Nature Conservation:
The Role of Remnants of Native Vegetation

A busload of Busselton boffins,
Trampled remnants with nails in their coffins,
But the forests fought back,
With a tree on the track,
Don't fool with your fragments too often!

Ian Herford

Nature Conservation:
The Role of Remnants of Native Vegetation

Edited by

Denis A. Saunders, Graham W. Arnold, Andrew A. Burbidge
and Angas J. M. Hopkins

Published by

Surrey Beatty & Sons Pty Limited

In association with

Commonwealth, Scientific and Industrial Research Organization,
Division of Wildlife and Rangelands Research,
Helena Valley Laboratory; and
Western Australian Department of Conservation and Land Management,
W.A. Wildlife Research Centre

Copyright Reserved

ISBN—0 949342 08 6
National Library Canberra

Published March, 1987

Printed and Published in Australia by
SURREY BEATTY & SONS PTY LIMITED
43 Rickard Road, Chipping Norton, NSW 2170

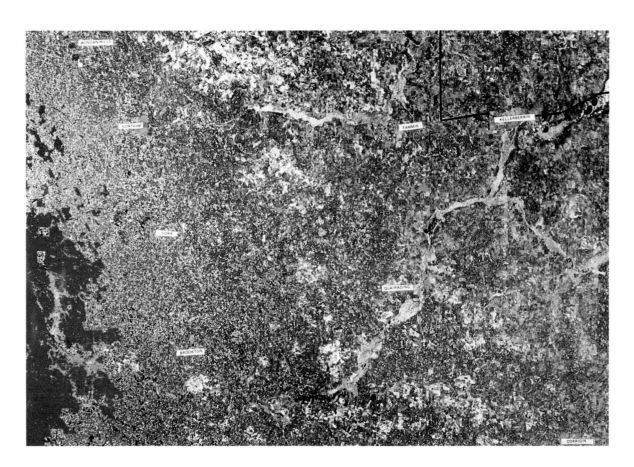

LANDSAT 'False Colour' image of the central wheatbelt of Western Australia. The image covers 14,500 square kilometres. The large belt of red represents part of the State Forest of the SW; the yellow and red spots, uncleared native vegetation; and the grey, farmland. The black line in the top right hand corner of the picture encloses the study area in the Kellerberrin district of the CSIRO, Division of Wildlife and Rangelands Research.

PREFACE

D. A. Saunders[1], G. W. Arnold[1], A. A. Burbidge[2] and A. J. M. Hopkins[2]

During the early 1980s the Western Australian Government asked the Commonwealth Scientific and Industrial Research Organization (CSIRO) to carry out research to help provide a scientific basis for the management of the many small nature conservation reserves in the agricultural areas of the south-west of the State. This main cereal growing area of Western Australia lies within the 300 and 800 mm annual mean rainfall isohyets (see Main, Chapter 1, Fig. 1, this volume) and, before development for agriculture, was extremely rich in species of plants and animals. Most of this biotic wealth now exists only in the remnants of natural vegetation scattered through the region, many of which are conservation reserves. If the species richness is to be maintained, much needs to be known about these remnants, their ecosystems and their management.

In mid-1984, in response to the Western Australian Government's request, the Helena Valley Laboratory of the CSIRO, Division of Wildlife and Rangelands Research began a study of the dynamics of plant and animal species in small patches of native vegetation in the Western Australian wheatbelt. The objective of this research programme is to establish the ecological principles on which the management of remnants of native vegetation should be based. In planning the programme it was considered desirable to organize a workshop to discuss the problems of remnants of native vegetation and to identify some of the key areas for research. It was recognized that such a workshop would only be worthwhile if it was attended by a broad spectrum of interest groups ranging from research to management, with the latter being particularly important.

In Western Australia, the Department of Conservation and Land Management (CALM) is responsible for the conservation of flora and fauna and for the management of National Parks, Nature Reserves, and State Forests. Consequently, Denis Saunders and Graham Arnold (CSIRO) approached Andrew Burbidge and Angas Hopkins (CALM) with the suggestion that the Helena Valley Laboratory (CSIRO) and the W.A. Wildlife Research Centre (CALM) jointly organize a workshop on the topic: 'NATURE CONSERVATION: THE ROLE OF REMNANTS OF NATIVE VEGETATION'.

Andrew and Angas agreed to the suggestion with the *proviso* that topics of concern to managers should be adequately covered at such a workshop. Together we developed a proposal to run a workshop dealing with four major themes:

1. Ecological studies as the basis for management;
2. Fragmentation and population genetics;
3. Measuring and monitoring dynamics of remnants; and
4. Management.

Notices of the intending workshop and of the four themes to be covered were sent to all tertiary institutions and conservation agencies in Australia and New Zealand, to individuals throughout the world known to have published papers on fragmentation of habitat and its effect on flora and fauna and to journals and newsletters of relevant professional societies. These notices solicited potential contributions from participants and invited suggestions for discussion topics to be covered.

The workshop was deliberately structured to provide maximum time for discussions and interaction. For this reason the number of participants was limited and each afternoon was devoted entirely to discussions. Unfortunately, it also meant we had to reject a number of titles which were submitted in response to the first circular.

[1] Division of Wildlife and Rangelands Research, CSIRO Helena Valley Laboratory, L.M.B. No. 4, P.O. Midland, Western Australia 6056.
[2] Department of Conservation and Land Management, Western Australian Wildlife Research Centre, P.O. Box 51, Wanneroo, Western Australia 6065.

The venue, Busselton, was chosen because it was well away from Perth, the capital of Western Australia. This encouraged people to stay together and to continue their discussions after the day's organized activities had ceased.

Each day was devoted to one of the workshop themes. The morning started with a review paper which was followed by six or seven relevant contributed papers. In the afternoon the participants were divided into five discussion groups, each of up to 19 people, discussing two of five topics listed for that day. Discussion topics were developed from ideas provided by people who responded to the preliminary notice, and discussion group leaders were selected from those who had indicated an interest in a particular topic. Each leader presented a brief introduction, chaired the ensuing discussion, presented a resumé at a plenary session that afternoon and prepared a summary for this book. As organisers, we allocated people to discussion groups with the aim of providing in each as broad a range of expertise and backgrounds as possible.

In the middle of the week a field trip was arranged to allow workshop participants to view local remnants, some of the State Forest to the south of Busselton and part of the Leeuwin-Naturaliste National Park. The forest, as well as demonstrating several local conservation issues and problems such as the replacement of eucalypt forest with pine plantations, the devastation caused by the root-rot fungus *Phytophthora cinnamomi*, and the highly saline ground water in some places, also provided some excitement in the form of a large recently fallen jarrah *Eucalyptus marginata* across a narrow track. However, with some help from two CALM trucks and the passengers, the bus drivers managed to turn around and unbog their vehicles (see Limerick at front of volume), and we finally made it for a late lunch at two local vineyards.

The range of backgrounds represented and the high quality of the papers and discussion sessions at Busselton were of such great interest and value that we decided to publish them so that the information is widely available and in one volume. We are extremely grateful to Ivor Beatty of Surrey Beatty & Sons, who readily agreed to publish the material for us and offered encouragement throughout the publishing process.

The layout of this volume is much as the material was presented at Busselton with the exception of the first and last chapters. The chapters are divided into the four themes in which they were presented and the review chapters are the first in each section, i.e. Chapters 2, 9, 16 and 23. These are followed by each of the contributed chapters which were presented in the four themes. These cover Chapters 3 to 30 with the exception of the above. For this volume each author was encouraged to expand the points presented at the workshop and refer to material presented by other people. Each contribution was reviewed by two referees and also underwent some editing. On that score, we are extremely grateful to the referees for the task they performed on our behalf and we are very appreciative of the speed with which the contributors prepared their chapters and the good grace displayed during the editorial phase.

In addition, several posters relating to the overall topic were presented at the workshop and these have been expanded and incorporated in this volume in Chapters 31 to 34.

As mentioned earlier, discussion group leaders were asked to prepare summaries of the discussions they led. Each summary was sent to two participants to verify that it did represent an accurate account of the findings of the discussion groups. In all, 20 topics were discussed at the Busselton workshop and 17 of these are published here in Chapters 35-51.

The two chapters which contain material not presented at Busselton are the first and last. We asked Professor A. R. Main (a member of the former W.A. Wildlife and National Park Authorities) if he would prepare an introductory chapter for this volume, drawing on the wheatbelt of Western Australia as an example. His chapter 'Management of remnants of native vegetation: a review of the problems and the development of an approach with reference to the wheatbelt of Western Australia' grew partly from his frustration at the seeming lack of a coherent approach to the management of remnants and partly out of our need for a suitable chapter to set this volume in perspective. We are extremely grateful to Bert Main for his efforts in presenting this contribution at such short notice.

As editors we have exercised our prerogative to have the last word and our chapter 'The role of remnants of native vegetation in nature conservation: future directions' represents our view of the main points which came from the Busselton experience. We should emphasize that they represent our view of the future, not necessarily the view of other contributors to this volume nor of other participants of the Busselton workshop.

ACKNOWLEDGEMENTS

The workshop at Busselton on the role of remnants brought together many people from a wide area (see list of participants). The success of the workshop was due in no small measure to the support we received from many sources and in particular we would like to thank: Alcoa; Ansett Airlines of Australia; Burswood Management; Cliffs Robe River; Murdoch University; Swan Portland Cement; Conservation Council of Western Australia; and West Australian Petroleum Pty Ltd. We would like to thank the management and staff of The Geographe Motel who provided the venue; Ken Tinley, Judith Brown and Greg Keighery who helped organize the tour during the workshop; Judith Brown, Ken Wallace, Ken Atkins, Jack Kinnear, Tony Friend and Bert and Barbara Main who helped organize the post-workshop tour; CSIRO and CALM for support; and Mrs Claire Taplin who assisted in the production of this volume.

The index was prepared by Andrew Burbidge from index cards supplied by the authors. It was compiled at the WA Department of Conservation and Land Management's Wildlife Research Centre utilizing dBASE III PLUS programmes written by Mike Choo, and we are most grateful for his help. Our thanks also go to Jill Pryde and Phil Fuller for inputting words and page numbers to the data-base.

CONTENTS

Preface		v
List of Colour Plates		xiii

INTRODUCTORY CHAPTER

1. Management of remnants of native vegetation: a review of the problems and the development of an approach with reference to the wheatbelt of Western Australia. By A. R. Main ... 1-13

ECOLOGICAL STUDIES AS THE BASIS FOR MANAGEMENT

2. Ecological studies as the basis for management. By A. J. M. Hopkins and D. A. Saunders ... 15-28
3. Persistence of invertebrates in small areas: case studies of trapdoor spiders in Western Australia. By B. Y. Main ... 29-39
4. Conservation of mammals within a fragmented forest environment: the contributions of insular biogeography and autecology. By A. F. Bennett ... 41-52
5. Local decline, extinction and recovery: relevance to mammal populations in vegetation remnants. By J. A. Friend ... 53-64
6. Effects of patch area and habitat on bird abundances, species numbers and tree health in fragmented Victorian forests. By R. H. Loyn ... 65-77
7. The incidence and conservation of animal and plant species in remnants of native vegetation within New Zealand. By C. C. Ogle ... 79-87
8. Assessing the conservation value of remnant habitat 'islands': mallee patches on the western Eyre Peninsula, South Australia. By C. R. Margules and A. O. Nicholls ... 89-102

FRAGMENTATION AND POPULATION GENETICS

9. Effects of fragmentation on communities and populations: actions, reactions and applications to wildlife conservation. By M. B. Usher ... 103-21
10. Responses of breeding bird communities to forest fragmentation. By J. F. Lynch ... 123-40
11. Consequences of faunal collapse and genetic drift to the design of nature reserves. By W. J. Boecklen and G. W. Bell ... 141-49
12. Geographic population structure of eucalypts and the conservation of their genetic resources. By G. F. Moran and S. D. Hopper ... 151-62
13. North American forests and grasslands: biotic preservation. By R. F. Whitcomb ... 163-76
14. Retaining remnant mature forest for nature conservation at Eden, New South Wales: a review of theory and practice. By H. F. Recher, J. Shields, R. Kavanagh and G. Webb ... 177-94
15. Connectivity: an Australian perspective. By P. B. Bridgewater ... 195-200

MEASURING AND MONITORING DYNAMICS OF REMNANTS

16. Monitoring populations on remnants of native vegetation. By P. R. Ehrlich and D. D. Murphy ... 201-10
17. The response of a small insectivorous bird to fire in heathlands. By I. Rowley and M. Brooker ... 211-18
18. Monitoring population densities of western grey kangaroos in remnants of native vegetation. By G. W. Arnold and R. A. Maller ... 219-25
19. Three decades of habitat change: Kooragang Island, New South Wales. By R. T. Buckney ... 227-32
20. Disturbance regimes in remnants of natural vegetation. By R. J. Hobbs ... 233-40
21. Characteristics of problem weeds in New Zealand protected natural areas. By S. M. Timmins and P. A. Williams ... 241-47
22. Factors affecting survival of breeding populations of Carnaby's cockatoo *Calyptorhynchus funereus latirostris* in remnants of native vegetation. By D. A. Saunders and J. A. Ingram ... 249-58

MANAGEMENT

23. Management of remnant bushland for nature conservation in agricultural areas of south-western Australia — operational and planning perspectives. By K. J. Wallace and S. A. Moore ... 259-68
24. The changing environment for birds in the south-west of Western Australia; some management implications. By G. T. Smith ... 269-77
25. Management of disturbance in an arid remnant: the Barrow Island experience. By W. H. Butler ... 279-85
26. Management of remnant habitat for conservation of the helmeted honeyeater *Lichenostomus melanops cassidix*. By G. N. Backhouse ... 287-94
27. The viability of planning control and reservation as options in the conservation of remnant vegetation in New South Wales. By W. B. Giblin and S. King ... 295-304
28. Planning for fire management in Dryandra forest. By N. D. Burrows, W. L. McCaw and K. G. Maisey ... 305-12
29. Conservation strategies for human-dominated land areas: the South Australian example. By S. G. Taylor ... 313-22
30. The use of fire as a management tool in fauna conservation reserves. By P. Christensen and K. G. Maisey ... 323-29

POSTER PAPERS

31. The impact of tree decline on remnant woodlots on farms. By F. R. Wylie and J. Landsberg ... 331-32
32. The conservation and study of invertebrates in remnants of native vegetation. By J. D. Majer ... 333-35

33. A monitoring system for natural area management in Western Australia. By A. J. M. Hopkins, J. M. Brown and J. T. Goodsell 337-39
34. Bird dynamics of Foster Road Reserve, near Ongerup, Western Australia. By B. J. Newbey and K. R. Newbey 341-43

WORKSHOP REPORTS

35. Relevance, accountability and efficiency of research for management. By R. McKellar 345-46
36. Achieving a balance between long and short term research. By K. L. Tinley 347-50
37. Nutrient cycles: their value in devising management strategies. By A. R. Main 351-52
38. Invertebrates as indicators of management. By J. D. Majer 353-54
39. Modelling: its role in understanding the position of the remnants in their ecosystems and the development of management strategies. By G. Beeston 355-56
40. The value of corridors (and design features of same) and small patches of habitat. By T. Dendy 357-59
41. Single large or several small reserves? By C. Margules 361
42. Ecotones, patchiness and reserve size. By E. Russell 363-64
43. Viability of small populations of animals and the value of introductions and translocations. By J. Kinnear 365
44. The use of surveys and data bases for conservation. By M. Austin 367-68
45. Monitoring of management practices. By G. R. Friend 369
46. Measuring and monitoring dynamics of remnants. Types of organisms that should be monitored: why and how. By T. J. Ridsdill-Smith 373
47. The integration of survey and monitoring. By R. Braithwaite 377
48. The role of government and the community. By G. J. Syme 379-80
49. Management options, practical constraints and the establishment of priorities. By A. A. Burbidge 381-82
50. Creation of ecotones and management to control patch size. By L. Mattiske 383
51. Management theory and optimum feedback. By A. J. M. Hopkins 385-86

CONCLUSION

52. The role of remnants of native vegetation in nature conservation: future directions. D. A. Saunders, G. W. Arnold, A. A. Burbidge and A. J. M. Hopkins 387-92
Appendix 1 — List of Participants 393-96
Index 397-410

LIST OF COLOUR PLATES

			Opposite Page
PLATE 1	CHAPTER 2:	Wandoo *Eucalyptus wandoo* woodland in the Mt Lesueur area (E. G. Griffin)	106
PLATE 2	CHAPTER 2:	Ecotone between woodland and heath in the Mt Lesueur area (E. G. Griffin)	106
PLATE 3	CHAPTER 9:	One of the largest of the limestone pavements that surround Ingleborough, North Yorkshire, England (M. B. Usher)	106
PLATE 4	CHAPTER 9:	Rigid buckler-fern *Dryopteris villarii montana* (M. B. Usher)	107
PLATE 5	CHAPTER 9:	Primrose *Primula vulgaris* (M. B. Usher)	107
PLATE 6	CHAPTER 12:	Flowers of *Eucalyptus caesia* subsp. *magna* (S. Hopper)	107
PLATE 7	CHAPTER 12:	*Eucalyptus caesia* subsp. *caesia* (S. Hopper)	154
PLATE 8	CHAPTER 12:	Buds, flowers and fruit of *Eucalyptus pendens* (S. Hopper)	154
PLATE 9	CHAPTER 12:	*Eucalyptus suberea* (S. Hopper)	154
PLATE 10	CHAPTER 13:	Iowa tall-grass prairie (R. Whitcomb)	155
PLATE 11	CHAPTER 13:	*Colladonus clitellarius*, a savanna insect (R. Whitcomb)	155
PLATE 12	CHAPTER 13:	Wheatgrass *Agropyron smithii* prairie (R. Whitcomb)	155
PLATE 13	CHAPTER 17:	Colour banded male *Malurus splendens* (G. S. Chapman)	226
PLATE 14	CHAPTER 17:	Gooseberry Hill study area four days after the fire of 30 January 1985 (M. Brooker)	226
PLATE 15	CHAPTER 17:	Territory V before the fire of 30 January 1985 (M. Brooker)	226
PLATE 16	CHAPTER 17:	Territory V 11 days after the fire of 30 January 1985 (M. Brooker)	227
PLATE 17	CHAPTER 20:	Experimental fire in a remnant area of *Banksia menzeissii/B. attenuata* woodland near Gingin, Western Australia (R. Hobbs)	227
PLATE 18	CHAPTER 20:	Small-scale disturbance caused by gopher activity in an area of annual grassland on serpentine soil in N. California (R. Hobbs)	227
PLATE 19	CHAPTER 22:	Typical road reserve in the northern wheatbelt of Western Australia (D. Saunders)	227
PLATE 20	CHAPTER 22:	Carnaby's cockatoos feeding in roadside native vegetation (L. Moore)	250
PLATE 21	CHAPTER 22:	Wandoo woodland at Coomallo Creek (D. Saunders)	250
PLATE 22	CHAPTER 22:	Carnaby's cockatoo (G. Chapman)	250
PLATE 23	CHAPTER 25:	Euros on Barrow Island (H. Butler)	251
PLATE 24	CHAPTER 27:	Brigalow remnant surrounded by farmland in northwest New South Wales (R. Dick)	251
PLATE 25	CHAPTER 4:	The Long-nosed Potoroo, *Potorous tridactylus* (G. Coulson)	251
PLATE 26	CHAPTER 24:	Female Noisy Scrub-bird *Atrichornis clamosus*, leaving the nest carrying a faecal sac (G. Chapman)	251

FRONTISPIECE:	LANDSAT 'False Colour' image of the central wheatbelt of Western Australia.		
CHAPTER 3	Fig. 3	Typical vegetation association of *Acacia* and *Allocasuarina* ('woodjil') occupied by *Anidiops villosus*. *Ecdeiocolea* tussocks and myrtaceous shrubs in foreground	p. 33
	Fig. 4	Nest of *Anidiops villosus*. Note fan of twiglines which increases foraging area	p. 33
	Fig. 5	Nest of *Anidiops villosus* in laterite, showing gravel pebbles which spider has carried from bottom of burrow and deposited at edge of twiglines	p. 33
	Fig. 6	Nest of *Aganippe* sp. D against shrub	p. 33
CHAPTER 15	Fig. 3	Three ecolines in a region of coastal Gippsland, Victoria	p. 199
	Fig. 4	Ecolines in the Pilbara region, Western Australia, near Turee Creek	p. 199
	Fig. 5	Ecolines in the Westernport Bay region, Victoria	p. 199
CHAPTER 30	Fig. 3a	Tammar wallaby *Macropus eugenii*	p. 325
	Fig. 3b	A heartleaf poison *Gastrolobium bilobum* thicket	p. 325

CHAPTER 1

Management of Remnants of Native Vegetation — A Review of the Problems and the Development of an Approach with Reference to the Wheatbelt of Western Australia

A. R. Main[1]

An approach to the management of remnants of native vegetation in the wheatbelt of Western Australia, a region subject to a variety of natural perturbations such as fire, flood, drought and windstorm, is developed. The effects of these perturbations upon ecosystem functions such as nutrient abstraction, retention and recycling, as well as nitrogen fixation and persistance of the biota of the remnant, are considered.

The putative ecosystem function related to management goals is tabulated, as are elements to be noted and questions asked in monitoring programmes, in the light of specific management goals and in the context of a new perturbation, namely the isolation of the remnant.

The difficulties of interpreting fluctuations within remnants are discussed in the light of theoretical considerations of stability, resilience and persistence. The approach developed may have wider application.

INTRODUCTION

REMNANTS of vegetation arise from causes ranging from unsuitability of land for agriculture, choice of the landholder or reservation for public purposes, e.g. roads, railway, water or conservation reserves. An argument can be developed that a role of remnants of native vegetation, especially those held as public land, is to retain places where native organisms and natural habitats can continue to be found. A counter argument can be developed which states that there is no such role because remnants will lose biotic elements and so cease to be examples of the habitats they are supposed to represent. Alternatively, or in addition, advocates of this second line of reasoning may argue that even if habitats and biotic diversity could be retained, the management costs would be prohibitively high and hence management is not worth pursuing. These are not arguments to which anyone committed to conservation can subscribe. Yet in order to counter them it is necessary to demonstrate that management is possible and there are principles which, if followed, are likely to lead to the retention of the native biota in remnants of vegetation both on public land set aside as conservation reserves and on private remnants.

The role of these remnants, in the context of conservation, is taken to be: (a) retention of an array of examples of native habitat which existed more widely before the fragmentation of the vegetation by farming, and (b) the provision of habitat for migratory, nomadic, or sedentary elements of the fauna. Clearly the foregoing roles cannot be fulfilled or the goal of persistence achieved without management to ensure that the remnants persist in a suitable state.

This chapter extends the approaches of earlier papers (Main 1981a, 1982, 1984) to managing and so conserving the biota included within remnants now held as conservation reserves in the wheatbelt of Western Australia. The conservation reserves of this region are chosen as an illustrative example of

[1]Department of Zoology, University of Western Australia, Nedlands, Western Australia 6009.
Pages 1-13 in NATURE CONSERVATION: THE ROLE OF REMNANTS OF NATIVE VEGETATION ed by Denis A. Saunders, Graham W. Arnold, Andrew A. Burbidge and Angas J. M. Hopkins. Surrey Beatty and Sons Pty Limited in association with CSIRO and CALM, 1987.

problems common to all reserves. I develop the argument in the following way: after a brief consideration of the historical and physical background of the array of remnants there is discussion of the matters which should be considered in formulating specific goals and strategies; and finally in the text and three tables there is mention of ecosystem functions which should be considered in detailed management and monitoring programmes. Thus the chapter develops an approach and advances principles rather than writes prescriptions for management.

BACKGROUND

In Western Australia the wheatbelt has passed from being essentially virgin land at the turn of the century to being more or less completely settled now. During the early settlement phase there was no policy of setting aside areas for nature conservation but many small areas were left uncleared. In contrast in recent years there has been an active policy of designating any remnants which were still public land as conservation reserves.

Twenty-two of these reserves have now been surveyed in detail for vertebrate fauna and flora. The results of these surveys have been analysed in terms of current hypotheses about species area curves and island biogeography (Kitchener *et al.* 1980a, 1980b, 1982; Kitchener 1982). The recently created wheatbelt reserves are not supersaturated, at least for lizard species (Kitchener *et al.* 1980a). The absence of supersaturation has been attributed to the archipelagic nature of the pristine environment induced by a mosaic of soil types with a diverse fire history (Kitchener *et al.* 1980a; Main 1979). Thus at the time of creation of the reserves the diversity of animal species was perhaps already reduced to what could survive in the post-fire refugia of unburnt habitat on various soil types.

From the foregoing papers of Kitchener and co-workers it might be concluded that despite the haphazard way in which the reservations were acquired there is a reasonable possibility that faunal elements can be retained. However, in view of their location within farmland it is also clear that the reserves will not survive without management. Furthermore these conservation areas have been acquired quickly and there has been no opportunity to develop management skills which are needed to: conserve rare species; retain as much as possible of the biotic assemblage now present within a reserve; or retain all the reserves in the regional setting so that mobile elements such as birds are not disadvantaged. Thus in a short time the emphasis has shifted from justifying the acquisition of the land to demonstrating that the reserves serve a conservation purpose which is being achieved through management.

THE ENVIRONMENT — CLIMATE

The wheatbelt is considered to have a low to moderate annual rainfall variability (Southern 1979). There are 80 or fewer rain-days per year. The average rainfall ranges from 580 mm on the western and southern boundaries to about 280 mm on the inland, eastern and northern boundaries. Evaporation ranges from 1500 to 2250 mm per year.

The number of months per year during which average rainfall exceeds effective rainfall ranges from five to six in the wetter areas to about three in the drier parts. Despite the acceptance that it is a region with a dry mediterranean climate and a reliable rainfall for wheatgrowing there are factors which cause considerable deviations from the broad climatological picture (Table 1).

The events listed in Table 1 may not be independent. Tornadoes can be associated with thunderstorms or intense cold fronts. Heavy rain can be from thunderstorms or associated with decaying tropical cyclones, while fire may be facilitated by rank growth following above average rainfall. Regardless of the associations, these events, even if rare, may be of significance in: initiating regeneration by destroying the canopy (fire, tornadoes, wind storms); preparing a seed bed (fire); promoting germination (fire, heavy rainfall); facilitating seedling establishment (summer rainfall); destruction of early stages of seedling regeneration (fire, drought).

It is to be expected that there will be climatic fluctuations and oscillations in the future. Yet even greater changes are possible, as indicated by Peters and Darling (1985) who have drawn attention to the possible results of the Greenhouse Effect on nature reserves and Chittleborough (1985) raises other issues which are relevant to Western Australia. Such effects should not be thought of as occurring only in the distant future.

SIZE OF REMNANTS

Kitchener *et al.* (1980a) reported that there were about 500 conservation reserves varying in size but three-quarters are less than 400 ha. Wallace and Moore (this volume) report the present number as being 626, ranging from 0.4 to 309,000 ha with a median area of 120 ha: the majority are thus small in area. A minimum adequate area for retaining desired habitat characteristics will vary depending on the purpose of the reserve. A reserve set aside to be a 'stepping stone' for nomadic birds, retain a vegetation assemblage, a particular plant species, a suite of reptiles, or an assemblage of invertebrate species need not be as large as one required to retain a large mammal (Main and Yadaov 1971), or reptiles (Kitchener *et al.* 1980a).

Table 1. Environmental factors which may significantly affect the native biota of the wheatbelt.

Event	Frequency	Authority
Drought	About once every 7 years	Fitzpatrick 1970, Southern 1979
Heavy rainfall	90 mm in 3 calendar days once every 20 years	Pierrehumbert 1973
	40 mm in 6 hours once every 5 years	
Tornado		Clarke 1962
	Comparable with tornado areas of USA	Minor, Peterson & Lourenz 1980 (Fig. 1)
Summer rainfall from tropical cyclone	Perhaps 3 times in decade	Lourenz 1981 Various records of cyclone tracks
Thunderstorms = thunder days	Up to 10 per year	Ashton 1960
Fire frequency	Vegetation will carry fire every 5 years	Tentative value of Leigh & Noble 1981 (Fig. 1)
	Frequency of past fires every 10 years	
	Absolute protection from fire in thicket formations has a detrimental effect on the vegetation	Gardner 1957, p. 173
	Woodland less frequently than once in 100 years	Hopkins & Robinson 1981

While a minimum adequate area for a remnant will depend on its perceived purpose it must be appreciated that small areas, while adequate for a stated conservation purpose, do impose additional management costs or require greater skills when:

1. the area is less than a complete landscape unit, i.e. it may occupy only part of a valley slope, or hilltop or valley floor so that, for example, water falling as rainfall may be distributed onto or off the reserve and control of the water is not under the control of the manager;

2. the area is so shaped that the boundary is large in relation to the area included;

3. the whole area is at risk from a single event (wildfire or wind storm);

4. the area does not readily lend itself to management for retaining the vegetation in something like the heterogeneity in terms of patches, patch size, post-fire regeneration and associated ecotones that is common under natural conditions; and

5. the area retains a species whose population is likely to be less than an effective breeding size.

Each of the above is of significance to management. Together they require the development of a very sophisticated set of management goals and principles. But each reserve and remnant on private property is part of the conservation resources of the region. Since some remnants will lose some of their biota the minimum goal of management is to see that somewhere in the suite of remnants the present biota is retained. The ideal is to retain all that is present on all remnants, yet this may not be possible since not only are the remnants of different sizes and shapes, in different parts of the landscape and under different controlling authorities, but they also possess different attributes with respect to soils, hydrological regime, nutrients, fire regime, plants and animals. If the broad goal of management is as stated above, then management must be cognizant of the fact that what can be retained on each remnant and within the region will depend as much on the size of the remnants as its physical attributes and included biota.

ATTRIBUTES OF REMNANTS

Soils. Different soil types form an interdigitating mosaic in any locality where the different kinds may be quite restricted in area. In a general way soils may be associated with landscape elements thus: laterites, sandy laterites, deep yellow sands and gritty poorly structured apron soils associated with granite outcrops are usually found on high ground; duplex soils (sands over clays) occur on slopes; and heavy loams in the valley floors (Mulcahy 1973).

Hydrological regime. Rainfall is the only source of water for soil and ground water recharge. Clearing of the Kwongan vegetation (sclerophyllous shrubland) has resulted in the development of permanent unconfined ground water in the deeper sands (Williamson 1973). The contribution of rainfall to soil water recharge depends on the intensity and duration of falls. Persistent steady rain makes the greatest contribution to soil water. The amount which penetrates depends on the nature of the soil, texture of soil, whether it is compact or open, the

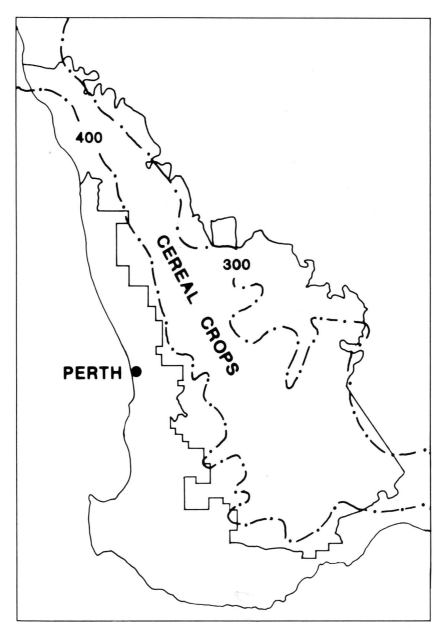

Fig. 1. Sketch map of southwestern Australia showing boundaries of cereal cropping areas (within continuous lines) and the approximate 400 and 300 mm isohyets (broken lines).

vegetation present and the site position with respect to the topography. Personal observations suggest that soil water recharge is best when normal winter rainfall follows summer rain from a tropical cyclone.

Farmed land is more prone to sheet flooding than areas still under native vegetation, hence a very significant factor for management is whether cleared land adjacent to the remnant is likely to lead to sheet flooding with redistribution of the flood water, debris and soils onto the remnant. However, even if sheet flooding does not occur, another problem could arise because the evapo-transpiration of crops and pastures is less than that of native vegetation. The consequence is that percolating water infiltrates the soil and ultimately raises the ground water table. The deep ground water is saline and if it is supplemented by percolation as related above, it will as it moves down the gradient intercept the surface and develop salt springs and seeps. Large areas of vegetation may be killed by this surface salt. The ground water table has risen markedly with increased clearing and salt-affected land is becoming more common (Burvill 1979). Peck (1975) discusses the effects of land use on salt distribution and cites the average salt storage when rainfall is less than 800 mm/yr as 8.1×10^5 kg/ha. Both sheet flooding and salinity pose serious problems for management of remnants.

Nutrients. Nutrients are low in all soils of the wheatbelt. Phosphorus is especially so. Sclerophylly and xeromorphy have been attributed to low phosphorus (Beadle 1966). Some native plants have adaptive root systems to cope with phosphorus shortage (Lamont 1981). Many native species display signs of toxicity when exposed to high or added levels of phosphorus (Ozanne and Specht 1981; Specht 1981).

The development of proteoid roots (Lamont 1981) and the association of many other plants with fungi to form mycorrhizae has led to a very efficient nutrient retention and recycling system. In addition, nitrogen fixation by a variety of organisms is common but not of a high order (Lamont 1984).

Fire regime. Wild fires started by lightning are always possible once the fuel bed provided by dried dead bark, leaves, twigs, branches and shrubs or grasses is both deep and continuous. In pre-settlement times such fires could be very destructive but they appear to have contributed to the patchiness or heterogeneity of the vegetation. Once burnt, an area would usually take five years before it could again carry a fire, but often remained unburnt for ten years (Leigh and Noble 1981); the earlier burns often acted as fire breaks and so restricted the extent of subsequent burns.

Prior to European settlement Aboriginals as well as lightning strike initiated fires. Hallam (1975) has documented the ubiquity of fire from this cause at the time of European discovery and settlement. The contrasts between the description of unsettled areas subject to Aboriginal burning cited by Hallam (1975) and areas of vegetation infrequently burned are remarkable. Nevertheless the frequency of Aboriginal burning is unknown. Hopkins (1985) discusses fire regimes of pre-European times and suggests that fires were rare in the wheatbelt.

It is interesting to note that Singh and Geissler (1985) propose that the sudden appearance of eucalypt pollen in the Lake George site in eastern Australia at about 130,000 years BP may be a consequence of the fire-based hunting technology of the Aboriginals which facilitated the spread of fire tolerant eucalypt-dominated vegetation. This presupposes an Aboriginal presence in Australia much earlier than present available archaeological evidence. Nevertheless it suggests that the expansion of fire tolerant plant associations is, in geological terms, recent.

Many of the present remnants are of an inappropriate size for wildfire to maintain the heterogeneity as occurred in the past. Moreover, wildfire in the present context is a danger to adjacent crops, pastures and farm infrastructure. Devising measures for protecting remnants from complete destruction by a single fire or from fires originating on adjacent farmland is a major task of management.

Plants. The common families are Myrtaceae, Proteaceae, Mimosaceae and Fabaceae. Casuarinaceae is not a large family but species of *Allocasuarina* are frequent and abundant elements. Many species of plants regenerate from seed after fire, including some that are also characteristc of disturbed ground. Many others re-sprout from below-ground parts. A considerable and significant component of many areas of bare soil are crusts of lichens and cyanobacteria. These are not only significant in fixing nitrogen but also bind the soil surface preventing wind erosion and scouring by flowing surface waters.

The specialized roots of Proteaceae and the symbiotic mycorrhizal associations of the Myrtaceae are significant in their contribution to nutrient availability.

In addition to cyanobacteria, nitrogen is fixed by both free-living micro-organisms and by symbionts associated with species of Casuarinaceae, Mimosaceae and Fabaceae (Lamont 1984).

A series of reviews of the biology of plants of the Kwongan are presented in Pate and Beard (1984).

Animals. It is more difficult to establish the presence of animals than plants. They vary immensely in size and number and may be sedentary or mobile. Absolute numbers of conspicuous or easily studied animals can be established but it is often difficult to relate these to a functional role other than the broad categories of predator, herbivore, detritus feeder, or scavenger. Nevertheless in another sense they all redistribute resources such as nutrients and energy from plant to animal biomass. Moreover, they shift these resources within and between areas. Nomadic birds and mammals, and breeding flights of termites (Isoptera) and ants (Formicidae) are possibly significant in redistributing resources. Redistribution of resources is not their sole role; invertebrates in conjunction with micro-organisms are the dominant elements in recycling nutrients. Furthermore animals and especially insects play a significant role in regulating the abundance and competitive status of many plant and animal elements of the biota (Whelan and Main 1979).

Many native animal species have changed in status by extinctions, range changes and through introductions of exotic species which are now feral. All changes have tended to alter ecosystems in ways which are essentially unstudied. The changes began with faunal extinctions of the late Pleistocene and Holocene and have continued to accelerate since European settlement. It is possible that no ecosystem has had time to adjust. The continuing effects of post-settlement changes such as clearing and introductions on distribution and abundance of cockatoos, which have been documented by Saunders *et al.* (1985) show how mobile faunal elements can be affected. However, changes in interspecies interactions generally remain unstudied: [for example, the effect of the loss of large herbivores on post-fire plant regeneration; the effect on plant regeneration of the loss of disturbed soil now that animals such as the boodie *Bettongia lesueur,* the dalgyte *Macrotis lagotis,* (the disturbed soil or rabbit burrows is not equivalent because rabbits graze on establishing plants) and the mallee fowl *Leiopoa ocellata* are absent; or the effect on the persistence of *Santalum* spp. now that dispersal by the emu *Dromaius novaehollandiae* no longer occurs and grazing of *Santalum* seedlings by the rabbit is intense]. In summary, changes which have occurred up to the time of settlement are now being exacerbated by the continued loss of native species and the acquisition of exotics.

FORMULATION OF GOALS AND STRATEGIES

Consideration of the matters already discussed is insufficient for the formulation of specific goals with regard to the management of remnants and the following should also be considered:

1. are any species unique to the remnant? In these cases the pivotal question is whether the present place of occurrence permits the deployment of the evolved life history traits. In terms of management for retention this means understanding the biology of the species and its relationship to the resources present in the remnant;

2. how many remnants in the regional set of remnants contain similar associations of plants and animals? Replication of areas with similar attributes and biological resources gives leeway in management decisions; and

3. are any areas less extensive than a landscape unit? If so they are likely to suffer reduction in species richness once they are isolated. This should be anticipated for two reasons: to ensure that if the species is lost from an area its functional role in the system is still filled, and to see if the species is conserved elsewhere.

Any hypotheses of likely losses of species from individual remnants will need to have regard to:

1. the topographic position of the remnant;

2. the adjacent land use and consequent soil disturbance and water distribution;

3. the adjacent land management, e.g., type of tillage, aerial application of fertilizer or use of herbicides, and the opportunity for fertilizers and herbicides to enter the remnant;

4. the soil mosaic within the remnant;

5. the plant associations present and their dynamics with regard to: longevity; seral stage; resilience or stability after perturbations in the sense of Sutherland (1981), i.e. having the potential to change community structure such as drought, flood, wind-storm, fire; tolerance to fertilizers and herbicides accidentally derived from adjacent farmland; the presence of ecotones on which animals may be dependent;

6. the contribution of plants and animals to nutrient cycling and availability;

7. the resilience or stability of any animal and plant population to perturbations;

8. the effective breeding population of any animal and plant species, its life history characteristics and breeding biology; and

9. the effect that the loss of a species has had or will have on the abundance of other elements.

When the foregoing have been considered it should be possible to list those remnants for which a goal of complete retention of the biota is unrealistic or requiring more resources than can be devoted to it. Inevitably there will be decisions involving judgements such as when all seral stages cannot be retained or when numbers of a dispersed species are below the effective breeding population size. This is so because when populations are small the effective breeding population (N_e) is likely to be such that alleles are lost and homozygosity increases. The significance of genetics in conservation practice has been discussed by Frankel (1982), Frankel and Soulé (1981) and Soulé (1985). The dangers of inbreeding are real; nevertheless many native Australian organisms have naturally small breeding populations, and even those which are abundant now may, because of habitat specificity, combined with effects of past fires and droughts, have been through a series of population reductions and thus been selected for a genetic mechanism which retains heterozygosity. Management and monitoring of populations in remnants would be helped if any genetic mechanism for maintaining heterozygosity were known (see James 1982).

MANAGEMENT

Management consists of those actions taken in the belief that they will contribute to the retention of the whole or designated biota in a remnant. This implies that we know the biology of each species, its role in the community and ecosystem, and its possible role when the ecosystem is perturbed by whatever means. Clearly this is not known now and is an almost unattainable goal.

However, a set of priorities can be established. Highest priority will be given to understanding the biology of species designated as being in special need of conservation because of their rarity or restricted occurrence. Next in priority should be those species which have changed status, i.e. become commoner or rarer. Changed status is a good indication that some of the interactions within the ecosystem have changed. Unfortunately it is easiest to study common species but increasing rarity is also a good indicator of changes within the ecosystem which need to be understood. For the remainder it may be prudent to do nothing, just wait and see what happens. Meanwhile, determining the longevity of various biotic elements, their reproductive strategies and whether recruitment is annual or episodic (particularly whether recruitment is related to some of the phenomena listed in Table 1) will contribute to an understanding of the overall dynamics of the biotic assemblage and enable judgements to be made as to when intervention becomes necessary.

Another set of data that can only be obtained from a long period of observation is related to woodland regeneration. In such cases it is necessary to know, in addition to conditions for recruitment and regeneration of trees, how long to maturity, when and how hollow limbs and nest sites develop, the survival and replenishment of hollow logs on the ground, the role of termites destroying fallen logs, and finally the role of wind-throw in the dynamics of the woodland.

Remnants may be perceived as a fire hazard by neighbours, and burning of an area for hazard reduction may be a procedure urged on managers. There are significant biological consequences, both advantageous and disadvantageous, of such action. Frequent fires will indeed reduce the fuel bed but they will also remove plant species which take longer than the interfiring period to produce seeds (see Hopkins 1985). Thus such a regime will also favour short life-span plants and remove fire sensitive species. Frequent burning will also remove or reduce invertebrate species dependent on deep litter and thus affect vertebrates utilizing them. Also affected will be animals dependent on plant species removed by burning. Many soil and litter animals, especially the young of burrowing forms, are very sensitive to the high summer temperatures which occur where the canopy has been destroyed or opened up by fire. Many termite species harvest litter and plant debris; in turn these termites are a food source for mammals such as the numbat *Myrmecobius fasciatus,* the echidna, *Tachyglossus aculeatus,* birds, lizards, and many predatory invertebrates, and as such are an irreplaceable element in the food chain.

Nevertheless some species will not regenerate without fire. In small remnants patchy burns may be very difficult to implement. Moreover small burnt areas may not regenerate adequately because of grazing of the post-fire regeneration by vertebrate and invertebrates resident on the unburnt areas adjacent (Main 1981b). Seedlings can be eliminated even when only grasshoppers are present (Whelan and Main 1979).

The relationships between management goals, their likely success, and the characteristics of the remnants which have a strong bearing on successful management are given in Table 2. The tabulation shows the complex ways in which remnant characteristics determine the reasonableness and likely success of management goals. Management of remnants vested in Statutory Authorities for the purpose of nature conservation or for related purposes can be directly related to Table 2. The relevance of the table to the management of remnants vested for other purposes or held as private land will depend upon the awareness by the controlling authorities of the conservation values of the remnants under their control as well as their capacity to fund the necessary action particularly in the case of small remnants or those not embracing a complete landscape unit.

MONITORING

Monitoring consists of those actions taken in order to establish that biota included within a remnant is being retained or conserved over the time period of interest. Efficient monitoring consists of a series of records from which trends in condition, recruitment or abundance can be established. Monitoring must not be expensive or excessively time consuming and must be useful to the manager. Monitoring is useless unless information flows to the manager for action and back to the monitoring programme to ensure that it is adequate for management purposes. Moreover, results of monitoring should be used to identify areas in which more biological information, such as longevity, reproductive strategies and nature of recruitment, is needed and hence where research should be directed. These results can be used to revise the management goals. All this information flow and the feed-back loops need to be considered and designed when the monitoring programme is set in place. Approaches to monitoring programmes are given in chapters in this volume (see Hopkins and Saunders) and in Table 3.

Apart from financial and other constraints, monitoring will be oriented by the anticipated or hypothetical outcome or consequences of the isolation of the remnant. Monitoring will then be directed to ensure that the anticipated goal is being reached, or, if it is not, to establish why this is so. The schema presented in Table 3 relates the goals of management as set out in this paper to the resources of the reserve or relict area in terms of soils, nutrients, plants and animals and the common perturbations, namely fire and drought. The treatment is not exhaustive but the questions raised in the table are likely to be those which will initiate effective monitoring.

Table 2. Management goals and their likely success in wheatbelt remnants related to the attributes of the remnant with respect to topography, boundaries, susceptibility to fire, patches and ecotones.

Attribute of remnant	Complete landscape unit on topographic high					Incomplete landscape unit not on topographic high (valley slope or floor)			
	Boundary		Risk from single wildfire	Vegetation patches, patch size and ecotones		Boundary		Risk from single wildfire	Patches, patch size and ecotones
Judgement of Management goal	Not excessive	Irregular and extensive				Not excessive	Irregular and extensive		
Retention of specific organism (may be plant or animal)	Success depends on size of area in relation to life history needs of organism	Success may be severely limited by nature of boundary	If risk high, prospects for retention not high without intense management; if risk low, possibility for retention very good	Success depends on maintaining patch characteristics and ecotones appropriate for species		Success may depend on use and management of adjacent land if placed on the slope	Success may be severely limited by nature and extent of boundary	If high fire risk, only intense management can lead to success; if fire risk low, then other factors more important.	Success only possible when patch characteristics can be maintained in the face of the impact of adjacent farm practices and their consequences and effects of fire regime
Retain present complexity	Depends on size of reserve-prospects of success not limited by boundary	Success depends on whether the boundary includes sufficient area of each soil type for all associations	As above	Success depends entirely on retention of factors which maintain patches		Success highly dependent on position in the landscape	Success will depend on how the adjacent land use interacts along the boundary	Success very difficult to achieve	Success depends in part on farm management practices, water runoff and flooding outside control of manager and chances of flooding and salination especially if near valley floor
Retain remnant with simplified associations	Success likely but boundary limiting in small areas; losses of biota only likely in very small areas where success may mean accepting anything other than an assemblage of exotics	Success higly likely when commonest association is acceptable	If high, decide whether fire disclimax is acceptable and manage accordingly; if not commit resources to reducing risk of destruction from single wild fire	Simplification likely if patch chacateristics cannot be maintained		Simplification most likely when area is small and adjacent land use highly deleterious or sited in or near the valley floor	Success depends on why boundary is irregular, i.e. the basis on which farm land was chosen; also sensitive to salination and flooding if in valley floor	If fire risk high, then need to accept that simplified fire disclimax may be outcome	Likelihood of success decreases as approaches the valley floor when prospects for flooding and salination increase

Table 3. Monitoring: elements to be noted and questions to be asked when assessing the status of a remnant in the light of specified management goals.

Management goal \ Elements of system	Soil	Plants — Population size	Plants — Age and structure	Plants — Recruits	Plants — Seed bank	Plants — Nutrients	Fire	Drought	Animals — Population size and structure	Animals — Migrants	Animals — Recruitment
Retention of specified organism	Bare soil: is lichen and algal crust intact? Is litter layer intact or subject to redistribution by flooding?	Is breeding and reproduction effective? Is diversity being lost? Is this likely to affect the designated organism?	Are the plants associated with the 'designated' organism being retained and in appropriate age structure?	Are other plants essential for designated species recruitment present? If not, is recruitment likely to be episodic?	Do essential plants have seed bank? What stimulates germination? Is this occurring?	Are all roles being filled?	Is designated organism dependent on post-fire succession, late seral stages, absence of fire, or ecotones? Are these being provided? Is fire detrimental to designated organisms?	Do favoured areas of water run-on act as refugia? What population size can be maintained and for how long?	If designated species is an animal, is N_e exceeded? If not, what should be done?	If N_e small or migration absent is supplementation necessary to maintain genetic diversity?	Is it regular or episodic? Is this expected?
Retention of whole biota	do.	Are rare species declining? Are they abundant elsewhere? Do they fill a role now or after perturbation?	Are species of different life expectancies all regenerating? If not, what will be effect on system?	Are dominant plants preventing recruitment? Are all habitats still being represented? Is the change successional? Is pollination successful, are the suite of pollinators and propagule dispensers present?	Is seed bank present? How long will it last? What stimulus for germination?	do.	Are patches and fire-induced heterogeneity being maintained? What is appropriate frequency for fire?	do. Are plants and animals re-establishing? Do they re-invade? Are there refugia in the reserve?	Are there refugia? If not, what degree of change or perturbation can be tolerated before management intervenes?	do.	Is there an obvious change in successful recruitment of any species? If so, is it likely to continue or reverse? Will it upset balance?
Retention of simplified biota	Is soil change as anticipated? Is flooding and salination worse than anticipated?	Is the simplified biota maintaining itself? If not, why?	Are the species replaced after maturity of the assemblage?	Is absence in recruitment of certain species as anticipated?	Is the seed bank needed for the expected simplification present? Has it been destroyed by salt or flooding?	How are the roles not present affecting the nutrition of the biota which is surviving?	As vegetation declines in abundance is reserve still capable of carrying a fire? If not, is it of concern? (e.g. weeds, exotics.)	Is the simplified system incapable of responding to drought?	Are the expected species persisting? If not, is it due to population size? How are populations regulated?	Is supplementation necessary for retention of simplified biota?	Is the expected persisting species still subject to population control? If not, will the carrying capacity be exceeded? What should be done?

EFFECTIVE POPULATION SIZE: ROLES AND FUNCTIONAL MODES WITHIN ECOSYSTEMS

Management and monitoring can most readily deal with taxonomic entities; however, the role or function of a population within a reserve may be a better guide to what is happening than anything else. Main (1981a) emphasized that the roles that organisms filled in ecosystems, and particularly in disturbed ones, were critical. Later (Main 1982), these relationships were extended to interpret the significance of rare species and so justify conservation efforts to retain them. This was so because of the possible significance of rare species when an ecosystem was subject to perturbations. In a functional sense this is a justification for retaining diversity. However, the contributions of the various elements making up the diversity need to be understood.

Should perturbations be experienced and have no effect on any of the roles, then the system can be said to exhibit stability. However, should this not be so, e.g. where nitrogen or minerals are lost, or recycling capacity is impaired or standing crop destroyed, the system will only display resilience (Holling 1973) and persist if the biotic elements needed to fill the role of replacing the lost or depleted function are present. Thus only if the appropriate diversity of role-filling organisms is present can the system enter a mode which restores the ecosystem function. If the biotic diversity and hence the capacity to enter the appropriate mode has been lost, the systems can only degrade following perturbation because nitrogen cannot be fixed, or nutrients recycled, or leaching prevented and so forth.

The principal roles are: nutrient abstraction from the deep soil profiles — (by plants, burrowing animals); nitrogen fixation (by micro-organisms, legumes, acacias, casuarinas); recycling those nutrients (by micro-organisms, litter invertebrates); scavenging nutrients (by hyphae of larger fungi, mycorrhizae, roots of quick growing ephemerals and plants with tubers and rhizomes which can quickly take advantage of favourable conditions and remove soluble nutrients before they are leached away); those redistributing nutrient resources from one site to another (litter fall, all animals). The functional modes resulting from the interplay of these roles with perturbations usually encountered are tabulated in Table 4.

Once experience is gained in thinking in terms of roles and functional modes following perturbation, it will be possible to ascribe biological significance to biotic elements in any system. Moreover, when changes in abundance of biotic elements are detected it should be possible to hypothesise as to what these changes mean:

Table 4. Nutrient response modes following the common natural perturbations.

Roles \ Source of perturbation	Soil disturbance such as after uprooting of trees caused by strong winds	Flood	Fire	Drought
Abstracting (making nutrients available from deeper soil profiles)	Dominant in early succession plants and early seres until soil formation allows some recycling		Deep-rooted seedlings and plants regenerating from underground parts	Capacity to enter mode depends on degree of soil disturbance, survival of biota, seed bank and regeneration from lignotubers.
N fixation (micro-organisms both free-living and symbionts)	Restricted to those N-fixing plants occupying early seres		Dominated by *Acacia* Fabaceae seedlings if these are in seed bank	Depends on seed bank and stimulus to germination; may be restricted to micro-organims and lichens
Recycling (making nutrients already incorporated into biological materials available again) invertebrates and micro-organisms	Not important until plant material is available and upper soil horizons have formed.	Dominant once ground is saturated depends on whether flooding inhibits aerobic organisms.		
Scavenging (prevention of leaching of soluble nutrients) tuberous and rhizostomaceous plants hyphae of fungi and mycorrhizas		Important if water movement is likely to lead to leaching	Dominant post-fire mode for nutrient minerals	Important at break of season and break of drought
Redistribution (the shifting of nutrient resources from one locality to another done mostly by animals)		Important if flooding has carried resources into a sink	Can be important if burnt area is small	Depends on what survives and whether area occupied by nomadic migrants or colonising populations

(a) to the present system;

(b) to the functional roles of the system if management does not intervene; and

(c) the capacity of the system to move to the appropriate mode following one or the other of the common perturbations.

RESEARCH

It is clear from the foregoing sections that research is needed at many levels. Research on rare and endangered vertebrates such as rock wallabies *(Petrogale)* and the numbat *(Myrmecobius)* is obvious and should proceed. As results come to hand all sorts of relationships to elements of the system in which they occur will become apparent; similarly if research is pursued on species whose status has changed. However, the most comprehensive understanding of the general function of systems within reservations is likely to arise from research naturally posed by management following the assessment of monitoring results interpreted in the context of remnant size, its attributes and place in the landscape, roles and functional modes when perturbed, and the likelihood of remnants showing either stability or resilience. Possible research will arise from questions posed in the various cells of Tables 2 and 3. This is an applied approach to research, yet it is likely to lead to much basic research and good science.

As relative abundances of some taxonomic entities change, research will be needed in order to establish the effects on ecosystem roles as well as on the populations and probability of persistence of other entities. Research posed on the basis of taxonomic categories, is likely to contribute less to achieving management goals than research based on role-centred questions, e.g. recycling (the role of micro-organisms, nematodes, insects); or individual growth or population control (disease organisms, nematodes, sap-sucking insects, folivorous insects, insects as vectors of disease); or redistribution (mammals, birds, insects, especially moths, termites, ants).

DISCUSSION

It is reasonable that the goals of management of conservation reserves containing rare and endangered species be devoted to the taxonomic entity about which there is concern. The presumed causes of rarity will, however, direct the detailed management (Main 1984). Also the aspects monitored will almost certainly be directly related to the rare species. When the management goal is wider and includes all taxonomic entities as a functional ecosystem, a taxonomic approach is not necessarily the best guide for management. This chapter has attempted to go beyond cases involving the rare or endangered and develop a general set of guidelines for management and monitoring. The guidelines are based on a theory of functional roles within the ecosystem which must be filled in order that the system can persist. Thus taxonomic and genetic diversity are regarded as important because the diversity reflects the potential for the system to adapt. In the past conditions have been different [arid or wet, colder or warmer (Bowler 1978)] and may again change in the future. When these changes occur, changes in the abundance and commonness of biotic elements will follow. Regardless of the change in abundance the system will only continue to function if the diversity prior to the environmental change can, through changes in the abundance of different taxonomic elements, provide elements to fill all the functional roles. The theory is clearly based on the assumption that if the diversity present now has coped with past changes, then it follows that future changes including perturbations can only be anticipated and handled by the system if the present diversity is retained.

As suggested earlier, there will be a class of remnant where simplification is to be accepted because of the location, size, configuration or other attribute of the remnant makes management unacceptably costly. However, there will be other remnants where complete retention of the biota is the goal. In these cases the comments of Holling (1973) are relevant: that the more homogeneous the environment in space and time, the more likely the system is to have low fluctuations and low resilience. This is of course the converse of the situation in Western Australia where the environment is variable in both space and time, and where resilience might be considered a normal response. So now the question is: how are changes to be interpreted? Are they merely manifestations of resilience, or are they caused by the new perturbation arising from being a remnant and thus conforming with the predictions of Island Biogeography Theory (IBT) as it applies to the wheatbelt situation (Kitchener 1982; Kitchener *et al.* 1980a, b, 1982). Or are they fluctuations reflecting manifestations of resilience (Holling 1973)? Or is it an example of instability arising from the specific assemblage included within the remnant (Sutherland 1981)? The problem is to decide whether fluctuations will be self correcting, i.e. part of a resilience response, or not. Clearly expectations can only be based on biological knowledge of the assemblage and expectations derived from theoretical considerations of the island-like state of the remnant (IBT) with considerations of diversity, stability, resilience and persistence (Holling 1973). Should perturbations turn out to have the Type III effect (be permanent) of Sutherland (1981), this means removal of the population and the biota becoming simpler.

Regardless of the cause, the consequence of simplification will depend on what is lost. Is it merely one of a set of alternatives for a function? Will it mean reduction in a critical function to such a level that the system will ultimately collapse? Changes are essentially unpredictable; they will depend on the severity of the initial change, the importance to the ecosystem of the elements removed or reduced in number, and the efficiency of the alternatives remaining. However, the presence of remnants and the need to manage them offers an opportunity to test theoretical expectations empirically. Only after some examples of simplification have been studied will we be in a position to make even tentative predictions about the consequences for other remnants of simplification of their biota. Meanwhile, this inability to predict or even speculate profitability gives an urgency to developing management skills within an adequate theoretical framework.

In essence I have taken the stance that functionally related localised groups of organisms which, taken together, are a mechanism for biological persistence. The interrelationships are most apparent in the roles occupied by different organisms. The listing of the frequencies of the common perturbations experienced in the wheatbelt in Table 1 shows that the likelihood of experiencing these perturbations (or sequence of perturbations) is so high that retention of diversity is essential if resilience is to be retained. But it must not be diversity for its own sake; it must encompass the kinds of organisms needed to fill:

1. the various roles during normal times;

2. the modal roles so that resilience is expressed following perturbation; and

3. the same roles under changed environmental conditions should they occur.

These are more explicit developments of the ideas advanced by Main (1982, 1984). The approach is more difficult than studies of energy flow but it aims at understanding the system rather than obtaining data which might be technically easier but unlikely to give the same insight into system function.

This chapter has not set out to develop a prescription for management, though there are indications of where management monitoring and research might be directed if ecosystem function is modelled around nutrient flow and retention. Data collecting and recording should commence immediately. This information by itself is not of much use; it must be interpreted within some hypothetical or conceptual framework such as the central theme expressed here. It is likely that it will need to be modified in the future. Meanwhile, because it has an operational orientation that is not achieved by a taxonomically oriented approach, it is possibly a more meaningful framework than a simple desire to protect or conserve what has been included within remnants in conservative reservations.

ACKNOWLEDGEMENTS

Grateful acknowledgement is made to A. J. M. Hopkins for his constructive comments on an early draft of the manuscript.

REFERENCES

Ashton, H. T., 1960. Thunder days in Australia. *Aust. Met. Magazine* 30: 44-51.

Beadle, N. C. W., 1966. Soil phosphate and its role in molding segments of the Australian flora and vegetation with special reference to xeromorphy and sclerophylly. *Ecology* 47: 992-1007.

Bowler, J. M., 1978. Glacial age aeolian events at high and low latitudes: A Southern Hemisphere perspective. Pp. 149-72 *in* Antarctic Glacial History and World Paleoenvironments ed by E. M. Van Zinderen Bakker. A. A. Balkema, Rotterdam.

Burvill, G. H., 1979. The natural environment. Pp. 91-105 *in* Agriculture in Western Australia ed by G. H. Burvill. University of Western Australia Press, Nedlands.

Chittleborough, R. G., 1985. Towards a state conservation strategy 1. Planning to meet climatic change. Bulletin 207 Department of Conservation and Environment Perth, Western Australia.

Clarke, R. H., 1962. Severe local wind storms in Australia. Technical Paper No. 13. CSIRO Division of Meteorological Physics, Melbourne.

Fitzpatrick, W. A., 1970. The expectancy of deficient winter rainfall and the potential for severe drought in the south-west of Western Australia. Institute of Agriculture Miscellaneous Publication 70/1. The University of Western Australia Agronomy Department, Nedlands.

Frankel, O. H., 1982. The role of conservation genetics in the conservation of rare species. Pp. 159-62 *in* Species at Risk: Research in Australia ed by R. H. Groves and W. D. L. Ride. Australian Academy of Science, Canberra.

Frankel, O. H. and Soulé, H. E., 1981. Conservation and Evolution. Cambridge University Press, Cambridge.

Gardner, C. A., 1957. The fire factor in relation to the vegetation of Western Australia. *West. Aust. Nat.* 5: 166-73.

Hallam, S. J., 1975. Fire and Hearth. Australian Institute of Aboriginal Studies, Canberra.

Holling, C. S., 1973. Resilience and stability of ecological system. *Ann. Rev. Ecol. and Syst.* 4: 1-23.

Hopkins, A. J. M., 1985. Fire in the woodlands and associated formations of the semi-arid region of southwestern Australia. Pp. 83-90 *in* Fire Ecology and Management of Western Australian Ecosystems ed by J. R. Ford. Western Australian Institute of Technology, Bentley.

Hopkins, A. J. M. and Robinson, C. J., 1981. Fire induced structural change in a Western Australian woodland. *Aust. J. Ecol.* 6: 177-88.

James, S. H., 1982. The relevance of genetic systems in *Isotoma petraea* to conservation practice. Pp. 63-71 *in* Species at Risk: Research in Australia ed by R. H. Groves and W. D. L. Ride. Australian Academy of Science, Canberra.

Kitchener, D. J., 1982. Predictors of vertebrate species richness in nature reserves in the Western Australian wheatbelt. *Aust. Wildl. Res.* 9: 1-7.

Kitchener, D. J., Chapman, A., Dell, J., Muir, B. G. and Palmer, M., 1980a. Lizard assemblage and reserve size and structure in the Western Australian wheatbelt — some implications for conservation. *Biol. Conserv.* 17: 25-62.

Kitchener, D. J., Chapman, A., Dell, J., Muir, B. G. and Palmer, M., 1980b. Conservation value for mammals of reserves in the Western Australian wheatbelt. *Biol. Conserv.* **18**: 179-207.

Kitchener, D. J., Dell, J., Muir, B. G. and Palmer, M., 1982. Birds in Western Australian wheatbelt reserves — implications for conservation. *Biol. Conserv.* **22**: 127-63.

Lamont, B. B., 1981. Specialized roots of non-symbiotic origin in heathlands. Pp. 183-95 *in* Heathlands and Related Shrublands: Ecosystems of the World, Vol. 9B ed by R. L. Specht. Elsevier Scientific Publishing Company, Amsterdam.

Lamont, B. B., 1984. Specialised modes of nutrition. Pp. 126-45 *in* Kwongan — Plant Life of the Sandplain ed by J. S. Pate and J. S. Beard. University of Western Austrlia Press, Nedlands.

Leigh, J. H. and Noble, J. C., 1981. The role of fire in the management of rangelands in Australia. Pp. 471-95 *in* Fire and the Australian Biota ed by A. M. Gill, R. H. Groves and I. R. Noble. Australian Academy of Science, Canberra.

Lourenz, R. S., 1981. Tropical Cyclones in the Australian Region, July 1909 to June 1981. Meteorological Summary. Bureau of Meteorology, Australia.

Main, A. R., 1979. The fauna. Pp. 77-99 *in* Environment and Science by B. J. O'Brien. University of Western Australia Press, Nedlands.

Main, A. R., 1981a. Ecosystem theory and management. *J. R. Soc. West. Aust.* **4**: 1-4.

Main, A. R., 1981b. Fire tolerance of heathland animals. Pp. 85-90 *in* Heathlands and related shrublands of the world, Vol. 9B ed by R. L. Specht. Elsevier Scientific Publishing Company, Amsterdam.

Main, A. R., 1982. Rare species: precious or dross? Pp. 163-74 *in* Species at Risk: Research in Australia ed by R. H. Groves and W. D. L. Ride. Australian Academy of Science, Canberra.

Main, A. R., 1984. Rare species: problems of conservation. *Search* **15**: 94-7.

Main, A. R. and Yadav, M., 1971. Conservation of macropods in reserves in Western Australia. *Biol. Conserv.* **3**: 123-32.

Minor, J. G., Peterson, R. E. and Louenz, R. S., 1980. Characteristics of Australian tornadoes. *Aust. Met. Magazine* **28**: 57-77.

Mulcahy, M. J., 1973. Land forms and soils of southern western Australia. *J. R. Soc. West. Aust.* **56**: 16-22.

Ozanne, P. G. and Specht, R. L., 1981. Mineral nutrition of heathlands: phosphorus toxicity. Pp. 209-13 *in* Heathlands and Related Shrublands: Ecosystems of the World, Vol. 9B ed by R. L. Specht. Elsevier Scientific Publishing Company, Amsterdam.

Pate, J. S. and Beard, J. S., 1984. Kwongan — Plant Life of the Sandplain. University of Western Australia Press, Nedlands.

Peck, A. J., 1975. Effects of land use on salt distribution in the soil. Pp. 77-90 *in* Plants in Saline Environments ed by A. Poljakoff-Mayber and J. Gale. Springer Verlag, Berlin.

Peters, R. L. and Darling, J. D. S., 1985. The greenhouse effect and nature reserves. *Bioscience* **35**: 707-17.

Pierrehumbert, C. I., 1973. Short duration rainfall over south-west Western Australia. Pp. 21-6 *in* Proceedings Hydrology Symposium 1973 Perth August 8-11. The Institution of Engineers, Australia, National Conference Publication No. 73/3, Perth.

Saunders, D. A., Rowley, I. and Smith, G. T., 1985. The effects of clearing for agriculture on the distribution of cockatoos in the south-west of Western Australia. Pp. 309-21 *in* Birds of Eucalypt Forests and Woodlands: Ecology, Conservation, Management ed by A. Keast, H. F. Recher, H. Ford, and D. Saunders. Surrey Beatty & Sons, Sydney.

Singh, G. and Geissler, E. A., 1985. Late Cainozoic history of vegetation, fire, lake levels and climate, at Lake George, New South Wales, Australia. *Phil. Trans. R. Soc. Lond. B.* **311**: 379-447.

Soulé, M. E., 1985. What is conservation biology? *Bioscience* **35**: 727-34.

Southern, R. L., 1979. The atmosphere. Pp. 183-226 *in* Environment and Science ed by B. J. O'Brien. University of Western Australia Press, Nedlands.

Specht, R. L., 1981. Conservation: Australian heathlands. Pp. 235-40 *in* Heathlands and Related Shrublands: Ecosystems of the World, Vol. 9B ed by R. L. Specht. Elsevier Scientific Publishing Company, Amsterdam.

Sutherland, J. P., 1981. The fouling community at Beaufort, North Carolina: A study in stability. *Amer. Nat.* **118**: 499-519.

Whelan, R. and Main, A. R., 1979. Insect grazing and post-fire plant succession in southwestern Australian woodland. *Aust. J. Ecol.* **4**: 387-98.

Williamson, D. R., 1973. Shallow ground water resources of some sands in southwestern Australia. Pp. 85-90 *in* Proceedings Hydrology Symposium 1973 Perth August 8-11. The Institution of Engineers, Australia, National Conference Publication No. 73/3, Perth.

CHAPTER 2

Ecological Studies as the Basis for Management

A. J. M. Hopkins[1] and D. A. Saunders[2]

Ideally, management decisions for nature conservation should be based on a sound knowledge of the resources of an area and the processes (geomorphological, ecological, evolutionary, etc.) that sustain those resources. Ecological studies are thus fundamental to good management. A case study is used to illustrate the merits of undertaking detailed and multi-faceted research at a few selected sites in order to develop a comprehensive theory for management. Unfortunately, management is often carried out in the absence of most, if not all, of the information necessary for effective planning. Ecological studies must then become a *part* of management *in addition* to being a precursor: every management activity must be planned and performed in a manner that incorporates information gathering to promote improvements in decision making. Monitoring and re-evaluation are basic to this. An experimental approach to management should also be adopted. These procedures will contribute to a better integration of research, planning and management.

INTRODUCTION

THE past 10-15 years has seen a massive increase in the amount of land dedicated for nature conservation in most parts of Australia. For example, in Western Australia, the area of National Parks and Nature Reserves has gone from 2.2 million ha in 1969 to 14.3 million ha in 1984 (Table 1). A similar trend for other States can be seen by inspection of the annual reports of relevant government departments. This acquisition phase is now losing momentum and the interest of authorities charged with conservation is turning towards the development of management practices (e.g. Ride and Wilson 1982a), including theory and techniques. There is increasing recognition that benign neglect (*sensu* Soulé *et al.* 1979) will result in a decline in the conservation values of land set aside as National Parks and Nature Reserves and that interventionist management will be necessary to halt this decline to ensure that the conservation objectives can be met.

Effective management of natural lands for nature conservation purposes must be based on sound knowledge of the biological and physical resources of those lands and the processes contributing to the continued existence of the resources. On that basis, the title of this chapter is a statement of the obvious. If it were to read 'Ecological Studies *should be* the Basis for Management' then there would be no dissent. As the title stands, however, we have the opportunity to comment on the sorts of studies that should be undertaken and how they should be linked to the conservation planning and management framework to ensure maximum effect.

Table 1. Trends in acquisition of land for National Parks and Nature Reserves in Western Australia over the period 1969-84. (Data from annual reports of the National Parks Board, National Parks Authority and Western Australian Wildlife Authority). Note that, in addition, in Forests Department Working Plan No. 86 (1977) some 360,000 ha of State Forest were designated as Special Management Priority Areas for Conservation.

As at June 30	National Parks No.	Area (ha)	Nature Reserves No.	Area (ha)
1969	34	478,932	278	1,721,788
1970	47	1,419,449	320	1,774,386
1971	51	1,445,333	363	4,958,363
1972	62	1,462,527	394	5,078,660
1973	63	1,770,822	440	5,013,287
1974	63	1,770,784	454	5,933,935
1975	64	2,271,702	491	5,013,037
1976	64	2,281,450	918	5,339,947
1977	68	3,871,505	946	6,927,627
1978	69	4,463,893	977	8,260,614
1979	61	4,551,593	1016	8,536,654
1980	62	4,274,465	1036	9,083,823
1981	63	4,363,968	1062	9,663,637
1982	63	4,366,255	1069	9,683,211
1983	65	4,416,715	1075	9,890,796
1984	63	4,426,811	1131	9,900,531

---

[1]Department of Conservation and Land Management, Western Australian Wildlife Research Centre, P.O. Box 51, Wanneroo, Western Australia 6065.
[2]CSIRO, Division of Wildlife and Rangelands Research, L.M.B. No. 4, P.O. Midland, Western Australia 6056.
Pages 15-28 *in* NATURE CONSERVATION: THE ROLE OF REMNANTS OF NATIVE VEGETATION ed by Denis A. Saunders, Graham W. Arnold, Andrew A. Burbidge and Angas J. M. Hopkins. Surrey Beatty and Sons Pty Limited in association with CSIRO and CALM, 1987.

OBJECTIVES FOR MANAGEMENT OF NATURE CONSERVATION AREAS

In the hierarchy of objective setting it is usual to go from the general, or principal, objectives for a region to more specific, issue-related or area-related objectives. For a nature conservation agency, an appropriate principal objective would be to conserve in perpetuity the full array of plant and animal species found within the region. 'In perpetuity' is an unlimited concept of time, so to better define the time-scale involved we suggest that a period of 10,000 years be used. This figure, drawn from studies of extinction kinetics (Slatyer 1975), is likely to incorporate a major part of any global climatic cycle (e.g. Bowler 1982) with its attendant implications for changes in relative importance of plant and animal species (Main 1984). The figure also serves to emphasize the long-term nature of the commitment to nature conservation.

At a regional level it is appropriate to focus on total numbers of species when setting objectives. At a more local level, however, the objectives should also incorporate consideration of combinations of species (communities) and habitats, the physical environments in which those species and communities exist and in which management at any reasonable level of intensity can contribute effectively to conservation. For some nature conservation areas, a particular species may be identified as having priority for management. This is often the case where rare species are concerned and, more particularly, where there is some special statutory protection of those rare species.

The objectives for management of a nature conservation area are often spelt out at the time of dedication of that area through the process of gazetting a purpose and vesting body. Where land is reserved for conservation of flora and fauna, a principal objective as described above would apply. Where the purpose includes recreation, as for many national parks, then objectives of management must include provision for that activity.

As will become apparent later in this chapter, many conservation areas are too small, too fragmented and too dispersed across the landscape for each to be fully effective in achieving conservation of the complete assemblage of plants and animals that existed there prior to the surrounding native vegetation being cleared. Such is the case with the reserves in the Western Australian wheatbelt, a region of about 14 million ha where there are about 500 nature reserves and about three-quarters of which contain less than 400 ha each (Kitchener *et al.* 1980a; see also Wallace and Moore, this volume). It is not realistic to try to conserve all species on all reserves; therefore, it becomes an important part of the planning process to interpret and redefine management objectives for each individual reserve in the context of the whole system of reserves and associated privately owned remnants of native vegetation. It is these local objectives that invariably become the focus of attention for reserve managers because they are regarded as achievable. Furthermore, they are used in the setting of regional nature conservation priorities. It is important, however, not to lose sight of the principal objectives because these provide a perspective that serves to emphasize the need for care and communication in the management process.

In the planning and management process it is necessary to set local objectives for each reserve and to establish priorities for actions. Invariably, however, only generalized objectives can be set because there is insufficient knowledge on the resources of any particular reserve, how they contribute to the overall conservation reserve system or how they are to be managed. It is generally only where special biological or ecological values have already been identified that specific objectives can readily be set. For most reserves then, the objective will be essentially 'to conserve as far as possible the complete array of organisms present on the reserve'. Refinement of objectives and priorities will depend heavily on ecological research being conducted.

WHEN IS MANAGEMENT NECESSARY?

Resource management, in the broadest terms, involves a series of actions resulting from conscious decisions about that resource. This definition incorporates the 'do-nothing' option as a valid form of management. Looking over the fence into a nature reserve and deciding to return in 12 months time constitutes a management action and should be properly noted as such within the appropriate records system.

The biological resources of a nature conservation area are the species present in that area. The populations of these and the communities of which they form part are dynamic, being sustained by a variety of processes (geomorphological, ecological, etc. and, on a longer time-scale, evolutionary) and interactions between species. While the objective of management is to conserve, the actions of the manager are directed towards maintaining these natural processes and interactions and, at the same time, minimising adverse impacts resulting from most human activities. As an example, in the Western Australian wheatbelt, the manager may seek to maintain the regimé of burning that existed prior to European colonization by prescribing fires for ecological purposes (Hopkins 1985a) while minimizing any tendency towards excessive burning by using a variety of fire protection techniques. The manager also has obligations to protect neighbouring landholders from fire. While this forms part of a secondary objective it is very important because failure to achieve it may result in loss of community support for nature conservation generally.

There are some adverse impacts that are beyond the capacity of managers to redress. These would include such things as global climatic change (Chittleborough 1985), effects of acid rain (e.g. Troyanowsky 1985) and gradual pollution of groundwaters (e.g. Malcolm 1983). The manager should be aware of these impacts and set up procedures to monitor their effects.

Remnants have two characteristics that engender in them a need for management. Firstly, they are usually isolated from nearby areas of intact native vegetation by clearing or overgrazing. Many conservation problems arise as a result of fragmentation; these and their management solutions are discussed elsewhere in this volume. The second characteristic is size; often remnants are too small to be viable without considerable management.

The issue of self-sustainability of nature conservation lands has surfaced most often in the context of the debate about reserve size and design (single large versus several small reserves) that has arisen as a result of island biogeographic studies (e.g. Boecklen and Bell, this volume). We do not wish to become embroiled in that debate but reserve size is also important when considering the level of management likely to be necessary to conserve the biota. Stated simply, if the faunal (and floral) collapse is faster in a small reserve than in a large reserve, then the level of management required to arrest that collapse will be greater for the smaller area. We consider that, as a general rule, any isolated reserve in Australia of less than 500,000 ha will require active, interventionist management if it is to maintain its full complement of species in the long term. (Larger areas may require control of exotic animals and plants as minimum management activity). This figure of 500,000 ha is based on an evaluation of the literature on reserve size and design and taking into account the special issues that arise in conserving some rare species. For example, Barrow Island (22,250 ha) is too small to have retained the red kangaroo *Megaleia rufa* and the emu *Dromaius novahollandiae* since it separated from the mainland 6-8000 years ago. Slatyer (1975) calculated that, in the arid zone, an area of 500,000 ha would be required for the conservation of a population of around 5000 red kangaroos; this area would probably support a viable emu population in the long term also. The figure of 500,000 ha provides a rough but useful guide to a minimum size for a self sustaining reserve for the purposes of discussion in this chapter. We do recognize however, that such figures are greatly affected by local factors.

Looking at data on National Parks and Nature Reserves for Western Australia (Table 2) it can be seen that only eight out of a total of 1100 parks and

Table 2. Sizes of Nature Conservation Areas in Western Australia in 1982 (compiled from Hinchey 1982). Areas of State Forest managed for nature conservation not included.

Size Class of Reserves (ha)	Number of reserves	
	National Parks	Nature Reserves
0-10	3	133
10-20	2	58
20-50	2	142
50-100	4	133
100-500	4	318
500-1000	2	87
1000-10,000	22	138
10,000-100,000	14	23
100,000-500,000	7	7
500,000-1,000,00	1	4
>1,000,000	1	2
Total Number of Reserves	62	1045
Total Area (ha)	4,366,248	9,667,371
Average Reserve Size (ha)	70,423	9251
Total Reserve Area 14,033,619 ha (5.56% of State)		

reserves exceed this size threshold. Even if the area capable of sustaining itself is reduced to, say, 100,000 ha the fact remains that by far the majority of conservation areas in Western Australia will require deliberate management.

ECOLOGICAL STUDIES FOR MANAGEMENT

If management is to be based on informed decisions, then three types of knowledge are required for any area — the what, where and how:

What species are present. This is generally gleaned from surveys supported by appropriate taxonomic research.

Where each species is located in that area, especially in relation to its occurrences elsewhere. A detailed ecological survey with appropriate analysis will provide the basis for developing predictive models (e.g. BIOCLIM, Busby 1985; GLIM, Austin *et al.* 1983) which will greatly enhance future survey work. For rare species with special statutory protection, detailed location data are required to avoid destruction by management activities including such things as road and fire trail construction and prescribed burning.

How each species persists on that site. Generally, insight into the processes and interactions that sustain a viable population will require studies of a broad range of autecological topics that include such things as reproductive biology and some aspects of ecophysiology. Attention should also be paid to studies of adverse impacts (e.g. trampling at visitor access points).

As was suggested earlier, it is necessary to have sufficient information to place each reserve in a regional context. The broader studies needed to achieve this include regional biological surveys (McKenzie 1984) as well as those just across the

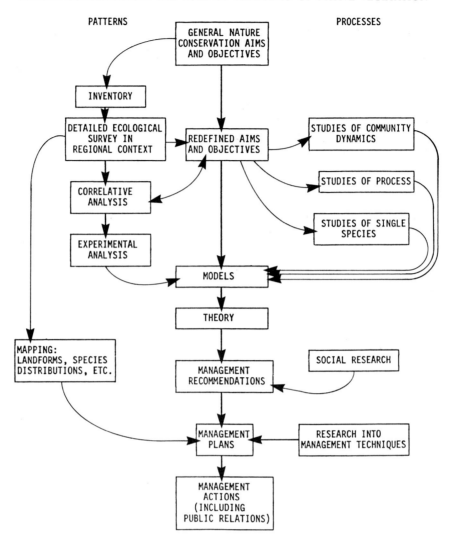

Fig. 1. A framework for ecological studies on nature conservation lands leading to the development of sound management plans (adapted from Austin 1984).

boundary of the reserve to examine the interactions between reserve and non-reserve land and the reserve and nearby areas of natural vegetation not formally designated reserve.

Where we have the luxury of undertaking research prior to management then studies can be developed according to a logical sequence as outlined in Figure 1. These studies start off with a general reconnaissance of the area in question, become increasingly detailed and lead to the development of models and theory. Studies on the left of Figure 1 under the heading 'patterns' provide the *what* and *where* information while studies under 'processes' provide insight into *how*. (The omission of studies of evolutionary biology from this figure is a reflection of the title of this chapter rather than a comment on their importance).

Clearly it will not be possible to undertake the full range of ecological studies outlined here on every nature conservation area. The most realistic approach is to undertake detailed studies of patterns and processes on a few selected sites and to work towards the development of a theory that explains the what, where and how for species at these sites. It is this theory that will provide the foundation for management activity across the landscape because it can be extrapolated from site to site, whereas the information on species and communities may be of local value only.

A CASE STUDY

An example of the detailed approach to the study of natural areas is provided by the programme of research in the so-called northern sandplains of southwestern Australia (the Irwin Botanical District, Beard 1980). This District (Fig. 2) encompasses the northern part of the Perth Sedimentary Basin which is composed of sandstones, siltstones and shales deposited in the Mesozoic (Playford *et al.* 1976). The District has an area of about 40,000 km^2 of which 96% supported shrub-dominated vegetation types (kwongan) and a mere 2% were woodlands prior to European occupation (Beard and Sprenger 1984). Approximately 60% of the district has been cleared for agriculture.

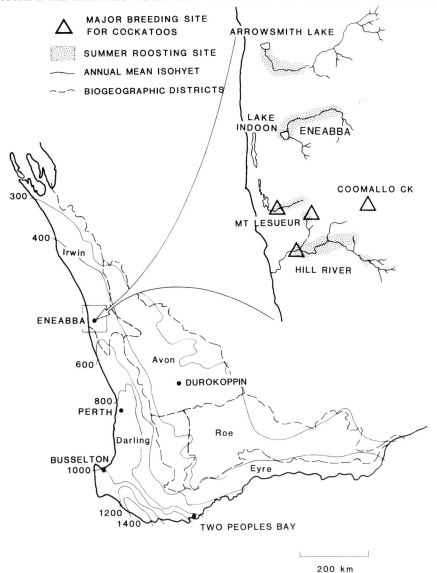

Fig. 2. Map of the biogeographic districts (Beard 1980) in southwestern Australia with the major study area indicated. The insert shows the major breeding and summer roosting sites for Carnaby's cockatoo in the Mt Lesueur-Eneabba area.

Research has been concentrated in the area around and between Mt Lesueur and Eneabba in the southern part of the Irwin District (Fig. 2) in the expectation that the general findings and the resulting theory would be applicable over a wider area. This study area was selected on the basis of the following attributes:

1. it was thought to be rich in plant species including some rare and restricted species;

2. there were a number of Nature Reserves and National Parks; but

3. some other areas of supposed biological importance were not included in the system of reserves; and

4. there were some major conflicts concerning the use of land in the area (mainly conservation versus mining and agriculture) that were inhibiting reservation of these additional areas.

This last attribute provided the initial impetus for some of the studies and subsequently led to an increased research effort because of the requirements for effective post-mining rehabilitation.

Apart from the early, often opportunistic, plant collections that were made in the Mt Lesueur-Eneabba area, the initial reconnaissance surveys of the region were by environmental consultants to mining companies and they were mostly little more than species lists. The seminal work of Speck (1958) was more systematic in that it covered the whole Irwin Botanical District but no attempt was made to quantify aspects of the much-vaunted botanical importance of the area until the work of George *et al.* (1979).

Detailed plant ecological studies in the Eneabba area (Lamont 1976; Hnatiuk and Hopkins 1981; Hopkins and Hnatiuk 1981) have been complemented by similar work in the Mt Lesueur area (Hopkins and Griffin 1984; Griffin and Hopkins

1985a). These studies were designed to identify groupings of plant species and important relationships between plants and their environment. Identification of floristic groupings was necessary because of the disparity between patterns defined on the basis of vegetation structure (physiognomy) and those defined on the basis of floristic composition (see Hopkins and Griffin 1984). In general, soil type was shown to be the most important local factor influencing floristic patterns but the correlation was not particularly strong; for example, the soils-related axis (Axis 2) in the ordination of Hnatiuk and Hopkins (1981) accounted for only 6% of the variability of the data. Some possible reasons for this poor correlation have been advanced by Hopkins and Griffin (1984) (including inadequacies in the existing soil classification, the presence of a complex mosaic of soils and the floristically rich and variable nature of the vegetation) and work is continuing with a view to better defining the relationships between plants and soils and to elicit the contribution of other factors to floristic patterns.

As a result of the difficulties encountered in adequately describing soil profiles during the local ecological studies, the subsequent regional study (Griffin et al. 1983) was designed to sample plants on one, clearly defined soil type (laterite in upper landscape positions). That study revealed a significant correlation between the distribution of floristic groups and climate patterns. What is probably more important in the context of this chapter is that the study provided a regional framework for interpreting the results of the more detailed, local studies. For example, it showed that the Mt Lesueur sites were all similar to each other but different from the sites at Eneabba.

Studies of community dynamics (processes, Fig. 1) have, thus far, been concentrated on the effects of fire and other disturbance on the vegetation (e.g. Hnatiuk and Hopkins 1980; Griffin and Hopkins 1981; Bell et al. 1984). These have now been supplemented by a detailed study of nutrient cycling (Low and Lamont 1985) and population studies of selected species (Cowling and Lamont 1984; Lamont 1985). A study of the role of mycorrhizae is also proposed (Brooks 1985).

A major contribution to understanding the dynamics of plant communities in the Mt Lesueur-Eneabba area came from work associated with programmes examining the rehabilitation of mined areas. The companies mining mineral sands on Crown land at Eneabba are required to rehabilitate to native vegetation after completion of mining and to undertake a continuous programme of research and monitoring as a condition of approval to mine. One of the key studies is a 4 × 2 × 2 × 2 × 5 factorial experiment involving a variety of possible rehabilitation treatments; e.g. topsoil storage, various levels of fertilizer applications and application of brush matting. The results of one analysis (Fig. 3) demonstrate the interaction between abundance and plant cover: as individual plants increase in size and thus cover increases, there is a decline in abundance as some individuals are affected by competition. Because the density of plants of each species is low, this thinning process is also reflected in a decline in species richness.

The need to review the success of rehabilitation has also, by necessity, produced useful insights into mechanisms contributing to community processes. For example, the absence of some species from rehabilitation sites led to the analysis of species by reproductive mode (Griffin and Hopkins 1985b). Where regeneration is by migration (*sensu* Grubb and Hopkins 1986) because there are no propagules remaining *in situ* after mining, then five categories of species based on propagule type need to be recognized. Each category requires a particular technique to ensure its contribution to the flora of the rehabilitated area.

Formal models of the plant communities of the Irwin Botanical District have yet to be developed but several approaches are currently being explored. A JABOWA-type model (Botkin et al. 1972) using a simplified array of species types and some of the vital attributes identified by Noble and Slatyer (1980) could provide a tool for investigating the effects of various management strategies on species composition (Shugart and West 1980, 1981). As such it could be incorporated into a computerised land management information system such as PREPLAN (Kessell et al. 1984). The use of models that involve investigation of single species distribution patterns (e.g. GLIM, Austin et al. 1983; BIOCLIM, Busby 1985) may provide the basis for extrapolating results from the JABOWA/PREPLAN models to other sites.

A unified theory of the plant communities of the Mt Lesueur-Eneabba area has not yet been developed. However, elements of such a theory are gradually emerging as specific questions are raised and the data are reviewed in the light of those questions (e.g. Hopkins et al. 1983; Bell et al. 1984; Hopkins and Griffin 1984; Lamont et al. 1984; Griffin and Hopkins 1985a and b; Hopkins 1985a and b; Grubb and Hopkins, in press). While much remains to be done to consolidate this work, results to date demonstrate conclusively the merits of detailed and many-faceted studies of a limited area: the present level of understanding of the communities and ecosystem processes is much greater than the sum of the contributions of the individual, component studies because of the cross-referencing of information that is possible. Furthermore, it is anticipated that the theory that will be developed as a result of this work will be relevant to sclerophyllous plant communities

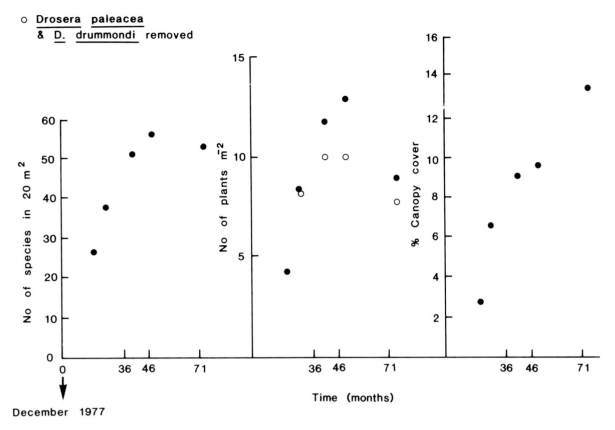

Fig. 3. Regeneration of native plant species in Block B4 (best treatment: fresh topsoil plus brush matting) of the Eneabba South East Factorial Experiment. Data are averaged for two 20 m² quadrats. (Data courtesy of Allied Eneabba Pty Ltd).

throughout the whole of southwestern Australia and will provide a solid foundation for continuing conservation and management.

While there have been extensive studies of the flora of the area, the fauna has not received the same attention. Parts of the Mt Lesueur area have been the subject of a vertebrate fauna survey (Chapman et al. 1977) and studies of sub-fossil deposits in a nearby cave (Lundelius 1960; Baynes 1979) have supplemented these data as well as providing insight into the palaeo-fauna. The Eneabba site, on the other hand, has not been comprehensively surveyed for vertebrates but the invertebrates have received attention because of their postulated role in the process of rehabilitation of mined areas (Majer et al. 1982). The only component of the fauna that has been studied on a regional basis is the cockatoos (see Saunders and Ingram, this volume; Saunders 1979, 1980, 1982). However, most of the cockatoo study sites are in areas other than those where detailed botanical studies have been conducted.

So far, little attempt has been made to draw together these rather disparate fauna studies and to integrate the results with those from the botanical studies. Integration is proposed as a part of the process of developing management plans for conservation lands in the region. One of the first steps in this will involve studying the use by cockatoos of the Mt Lesueur area. This information can be readily placed in a regional context that will permit management orientated inferences to be drawn. A pointer to the types of inferences that can be anticipated is provided by results of the work on Carnaby's cockatoo *Calyptorhynchus funereus latirostris*, summarized below.

Carnaby's cockatoo is the largest cockatoo found in the region. It has four major requirements from the environment in which it lives:

1. free water which, because of the mediterranean-type climate, is only found all year in small pools and soaks in the local, sluggish drainage system and in the few wetlands of the region;

2. nest hollows of particular dimensions (Saunders 1979). In the Mt Lesueur area suitable hollows are generally confined to *Eucalyptus wandoo* formations;

3. food resources (Saunders 1980). The birds feed on seeds and seed-eating invertebrates which are found in shrub formations dominated floristically by species of Proteaceae; and

4. trees around watercourses. During the hot summer the birds are not able to forage during the heat of the day and they must shelter in trees, close to water and food.

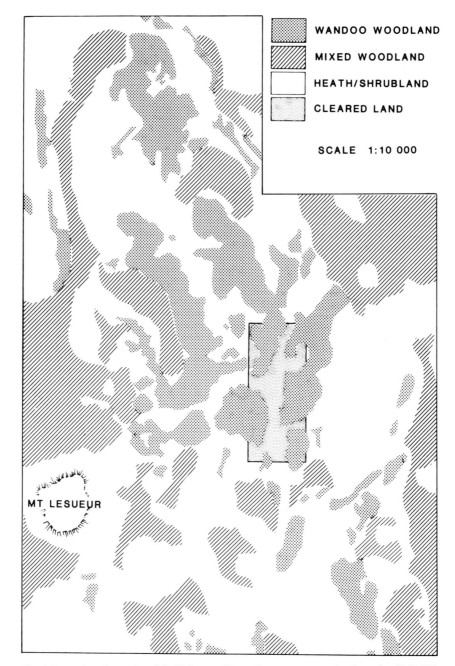

Fig. 4. Vegetation of a portion of the Mt Lesueur Nature Reserve as mapped at the scale of 1:25,000 (E. A. Griffin, Unpublished) showing the mosaic of *Eucalyptus wandoo* woodland, mixed woodlands, mainly dominated by *E. accedens, E. todtiana* and *Banksia* spp. and shrub dominated communities (kwongan).

Prior to the large scale clearance of native vegetation for agriculture that has occurred over the past 25 years (see Saunders and Ingram, this volume; Saunders 1986, Table 1, data for Coomallo Creek) the landscape of the region consisted mainly of kwongan with a few small pockets of woodlands (Beard 1979). Even within these woodland pockets, however, shrublands are very important as is illustrated in Figure 4: the area to the east of Mt Lesueur was mapped at the 1:250,000 scale by Beard (1979) as woodland whereas more detailed mapping at 1:25,000 reveals a mosaic of shrublands and woodlands. This mosaic provides ideal habitat for Carnaby's cockatoo because of the juxtaposition of suitable nest sites with abundant food resources (see Saunders and Ingram, this volume).

Within the Mt Lesueur-Eneabba area there are only a few pockets of woodland which satisfy the requirements for the successful breeding of Carnaby's cockatoo (Fig. 2). Once breeding has finished, the birds from all of these breeding areas range nomadically over much of the area but during the hot weather they must concentrate in large flocks on those areas where there is both free water and shade trees (Saunders 1980). It is only after the rains come in April and free water is more widespread that the birds can move away from these limited areas and disperse into smaller flocks. As a result, for much of the year the birds are tied to a few locations which supply the resources essential to their survival.

An examination of the areas which satisfy the animals' needs shows that many are remnants of native vegetation which are on private property or vacant Crown land and, therefore, outside the designated conservation areas. This study of Carnaby's cockatoo illustrates the importance of considering the *system* of remnants (private and reserve) when looking at management for conservation rather than any one remnant in isolation. In the case of this species, it is clear that any major disturbance to one site (e.g. a fire that consumes all the proteaceous shrubs, clearing that removes the trees close to water sources) will have an impact on the other sites in the region because the birds will tend to congregate more densely at those other sites.

It is also important to consider the effects of other changes in the use of land in the region when devising management strategies for conservation lands. Largely as a result of land clearing for agriculture in the Mt Lesueur-Eneabba region, two other species of cockatoo, the galah *Cacatua roseicapilla* and the long-billed corella *C. pastinator pastinator* have increased in abundance (Saunders *et al.* 1985). These two species feed on agricultural products and agricultural weeds. It is likely that both of them will compete with Carnaby's cockatoo for nest hollows and thus may jeopardise the species survival in the absence of management.

IDENTIFICATION OF SHORT-CUTS: STUDIES OF VULNERABLE SPECIES

Clearly it is not possible to commit resources to the study of every part of Australia to the extent that has been done for the Mt Lesueur-Eneabba region; nor is it really necessary. Detailed studies at a few sites provide the basis for development of theories that can have widespread applicability. Site validation of these theories may be achieved through brief surveys followed by well designed monitoring programmes.

In order to develop management guidelines in the absence of broad theory, it is possible to utilize existing data more effectively than has often been the case in the past. One such short-cut involves the identification of vulnerable species (*sensu* Ride and Wilson 1982b): species that could be endangered by a change in environmental processes or loss of environmental requirements. Once the vulnerable species are identified then management prescriptions can be modified to ensure continued persistence of those species.

A recent survey of the effects of clearing for agriculture on the distribution of nine species of cockatoos in southwestern Australia, based in part on the above-mentioned work in the Mt Lesueur-Eneabba area (Saunders *et al.* 1985) revealed that three of the nine are vulnerable because of their dependance on native vegetation and one further species is vulnerable as a consequence of its total dependence now on an agricultural weed (*Emex australis*) as a food resource. The simple dietary analysis forming the basis of these conclusions could have been made on the basis of information published over ten years ago and did not require the level of detail that went into it (Table 3). In a similar approach, Kitchener *et al.* (1980a, b, 1982) have analyzed presence/absence data of vertebrate fauna on selected reserves throughout the Western Australian wheatbelt to assess the conservation value of these reserves to birds, mammals and lizards. At the same time they identified vulnerable species of birds based on a variety of life history and habitat useage characteristics [see also Terborgh (1974) for an example of an international approach to this issue].

A group of plant species that is vulnerable to altered disturbance regimes (including fire) has been recognized on the basis of reproductive and life history characteristics (Bell *et al.* 1984; Hopkins 1985b; see also Grubb and Hopkins 1986). These species were grouped by virtue of being fire sensitive obligate seed regenerators with a seed store in woody or papery fruits held above-ground. For continued persistence at any site, each of these species requires a period free from disturbance of any kind likely to cause widespread mortality (e.g. fire, drought, flooding, trampling, etc.). This period must be greater than the primary juvenile period in order to re-establish the seed bank. Recurrent disturbance at any interval less than, equal to, or even slightly longer than, the primary juvenile period can cause attrition of the population leading to eventual, local extinction. The life history details of species in this group have been used to devise fire management guidelines for conservation lands in southwestern Australia (Hopkins 1985b).

INTEGRATION OF RESEARCH WITHIN THE PLANNING AND MANAGEMENT PROCESS

The discussion so far has largely reflected the traditional view of the role of research whereby research results are regarded as a prerequisite to management planning which, in turn, precedes management; i.e. the situation in Figure 1. But the situation in fact is that decisions about management will almost always be made in the absence of adequate knowledge derived from research. Most management decisions cannot be postponed to allow time for research to be undertaken *even if* sufficient resources were to be allocated to the research effort.

Table 3. Major food items of the nine species of cockatoo found in southwestern Australia (modified from Table 1, Saunders et al. 1985). (xx indicates that item comprised the bulk of the diet).

	Agricultural grains	Grass and herb seeds	Major agricultural weeds	Proteaceae seed	Marri *Eucalyptus calophylla* seed	Seed or Invertebrates from native vegetation other than that listed in other columns
Galah *Cacatua roseicapilla*	xx	x		x		x
Major Mitchell's cockatoo *C. leadbeateri*	xx	x	xx	x		x
Sulphur crested cockatoo *C. galerita*	xx	x			x	x
Little corella *C. pastinator gymnopis*	xx	x	x			x
Long-billed corella *C. p. pastinator*	xx	x	x			x
Forest red-tailed black cockatoo *Calyptorhynchus magnificus naso*					xx	x
Inland red-tailed black cockatoo *C. m. samueli*			xx			
Baudin's cockatoo *C. baudinii*				x	xx	x
Carnaby's cockatoo *C. funereus latirostris*				xx	x	x

In view of the need for management to proceed yet, at the same time, for research data to be gathered to provide for improvements in management decisions, it is necessary to consider an integration of these two activities. An organizational framework which should achieve integration is shown in Figure 5. The key feature of this framework is the sections involving monitoring and re-evaluation: each management action includes appraisal, the results of which feed back into the information base to generate a gradual improvement in the knowledge of the system.

The procedure of monitoring and re-evaluation can best be incorporated into management by those responsible for planning. The planning process involves assessment and analysis of available information and prescription of management actions. These management actions should be described in terms of hypotheses such that an experimental procedure for testing each hypothesis (e.g. a monitoring programme) is designed into the management programme. In addition, the planners are in a strong position to identify critical gaps in the information base and to design appropriate research programmes to supply the information.

In this scheme, the managers not only become responsible for implementing the management programme, they also establish monitoring sites and carry out the monitoring of those sites. Because of their close involvement with the monitoring process, managers must also be involved in interpreting the results of the monitoring programmes. This is an efficient and effective means of collecting research data but the process has the added advantage that it fosters in the managers an awareness of, and interest in, the results of management actions.

The research into patterns and processes as previously outlined is necessary and should continue. But research teams should also become involved in the process of planning for management. Researchers must interact closely with the planners in the design of experimental and monitoring programmes. They must guide and assist the managers in setting up the monitoring programmes and then take a major role in interpreting the results of those programmes. As a result, the researchers become more responsive to the needs of planners and managers and forge close links with those operational personnel.

Adoption of an organizational framework like the one proposed in Figure 5 can provide many benefits. It certainly allows for a gradual improvement in knowledge while management continues. It provides safeguards on management actions because results of those actions are subject to evaluation. It can improve efficiency of data gathering through having that work done by on-site operational staff, and it brings together the personnel from the three functional areas of nature conservation agencies (research, planning, management) to work

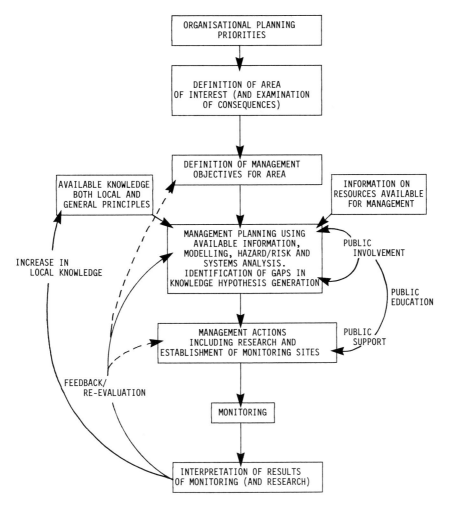

Fig. 5. A conceptual framework for the integration of research, planning and management activities in a nature conservation agency with emphasis on monitoring and re-evaluation.

collaboratively. Unfortunately we are unable to point to a situation where such institutional arrangements have been applied and to report on their success (or otherwise). Furthermore it is likely that other arrangements could equally well advance the cause of nature conservation. It just happens that the organizational framework given in Figure 5 is one that appeals to us as a logical one. It should be noted, however, that no arrangement is a panacea, or a substitute for motivated and dedicated staff.

EXPERIMENTAL MANAGEMENT AND UNCERTAINTY

The integration of research and management functions provides opportunities to undertake experimental management as a means of achieving both research and management objectives. In this context, experimental management means implementing management programmes as controlled field experiments: where appropriate, the programmes are designed to fit within an experimental framework so that the effects of those programmes can be evaluated and the results clearly interpreted. Ideally, hypotheses should be generated and rigorously tested. However, more open-ended inquiry is also possible within the general framework of experimental management because that framework serves to ensure that research is relevant.

An elaboration of the procedures outlined above has been given by Holling (1978) who has also drawn attention to the many unforeseen events that can impinge on programmes of management. These include natural events such as drought, fire, floods and epidemics. But, in addition, changes in economic or political factors or social attitudes may have dramatic effects on management programmes. Natural resource management operates not only in a field of inadequate knowledge but also in an atmosphere of uncertainty. To cope with the uncertainty, Holling has advocated inclusion in programmes of experimental management some programmes that are specifically established to explore the boundaries of stability of the communities and ecosystems. They should be designed to increase our understanding and awareness of the way these communities and ecosystems function under stress such that, should an unforeseen event arise, there is at least some appropriate knowledge available to enable managers to cope with it. This recommendation comes with two *caveates:*

1. there should be a structured learning process associated with the experimental management programme (e.g. a monitoring programme); and

2. the programme should not destroy the system or the observer.

CONCLUSIONS

The development of sound management practices that will ensure the conservation of the indigenous biota is one of the most challenging issues facing conservationists, land managers and governments today. The problems are particularly acute when:

1. remnants are small and thus prone to deterioration and loss of conservation values; and

2. publicly owned conservation lands inadequately sample the landscape such that reliance must be placed on conservation values of private lands.

This is the situation that exists throughout most of the Western Australian wheatbelt where only small, scattered remnants of native vegetation remain in an agricultural landscape.

Sound management depends on a knowledge of the resources to be managed and an understanding of the processes that sustain those resources and the effects of management practices on the resources and processes. To that extent, research is essential to management. But the research must be relevant. We have suggested two strategies that might be employed to ensure relevance. Firstly, studies should be structured within a clearly defined framework leading to the development of models and theory that become the building blocks of management practice. Secondly, the management questions can be identified in the planning process and, when couched in terms of hypotheses, testing procedures can be readily determined and implemented.

Inevitably, most management of remnants will occur in the absence of perfect knowledge and understanding. In this event it is often possible to make some intelligent choices between management options based on existing information. Here we have emphasized the value of basing such decisions on known responses of vulnerable species. This view is based on the premise that, if persistence of vulnerable species can be ensured, then other, more resilient species may also be conserved.

Most nature conservation agencies have insufficient research capability to remedy the apparent deficiencies in resource management information. To cope with this, we have proposed changes in organizational arrangements whereby management personnel become more closely involved in data gathering through the process of monitoring of management actions and their effects. The short-term result of implementing the changes we propose should be to produce a more efficient and effective, integrated organization. In the longer term, the monitoring procedures should lead to iterative improvements in management decisions and practices as information accumulates.

Finally, it is noted that, once management becomes more closely associated with research, the opportunities to undertake experimental management will become apparent. Programmes of experimental management should include not only application of disturbance regimés under controlled conditions with appropriate evaluation but also the application of some extreme disturbances in order to explore the stability limits of the communities and ecosystems.

ACKNOWLEDGEMENTS

Many of the ideas presented in this chapter have their origins in the many lengthy discussions that we have had with colleagues, reserve managers and reserve neighbours over the past decade or so or have appeared elsewhere in print during that period. In some cases we are unable to attribute credit directly to the people who introduced the ideas to us, because of the passage of time. We do wish to thank Richard McKellar for giving us the benefit of his insights into the planning process. Andrew Burbidge, Peter Bridgewater and Mike Austin provided valuable comments on an earlier draft of the manuscript, while preparation of the chapter was greatly assisted by Jill Pryde, Raelene Hick, Ted Griffin, Judith Brown and Perry de Rebeira.

REFERENCES

Austin, M. P., 1984. Problems of vegetation analysis for nature conservation. Pp. 101-29 *in* Survey Methods for Nature Conservation ed by K. Myers, C. R. Margules and I. Musto. Vol. I. CSIRO Division of Water and Land Resources, Canberra.

Austin, M. P., Cunningham, R. B. and Good, R. B., 1983. Altitudinal distribution of several eucalypt species in relation to other environmental factors in southern New South Wales. *Aust. J. Ecol.* **8**: 169-80.

Baynes, A., 1979. Analysis of a late Quaternary mammal fauna from Hastings Cave, Jurien, Western Australia. Ph.D. Thesis, University of Western Australia, Perth.

Beard, J. S., 1979. Vegetation Survey of Western Australia. The Vegetation of the Moora and Hill River Areas, Western Australia. Map and Explanatory Memoir. 1:250,000 series. Vegmap Publications, Perth.

Beard, J. S., 1980. A new phytogeographic map of Western Australia. *West. Aust. Herb. Research Notes* **3**: 37-58.

Beard, J. S. and Sprenger, B. S., 1984. Geographical data from the vegetation survey of Western Australia. Vegetation Survey of Western Australia. Occasional Paper No. 2. 62 pp. Vegmap Publications, Perth.

Bell, D. T., Hopkins, A. J. M. and Pate, J. S., 1984. Fire in the kwongan. Pp. 178-204 *in* Kwongan — Plant Life of the Sandplain ed by J. S. Pate and J. S. Beard. University of Western Australia Press, Nedlands.

Botkin, D. B., Janak, J. F. and Wallis, J. R., 1972. Some ecological consequences of a computer model of forest growth. *J. Ecol.* 60: 849-72.

Bowler, J. M., 1982. Aridity in the later Tertiary and Quaternary of Australia. Pp. 35-45 *in* Evolution of the Flora and Fauna of Arid Australia ed by W. R. Barker and P. J. M. Greenslade. Peacock Publications, Adelaide.

Brooks, D. R., 1985. Mycorrhizal investigations in rehabilitated areas — implications for Eneabba. Pp. 27-8 *in* Proceedings of a Seminar on the Plant Ecology of the Eneabba Heathlands ed by B. Lamont and B. Low. Western Australian Institute of Technology School of Biology, Bulletin No. 10.

Busby, J. R., 1985. Bioclimate prediction system. Users manual, version 1.2. Bureau of Flora and Fauna, Canberra.

Chapman, A., Dell, J., Johnstone, R. E. and Kitchener D. J., 1977. A vertebrate survey of Cockleshell Gully Reserve, Western Australia. *Rec. W. Aust. Mus. Suppl.* No. 4.

Chittleborough, R. G., 1985. Towards a State conservation strategy. 1. Planning to meet climatic changes. *Dept. Cons. Envt. Bull. West. Aust.* 207: 1-8.

Cowling, R. M. and Lamont, B. B., 1984. Population dynamics and recruitment of four co-occurring *Banksia* spp. after spring and autumn burns. Pp. 31-2 *in* Proceedings of the 4th International Conference on Mediterranean Ecosystems ed by B. Dell. Botany Department, University of Western Australia, Nedlands.

George, A. S., Hopkins, A. J. M. and Marchant, N. G., 1979. The heathlands of Western Australia. Pp. 211-29 *in* Ecosystems of the World. Vol. 9A. Heathlands and Related Shrublands. Descriptive Studies ed by R. L. Specht. Elsevier, Amsterdam.

Griffin, E. A. and Hopkins, A. J. M., 1981. The short-term effects of brush harvesting on the kwongan vegetation at Eneabba, Western Australia. *Dept. Fish. Wild. West. Aust. Report.* 45: 1-38.

Griffin, E. A. and Hopkins, A. J. M., 1985a. The flora and vegetation of Mt Lesueur, Western Australia. *J. Roy. Soc. West. Aust.* 67: 45-57.

Griffin, E. A. and Hopkins, A. J. M., 1985b. The flora and vegetation of the Crown land south of Eneabba. Pp. 3-11 *in* Proceedings of a Seminar on the Plant Ecology of the Eneabba Heathlands ed by B. Lamont and B. Low. Western Australian Institute of Technology School of Biology, Bulletin No. 10.

Griffin, E. A., Hopkins, A. J. M. and Hnatiuk, R. J., 1983. Regional variation in mediterranean-type shrublands near Eneabba, southwestern Australia. *Vegetatio* 52: 103-27.

Grubb, P. J. and Hopkins, A. J. M., 1986. Resilience at the level of the plant community. Pp. 21-38 *in* Resilience of Mediterranean Type Ecosystems ed by B. Dell, A. J. M. Hopkins and B. B. Lamont. Dr. W. Junk, Dordrecht.

Hinchey, M. D. (ed.), 1982. Nature Conservation Reserves in Australia (1982). Australian National Parks and Wildlife Service. Occasional Paper No. 7.

Hnatiuk, R. J. and Hopkins, A. J. M., 1980. Western Australian species-rich kwongan (sclerophyllous shrubland) affected by drought. *Aust. J. Bot.* 28: 573-85.

Hnatiuk, R. J. and Hopkins, A. J. M., 1981. An ecological analysis of kwongan vegetation south of Eneabba, Western Australia. *Aust. J. Ecol.* 6: 423-38.

Holling, C. S. (ed.), 1978. Adaptive Environmental Assessment and Management. Wiley-Interscience, Chichester.

Hopkins, A. J. M., 1985a. Planning the use of fire on conservation lands in southwestern Australia. Pp. 203-8 *in* Fire Ecology and Management in Western Australian Ecosystems ed by J. R. Ford. Western Australian Institute of Technology Environmental Studies Group Report No. 14. Western Australian Institute of Technology, Perth.

Hopkins, A. J. M., 1985b. Fire in the woodlands and associated formations in the semi-arid region of southwestern Australia. Pp. 83-90 *in* Fire Ecology and Management in Western Australian Ecosystems ed by J. R. Ford. Western Australian Institute of Technology Environmental Studies Group Report No. 14. Western Australian Institute of Technology, Perth.

Hopkins, A. J. M. and Griffin, E. A., 1984. Floristic patterns. Pp. 69-83 *in* Kwongan — Plant Life of the Sandplain ed by J. S. Pate and J. S. Beard. University of Western Australia Press, Nedlands.

Hopkins, A. J. M. and Hnatiuk, R. J., 1981. An ecological survey of the kwongan south of Eneabba, Western Australia. *Wildl. Res. Bull. West. Aust.* 9: 1-33.

Hopkins, A. J. M., Keighery, G. J. and Marchant, N. G., 1983. Species-rich uplands of southwestern Australia. *Proc. Ecol. Soc. Aust.* 12: 15-26.

Kessell, S. R., Good, R. B. and Hopkins, A. J. M., 1984. Implementation of two new resource management information systems in Australia. *Environ. Manage.* 8: 251-70.

Kitchener, D. J., Chapman, A., Dell, J., Muir, B. G. and Palmer, M., 1980a. Lizard assemblage and reserve size and structure in the Western Australian Wheatbelt — some implications for conservation. *Biol. Conserv.* 17: 25-62.

Kitchener, D. J., Chapman, A., Muir, B. G. and Palmer, M., 1980b. The conservation value for mammals of reserves in the Western Australian Wheatbelt. *Biol. Conserv.* 18: 179-207.

Kitchener, D. J., Dell, J., Muir, B. G. and Palmer, M., 1982. Birds in Western Australian wheatbelt reserves — implications for conservation. *Biol. Conserv.* 22: 127-63.

Lamont, B. B., 1976. A Biological Survey and Recommendations for Rehabilitating a Portion of Reserve 31030 to be Mined for Heavy Minerals During 1975-81. WAIT-AID, Perth.

Lamont, B., 1985. Fire responses of sclerophyll shrublands — a population ecology approach, with particular reference to the genus *Banksia*. Pp. 41-6 *in* Fire Ecology and Management in Western Australian Ecosystems ed by J. R. Ford. Western Australian Institute of Technology Environmental Studies Group Report No. 14. Western Australian Institute of Technology, Perth.

Lamont, B. B., Hopkins, A. J. M. and Hnatiuk, R. J., 1984. The flora — composition, diversity and origins. Pp. 27-50 *in* Kwongan — Plant Life of the Sandplain ed by J. S. Pate and J. S. Beard. University of Western Australian Press, Nedlands.

Low, A. B. and Lamont, B., 1985. Nutrient allocation in native heath and rehabilitated communities. Pp. 12-9 *in* Proceedings of a Seminar on the Plant Ecology of the Eneabba Heathlands ed by B. Lamont and B. Low. Western Australian Institute of Technology School of Biology Bulletin No. 10.

Lundelius, E. L., 1960. Post Pleistocene faunal succession in Western Australia and its climatic interpretation. *Report Int. Geol. Congress 21st Session Part IV:* 142-53.

Main, A. R., 1984. Rare species: precious or dross? Pp. 163-74 *in* Species at Risk: Research in Australia ed by R. H. Groves and W. D. L. Ride. Australian Academy of Science, Canberra.

Majer, J. D., Sartori, M., Stone R. and Perriman, W. S., 1982. Recolonization by ants and other invertebrates in rehabilitated mineral sand mines near Eneabba, Western Australia. *Reclam. Reveg. Res.* 1: 63-81.

Malcolm, C. V., 1983. Wheatbelt salinity. A review of the salt land problem in southwestern Australia. *West. Aust. Dept. Ag. Tech. Bull.* No. 52.

McKenzie, N. L., 1984. Biological surveys for nature conservation by the Western Australian Department of Fisheries and Wildlife — a current view. Pp. 88-117 *in* Survey Methods for Nature Conservation Vol. 2 ed by K. Myers, C. R. Margules and I. Musto. CSIRO Division of Water and Land Resources, Canberra.

Noble, I. R. and Slatyer, R. O., 1980. The use of vital attributes to predict successional changes in plant communities subject to recurrent disturbances. *Vegetatio* 43: 5-21.

Playford, P. E., Cockburn, A. E. and Low, G. H., 1976. Geology of the Perth Basin, Western Australia. *Geol. Surv. West. Aust. Bull.* 124: 1-311.

Ride, W. D. L. and Wilson, G. R., 1982a. Towards informed management. Pp. 181-9 *in* Species at Risk: Research in Australia ed by R. H. Groves and W. D. L. Ride. Australian Academy of Science, Canberra.

Ride, W. D. L. and Wilson, G. R., 1982b. The conservation status of Australian animals. Pp. 27-44 *in* Species at Risk: Research in Australia ed by R. H. Groves and W. D. L. Ride. Australian Academy of Science, Canberra.

Saunders, D. A., 1979. The availability of tree hollows for use as nest sites by white-tailed black cockatoos. *Aust. Wildl. Res.* 6: 205-16.

Saunders, D. A., 1980. Food and movements of the short-billed form of the white-tailed black cockatoo. *Aust. Wildl. Res.* 7: 257-69.

Saunders, D. A., 1982. The breeding behaviour and biology of the short-billed form of the white-tailed black cockatoo *Calyptorhynchus funereus. Ibis* 124: 422-55.

Saunders, D. A., 1986. Breeding season, nesting success and nestling growth in Carnaby's cockatoo, *Calytorhynchus funereus latirostris,* over 16 years at Coomallo Creek, and a method for assessing the viability of populations in other areas. *Aust. Wildl. Res.* 13: 261-73.

Saunders, D. A., Rowley, I. and Smith, G. T., 1985. The effects of clearing for agriculture on the distribution of cockatoos in the south-west of Western Australia. Pp. 309-21 *in* Birds of Eucalypt Forests and Woodlands: Ecology, Conservation, Management ed by A. Keast, H. F. Recher, H. Ford and D. Saunders. Surrey Beatty and Sons, Sydney.

Shugart, H. H. and West, D. C., 1980. Forest succession models. *Bioscience* 30: 308-13.

Shugart, H. H. and West, D. C., 1981. Long-term dynamics of forest ecosystems. *Amer. Sci.* 69: 647-52.

Slayter, R. O., 1975. Ecological reserves: size, structure and management. Pp. 22-38 *in* A National System of Ecological Reserves in Australia ed by F. Fenner. Australian Academy of Science Report No. 19.

Soulé, M. E., Wilcox, B. A. and Holtby, C., 1979. Benign neglect: a model of faunal collapse in the game reserves of East Africa. *Biol. Conserv.* 15: 259-72.

Speck, N. H., 1958. The vegetation of the Darling-Irwin Botanical Districts and an investigation of the distribution of the family Proteaceae in southwestern Australia. Ph.D. Thesis, University of Western Australia.

Terborgh, J., 1974. Preservation of natural diversity: the problem of extinction prone species. *Bioscience* 24: 715-22.

Troyanowsky, C. (ed.), 1985. Air Pollution and Plants. Vch Verlagsgesellschaft, Berlin.

CHAPTER 3

Persistence of Invertebrates in small areas: Case Studies of Trapdoor Spiders in Western Australia

Barbara York Main[1]

A diversity of spiders in isolated reserves e.g. The Wongan Hills (15 families, 40 species), King's Park (28 families) and Torndirrup National Park (24 families) indicates that they can exist in relatively small areas. It is proposed that mygalomorphs (trapdoor spiders) are admirably fitted to persist in small isolated areas because of their low dispersion powers, long life cycle and sedentary life style. This contention is supported by occurrence of trapdoor spiders on offshore continental islands. As predators the occurrence of trapdoor spiders also indicates the persistence of other terrestrial invertebrates. Predictability of long term persistence of trapdoor spiders is postulated as dependent on knowledge of life style strategies of individual species combined with availability of their preferred microhabitats and other environmental features. A study of trapdoor spiders at North Bungulla Nature Reserve (104 ha) demonstrates the occurrence of 15 species with status ranging from rare (7 species), uncommon (3), sparse (3) to common (2). A model is presented suggesting that only the two common species are likely to persist indefinitely. A propensity to form matriarchal clusters on which other advantageous features are dependent, including maximisation of foraging and aspects of the reproductive strategies, combined with microhabitats in close proximity are paramount in determining persistence. It is suggested that a habitat supporting about 20 matriarchs (adult females which have reproduced at least once) of *Anidiops villosus* (Rainbow) in close proximity is capable of maintaining a viable population. However, it is suggested that a surrounding area of at least 25 ha (with additional subpopulations) and with a diversity of microhabitats which support prey species (predominately ants and termites) is required to buffer and augment the microhabitat area of the spiders.

INTRODUCTION

INVERTEBRATES, particularly insects, are frequently collected from isolated small habitat patches. As predators, occurrence of spiders indicates the presence also of sufficient insects or other invertebrates for their sustenance. Since they are at the apex of food pyramids, spiders are good indicators of the general balance of communities. In assessing the status of invertebrates in an ecosystem spiders are thus a useful taxon.

Recent surveys of several reserves and other isolated areas in Western Australia have demonstrated the occurrence of a diversity of spiders. For example visual search collecting in The Wongan Hills (with an area of approximately 1750 ha) on several days in 1975, demonstrated the presence of 15 families of spiders, about 28 genera and 40 species (Main 1977). Search was concentrated on Araneomorphs but four species in two Mygalomorph families were collected. Twenty-eight families of spiders have been recorded from King's Park (400 ha) (Main unpublished; Barendse *et al.* 1981). Notable amongst the latter collection was the occurrence of species of *Mimetus* (Mimetidae), *Laestrygones* (Toxopidae) and *Micropholcomma* (Micropholcommatidae), all small spiders with low powers of dispersion and characteristic of eastern Australian rainforest habitats. Their presence is also surprising in view of the frequent burning and disturbance of a bushland park in the heart of a city over 150 years old. Some species were restricted to ground cover plants not susceptible to burning e.g. *Jacksonia sericea* and deep *Casuarina* litter

---

[1] Zoology Department, University of Western Australia, Nedlands, Western Australia 6009.

undisturbed for about 17 years. In 1983 pitfall-trap sampling in heathland in Torndirrup National Park, southwest of Albany, demonstrated the presence of about 24 families and eight species of mygalomorphs. Of the spiders recorded, species of Archaeidae, *Toxops* (Toxopidae) and *Tasmanoonops* (Orsolobidae), and several others are relics of an early Tertiary environment. Their presence in a relatively restricted area testifies to the value of retaining small reserves in conservation programmes.

Nevertheless occurrence of a species at any one time is not necessarily a guarantee of persistence. Several workers, notably Majer (1980), Watson (1981), Greenslade and Greenslade (1984) and New (1984 p. 112) have emphasised the likely biological significance of invertebrates in Australian terrestrial ecosystems and their potential as 'bio-indicators'. However, there is little if any documentary evidence from Australia, which either confirms assumptions of persistence or from which predictions of potential persistence can be made. To predict potential persistence of taxa, the *capacity to persist* must first be assessed. The capacity to persist is dependent on a combination of environmental factors and life style strategies. Until the components of this interaction are understood, management plans of areas cannot be devised to facilitate or ensure persistence of particular species.

The primary environmental factors and life style strategies that appear to be relevant for persistence of all invertebrates in isolated areas are as follows.

1. *Environmental factors.* Size of area; diversity of area — topography, soil, vegetation i.e. microhabitat structure; proximity to other similar natural areas; regularity of weather patterns; fire proneness and degree of interference e.g. nonindigenous 'use' (humans, stock etc.).

2. *Life style characteristics.* Sedentary, vagarious, volant; foraging strategies (all predatory in present context); life cycle pattern — whether critical life history stages are seasonally dependent; life history term — annual, biennial or perennial (perennial in present context) and reproductive potential.

The present study addresses the possibility of persistence of trapdoor spiders (Mygalomorphae) in small wheatbelt reserves in Western Australia. As already mentioned, mygalomorphs are a good biological indicator group, thus from the case studies reported here, generalisations can be made regarding persistence of other invertebrates.

Because of their sedentary behaviour, long life cycle and poor powers of dispersal (Main 1976, 1985), mygalomorphs seem admirably fitted to persist in fragmented natural areas. Records of mygalomorphs from some continental islands demonstrate persistence of viable populations in small areas for at least 7000 years [e.g. five species on Rottnest Island of 9.6 square kms^2 (Main 1976)]. Extrapolating from such examples it is assumed that at least some mygalomorphs which have been artificially confined to fragments of bushland on the mainland, in the wheatbelt and other small reserves, may well persist. Nevertheless chance observations of trapdoor spiders on road verges or such places is no guarantee that a viable population is present. Spiders may live over 20 years and thus individuals can 'hang on' in precarious sites. Analysis of the population structure and assessment of habitat requirements can, however, indicate viability.

TRAPDOOR SPIDERS IN WHEATBELT 'RESERVES': CASE STUDIES

North Bungulla Nature Reserve

Mygalomorph spiders have been observed and some species have been intermittently monitored in several wheatbelt reserves. A long term study of the trapdoor spiders, including documentation of case histories of individuals of *Anidiops villosus* (Rainbow) was begun in the North Bungulla Nature Reserve (NBR) (31°32' S, 117°35' E) in 1973 (Anon. 1977; Main 1978). Prior to this, cursory observations had been made over many years and pitfall trap collections revealed presence of otherwise unobserved species. The nearest natural bushland is Heitman's Scrub about 1.5 kms northwest of the Reserve on private property (Main 1967) (see Fig. 1 in Chapman *et al.* 1980).

The reserve originally comprised two separate reserves (Numbers 17732 and 19950) set aside for the purposes of recreation, racecourse and hall site (collectively known as North East Tammin). Vesting was transferred from the Tammin Shire to the Western Australian Wildlife Authority in 1977 when the purpose of the combined reserves was changed to conservation of flora and fauna.

The area, of 104 ha, was surveyed by the Biological Survey Department of the Western Australian Museum during several periods in 1974, 1975 and 1977 (Fig. 1). As a result the general features, including a detailed account of the botany, and vertebrate fauna have been documented (Chapman *et al.* 1980). The area is within the region with an average and median annual rainfall of 339 mm and 336 mm respectively. Most rain falls between May and August. At least one summer thunderstorm generally ameliorates the summer drought. July and August are the coldest months, January the hottest. The reserve has a high botanical diversity with 150 species recorded from 16 vegetation associations distributed in three formations of mallee, shrubland and heath (Muir 1980). The area has not been burnt for about seventy years. Hearsay reports suggest that

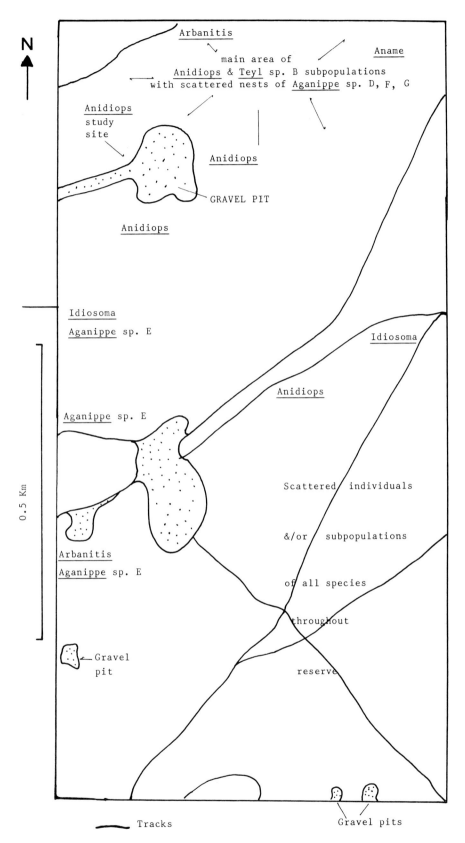

Fig. 1. Map of the North Bungulla Nature Reserve indicating main locations of populations of *Anidiops villosus, Teyl* sp. B, *Aganippe* spp., *Arbanitis, Idiosoma* and *Aname*. The large cleared areas (stippled) in the mid-Western region formerly included tennis courts, hall site and racecourse (see Chapman *et al.* 1980 for vegetation).

Fig. 2. Anidiops villosus (Rainbow), female. Scale bar = 5.0 mm.

it may have been burnt in 1912. There is charcoal in the soil and regrowth from mallee stumps indicates fire from about that period.

The reserve has a gentle slope from southeast to northwest. The soil is predominantly a coarse-grained yellow sandy loam overlying laterite and with patches of weathered granite exposures. The latter supports some mixed *Callitris* groves and patches of *Borya nitida* (pincushion or resurrection plants). On the lower parts, saucer-like depressions are filled with shallow, silt-like soil comprising a denser clay component than in most of the reserve. These depressions or slopes are subjected to sheet flooding in winter and after summer thunderstorms. In the higher areas, particularly in the southern end of the reserve there are exposures of laterite.

Botanical diversity, combined with different edaphic types suggests a diversity of microhabitats for trapdoor spiders and this is confirmed by my observations. These spiders are less abundant in habitats associated with laterite (Fig. 5) and weathered granite exposures where the soil is difficult to burrow in and becomes hot and dry in the summer.

The reserve suffered non-natural disturbance and partial clearing between 70 and 45 years ago for hall site, racecourse and tennis courts. One of the gravel pits (Fig. 1) was used until about five years ago. Only one of the tracks which transects the reserve is now maintained. Natural regrowth has occurred in the gravel pits and cleared areas except on heavily compacted areas including the tennis courts, hall site and some of the old tracks.

Table 1. The fifteen species of mygalomorph spiders in N Bungulla Nature Reserve and their relative commonness or rareness status. R, rare (less than 5 specimens collected or burrows ever observed). U, uncommon (5 to 10 specimens or viable burrows in any one year). S, sparse but apparently stable population (10 to 25 specimens or burrows in any one year). C, common, abundant and likely to persist indefinitely (100+ burrows in any one year). Numbers of specimens refers to adults or relatively mature specimens. Species marked with an asterisk are absent from "Fairfields" (see Table 2).

	R	U	S	C
ACTINOPODIDAE				
Missulena hoggi Womersley	x			
DIPLURIDAE				
Aname sp. A	x			
Chenistonia tepperi Hogg			x	
Genus A sp. A*	x			
Genus A sp. B*	x			
Teyl luculentus Main		x		
Teyl sp. A*	x			
Teyl sp. B*				x
CTENIZIDAE				
Aganippe sp. D*			x	
Aganippe sp. E (*cupulifex* group)		x		
Aganippe sp. F			x	
Aganippe sp. G*	x			
Anidiops villosus (Rainbow)*				x
Arbanitis sp. A	x			
Idiosoma nigrum Main		x		

Other possible species: *Kwonkan eboracum* Main, *Aname armigera* Rainbow and Pulleine.

Taxonomic note: Code names are consistent with usage elsewhere (see Main 1982). All 'coded' species are now designated in manuscripts (in preparation).

Trapdoor spiders at NBR and their status. A simple survey of the reserve by visual search for nests (intermittently over the last 20 years) and continuous pitfall trapping for wandering males over two years, 1969-70 and 1983, revealed the occurrence of 15 mygalomorph species from three families and belonging to nine genera (Table 1). One other genus, *Kwonkan* Main and two species, *K. eboracum* Main and *Aname armigera* Rainbow and Pulleine, possibly did and may still occur in the reserve. An assessment of the relative rareness/commonness of the spiders has been deduced according to an arbitrary scaling of numbers of nests or spiders observed and/or collected at various times over about 20 years. This assessment demonstrates that of the 15 species, two are common, three sparse, three uncommon and seven are rare (Table 1). Following discussion of the NBR spiders a comparison is made with the trapdoor spiders of heavily disturbed bushland surrounding a homestead and sheds ('Fairfields' settled in 1919) 3.2 kms north of the reserve (Table 2 and later).

By noting characteristics of foraging strategies, preferred habitats of spiders, dispersion pattern and phenology, life cycle and reproductive strategies and effect of predators and other animals (all in

Fig. 3. Typical vegetation association of *Acacia* and *Allocasuarina* ('wodjil') occupied by *Anidiops villosus*. *Ecdeiocolea* tussocks and myrtaceous shrubs in foreground.

Fig. 4. Nest of *Anidiops villosus*. Note fan of twiglines which increases foraging area.

Fig. 5. Nest of *Anidiops villosus* in laterite, showing gravel pebbles which spider has carried from bottom of burrow and deposited at edge of twiglines. Normal spoil is pushed through a temporary tunnel beneath hinge of door (visible in picture as a sinuous heap of soil to right of hinge).

Fig. 6. Nest of *Aganippe* sp. *D* against shrub; door opens outward (i.e. away from butt of shrub) and twiglines are draped moustache fashion from rim of burrow to the ground. Prey, primarily ants, is thus directed past the mouth of the burrow where the spider positions itself in foraging stance (Sp.D now *Aganippe castellum*).

Table 2. Species of mygalomorph spiders in bushland (about 7 ha) around farmhouse and sheds at "Fairfields", during the period 1952-57 and in 1980. Symbols as in Table 1.

	1952-57	1980
ACTINOPODIDAE		
Missulena hoggi	U	R
DIPLURIDAE		
Aname sp. A	S	–
Aname armigera	R	–
Chenistonia tepperi	U	–
Genus A. sp. A	–	–
Genus A. sp. B	–	–
Teyl luculentus	S	R
Teyl sp. A	–	–
Teyl sp. B	–	–
CTENIZIDAE		
Aganippe sp. D	–	–
Aganippe sp. E (*cupulifex* gr.)	C	S
Aganippe sp. F	U	R
Aganippe sp. G	–	–
Anidiops villosus	–	–
Arbanitis sp. A	R	R
Idiosoma nigrum	C	U

Kwonkan eboracum unlikely ever to have been present. In 1952-57, nine species only present of which eight also present in NBR (*Aname armigera* not present at NBR); the two COMMON species of NBR absent; the two COMMON "Fairfields" species both UNCOMMON at NBR. Status of "Fairfields" species changed as follows: two COMMON species now SPARSE or UNCOMMON, the SPARSE species now RARE or absent, the three UNCOMMON species now RARE (two spp.) or absent, the two RARE species are now RARE or absent.

relation to extent of microhabitats available, seasonal weather and incidence of summer rain and drought), it is possible to indicate the likely persistence or otherwise of each species.

1. *Foraging strategies*. Main (1982) presented a schema of foraging strategies of arid-adapted mygalomorphs by using a combination of characters relating to burrow structure (entrance, door and adjuncts) and associated methods of prey-attack. Spiders were arranged into six groups. Each category or group was rated as having a Low, High or Very High level of foraging efficiency. By using this schema the species at NBR are rated as follows.

Low: *Aganippe* spp. *E, F* and *G, Arbanitis* sp. and *Missulena hoggi*.

High: *Chenistonia tepperi, Aname* sp., *Teyl luculentus, Teyl* spp. *A, B,* diplurine Genus A spp. *A* and *B.*

Very High: *Anidiops villosus, Idiosoma nigrum, Aganippe* sp. *D.*

Apart from the low raters, the particular foraging strategies differ within the groups. Characteristics of such strategies are summarized as follows.

Low Raters build soil doors and have no accessory prey-intercepting structures; the spiders adopt a sit-and-wait posture at the entrance and lunge at passing prey *without completely emerging* from the nest. Thus the foraging area is restricted to 180°, or usually, a smaller segment of a semicircle anterior to the nest and with a radius roughly equal to the 'reach' of the spider from the burrow rim, while retaining a hold with posterior tarsal claws.

High Raters may emerge completely from the nest in the lunge-attacks on prey, and the foraging area is a complete circle with a radius exceeding the 'body-reach' of the spider. *Chenistonia, Aname, Teyl luculentus* and *Teyl* sp. *B* and both species of *Genus A* all have open holes; *Teyl* sp. *A* has a trapdoor which when open, lies flat on the ground and over which the spider spreads its legs and body while waiting for prey (Main 1982, Fig. 7C).

Very High Raters affix twigs and leaves by their ends to the rim of the burrow (Figs 4, 5). Such twig-lines have been demonstrated to increase the foraging area (Main 1957, 1976, 1982; Gray 1968). In addition some species further maximise predatory possibilities by aggregating in clusters around matriarchal females in litter mats which termites use as forage and across which ants pass to feed on trees. *Missulena* sometimes extends silk strands from the lips of the double-doored entrance which possibly act as interceptors and thus enlarge the foraging area. *Aganippe* sp. *D,* which sites its nests against butts of shrubs, is effectively an arboreal forager (see Fig. 6), catching prey (mainly ants) which travel between the ground layer and foliage.

In spite of catholic dietary taste, there are some notable differences in prey components of the various species which are related to the microhabitats occupied and predatory techniques. For example the twiglining species are predominantly ant feeders which is due in part to ants being directed along twiglines to burrow openings. Litter foraging termites also rate high with *Anidiops* and *Idiosoma*. In addition, and because of its large size *Anidiops* is able to subdue large and heavily armoured insects such as curculionid beetles. Diplurines which have open holes and chase prey, capture moths which are prevalent in areas with native grasses and *Ecdeiocolea monostachya* tussocks.

2. *Preferred habitats*. Associated with their twig-lining foraging technique, *Anidiops* and *Idiosoma* are dependent on stable litter mats for burrow sites — *Anidiops* in 'wodjil' i.e. mixed *Acacia/ Allocasuarina* dominated associations (Fig. 3) and *Idiosoma* in *Eucalyptus loxophleba* and/or *Acacia acuminata* associations. Similarly *Aganippe* sp. *D* utilizes twiglines but its preferred habitat is in flood-prone depressions and flats which support a heath of myrtaceous shrubs. *Teyl* sp. *B* and both species of the undescribed diplurine genus all prefer open soil patches within the litter matrices. All other species burrow in open ground or lightly littered (unstable) areas.

3. *Dispersion patterns and phenology*

(a) *Dispersion*. Juveniles of all species disperse (Main 1981; Platnick 1976; Udvardy 1969) and burrow after the secondary bouts of autumn/winter rain. *Anidiops, Idiosoma* and *Teyl* sp. *B* have limited dispersion powers and tend to aggregate in 'clusters' around matriarchal nests wherever space permits; other species are more prodigal, disperse widely and settle solitarily at a distance from parent nests. *Missulena* juveniles disperse aerially during mild, warm mid-winter weather. It is one of only two mygalomorph genera which disperse aerially in Australia. Although it is not known how far spiderlings can travel, observations show that the spiders' nests are always widely spaced. This habit may be advantageous in preventing competition for food and facilitating recolonization of damaged habitats e.g. following fire, drought or other disturbance. However, it also means that males have a more hazardous search for mates than in aggregated populations. In the conservation context it may mean that only large reserves or conversely small reserves in close proximity can support indefinitely *Missulena* populations.

(b) *Foraging*. *Teyl* species are winter-active for only about three months and plug their burrows for the rest of the year. At least some members of the populations of other diplurines forage throughout the year although juveniles may aestivate. *Missulena* and all the ctenizids apart from *Anidiops* feed opportunistically throughout the year except when moulting. However, *Anidiops* which is a large spider with maximum body length of 5.0 cms (Fig. 2) digs a deep burrow (up to 70 cms) and is markedly seasonal in its general behaviour. All age classes younger than six years and penultimate males and brooding females aestivate in plugged burrows. Non-breeding females feed throughout the year and take advantage of prey flushes associated with summer thunderstorms.

(c) *Wandering of males*. Most species in the reserve are autumn/winter wanderers with some modification e.g. *Anidiops* runs with late summer thunderstorms if they occur, otherwise with the first bout of autumn/winter rain. *Teyl*. sp. *B* runs with the first autumn/winter rains, *Missulena* in midwinter on clear days. It is noteworthy that *Missulena* is the only Australian mygalomorph known to wander during daylight. *Chenistonia tepperi* is completely opportunistic and runs at any time of the year.

4. *Life cycle and reproductive strategies*. All mygalomorphs are long lived and take several years to mature. However, the only comprehensive field study of the life history of a mygalomorph is the current one on *Anidiops villosus* at NBR where a clinician's case-history approach has been adopted (Main 1978 and unpublished records). Males take at least seven or eight years to mature, females at least eight years and they may not reproduce until they are much older. Males have a short adult life and die after mating with one or more females. Females can live for at least 23 years and can reproduce every second year. Thus some females under observation have reproduced at least five times (Table 3). Main

Table 3. *Anidiops villosus*. Number of times 17 females with two or more opportunities have reproduced in the 11 year period 1974-1984. Asterisks include spiders' first reproductive phase. Other spiders were adults when first observed and some (e.g. 1,31) had reproduced before the study period; 1, 4 and 31 died before the end of the study period.

Number of reproductive opportunities	5	4	3	2
Identification (tag) numbers of individual spiders; (number of reproductive phases in parentheses).	3(5)*	1(4)	31(3)	13(2)*
	8(5)	126(4)	29(2 or 3)*	45(2)*
	4(3)	12(3)*		46(2)*
	7(2)	6(2)		56(1)*
	11(2)	64(2)		57(1)*

Note: There is no way of determining whether spiders forfeit reproductive phases between maturity and the first reproductive phase experienced — forfeiture could be due to lack of mating opportunity. The earliest age at which spiders of known age have reproduced is assumed to be the base maturation age — some spiders may have matured later than others or missed their 'first' opportunity.

(1978) indicated a high mortality with less than 4% of established emergents of any one year likely to mature. Any group of clusters exhibits a peculiarly imbalanced population structure due to the longevity of females: about one-sixth to one-quarter of the population consists of adult females (i.e. matriarchs — see Main 1978). Of these not all are reproductively available in any one year and of these only some actually reproduce. In fact only about a quarter to less than half of the matriarchs actually reproduce in a season and in any one season there are less males than available females (Table 4). Nevertheless more females reproduce than there are males available so either males mate with more than one female or females store sperm for several reproductive phases. It seems that the longevity of females and their iteroparous reproduction, and variable maturation time of males (7-11 years) ensures persistence in the face of drought. This

Table 4. *Anidiops villosus*. Reproductive behaviour (for the five year period 1980-1984) within the ambit of a group of clusters (A,B,C,F) in the study site at NBR (see Main 1978) in an area of less than 1000 m². Total number of matriarchs, number (and % of total) of reproductively available matriarchs (i.e. those that did not reproduce in previous year), number (and % of total) of matriarchs which actually reproduced and number of males which ran in each year. Note: The apparent anomaly of numbers of spiders which have reproduced relative to number of matriarchs and available matriarchs for the years 1983, 1984 is due to the natural inconsistency of mortality and recruitment to the matriarchy. (For terminolgy relating to mygalomorph biology and reproductive behaviour see Main 1976, 1978, 1985).

Year	Matriarchs	Reproductively available No.	%	Reproduced No.	%	No. of Males which ran
1980	15	6	40	5	33	3
1981	19	14	74	6	31.5	2
1982	23	17	74	11	48	3
1983	23	14	61	5	22	2
1984	24	20	83	8 or 9	33	3

capacity fortuitously insures the spiders also against naturally restricted microhabitats, and in the reserve system context, assures a capacity to persist in artificially constricted habitats. From case history records from clusters within an area of less than 1,000 m² it appears that a population supported by about 20 matriarchs can be regarded as viable. Nevertheless, areas (in close proximity) supporting similar viable subpopulations may be necessary to sustain a population through adverse seasons or events. At NBR such close-by subpopulations are scattered throughout an area of about 25 ha.

It is assumed that other species have a similar pattern of reproductive strategies. It is unlikely, however, that they have such a long life cycle or take so many years to mature. Apart from the Theraphosidae or so-called 'bird-eating spiders', *Anidiops* is one of the largest mygalomorphs in Australia. The large size gives it certain physiological advantages, as well as a capacity to dig very deep burrows and thus avoid both inimical summer conditions and long droughts.

5. Effect of predators and other animals. The natural predators noted at NBR are ground-feeding birds, the goanna *Varanus gouldii* and the scorpion *Isometroides vescus*. Birds scratch out juvenile spiders when they are pushing out spoil while deepening their burrows; goannas dig up larger spiders in their burrows and *Isometroides* enters burrows to eat the resident spiders. Possibly nocturnal birds predate wandering male spiders. The effects of natural predators on already reduced populations of spiders are not known. Although echidnas *Tachyglossus aculeatus* and quails *Turnix* spp. disturb bare ground and litter when feeding, permanent damage to spider nests rarely if ever occurs and such disturbance sometimes provides new sites for spider burrows (Main 1978). There is evidence that rabbit *Oryctolagus cuniculus* warrens caused habitat interference in the past (pre mid 'fifties). Old warren sites have been recolonized by *Anidiops, Teyl* sp. *B* and *Aganippe* sp. *D*. However, disturbance by rabbits is again occurring in some areas.

Diminution of microhabitat sites (the result of fragmentation of the natural landscape) combined with recurrent drought is now the primary obstacle to persistence of the spiders at NBR. There is also the added potential hazard of threat from fire, which in a restricted area could be total instead of patchy. The additional hazard of regular spraying of insecticides and weedicides in surrounding paddocks has not been considered here. Inevitably there must be some effect on terrestrial arthropods as well as on insects on vegetation.

From the foregoing it seems likely that only two species, possibly three, have the capacity to persist. It is therefore pertinent to summarise for at least these two species the combination of biological strategies which, given that only minimal change to the present environment is likely, seem to ensure persistence.

Anidiops villosus

1. Limited dispersion and formation of matriarchal clusters i.e. aggregations of nests of successive cohorts around respective matriarchs ensures aggregation in 'proven' microhabitats.

2. Maximisation of foraging:
   (a) very high predatory strategy i.e. twiglining behaviour;
   (b) siting of burrows in stable litter mats i.e. areas with high prey potential; and
   (c) non-brooding matriarchs forage throughout the year thus taking advantage of spasmodic flushes during the summer.

3. Reproductive strategies:
   (a) mate finding is easier due to clustering;
   (b) longevity of females and repetitive brood production (perhaps at least five broods per female) offsets apparent low (individual brood) fecundity; iteroparity allows for some brood failures and thus spreads the risk over a number of years;
   (c) long, variable maturation time of males (7-11 years) accommodates forfeiture of mating in some unfavourable seasons and again spreads risk over number of years; and
   (d) opportunistic seasonal wandering of males, with late summer thunderstorms OR early autumn/winter rains, males running at first opportunity and so ensuring against the possibility of winter drought.

4. Large body size and deep burrows.

5. Aestivation and nest plugging by immature spiders, brooding matriarchs and penultimate instar males affording protection primarily against drought and secondarily against fire.

Teyl sp. *B*

1. Minimal distance (cautious) dispersion and thus semi-aggregative groups in 'proven' microhabitats.

2. Maximization of foraging:
   (a) high predatory strategy (chase-attack on prey);
   (b) siting of burrows in bare interstices in stable litter zones — catch prey 'strays' from litter; and
   (c) intensive predation by total population during season of high prey abundance in early-mid winter which offsets prolonged aestivation.

3. Reproductive strategies:
   (a) mate finding is easier due to concentration of populations in aggregations;

(b) longevity and iteroparity of females; and
(c) synchronization of male wandering with the short activity period of females (early/mid winter).

4. Aestivation and burrow-plugging by total population affords protection against summer drought and fire.

Conversely (see Fig. 7) those species which exhibit prodigal dispersion and thus do not form clusters but scatter widely, settle distantly from parents and in isolation *and* which site burrows in open ground are at a disadvantage in relation to: foraging, even if they have a high or very high rating predatory technique the prey is less concentrated; reproduction, since males have a longer and more hazardous search for females; habitat, in that a relatively large area of habitat is required to accommodate a viable population.

"Fairfields" farmstead bushland — a Comparison With NBR (see table 2)

Unlike the situation at 'Fairfields', NBR has suffered minimal ground and litter disturbance over the last 40 years because of the absence of non-indigenous animals apart from feral ones e.g. rabbits and foxes *Vulpes vulpes*. The farmstead bushland at 'Fairfields' has been continuously disturbed by poultry, sheep (during shearing activities) and rabbits. Subsequent invasion by introduced weeds has prevented regeneration of the senescent vegetation, except for jam *Acacia acuminata*. It is now dominated by old York gum trees *Eucalyptus loxophleba* and *A. acuminata*. Senescent trees of *Allocasuarina acutivalvis* and *Melaleuca uncinata* persist, along with tussocks of *Ecdeiocolea monostachya*. However, the shrub understorey and ephemerals previously characteristic of the area have now been replaced with introduced seasonal

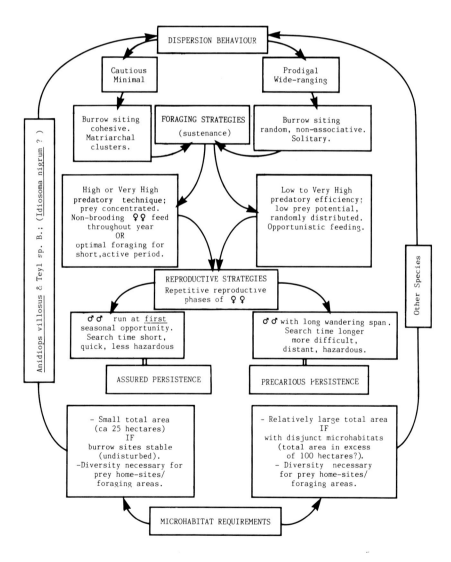

Fig. 7. Model for persistence of trapdoor spiders at North Bungulla Reserve, see text. Persistence of *Idiosoma* is only tentative; its preferred habitat here (eucalypt litter) is fragmented and minimal.

weeds. The overall effect is that spider burrow sites are destroyed and the abundance of prey, particularly termites and ants, is reduced. Conversely at NBR some natural regeneration is occurring in small patches. Nevertheless regeneration has been partially inhibited by successive droughts in recent years.

CONCLUSIONS

The mosaic pattern of vegetation, soil types and subtle topographic variations of the former wheatbelt landscape provided a matrix of naturally fragmented or tenuously continous microhabitats. Thus before European settlement, much of the terrestrial and subterranean invertebrate fauna was already adapted to surviving in small semi-isolated sites. A species could withstand occasional population collapses due to natural catastrophes such as fire and drought in some 'cells' of the distribution lay-out while other 'cells' flourished. Recolonization from undisturbed pockets may have been slow but nevertheless possible.

In the present artificially fragmented landscape with unnegotiable barriers between favourable 'cells', small areas cannot be expected to retain all species of the already diminishing diversity. But at least we can ascertain which species have some capacity to persist and thus design protective management practices to ensure their actual persistence.

In summary a model (Fig. 7) is proposed for the predicted persistence of trapdoor spiders at North Bungulla. *Anidiops villosus*, one of the three most likely species to persist is already adapted to drought. It is also widely distributed and occurs in suitable habitats throughout the northeastern wheatbelt and extends into the mulga zone. It is probably a 'safe' species, as is also possibly *Teyl* sp. *B*.

CONSERVATION AND MANAGEMENT

The objectives now for conservation are to:

1. *Ensure long term persistence of the two common, potentially persistent species* and hold as well as possible the *status quo* position of the other species at NBR. This may require several approaches:

    (a) *Active control through* minimial disturbance of litter and soil, reduction and/or extermination of rabbits and minimization of vistors.

All of the above activities if unchecked destroy spider nest sites and inhibit cyclic regeneration of vegetation. Some regeneration of vegetation is necessary for the persistence of prey species (ants, termites, other arthropods) and shade and litter supplementation for spider nests.

(b) *Management procedures to promote rejuvenation of vegetation.* It may be necessary sometime in the future to burn small selected areas of senescent clumps of vegetation to encourage regrowth, particularly in the southern end of the reserve. It appears that on the sandy loam soils e.g. in the northern part of the reserve, activity of echidnas is in part promoting regrowth of at least *Allocasuarina*, hakeas, *Grevillea paradoxa* and *Acacia* species.

(c) *Examination of interaction of ants and termites with the habitat.* The long term requirements of the termites are not known. Two species are present, *Drepanotermes perniger* a hearth (= pavement) building species, and a mound builder *D. tamminensis*. At present there appears to be excessive predation of the mound builder by the meat ant *Iridomyrmex purpureus*. It is perhaps noteworthy that the meat ant nests are generally sited on disturbed areas such as cleared patches, edges of gravel pits, old tracks and edges of the track still in use and sites of defunct termite mounds.

In view of the dependence of trapdoor spiders on termites and ants, monitoring of these insects in relation to their habitat shifts, general requirements and persistence needs to be undertaken.

2. *Look at the distribution of the less common and rare species* e.g. *Aganippe* sp. *D* and *Teyl sp. A* which have unusual specializations and require larger areas of habitat. It is hoped that the other species are included somewhere in the larger reserves or if only in small reserves of comparable size to NBR, that such reserves contain a more extensive proportionate area of preferred microhabitats.

ACKNOWLEDGEMENTS

The Wildlife section (formerly of the Department of Fisheries and Wildlife) now of the Department of Conservation and Land Management has permitted continuation of this long term study in the reserve. The Lions Club of Tammin constructed the fence along the western boundary. Grateful thanks are due to neighbouring farmers and other local residents for their continued interest and 'watchdog' support of the reserve. Facilities and equipment provided by the Zoology Department of the University of Western Australia are gratefully acknowledged.

REFERENCES

Anon, 1977. Trapdoor spider study. *S.W.A.N.S. Department of Fisheries and Wildlife, Western Australia.* 7(1): 3-4.

Barendse, W. J., Bolton, A. J., Collins, L. M., Craven, L., Pusey, B. J., Sorokin, L. M. and Ward B. H. R., 1981. Spiders in King's Park — an appraisal of management. B.Sc. Honours Thesis, Zoology Library, University of Western Australia. 140 pp.

Chapman, A., Dell, J., Kitchener, D. J. and Muir, B. G., 1980. Biological survey of the Western Australian Wheatbelt. Part II: Yorkrakine Rock, East Yorkrakine and North Bungulla Nature Reserves. *Rec. West. Aust. Mus.* Supp. **12**: 1-76.

Gray, M. R., 1968. Comparison of three genera of trapdoor spiders (Ctenizidae, Aganippini) with respect to survival under arid conditions. Unpublished M.Sc. Thesis, University of Western Australia.

Greenslade, P. J. M. and Greenslade, P., 1984. Invertebrates and environmental assesment. *Environment and Planning* **3**: 13-5.

Main B. Y., 1957. Biology of aganippine trapdoor spiders (Mygalomorphae: Ctenizidae). *Aust. J. Zool.* **5**: 402-73.

Main, B. Y., 1967. Between Wodjil and Tor. Jacaranda, Brisbane.

Main, B. Y., 1976. Spiders. Collins, Sydney.

Main, B. Y., 1977. Spiders. Pp. 100-107 *in* The Natural History of the Wongan Hills ed by K. F. Kenneally. Western Australian Naturalist's Club, Perth, WA.

Main, B. Y., 1978. Biology of the arid-adapted Australian trapdoor spider *Anidiops villosus* (Rainbow). *Bull. Br. Arachnol. Soc.* **4**: 161-75.

Main, B. Y., 1981. A Comparative account of the biogeography of terrestrial invertebrates in Australia: some generalizations. Pp. 1055-1077 *in* Ecological Biogeography of Australia ed by A. Keast. Junk, The Hague.

Main, B. Y, 1982. Adaptations to arid habitats by mygalomorph spiders. Pp. 273-83 *in* Evolution of the Flora and Fauna of Arid Australia ed by W. R. Barker and P. J. M. Greenslade. Peacock Publications, Frewville, SA.

Main, B. Y., 1985. Mygalomorphae. Pp. 1-48 *in* Zoological Catalogue of Australia, Vol. 3. Australian Government Publishing Service, Canberra.

Majer, J. D., 1980. Report on the study of invertebrates in relation to the Kojonup Nature Reserve fire management plan. *Department of Biology Bulletin,* Western Australian Institute of Technology. **2**: 1-22.

Muir, B. G., 1980. Vegetation of Yorkrakine Rock, East Yorkrakine and North Bungulla Nature Reserves. Pp. 15-48 *in* Biological Survey of the Western Australian Wheatbelt, Part II. *Rec. West. Aust. Mus.* Supp. **12**..

New, T. R., 1984. Insect Conservation. Junk, Netherlands.

Platnick, N. I., 1976. Concepts of dispersal in historical biogeography. *Syst. Zool.* **25**: 294-5.

Udvardy, M. D. F., 1969. Dynamic Zoogeography, with special reference to land animals. Van Nostrand Reinhold, New York.

Watson, J. A. L., 1981. Odonata (dragonflies and damselflies). Pp. 1139-67 *in* Ecological Biogeography of Australia ed by A. Keast. Junk, The Hague.

TAXONOMIC NOTE: Descriptions of two of the 'coded' taxa have now been published and the names are thus available as follows: *Genus A* = *Yilgarnia* Main and *Aganippe* Sp.D = *Aganippe castellum* Main.

CHAPTER 4

Conservation of Mammals within a Fragmented Forest Environment: The Contributions of Insular Biogeography and Autecology

A. F. Bennett[1]

At Naringal in southwestern Victoria, small forest fragments (<80 ha in size) comprise the remaining natural habitat for native mammals. To examine the consequences of habitat fragmentation for the mammal fauna two approaches were followed: (i) a study of the insular biogeography of the mammal fauna within 39 forest patches; and (ii) a study of the population ecology of the macropodid marsupial, the long-nosed potoroo *Potorous tridactylus* within a fragmented forest system.

Insular biogeographic analysis demonstrated that habitat fragmentation resulted in changes in species richness and in the composition of mammal assemblages within forest patches. Important determinants of mammal species richness were: (i) forest area and habitat richness; (ii) disturbance to forest vegetation from domestic stock; and (iii) the time since isolation from contiguous forest. Composition of mammal assemblages showed characteristic variation in relation to forest patch size.

The successful persistence of *P. tridactylus* within the fragmented environment at Naringal is attributed to the survival of suitable habitat patches and to the population ecology of this species. Small body size, restricted home range, extensive range overlap, stable population structure, dispersal of both sexes, and a continuous pattern of reproduction are all attributes which facilitate the survival of populations of *P. tridactylus* within the relatively small and patchy habitats which it favours.

Insular biogeography (as distinct from equilibrium island biogeography) and autecology are viewed as complementary approaches which both provide useful information relevant to the conservation of fauna within fragmented environments. Insular biogeography primarily documents the observed changes occurring in the faunal assemblage following habitat fragmentation, whereas autecological studies attempt to understand the basis for the status and performance of individual species with the fragmented system.

INTRODUCTION

THROUGHOUT the world clearing and fragmentation of natural environments is having a profound impact on the status of many wildlife species. In Australia the loss of natural vegetation has been most severe in those areas intensively developed for agriculture, such as the wheatbelt region of Western Australia (Kitchener *et al.* 1980a) and the plains of western Victoria (Willis 1964). In these areas the majority of the natural environment has been cleared, leaving only disjunct remnants of the original forest and woodland vegetation within an expanse of developed agricultural land. As most Australian animals have not adapted successfully to environments severely modified by humans (Frith 1979), the remaining patches of natural vegetation are frequently of crucial importance to the regional conservation of the indigenous fauna. In view of the increasing loss and isolation of habitat for wildlife, the challenge to ecologists and wildlife managers is twofold: (i) to determine the consequences of habitat fragmentation for faunal communities; and (ii) to develop strategies for the conservation of fauna before their natural environments are entirely alienated.

---

[1]Department of Zoology, University of Melbourne, Parkville, Victoria 3052. Present Address: Arthur Rylah Institute for Environmental Research, Fisheries and Wildlife Service, 123 Brown Street, Heidelberg, Victoria 3084.
Pages 41-52 *in* NATURE CONSERVATION: THE ROLE OF REMNANTS OF NATIVE VEGETATION ed by Denis A. Saunders, Graham W. Arnold, Andrew A. Burbidge and Angas J. M. Hopkins. Surrey Beatty and Sons Pty Limited in association with CSIRO and CALM, 1987.

The equilibrium theory of island biogeography (MacArthur and Wilson 1967), originally formulated to account for patterns of species richness on oceanic islands, was initially welcomed as a model that could also provide an understanding of the distributional patterns of fauna in mainland habitat isolates and thus serve as a theoretical basis for conservation strategies (Simberloff 1974; Diamond 1975; Wilson and Willis 1975). Widespread interest in the application of island biogeography to terrestrial environments stimulated numerous studies of the biogeography of fauna within mainland habitat isolates (Moore and Hooper 1975; Galli *et al.* 1976; Kitchener *et al.* 1980a, b,1982; Matthiae and Stearns 1981; Whitcomb *et al.* 1981; Howe *et al.* 1982; Ambuel and Temple 1983; Howe 1984). These studies of 'insular biogeography' have provided little support for the equilibrium theory of island biogeography, but they have provided valuable quantitative information concerning the consequences of habitat fragmentation for faunal communities. Concurrently, debate over the validity of the equilibrium theory (Gilbert 1980) and particularly its application to terrestrial conservation strategies (McCoy 1982; Simberloff and Abele 1982; Boecklen and Gotelli 1984) led to an increasing call for a greater emphasis on autecological and synecological studies as the basis for conservation planning.

At Naringal in southwestern Victoria, native mammals are restricted to small patches of forest which are remnants from the extensive clearing and forest fragmentation that accompanied agricultural settlement. To examine the consequences of habitat fragmentation on the mammal fauna at this locality, two approaches were followed: (i) a study of the insular biogeography of mammals in remnant forest patches; and (ii) a study of the population ecology of the long-nosed potoroo *Potorous tridactylus* in a fragmented forest system. *P. tridactylus* is a member of the Family Potoroidae (Marsupialia: Macropodoidea), a group of nine species known colloquially as 'rat-kangaroos' (Strahan 1983). The Potoroidae are among those native mammals which have been most severely disadvantaged by the past 200 years of European settlement in Australia (Calaby 1971). In Victoria, three of the five species historically present no longer occur in the state (*Bettongia penicillata, B. gaimardi, Aepyprymnus rufescens*), one is rare (*Potorous longipes*), and the remaining species (*Potorous tridactylus*) is uncommon.

This chapter briefly summarizes the information provided by the two approaches, insular biogeography and autecology, and compares the relative contribution of each approach towards understanding the consequences of forest fragmentation and the requirements for conservation of the native mammal fauna at Naringal. The ecology of *P. tridactylus* within the fragmented forest environment will be discussed in greater detail elsewhere.

METHODS

Study Area

The study area comprised a region of some 194 km^2, at Naringal in southwestern Victoria (38° 24' S, 142° 45' E). The natural environment of southwestern Victoria has been fully described elsewhere (Land Conservation Council 1976). Briefly, the climate is temperate with cool winter months (June-August) and a warm dry summer period (December-February). Rainfall is reliable with some 54% of the annual rainfall (780-890 mm mean annual rainfall) occurring between May and September. The natural vegetation in the study area comprises open forests dominated by *Eucalyptus obliqua, E. ovata* and *E. viminalis*, with dense sclerophyllous understories.

European settlement, commencing in the latter half of the 19th century, brought about dramatic changes to the forest environment. Clearing of forests to create agricultural lands resulted in the progressive fragmentation of the forest vegetation. In 1942 some 51% of the study area remained as forest, but by 1980 this had declined to less than 10% of the study area. The remaining forest patches, all less than 100 ha in size, are scattered amidst cleared farm paddocks and are linked only by narrow strips of forest along roadsides and creeks. Forest fragmentation in this district was also accompanied by extensive disturbance from timber extraction, frequent fires and, in many forest patches, grazing by domestic stock.

Insular Biogeography

Surveys were conducted to determine the occurrence of mammals (excluding bats and aquatic species) within 39 forest remnants ranging in size from 0.3 to 82 ha. The primary survey methods employed were live-trapping, nocturnal spotlighting and diurnal observations of animals or their tracks, signs, faeces or skeletal remains. Constant survey effort in each forest remnant was impractical; consequently, survey intensity was varied in relation to the forest patch area. Minimum levels of trapping effort and of spotlighting effort in each forest patch surveyed were defined by the equations:

$T = 140 \log A - 13$; and

$SH = 1.4 \log A - 0.15$;

where T represents the number of trap-nights and SH represents the number of spotlight-hours in a forest remnant of area A. Species richness was represented by two variables, total mammal species richness and native mammal species richness.

Variables quantifying size, relative isolation, disturbance and habitat richness for each forest patch were measured as listed below.

1. Size. The AREA (ha) and PERIMETER (km) of each forest patch were measured from aerial photographs (1980, Department of Crown Lands and Survey, Melbourne).

2. Isolation. The distance (km) to the nearest forest patch of at least 25% comparable size (DISNEAR), the distance to the nearest larger forest patch (DISLARGE) and the percentage of forested land within a radius of 2.0 km (FOREST%) were calculated from aerial photographs for each forest patch. The time that each patch had been isolated from contiguous forest, or since it was reduced to its present size (TIME), was determined as accurately as possible from sequential aerial photographs (available for 1946, 1966, 1971 and 1980) and by interviewing local land-owners. As absolute values could not be obtained for all patches, each patch was assigned to one of five time periods: (1) <5 years; (2) 5-10 years; (3) 11-20 years; (4) 21-40 years; and (5) >40 years.

3. Disturbance. A subjective estimate of the intensity of grazing by domestic stock (GRAZING) and the length of time since the forest patch was last burnt (FIRE) were quantified by using the following scales, based upon information from local landowners and from field observations.

   Grazing: (0) no grazing for more than 40 years; (1) no grazing for more than 15 years; (2) periodic light grazing during the past 15 years; (3) sustained light grazing; (4) periodic heavy grazing during the past 15 years; and (5) sustained heavy grazing. 'Light' and 'heavy' grazing are subjective assessments based upon the type of domestic stock (sheep, cattle, horses) and the relative stocking intensity.

   Time since last fire was categorized as follows: (1) <5 years; (2) 5-10 years; (3) 11-20 years; (4) 21-40 years; and (5) >40 years.

4. Habitat. Vegetation was used as a measure of habitat availability. In each forest patch the number of alliances of canopy tree species (n = 7) was recorded and used as an estimate of habitat richness (VEGCLASS).

Correlation coefficients were calculated between mammal species richness and each of the forest patch variables. Transformed values ($\log_{10}$) of variables were used where these provided greater linearity in the relationship with species richness. Correlation coefficients were also calculated between the so-called 'independent' variables and these showed a high level of multicollinearity between five of the nine variables (AREA, PERIMETER, DISLARGE, GRAZING and VEGCLASS). To reduce this complexity of interrelationships, principal components analysis (Dixon 1983) was employed. This procedure creates a smaller number, or subset, of new orthogonal variables which are linear combinations of the original variables. The derived principal components were then incorporated with the remaining variables in a stepwise multiple regression to obtain an equation accounting for mammal species richness. Stepwise multiple regression sequentially selects variables which make a significant contribution to the explanation of variance in the dependent variable (i.e. species richness).

Population Ecology of Potorous tridactylus

Seven species of small mammal, including *Potorous tridactylus*, were regularly captured on three trapping grids. Reads Bush (RB) grid was a square trapping grid comprising nine rows of nine sites with 24 m spacing between trap sites, situated in a portion of a 10 ha forest remnant. Wire-mesh cage traps and Elliot aluminium traps, baited with a mixture of peanut butter, rolled oats and honey, were set alternatively along rows and switched at the subsequent trapping session. Unmade Road (UR) grid and Cobden Road (CR) grid were linear grids of 10 sites spaced 24 m apart, which were located in narrow strips of remnant forest vegetation. Unmade Road was an unused road reserve 20 m in width linking Reads Bush to another forest remnant one km distant; the UR grid was adjacent to the RB grid. Cobden Road was a forested road verge 30 m in width adjacent to a bituminised road; the CR grid was situated 0.6 km from the closest forest patch. All three grids were trapped concurrently for four nights at approximately six-week intervals from March 1980 to March 1982.

Captured *P. tridactylus* were weighed and the identity, sex, pouch condition (for females) and trapping location noted before the animal was released at the point of capture. Animals were individually identified by inserting numbered metal tags in each ear. Females were regarded as reproductively immature until the time of birth of their first young. Males were regarded as immature until a body weight of 600 g was achieved, based upon observations by Hughes (1964) that male *P. tridactylus* >600 g were capable of spermatogenesis. This criterion is a conservative estimate as the population studied by Hughes (1964) attained a greater adult body weight (1.3 kg) than did male *P. tridactylus* at Naringal (0.8 kg).

Population size for *P. tridactylus* was determined using the known-to-be-alive (KTBA) estimate. This measure includes all individuals captured during a trapping session plus any individuals not captured but recorded from at least one previous and one subsequent session. An estimate of population density for the RB grid was calculated by dividing the population size by the effective trapping area, which was defined as the area of the trapping grid plus a boundary strip equal to the average distance moved between captures (Flowerdew 1976). The area of an individual home range was approximated using the exclusive boundary strip method (Stickel 1954). Home ranges were estimated only for those individuals whose capture points indicated a range entirely, or almost entirely, within the RB trapping grid. A simple measure of home range overlap for individual *P. tridactylus* on the RB grid was calculated as the proportion of an individual's capture sites at which other individuals were also captured. For example, a male which was captured at 10 sites, at four of which other males were also captured, would be regarded as having a 40% range overlap with other males. Values for range overlap were calculated only for individuals captured at least 10 times during a given eight-month period (March-October 1980, November 1980-July 1981, August 1981-March 1982), and transient and juvenile animals were not included in the calculations.

RESULTS

Insular Biogeography

A total of 19 species of mammal, comprising 13 native mammals and six introduced mammals was recorded from remnant forest vegetation in the study area during this survey. Mammal species richness was significantly correlated with five of the forest patch variables (Table 1). The highest correlations for both total mammal species richness and native mammal species richness were with $\log_{10}$ AREA ($r = 0.91$, $p<0.001$ for both). The regression equations best describing the relationship between species richness and area were:

$$S_t = 3.16 + 4.60 \log_{10} \text{AREA} \quad (F = 177.1, p<0.001, r^2 = 0.83)$$

$$S_n = 1.46 + 3.70 \log_{10} \text{AREA} \quad (F = 169.2, p<0.001, r^2 = 0.82)$$

Table 1. Correlation coefficients for the relationship between mammal species richness and variables representing size, isolation, disturbance and habitat attributes of forest patches at Naringal, Victoria. (**$p<0.01$, ***$p<0.001$).

	Total Mammals	Native Mammals
Area ($\log_{10}$)	0.91***	0.91***
Perimeter ($\log_{10}$)	0.86***	0.86***
Disnear	0.19	0.12
Dislarge ($\log_{10}$)	0.47**	0.43**
Forest%	0.18	0.23
Time ($\log_{10}$)	-0.18	-0.18
Fire ($\log_{10}$)	-0.22	-0.22
Grazing	-0.49**	-0.47**
Vegclass ($\log_{10}$)	0.75**	0.78**

where S_t and S_n represent total mammal and native mammal species richness respectively (Fig. 1). Other variables, although also highly significantly correlated with mammal species richness, accounted for a smaller proportion of the variance.

Because of the high level of intercorrelation between five of the so-called independent variables, principal components analysis was employed in an attempt to separate the respective contributions of each variable towards species richness. This analysis yielded two components which together accounted for some 82% of the variation in the original

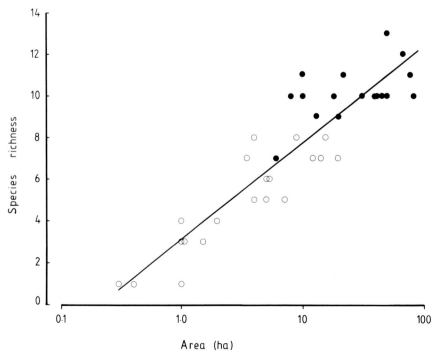

Fig. 1. Relationship between total mammal species richness and area for forest patches at Naringal. The equation for the regression line is: Species richness = 3.16 + 4.60 log Area. Solid points represent forest patches from which *Potorous tridactylus* was recorded.

Table 2. Principal components analysis (VARIMAX rotation) of five intercorrelated variables measured for forest patches at Naringal. Vertical lines indicate the variables primarily contributing to each component.

Variables	Loadings for	
	PC1	PC2
Area ($\log_{10}$)	0.828	-0.501
Perimeter ($\log_{10}$)	0.826	-0.491
Dislarge ($\log_{10}$)	0.807	0.189
Vegclass ($\log_{10}$)	0.749	-0.444
Grazing	-0.095	0.888
Variance (%)	65.3	16.8
Cumulative Variance (%)	65.3	82.1

variables (Table 2). The first principal component, PC1, primarily represents size and habitat richness within each patch. The greatest contributor to the second component, PC2, was the variable GRAZING (Table 2), suggesting that this component reflects the extent of disturbance to the forest vegetation. In forest patches which had experienced sustained grazing pressure by domestic stock, the shrub and field stratum was greatly reduced in complexity and introduced grasses and weeds were common. Smaller forest patches which are unfenced and surrounded by cleared pasture land are particularly vulnerable to such disturbance.

Stepwise multiple regression analyses were carried out using these principal components in place of the five intercorrelated variables. These analyses (Table 3) provided equations with three significant steps, together accounting for 82.6% and 82.2% of the variance in the total mammal and native mammal species richness respectively. The three variables incorporated in each equation were identical, namely PC1, PC2 and TIME.

Clearly, although the regression between species richness and area alone accounted for a similar percentage of the variation in species richness, this analysis suggests that it is not only the size of the forest patch that is an important contributor to species richness, but also other attributes which are significantly intercorrelated with AREA. These include habitat richness and the level of disturbance to the forest vegetation associated with grazing by domestic stock. The inclusion of TIME as a significant variable in the regression equation predicts that with an increasing period of isolation mammal species richness will gradually decline from present levels.

To examine the composition of assemblages of mammals present in forest remnants, forest patches were grouped into five classes: <2 ha; 3-7 ha; 8-15 ha; 16-40 ha; and 41-100 ha. The percentage of forest patches in each size class from which each species was recorded were calculated. Species occurring with a frequency >50% were termed 'core species' (Table 4). The composition of mammal assemblages

Table 3. Stepwise multiple regression analyses for mammal species richness within remnant forest patches at Naringal. (*p<0.05, ***p<0.001).

Dependent variable	Independent variable, r^2 and coefficient (B) for significant steps			Total variance (%)
	Step 1	Step 2	Step 3	
Total Mammals	PC1*** $r^2 = 0.544, B = 2.37$	PC2*** $r^2 = 0.250, B = -1.61$	Time ($\log_{10}$)* $r^2 = 0.032, B = -2.93$	82.6
Native Mammals	PC1*** $r^2 = 0.532, B = 1.97$	PC2*** $r^2 = 0.260, B = -1.32$	Time ($\log_{10}$)* $r^2 = 0.030, B = -2.29$	82.2

Table 4. Percentage occurrence of mammal species in forest patches within five size-classes at Naringal. Core species (frequency of occurrence >50%) are enclosed with blocks (+ introduced species).

Size class (ha)		<2	3-7	8-15	16-40	41-100
Number of patches		8	8	8	8	7
+ *Oryctolagus cuniculus*	Rabbit	63	100	100	100	100
Rattus fuscipes	Bush rat	38	75	100	100	100
Pseudocheirus peregrinus	Common ringtail possum	25	88	100	88	100
+ *Vulpes vulpes*	Fox	0	63	100	100	100
Tachyglossus aculeatus	Short-beaked echidna	13	50	100	100	100
Antechinus stuartii	Brown antechinus	13	50	100	100	86
Wallabia bicolor	Swamp wallaby	13	13	75	63	86
Potorous tridactylus	Long-nosed potoroo	0	13	50	63	100
Macropus giganteus	Eastern grey kangaroo	13	50	13	63	57
+ *Mus musculus*	House mouse	25	25	38	38	57
+ *Felis catus*	Cat	25	38	50	38	50
Rattus lutreolus	Swamp rat	0	13	13	25	29
Perameles nasuta	Long-nosed bandicoot	0	13	13	0	43
Macropus rufogriseus	Red-necked wallaby	0	0	0	38	29
Petaurus breviceps	Sugar glider	0	0	13	25	29
Isoodon obesulus	Southern brown bandicoot	0	0	13	13	14
Trichosurus vulpecula	Common brushtail possum	0	0	13	0	14
+ *Rattus rattus*	Black rat	13	25	0	0	0
+ *Lepus capensis*	Brown hare	13	0	0	13	0

in forest remnants showed characteristic variation related to forest patch size. The occurrence of core species formed a consistent 'nested' pattern, with those present in the smaller size classes also present as core species in successively larger size classes (Table 4). Notably, core species in the smallest size-classes (e.g. *Oryctolagus cuniculus, Rattus fuscipes, Pseudocheirus peregrinus, Vulpes vulpes, Antechinus stuartii*) were either introduced mammals or the most widespread and abundant native mammals. The least common native mammals (e.g. *Perameles nasuta, Macropus rufogriseus, Isoodon obesulus*) were more likely to be present in the larger forest patches, and were generally absent from the smaller forest patches.

Potorous tridactylus was recorded from 17 of the 39 forest remnants surveyed, being captured on 115 occasions from 4946 trap-nights of survey effort. The majority of these captures were from the larger forest patches (20-80 ha) (Fig. 1.); *P. tridactylus* was recorded from only two patches less than 10 ha in size (6 and 8 ha respectively). The smallest size class in which *P. tridactylus* was a core species was that with the size range of 16-40 ha, in which it was recorded from five of the eight patches (63%). In forest patches 41-100 ha in size, *P. tridactylus* also occurred as a core species with a frequency of occurrence of 100% (Table 4). Characteristically, individuals were captured in forest vegetation having dense cover in the shrub and field strata, or from sites close to (within 20 m) dense ground cover. *P. tridactylus* was not captured in forest patches which had experienced sustained heavy grazing by domestic stock. Low-lying areas with poor drainage and supporting dense growths of the sedge *Lepidosperma laterale* provided suitable habitat, as did dense vegetative cover provided by *Xanthorrhea australis, Pteridium esculentum* or *Tetrarrhena juncea* in combination with sclerophyllous shrubs and herbs. Small runways were utilized by *P. tridactylus* as pathways for rapid movement through the dense ground cover.

Population Ecology of Potorous tridactylus

Over the 25-month duration of this study (8160 trap-nights), 61 individual *P. tridactylus* (27 males, 34 females) were captured on more than 825 occasions. Body weights of reproductively mature male *P. tridactylus* ranged up to a maximum of 950 g, but the mean value (± standard deviation) from all trapping session occurrences was 789 ± 77 g (n = 120). The body weights of females were, in many instances, biased by the presence of pouch young; but for females without pouch young and those with a very small pouch young (<10 g) the mean body weight was 777 ± 86 g (n = 61).

The number of *P. tridactylus* known to be alive on the trapping grids remained relatively constant throughout the 25-month period, ranging between a minimum of 20 and a maximum of 29 (Fig. 2). There was no marked seasonal fluctuation in population size, although annual peaks occurred during spring (September, October). The mean density of *P. tridactylus* on the RB grid, calculated by the addition of a boundary strip 80 m in width, was 2.55 individuals per ha (range 1.67-3.08). The composition of the population remained relatively constant with the only variation evident being an increase in the number of reproductively mature males during 1980 followed by a gradual decline throughout 1981 to the end of the study (Fig. 2). At all times the population was composed largely of reproductively mature animals. The number of females known to be

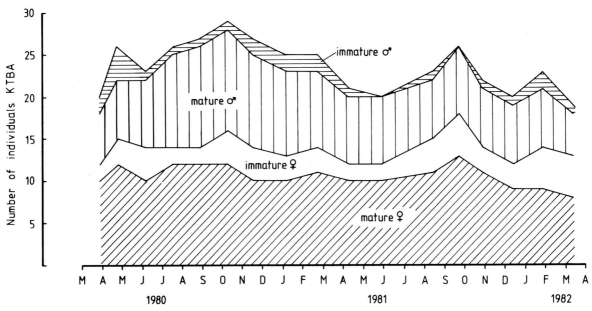

Fig. 2. Population size and structure of *Potorous tridactylus* at Naringal.

alive was always greater than the number of males and consequently the sex ratio (male:female) of the trappable population was less than unity in all trapping sessions. However, the sex ratio for 21 pouch young close to the end of pouch life was 13:8, a ratio not significantly different from parity. The level of transience in the population, here defined as the occurrence of an animal during only a single trapping session, was relatively low. Three of 34 females (9%) and three of 27 males (11%) captured during the study were transients.

Individual *P. tridactylus* occupied stable home ranges, although minor shifts in orientation and location occurred with time. The magnitude of individual home ranges increased with increasing number of captures (Fig. 3). Female home ranges approached an asymptotic value at approximately 1.4 ha, while for males the mean size of home ranges increased only marginally past 2.0 ha (Fig. 3). After identical numbers of captures the mean home range size for males was consistently greater than that for females. Overlap between home ranges is summarized in Table 5. Males and females both displayed a high level of range overlap with few sites occupied exclusively by a single animal. There was a particularly high level of range overlap between sexes (Table 5). Individual males overlapped with females at a mean of 75.5% of sites, a significantly greater level of range overlap (Students t-test; t = 3.33, p<0.01) than with other males (54.7% of sites). Female *P. tridactylus* also shared their home range

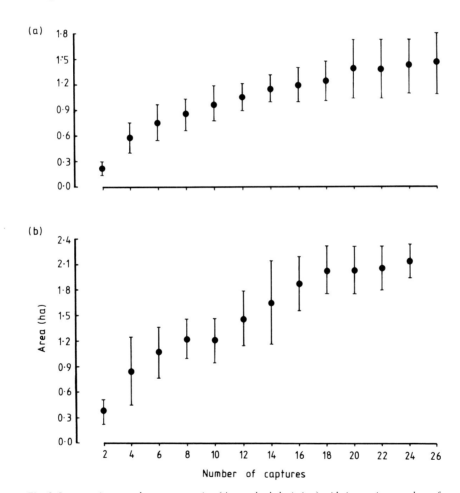

Fig. 3. Increase in mean home range size (± standard deviation) with increasing number of captures for (a) female and (b) male *Potorous tridactylus* at Naringal.

Table 5. Trap-revealed measures of home range overlap for *Potorous tridactylus* at Naringal. Values represent the mean ± standard deviation.

	Sample size	Overlap with males			Overlap with females			Sites occupied exclusively (%)
		Total overlap with ♂ (%)	Maximum overlap with single ♂ (%)	Number of ♂ overlapping	Total overlap with ♀ (%)	Maximum overlap with single ♀ (%)	Number of ♀ overlapping	
♂	9	54.7±13.3	31.0±18.6	4.6±1.0	75.5±18.6	40.7±13.4	6.1±1.6	10.6± 8.0
♀	12	60.7±18.1	37.8±15.5	4.1±1.4	54.8±13.8	27.9± 6.7	4.7±2.1	15.8±14.2

with other individuals of both sexes, but there was no significant difference (t = 0.89, p>0.2) between the percentage of capture sites shared with males ($\bar{x}$ = 60.7%) or with other females ($\bar{x}$ = 54.8%).

After a gestation period of approximately 38 days, or a delayed gestation of some 29 days (Hughes 1962; Shaw and Rose 1979) female *P. tridactylus* give birth to a single young which is carried and suckled in the pouch for approximately 120-130 days. At Naringal reproduction occurred throughout the year with no apparent seasonal variation in the percentage of reproductivly mature females carrying pouch young (Fig. 4). During all trapping sessions at least 70%, but frequently 90% or more, of females were carrying and suckling young in the pouch (Fig. 4). For individual females reproduction usually followed a continuous pattern with the birth of a new young apparently coinciding with pouch vacation by the previous occupant.

An estimate of the level of recruitment of pouch young into the trappable population was made by individually ear-tagging animals close to the end of pouch life and noting the percentage later captured independently. Of 19 pouch young, nine (47%) were later captured as individual animals. This is likely to be a minimum estimate of recruitment as juvenile *P. tridactylus* are markedly trap-shy during their first few months of independence.

Capture sites of juvenile *P. tridactylus* were either within or on the margin of the maternal home range. Young animals remained within the maternal range until close to, or at the time of, achieving reproductive maturity, at which time dispersal of both sexes apparently occurred. Sample sizes are small, but the available data (Table 6) indicate that female *P. tridactylus* leave the maternal home range before the time of first birth, while males leave either before or soon after reaching a body weight of 600 g. All nine individuals first tagged as pouch young and successfully recruited to the trappable population were no longer present in the maternal home range after 12 months. Most individuals disappeared from the trapping record, but in two instances dispersal away from the maternal range was confirmed by trap captures. Disappearance from the trapping record may be due to dispersal or mortality.

DISCUSSION

Insular Biogeography

A highly significant relationship between faunal species richness and insular area is characteristic of most insular biogeographic studies (Moore and Hooper 1975; Kitchener *et al.* 1980b; Suckling 1982; Howe 1984). However, where other attributes such as isolation and habitat richness are measured these are also frequently highly correlated with faunal

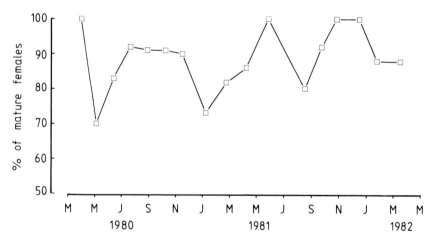

Fig. 4. Percentage of reproductively mature female *Potorous tridactylus* carrying a pouch young during each trapping session at Naringal.

Table 6. Status of young individuals of *Potorous tridactylus* at Naringal at the time of their last capture or known dispersal from the maternal home range. Individuals present at the close of the study are excluded. Mean adult body weight of *P. tridactylus* at Naringal was 789 g (male) and 777 g (female).

		Body weight (g)					Reproductive status (♀)	
	<300	300-399	400-499	500-599	600-699	700-799	given birth	not given birth
♂	2	1	–	1	3	–		
♀	1	1	–	1	–	1	0	4

species richness and with insular area (Kitchener *et al.* 1980a,b, 1982; Suckling 1982; but see Howe 1984). Kitchener *et al.* (1980a) found that the number of vegetation associations was a better predictor of lizard species richness in nature reserves in Western Australia than was reserve area. Other authors have included measures of habitat diversity (Kitchener *et al.* 1982; Ambuel and Temple 1983) and isolation (Howe *et al.* 1982; Opdam *et al.* 1984), togther with isolate area, in multiple regression equations describing faunal species richness.

In this study mammal species richness was highly significantly correlated with a number of variables representing patch size, habitat richness and disturbance. Area was the best single correlate of species richness accounting for some 82% of the variance, but it was evident that the contributions of other variables were being masked by area. Principal components analysis assisted in separating the individual contributions of these highly intercorrelated variables but was not able to distinguish between the variance in species richness attributable to area and to habitat richness respectively. Spatial isolation was not found to be an important factor influencing species richness in this locality, but temporal isolation was included as a significant step in the multiple regression equation. Most forest patches in the Naringal area have been isolated or reduced to their present size for less than 20 years following extensive forest clearing during recent decades (1940-1970). It may be expected that species richness will gradually decline in many forest patches as the duration of isolation increases.

Numerous studies have demonstrated that habitat fragmentation results in changes to the composition of faunal assemblages (e.g. Whitcomb *et al.* 1981; Ambuel and Temple 1983; Blake and Karr 1984; Howe 1984). This is due to the characteristically differing responses of different species to habitat fragmentation. At Naringal, the frequency of occurrence of mammal species in forest patches displayed marked variation in relation to forest patch area (Table 4), a response which has been documented for other faunal assemblages (Moore and Hooper 1975; Galli *et al.* 1976; Whitcomb *et al.* 1981; Humphreys and Kitchener 1982; Blake and Karr 1984). Differential occurrence of species in forest remnants has also been linked with spatial isolation of habitat patches (Opdam *et al.* 1984) and land use in the surrounding matrix (Butcher *et al.* 1981; Matthiae and Stearns 1981).

The differing responses of fauna to fragmentation at Naringal has not been explored in this chapter. Nevertheless, it may be noted that four primary factors have been associated with sensitivity to habitat fragmentation; these are relative body size (Brown 1971; Willis 1974), dietary type and trophic position (Willis in Terborgh and Winter 1980; Kitchener *et al.* 1982; Blake 1983), habitat specialization (Kitchener *et al.* 1980b, 1982; Whitcomb *et al.* 1981; Humphreys and Kitchener 1982) and general rarity (Terborgh and Winter 1980).

Faunal surveys which form the basis for insular biogeographic analyses also provide data which may be used to assess the relative status of individual species. Survey of the mammalian fauna at Naringal revealed that *P. tridactylus* was widespread and locally common within the study area. The frequency of occurrence of *P. tridactylus* in forest remnants, like most other native mammals, was clearly related to forest patch area (Table 4). The 'nested' pattern of occurrence of mammal species in forest remnants in the study area demonstrates the importance of the larger patches, both for the preservation of species-rich mammal assemblages and for the survival of the less common species. Small forest patches, of less than 10 ha for example, may provide valuable habitat for mammals, but such patches generally supported a smaller number of species dominated by widespread common native species or introduced mammals. Those native mammals whose status was best described as uncommon were seldom recorded in such small patches (Table 4).

Population Ecology of P. tridactylus

Potorous tridactylus has been recorded from a wide variety of forest types ranging from the fringes of rainforest to coastal woodlands (Schlager 1981; Seebeck 1981) but the feature common to all sites is the presence of dense effective cover in the shrub or field stratum (Heinsohn 1968; Schlager 1981; Seebeck 1981). In southern Victoria, forest vegetation providing suitable habitat for *P. tridactylus* is generally patchy in distribution and limited in size. It may be present in poorly-drained swampy areas, along drainage lines or watercourses, or be temporally available as a seral successional stage following fire or other disturbance (Seebeck 1981). It is proposed that the population ecology of *P. tridactylus* allows this species to exploit these relatively small and patchy habitats. Small body size, restricted home range, extensive range overlap, stable population structure, dispersal of both sexes, and a continuous pattern of reproduction are all attributes facilitating the maintenance of populations of *P. tridactylus* within relatively small areas.

Home range sizes calculated for *P. tridactylus* at Naringal are small in comparison with those recorded for the same species elsewhere (Kitchener 1973), and with other members of the Potoroidae. Kitchener (1973) estimated mean home range sizes of 19.4 ha and 5.2 ha for male and female *P. tridactylus* respectively in Tasmania (body weight ~ 1.2 kg). Christensen (1980) calculated mean home range sizes of 27 ha and 19 ha for male and female *Bettongia penicillata* respectively (body weight 1.3 kg), and Johnston and Rose (1983) described a home range of 65-135 ha for *B. gaimardi* (body weight 1.7 kg). Extensive overlap of home range areas (Table 5) also contributes to the maintenance of a relatively high population density, with individuals exclusively occupying either only a small part or none of their home ranges. Radio-tracking studies of *B. penicillata* (Christensen 1980) revealed that the home range of this species is composed of two parts; a small nest area in which there is little range overlap ($\bar{x}$ = 19.5%) and a much larger feeding area which is substantially shared with other animals ($\bar{x}$ = 63.9% overlap). *P. tridactylus* may have a similar pattern of home range utilization.

The major constraint on the reproductive potential of *P. tridactylus* is the limitation to a single young per litter, a feature characteristic of the Macropodoidea (with the sole exception of *Hypsiprymnodon moschatum*) (Strahan 1983). Within this limitation, reproductive potential is maximised by the mechanism of delayed gestation (Shaw and Rose 1979), a relatively short pouch life (in comparison with larger macropods) and a continuous year-round pattern of breeding by all reproductively mature females (Fig. 4). A similar reproductive pattern has been described for other potoroids (e.g. Christensen 1980). Despite these attributes, the maximum reproductive potential which may be achieved by *P. tridactylus* is between two and three young per year (Heinsohn 1968).

Dispersal is an important life-history feature for a species occupying patchy environments as it may lead to the re-colonization of unoccupied habitat patches or the discovery of newly created patches of suitable habitat. At Naringal, strips of forested vegetation along roadsides and creeks, such as those in which the CR and UR trapping grids were located, provide continuity between many forest remnants and facilitate the dispersal of *P. tridactylus* between the otherwise isolated habitat patches. The apparent timing of the major dispersal phase for *P. tridactylus*, close to the age of achieving reproductive maturity, also enhances the chance of individuals successfully establishing elsewhere as by this age they have passed the greatest period of mortality, namely, the transition from pouch life to independence.

In summary, the successful persistence of *P. tridactylus* within the fragmented and disturbed forest environment at Naringal may be attributed to the availability of numerous patches of suitable habitat (such as forest vegetation occurring in low-lying poorly drained areas, and dense regeneration following fire) and to the ability of *P. tridactylus* to disperse between and maintain populations within these habitats.

Insular Biogeography and Autecology: Relative Contributions to Fauna Conservation

Insular biogeography and autecological studies both provide useful information concerning the effects of habitat fragmentation on fauna, but each approach has a different focus. Insular biogeography primarily documents the observed changes that have occurred in the faunal assemblage and attempts to interpret them in relation to physical parameters such as area, isolation or habitat attributes. In contrast, autecological studies seek to investigate the underlying basis for the status and performance of individual species within a fragmented environment. Consequently insular biogeography and autecology do not represent optional alternatives, but rather complementary approaches to solving the same problem.

Insular biogeography makes its greatest contribution to conservation strategies directed toward conserving faunal communities rather than individual species. For the Naringal area, this approach identified the consequences of habitat fragmentation in terms of changes to species richness and changes to the species composition of mammal assemblages in forest remnants. These results suggest that the conservation objective of attaining maximum species richness and a more-complete faunal assemblage are most likely to be achieved in forest patches having large size, a variety of vegetation types and an absence of grazing by domestic stock. Accordingly, management strategies should be directed toward preventing further fragmentation, enhancing forest continuity between existing remnants to create effectively larger patches, and minimising disturbance to forest vegetation from stock. Larger forest size appears to be particularly important if the mammal assemblage is to include the least common native mammals.

Faunal surveys within forest remnants also provide a useful overview of the distribution and abundance of individual species and an estimate of their comparative status and sensitivity to habitat loss. However, insular biogeographic studies generally provide few clues concerning the reasons for the persistence or sensitivity of individual species to habitat fragmentation.

Autecological studies seek to determine the resource requirements and to understand the population dynamics of single species. These studies form a sound basis for single-species conservation strategies in disturbed environments as they provide a basis for understanding a species response to habitat fragmentation. Ecological study of *P. tridactylus* points to the important link between the population dynamics of this species and the distribution of suitable habitat. A conservation strategy for this species should emphasize the patchy distribution of favoured habitat, and the importance of forest continuity to allow dispersal between habitats. The potential need for management to maintain suitable dense vegetative cover (e.g. exclusion of grazing, or burning to promote vegetation succession) should also be considered.

The contribution of autecological studies to conservation strategies for which the goal is to conserve the entire faunal assemblage is more limited. It will chiefly depend upon the extent to which a strategy based upon the resource requirements of the target species will also fulfil the needs of other faunal species. Other species present may have differing or more specialized conservation requirements. Autecological studies can, however, identify principles which have wider relevance to the faunal community. For example, recognition of the heterogeneity of habitats and the importance of forest continuity to facilitate dispersal may be important not only to *P. tridactylus* but also to a range of other native mammals at Naringal.

ACKNOWLEDGEMENTS

I am grateful to Dr M. J. Littlejohn for supervision of this study and for comments on the manuscript. I also thank J. H. Seebeck, P. W. Menkhorst, C. E. Silveira, L. F. Lumsden, R. M. Bennett and R. J. Bennett for discussion and useful comments; and numerous residents of the Naringal area for allowing me to work on their properties. Financial assistance towards travelling costs is gratefully acknowledged from the M. A. Ingram Trust, the Ian Potter Foundation, and the Queen Elizabeth II Silver Jubilee Trust for Young Australians. Mammals were trapped under the provisions of Permit Nos. 79-167, 80-28, 81-53, and 82-48 from the Fisheries and Wildlife Service, Victoria.

REFERENCES

Ambuel, B. and Temple, S. A., 1983. Area dependent changes in the bird communities and vegetation of southern Wisconsin forests. *Ecology* 64: 1057-68.

Blake, J. G., 1983. Trophic structure of bird communities in forest patches in east-central Illinois. *Wilson Bull.* 95: 416-30.

Blake, J. G. and Karr, J. R., 1984. Species compositions of bird communities and the conservation benefit of large versus small forests. *Biol. Conserv.* 30: 173-88.

Boecklen, W. J. and Gotelli, N. J., 1984. Island biogeographic theory and conservation practise: species-area or specious-area relationship? *Biol. Conserv.* 29: 63-80.

Brown, J. H., 1971. Mammals on mountain tops — non equilibrium insular biogeography. *Am. Nat.* 105: 467-78.

Butcher, G. S., Niering, W. A., Barry, W. J. and Goodwin, R. J., 1981. Equilibrium biogeography and the size of nature reserves: an avian case study. *Oecologia* 49: 29-37.

Calaby, J. H., 1971. The current status of Australian Macropodidae. *Aust. Zool.* 16: 17-29.

Christensen, P. E. S., 1980. The biology of *Bettongia penicillata* (Gray 1837), and *Macropus eugenii* (Desmarest 1817) in relation to fire. *Forests Dep. W.A. Bull.* 90.

Diamond, J. M., 1975. The island dilemma: lessons of modern biogeographic studies for the design of natural reserves. *Biol. Conserv.* 7: 129-46.

Dixon, W. J. (ed.), 1983. BMDP Statistical Software. University of California Press, Berkeley.

Flowerdew, J. R., 1976. Techniques in mammalogy. Chapter 4. Ecological methods. *Mamm. Rev.* 6: 123-58.

Frith, H. J., 1979. Wildlife Conservation. Angus and Robertson, Sydney.

Galli, A. E., Leck, C. F. and Forman, R. T. T., 1976. Avian distribution patterns in forest islands of different sizes in central New Jersey. *Auk* 93: 356-64.

Gilbert, F. S., 1980. The equilibrium theory of island biogeography: fact or fiction. *J. Biogeog.* 7: 209-35.

Heinsohn, G. E., 1968. Habitat requirements and reproductive potential of the macropod marsupial *Potorous tridactylus* in Tasmania. *Mammalia* 32: 30-43.

Howe, R. W., Howe, T. D. and Ford, H. A., 1982. Bird distributions on small rainforest remnants in New South Wales. *Aust. Wildl. Res.* 8: 637-51.

Howe, R. W., 1984. Local dynamics of bird assemblages in small forest habitat islands in Australia and North America. *Ecology* 65: 1585-601.

Hughes, R. L., 1962. Reproduction in the macropod marsupial *Potorous tridactylus* (Kerr) *Aust. J. Zool.* 10: 193-224.

Hughes, R. L., 1964. Sexual development and spermatozoan morphology in the male macropod marsupial *Potorous tridactylus* (Kerr). *Aust. J. Zool.* 12: 42-51.

Humphreys, W. F. and Kitchener, D. J., 1982. The effect of habitat utilization on species-area curves: implications for optimal reserve area. *J. Biogeog.* 9: 391-6.

Johnson, K. A. and Rose, R., 1983. Tasmanian Bettong *Bettongia gaimardii*. Pp. 186 *in* Complete Book of Australian Mammals ed by R. Strahan. Angus and Robertson, Sydney.

Kitchener, D. J., 1973. Notes on home range and movement in two small macropods, the potoroo *(Potorous apicalis)* and the quokka *(Setonix brachyurus)*. *Mammalia* 37: 231-40.

Kitchener, D. J., Chapman, A., Dell, J., Muir, B. G. and Palmer, M., 1980a. Lizard assemblage and reserve size and structure in the Western Australian wheatbelt — some implications for conservation. *Biol. Conserv.* 17: 25-62.

Kitchener, D. J., Chapman, A., Muir, B. G. and Palmer, M., 1980b. The conservation value for mammals of reserves in the Western Australian wheatbelt. *Biol. Conserv.* 18: 179-207.

Kitchener, D. J., Dell, J., Muir, B. G. and Palmer, M., 1982. Birds in Western Australian wheatbelt reserves — implications for conservation. *Biol. Conserv.* 22: 127-63.

Land Conservation Council, 1976. Report on the Corangamite Study Area. Land Conservation Council, Victoria.

MacArthur, R. H. and Wilson, E. O., 1967. The Theory of Island Biogeography. Princeton University Press, Princeton, New Jersey.

Matthiae, P. E. and Stearns, F., 1981. Mammals in forest islands in southeastern Wisconsin. Pp. 55-66 *in* Forest Island Dynamics in Man — Dominated Landscapes ed by R. L. Burgess and D. M. Sharpe. Springer-Verlag, New York.

McCoy, E. D., 1982. The application of island biogeographic theory to forest tracts: problems in the determination of turnover rates. *Biol. Conserv.* 22: 217-27.

Moore, N. W. and Hooper, M. D., 1975. On the numbers of bird species in British woods. *Biol. Conserv.* 8: 239-50.

Opdam, P., van Dorp, D. and ter Braak, C. J. F., 1984. The effect of isolation on the number of woodland birds in small woods in the Netherlands. *J. Biogeog.* **11**: 473-8.

Schlager, F. E., 1981. The distribution and status of the Rufous Rat-kangaroo, *Aepyprymnus rufescens*, and the Long-nosed Potoroo, *Potorous tridactylus*, in northern New South Wales. Project Report No. 18, Department of Ecosystem Management, University of New England, Armidale, NSW.

Seebeck, J. H., 1981. *Potorous tridactylus* (Kerr) (Marsupialia: Macropodidae): its distribution, status and habitat preferences in Victoria. *Aust. Wildl. Res.* **8**: 285-306.

Shaw, G. and Rose, R. W., 1979. Delayed gestation in the Potoroo. *Aust. J. Zool.* **27**: 901-12.

Simberloff, D. S., 1974. Equilibrium theory of island biogeography; and ecology. *Ann. Rev. Ecol. Syst.* **5**: 161-82.

Simberloff, D. S. and Abele, L. G., 1982. Refuge design and island biogeographic theory: effects of fragmentation. *Am. Nat.* **120**: 41-50.

Stickel, L. F., 1954. A comparison of certain methods of measuring ranges of small mammals. *J. Mamm.* **35**: 1-15.

Strahan, R. (ed.), 1983. Complete Book of Australian Mammals. Angus and Robertson, Sydney.

Suckling, G. C., 1982. Value of reserved habitat for mammal conservation in plantations. *Aust. For.* **45**: 19-27.

Terborgh, J. and Winter, B., 1980. Some causes of extinction. Pp. 119-33 *in* Conservation Biology: An Evolutionary-Ecological Perspective ed by M. E. Soulé and B. A. Wilcox. Sinauer Associates, Sunderland, Mass.

Whitcomb, R. F., Robbins, C. S., Lynch, J. F., Whitcomb, B. L., Klimkiewicz, M. K. and Bystrak, D., 1981. Effects of forest fragmentation on avifauna of the eastern deciduous forest. Pp. 123-205 *in* Forest Island Dynamics in Man — Dominated Landscapes ed by R. L. Burgess and D. M. Sharpe. Springer-Verlag, New York.

Willis, J. H., 1964. Vegetation of the basalt plains in western Victoria. *Proc. Roy. Soc. Vic.* **77**: 397-468.

Willis, E. O., 1974. Populations and local extinctions of birds on Barro Colorado Island, Panama. *Ecol. Monogr.* **44**: 153-69.

Wilson, E. O. and Willis, E. O., 1975. Applied biogeography. Pp. 522-34 *in* Ecology and Evolution of Communities ed by J. A. Diamond and M. Cody. Belknap Press, Cambridge, Mass.

CHAPTER 5

Local Decline, Extinction and Recovery: Relevance to Mammal Populations in Vegetation Remnants

J. A. Friend[1]

The surviving populations of a number of species of mammal in Western Australia are largely limited to isolated remnants of native vegetation. It is postulated that in pre-European times, sharp declines, and sometimes extinctions of species occurred in local areas, due to climate-related events (including wildfire). Population recovery was due to immigration as well as reproductive increase by the remaining individuals.

Today, the opportunity for immigration after local declines (and extinctions) is often effectively lost, so recovery is limited to reproductive increase. The long-term survival of populations in some of these remnants may depend on manipulation of gene pools. Before decisions in this area may be made, it is necessary to have certain minimum biological information about the species involved. The case of the numbat *Myrmecobius fasciatus* is used to illustrate this problem.

A general model is developed to predict the size of the influx of a species by natural dispersal into a given area which is surrounded by continuous habitat. This may be used to determine the desirable size for a translocation group, should this be deemed necessary after a severe decline in an isolated population.

INTRODUCTION

AMONGST the faunal groups present in a largely cleared landscape dotted with fragments of native vegetation, evidence indicates that the mammals will show the greatest departure from their original richness (Kitchener 1982). The scale at which habitat fragmentation has been carried out by European man, in terms of the patch sizes of cleared and uncleared land, appears to have had the most devastating interaction with the scale of dispersal and home range size of this particular group of animals. This chapter compares the effects of natural population fluctuations in pre- and post-fragmentation scenarios and outlines some implications for the management of mammal populations persisting in remnants smaller than those considered of minimum useful size (e.g. Main and Yadav 1971).

The Western Australian wheatbelt is an area previously rich in mammals, which has been extensively cleared for agriculture. Seventeen of the 43 species of mammal recorded from the region since white settlement no longer exist there (Kitchener *et al.* 1980, but note that *Antechinus flavipes* is still common in larger western wheatbelt reserves). The medium-sized mammals (body weight 100-5000 g) have been most severely affected; three species in this size range historically known from the region have become extinct altogether (*Onychogalea lunata, Chaeropus ecaudatus, Potorous platyops*), while some species are now represented only by populations elsewhere (*Lagostrophus fasciatus, Lagorchestes hirsutus, Bettongia lesueur, Macrotis lagotis, Perameles bougainville, Leporillus* sp.). A number of species have undergone decline but still occur in the region. Three of the four surviving populations of *Bettongia penicillata*, and all known populations of *Phascogale calura* (body weight ~ 50 g) occur in remnants of native vegetation in the wheatbelt. Several rare or endangered species/subspecies surviving there are also found in

---

[1]Western Australian Wildlife Research Centre, Department of Conservation and Land Management, P.O. Box 51, Wanneroo, Western Australia 6065.
Pages 53-64 *in* NATURE CONSERVATION: THE ROLE OF REMNANTS OF NATIVE VEGETATION ed by Denis A. Saunders, Graham W. Arnold, Andrew A. Burbidge and Angas J. M. Hopkins. Surrey Beatty and Sons Pty Limited in association with CSIRO and CALM, 1987.

parts of the forest belt to the southwest, where the likelihood of movement between populations is greater (*Macropus eugenii, Dasyurus geoffroii, Myrmecobius fasciatus, Pseudocheirus peregrinus occidentalis*). *Petrogale lateralis lateralis* has declined in many parts of its former range, but still occurs on isolated granite rock reserves of the wheatbelt.

The persistence of some populations of these species is dependent on remnants of natural vegetation which are generally so widely spaced that exchange of individuals through dispersal is virtually impossible. The complete isolation of many of these remnants, however, is still a very recent event. Most clearing in the wheatbelt has occurred since the 1940s with the advent of the bulldozer, and with the application of new agricultural technology which has allowed previously unsuitable land to be farmed. Consequently, bushland adjacent to some important reserves has been cleared only in the last twenty years. With so few generations even of shorter-lived mammal species having passed since the fragmentation of their populations, it is likely that some of the genetic consequences are still developing. The early and extremely rapid extinction of several of the species listed above (Shortridge 1910; Kitchener *et al.* 1978) indicates that those most sensitive to habitat disturbance have already gone. The mammal fauna which remains apparently comprises species relatively tolerant to the ecological changes which have occurred. The longer-term effects of disrupted gene flow within species, however, are likely to take a further and continuing toll (Frankel and Soulé 1981). It is therefore imperative that this problem be examined and necessary action taken before further populations and genotypes are irretrievably lost.

POPULATION FLUCTUATIONS IN HABITAT ISOLATES

Local decline and even local extinction, and subsequent recovery or recolonization, are natural processes which clearly contribute significantly to the genetic dynamics of any population. Natural causes of declines or extinctions in local areas are generally related to climate, and include drought and wildfire.

Drought. There is evidence that declines and subsequent recoveries in several mammal species in southwest Western Australia have coincided with periods of drought followed by good rains (drought of 1895-1902, population crashes and subsequent recovery recorded by Leake 1962; drought of the 1930s — subsequent recovery documented by Serventy 1954; drought of the 1970s — declines of some species shown by Christensen 1980a). Patterns of rainfall fluctuations in Western Australia show strong local variation (Fig. 1), so drought-induced population declines may conceivably occur adjacent to areas experiencing adequate rainfall. The mechanisms by which rainfall fluctuations cause population crashes may be diverse; however lower food availability, and consequently greater exposure to predation, and lower resistance to pathogens are likely to be largely responsible.

As a specific example, the numbat, *Myrmecobius fasciatus* has suffered a recent decline and recovery, although over a wide area covering both Perup and Dryandra Forests, southwest Western Australia (Figs 2 and 3). This decline coincided with drought during the mid-1970s, and with an apparent rise in numbers of the introduced red fox *Vulpes vulpes*. It has been suggested that increased predation over this period was the major cause of the marked decline of several medium-sized mammals, including the numbat (Christensen 1980a).

Wildfire. Dramatic declines and temporary local extinctions of some species of small mammal due to a catastrophic wildfire were recorded by Newsome *et al.* (1975) in southern New South Wales. Other Australian studies have also demonstrated this effect (see Recher and Christensen 1981). Similar crashes in small mammal populations after fire are reported in chapparal fires in California (Chew *et al.* 1959; Cook 1959). Even controlled burns can cause a decline in numbers of small- and medium-sized mammals (Christensen and Kimber 1975; Christensen 1980b), if not through the acute effects of the fire, then as a result of the subsequent shortage of food and shelter.

Where native vegetation is largely intact and mammal populations are not separated by impassable tracts of uninhabitable land, recovery after a decline is by both reproductive increase (R) and immigration by dispersing individuals from surrounding areas (I) (Fig. 4). The relative contributions of these two processes to recovery are determined by the size of the population surviving in the affected area, and the density of the population remaining in adjacent unaffected areas. A high contribution of immigration to I+R-recovery was shown, for instance, in a study of a population of the common brushtail possum, *Trichosurus vulpecula*, in New Zealand (Clout and Efford 1984). Within one year of a heavy poisoning programme which almost totally removed the species from the study area, numbers had recovered to 50% of their pre-poisoning level. This was achieved almost entirely through immigration of dispersing individuals. In this case there was an artificially sharp boundary between the affected and unaffected areas.

In a habitat isolate, on the other hand, recovery after decline is due only to reproductive increase (Fig. 4). Let us assume that habitat recovers more rapidly than the population can by reproduction alone, so that its instantaneous carrying capacity is

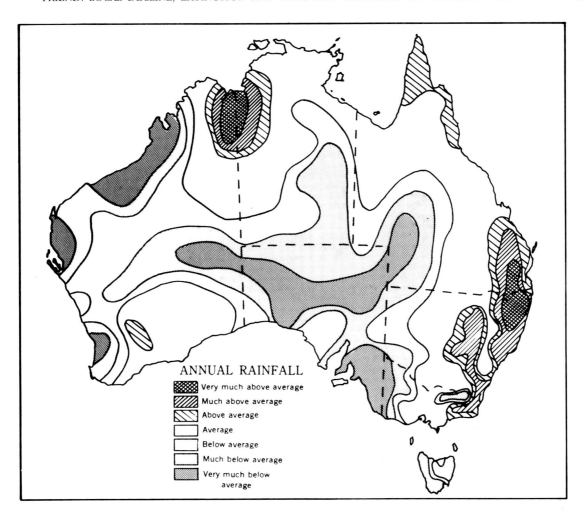

Fig. 1. Variation from average annual rainfall in Australia during 1959 (after Gibbs and Maher 1967).

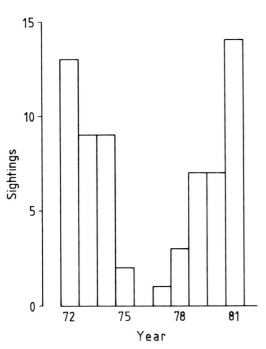

Fig. 2. Number of recorded sightings of numbats in the Perup Forest, Western Australia, each year from 1972 to 1981 (from Christensen *et al.* 1984).

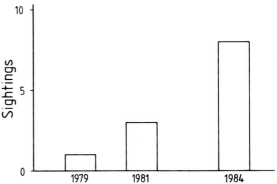

Fig. 3. Number of sightings of adult numbats during the surveys in Dryandra Forest, Western Australia, over seven weeks between August and November, 1979, 1981 and 1984. 1979 data from Turner and Borthwick (1980). The 1984 survey duplicated the 1981 survey routes.

not approached during the early stages of recovery. There are several important genetic consequences of R-recovery, as opposed to I+R-recovery. Firstly, the effect of the reduction of the population to low numbers *per se* (the 'founder effect') is that there is a high likelihood of significant loss of genetic variation within the population through sampling error. In the case of an isolated population, this loss

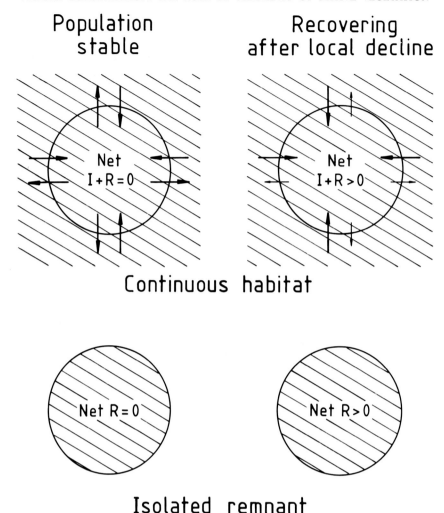

Fig. 4. Diagrammatic representation of modes of population recovery after decline in circular areas of habitat either surrounded by, or isolated from, sources of dispersing animals (shading represents habitat). **Net I + R** refers to the increase in numbers within an area due to immigration and reproduction, after allowing for emigration and mortality. **Net R** refers to the increase in numbers in an isolated area after allowing for mortality, assuming that dispersal movement away from isolated areas of habitat is negligible.

is not relieved by the addition of genetic variability through immigration, and the population goes through a 'genetic bottleneck'.

In addition to the loss of genetic variability, founder effects may also cause destabilization of polygenic balances, resulting in the selection of new balances (Carson 1983), even without significant loss of genetic variability (Templeton 1980). As this type of genetic system is believed to control integrated developmental, physiological and behavioural traits, these changes may affect profoundly the fitness of offspring.

The second consequence of R-recovery is that there is a risk that return to previous population levels will be slow. In this case, further loss of genetic variability is likely through genetic drift. The theoretical study of Nei *et al.* (1975) showed that the proportion of heterozygosity, a measure of genetic variation, remaining in a population after an extreme bottleneck is negligible when the rate of natural increase is very low ($r < 0.1$). When $r > 1.0$, the loss of genetic variation after a bottleneck is, by comparison, insignificant. In addition to a reduction in heterozygosity, alleles may be lost through genetic drift; the most likely to be lost, of course, are rare alleles, which have little effect on the short-term fitness of the population, but can be of major importance in adaptive evolution in the event of environmental change (Frankel and Soulé 1981). Immediately threatening the survival of a very small population, however, is the phenomenon known as inbreeding depression. This is the loss of fecundity, loss of fertility and increased juvenile mortality due either to loss of heterozygosity or to fixation of deleterious genes due to a high incidence of mating between close relatives (Franklin 1980). Inbreeding depression is well known in small breeding groups of domestic animals, and has been documented particularly for captive colonies in zoos (e.g. Ralls and Ballou 1983).

Thus the loss of genetic variation caused by the small size of the surviving group is exacerbated subsequently if recovery is very slow. Through I+R-recovery, these problems are much less likely to arise, due both to the more rapid rise in population numbers and to the import of new genetic material.

GENETIC MANAGEMENT OF ISOLATED POPULATIONS

The detrimental effects of fluctuations in small populations in habitat isolates should clearly be of concern to managers, as mere survival after decline cannot be taken as a sign that all is well. The definition of 'small' in this context can be related to estimates of minimum viable population (MVP) which have been attempted (Franklin 1980; Frankel and Soulé 1981; Lehmkuhl 1984). It has been suggested that an increase in inbreeding coefficient of 1% per generation is the maximum permissible to avoid inbreeding depression; this corresponds to an *effective* population size of 50, which should be regarded as a short-term MVP (Franklin 1980). The effective population size of any breeding group refers to an idealised population in which each individual contributes equally to the pool from which the next generation is formed, which has the same sampling variance as the real population. The difference between the actual population and the effective population of a particular species depends on such factors as variance of litter size, non-participation by individuals in reproduction, unequal sex ratio, overlapping generations and population fluctuations (Kimura and Crow 1963; Franklin 1980). The operation of these factors may mean that the actual population size is much greater than the corresponding effective population. By use of the equations of Kimura and Crow (1963) following Lehmkuhl (1984), Franklin's *short-term* MVP estimate of 50 effective individuals corresponds to an actual population size of 320 for the numbat. This figure should be reduced, however, due to the particular dispersal behaviour displayed by this species (see below).

Estimates of the minimum size of genetically-isolated populations which will allow *long-term* viability have also been put forward. Franklin (1980) proposed that at an effective population size of 500, the loss of additive genetic variance (in quantitative traits) by genetic drift is approximately balanced by the rate of gain by mutation. Below this number, he suggested, a population will suffer a net loss of variation which would reduce the chance of adaptive change occurring in the future. Franklin (1980) admitted that this number was chosen on the basis of slim evidence, but that it was of the right order of magnitude. Frankel and Soulé (1981) used arguments based on the relative strength of selection and drift on the frequency of a single gene in small populations to arrive at the same figure as Franklin (1980). This precise agreement appears to be due to their choice of values of certain parameters as much as to the accuracy of the estimate (Brown 1983). There is a severe shortage of directly useful experimental evidence in this discussion. Of particular interest, however, is the study by Main and Yadav (1971) relating the occurrence of macropods on islands off the Western Australian coast to island size and availability of habitat. They found that a carrying capacity of 200-300 appears to be the minimum required to allow the long term survival (10,000 years) of species in this family on an island.

In deciding which populations should be of concern to managers after a serious decline in numbers has occurred, the short-term MVP estimate above (effective population size of 50) is of most relevance. It is clear from the work of Nei *et al.* (1975) that a rapid recovery in numbers will minimize the long-term effects of a bottleneck. Efforts to reduce mortality, particularly amongst juveniles, by management of habitat (e.g. to maintain the availability of refuges and cover and, in Australia, to control introduced predators) are of prime importance at this stage. If the surviving population is very small, however, these measures may not be sufficient to prevent the development of undesirable genetic effects.

An obvious remedy for the ills of a small, isolated population which has recently suffered a decline is the injection of individuals from another source into the population during the early stages of recovery. This then mimics the immigration of dispersing animals which would happen naturally if the depleted area were adjacent to unaffected populations. If a sufficient number of animals are translocated and establish themselves as part of the breeding population, the effective minimum population size at the bottleneck will be increased. Consequently, the risk of inbreeding depression and of significant loss of genetic variation will be greatly reduced.

The effectiveness of any such introduction will depend on the biology of the species concerned. Effort might be wasted, for instance, if translocated males were excluded from breeding by the presence of an extremely aggressive resident male, and new females were merely added to his harem. This would result in a smaller contribution to the genetic diversity of the next generation. Introduction of new animals should be carried out at a time of year when the risk of predation before the next breeding season is least. Seasonal periods of food shortage should also be avoided. It is important, therefore, to have a good understanding of the ecology and behaviour of the species concerned.

Another important consideration in planning a translocation to enhance genetic diversity is the possibility that the source population shows significant genetical differences from the depleted population. Mixing of distinct genotypes in this way may have deleterious effects on progeny due to genetic incompatibilities, but more importantly, it may result in loss of overall genetic diversity by the species. It is most desirable, therefore, that some comparison (e.g. electrophoretic studies) of source and destination populations be made before animals are transferred. With respect to mammals, the likelihood of interpopulation differences may be predicted through knowledge of the particular species' history of isolation. For instance, vagile species whose habitat was previously more or less continuous, like the numbat, are unlikely to display large genetic differences within a region. On the other hand, in species whose habitat was always divided into remote patches between which individuals rarely moved, distinct genotypes might be expected to have evolved in different populations. For example, the Western pocket gopher, *Thomomys bottae*, which occupies a discontinous grassland habitat in North America, displays strong interpopulation genetic differences due to inhibited gene flow (Patton and Yang 1977).

As the artificial introduction of new animals into an isolated population is an attempt to mimic the immigration of dispersing individuals, it is important to have some idea of the magnitude of the movement which would have occurred if there had been adjacent occupied habitat. It is clearly desirable to know also the prevailing conditions or season during which dispersal occurs under natural conditions. The case of the numbat will be used (below) to illustrate the development of a strategy for restoring lost genetic variation to a hypothetical isolated population which has experienced a severe decline.

THE NUMBAT — A CASE STUDY

The numbat is a small marsupial belonging to the monotypic family Myrmecobiidae. It is of particular scientific interest, being the only myrmecophagous (ant- or termite-eating) marsupial, as it feeds almost exclusively on termites and possesses a number of morphological adaptations in keeping with this habit. The attractive appearance and appealing nature of the numbat also ensure high public interest in its conservation.

At the time of European settlement of Australia, the distribution of the numbat extended across most of the southern half of the continent (Friend and Kinnear 1983). The decline of the species apparently began almost immediately, and by the 1970s the species was restricted to forest and woodland at the southwestern extremity of its former range (Friend 1982). Calaby (1960) suggested that the numbat's remaining stronghold in the mid-1950s was woodland dominated by wandoo, *Eucalyptus wandoo*, in the Western Australian wheatbelt. Since that time, most of this vegetation type has been selectively cleared for agriculture, as the heavier soils on which it occurs are particularly suitable for cultivation. Very little land in the wheatbelt has been set aside for conservation, and those reserves which exist tend to comprise the least arable land; rocky upland sites, salt lakes and areas of sandy soils feature prominently amongst nature reserves. While the numbat has been found to persist in certain parts of the extensive jarrah, *E. marginata*, forests west of the wheatbelt, populations have survived in the agricultural region only on two of the larger conservation areas, Dryandra Forest (28,000 ha), Tutanning Nature Reserve (2000 ha) and small remnants of woodland near them (Connell and Friend 1985). Dryandra Forest has been the site of an intensive study of the species since 1982 (Friend and Burrows 1983). In particular, dispersal movements, as well as the daily activity of individuals have been followed, and these data allow practical recommendations for the introduction of new animals into isolated populations after a local decline. The detailed results of this work will be reported elsewhere, but those immediately relevant to this discussion are given below.

Reproduction, Home Range and Dispersal

Numbats produce young during summer, although at Dryandra most litters are born in mid-January. The young remain in the pouch until late June or July, when the female deposits them in a burrow. By late October the juvenile numbats are active and feeding on termites, but remain at first within their mother's home range (Friend and Burrows 1983). Dispersal of the young occurs in November and December.

Individual numbats in Dryandra Forest were fitted with radio-transmitters (AVM Instrument Company, California) to allow the measurement of home range and to follow dispersal of juveniles. Tracking was generally carried out on foot, but to record long-range dispersal movements it was necessary to use a light aircraft. The following findings are relevant to this discussion.

Dispersal distance. Distances of dispersal movements in 1982, 1983 and 1984 (n = 11) varied from 0.6 km to 10.9 km, with a mean and standard deviation of 3.59 ± 2.82 km. As the magnitude of the longer movements was apparently restricted by the size of the forest (the largest block measures 18 × 10 km), this mean is probably an underestimate; however, as the distribution of dispersal distance appears skewed towards lower values, as might be expected (Fig. 5), it is probably only the few highest values which are affected.

Dispersal direction. Zar (1984) defined the value r, which is a measure of the concentration of bearings around a circle. This value may vary from 0, when there is so much dispersion that a mean angle cannot be described, to 1.0, when all the data are concentrated in one direction. Calculating r for the direction of dispersal of the young (n = 11) gives a value of 0.46. This is not significantly different from zero at the 5% level (Rayleigh's test; Zar 1984), suggesting that the direction of dispersal is random, or at least that there is no general trend in the direction of dispersal.

Numbers of dispersing young. Over the three years, all females captured during the breeding season had produced between 1 and 4 young, but the mean number per litter (n = 7) surviving to dispersal was 2.0 young.

During the period of the study of home range size and juvenile dispersal, the Dryandra numbat population was recovering after the decline of the 1970s (Fig. 3). It is conceivable that low population densities early in the study may have had some effect on size of home ranges, dispersal distance and rate of juvenile survival. Most of these data, however, were collected in 1983 and 1984, by which time the frequency of sightings had returned to levels encountered in the early 1970s. It is felt that the values given here present a reasonable approximation given the non-pristine environment, where higher levels of predation exist due to introduced predators, and restrictions on movement are imposed by the finite bounds of the forest.

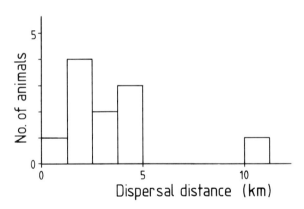

Fig. 5. Frequency of distance (1.25 km classes) between natal area and final home range of dispersing numbats in Dryandra Forest, 1982-1984.

Long-term studies of numbat movement in Dryandra Forest have shown that females occupy home ranges which are exclusive with respect to other females. The home ranges of males overlap those of females but are not shared by other males; however, in the breeding season (summer) and about two months prior to it, males roam well beyond their winter home range. Male and female winter home ranges are of similar mean size, that is, approximately 50 ha in area. A fully-inhabited area may thus be regarded as containing a pair of numbats in each 50 ha (although each pair does not coexist in exactly the same patch of habitat).

Dispersal of the young involves movement by individuals away from their natal areas and their subsequent establishment in new locations. This movement by an individual numbat rarely takes more than a week, and the approximate home range established at that stage appears to be occupied by that animal for the rest of its life. Mating occurs after dispersal of the young; the large distances travelled are obviously important in reducing the chance of breeding between close relatives. As a consequence, the size of a numbat population corresponding to a 1% increase of inbreeding coefficient per generation is probably rather smaller than is suggested by using the method of Lehmkuhl (1984) above.

Immigration Model

In order to recommend a strategy for the reintroduction of new animals into a population after a decline, a model is developed below which describes expected immigration into a defined area, on the basis of the findings already outlined. To summarize, these are as follows:

1. adult home ranges average 50 ha in area;

2. an average of 2.0 young per female survive to dispersal; and

3. dispersal direction is random.

The following assumptions are used in addition to these data:

1. numbat habitat is uniformly distributed over the region;

2. numbat habitat occupies 50% of the total area; and

3. all young disperse a uniform distance (3.6 km), irrespective of the density of adults.

The aim of the model is to calculate the number of individual numbats which might be expected to disperse each year, into a hypothetical nature reserve of given size and shape surrounded by continuous bushland, given the above assumptions. Thus, after a population decline in such an area, this is the size of the influx of animals which will contribute to the recovery of the population. The calculated influx may then be used as a guide to the number of animals which should be used in any programme to restock a population on an isolated reserve after a major decline, in order to prevent a critical loss of genetic variability.

We may regard the region surrounding the reserve as being uniformly occupied by adult pairs of numbats, in the first instance at the carrying capacity of the environment. The density of the population, then, corresponding to our assumptions of 50% useful habitat and 50 ha home ranges, would equal 1 pair per 100 ha. While the population is uniformly distributed, dispersing numbats will also be evenly distributed, if a uniform dispersal distance is assumed. Also, if two young from each litter survive to disperse, this density will equal two young per 100 ha, to 2 km^{-2} at carrying capacity.

We will suppose that the reserve in question is circular, and ignore the dispersal of young from inside the reserve, to allow calculation of the influx of new animals. An annular zone of width equal to the dispersal distance will exist inside the boundary which will be penetrated by young from outside. As young dispersing into the reserve will not be replaced by others moving out, there will be an annular zone of the same width surrounding the boundary, in which the density of dispersing young is below that in the surrounding region. A gradient of decreasing density of dispersing young will occur from the outside of the outer annulus to the inside of the inner one. Over this gradient, the density will fall from 2 km^{-2} to zero and the density of dispersing young will be proportional to the area under the curve. If the boundary were a straight line, the decrease of density would be linear, so as the radius of the reserve increases relative to the dispersal distance, this relationship approaches linearity. If we know this relationship, calculation of the number of animals entering a reserve of given radius may be performed by integral calculus. The density of dispersing animals at any point (animals.km^{-2}) appears along the y-axis, while x represents the distance from the centre of the reserve. The total number of young entering the reserve is then given by the volume of the solid of revolution of the (shaded) area under the curve, around the y-axis (Fig. 6). If we assume that the decrease in density (above) is always linear, then this line is described by the equation:

$$y = ax + B$$

$$\text{or, } x = \frac{1}{a} \cdot y + b \qquad \text{Where } b = \frac{-B}{a}$$

At the boundary of the reserve, $x = r$ and $y = D_y/2$. At the limit of penetration of dispersing young, $x = b = r-d$ and $y = 0$ (Fig. 6),

where: D_y = density of dispersing young outside reserve (animals.km^{-2})
r = radius of reserve (km)
d = dispersal distance of young (km)

The slope of the line $y = ax + b$ is

$$a = \frac{D_y}{2} \cdot \frac{1}{d} \qquad (1)$$

Thus the volume of the solid of revolution of the shaded area around the y-axis = the difference between the volumes of the solids of revolution

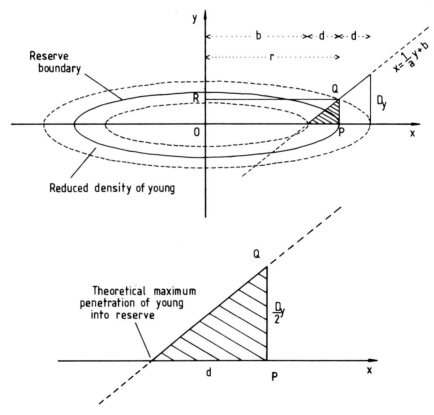

Fig. 6. Change in density of dispersing animals over annular areas near the boundary of a hypothetical circular reserve, given the assumptions of the model. Distance from the centre of the reserve is represented by x, while y represents the density of dispersing young. See text for fuller explanation.

around the y-axis of the rectangle OPQR (Fig. 6) and the area between the line $x = \frac{1}{a}y + b$ and the y-axis over the interval $y = 0$ to $y = \frac{D_y}{2}$

$$\therefore \text{Volume} = \pi r^2 y - 6\pi x^2 . dy$$

In this case $x = \frac{1}{a}y + b$, so substituting,

$$= \pi r^2 y - \pi \frac{(y + b)^2}{a} dy$$

Substituting for y,

$$= \pi r^2 . \frac{D_y}{2} - \pi \int_0^{\frac{D_y}{2}} (\frac{y^2}{a^2} + 2\frac{by}{a} + b^2) \, dy$$

$$= \pi r^2 . \frac{D_y}{2} - \pi [\frac{y^3}{3a^2} + \frac{by^2}{a} + b^2 y]_0^{D_y/2}$$

Substituting for a from (1), and $b = (r-d)$,

Volume =

$$\pi r^2 . \frac{D_y}{2} - \pi [\frac{4d^2 . y^3}{3D_y^2} + \frac{2d . (r-d) . y^2}{D_y} + (r-d)^2 . y]_0^{D_y/2}$$

Substituting for y,

Volume =

$$\pi r^2 . \frac{D_y}{2} - \pi . \frac{D_y}{2} [\frac{d^2}{3} + d(r-d) + (r-d)^2]$$

$$= \pi . \frac{D_y}{2} \left(dr - \frac{d^2}{3} \right) \quad (2)$$

This expression may be used to calculate the expected influx of dispersing young into a circular area of radius r km, within continuous habitat in which the density of dispersing young is D_y animals.km^{-2} and the dispersal distance of individual young is d km.

In the particular case of the numbat population at carrying capacity, the following values may be substituted:

$$D_y = 2 \text{ and } d = 3.6$$

Therefore, substituting in (2), the expected influx of young animals into the reserve (N) is

$$N = \pi(3.6r - 4.32) \quad (3)$$

Equation (2) may be used to calculate the theoretical immigration into circular reserves of given area when the surrounding habitat contains a given density of adult numbats. The calculated influx of young numbats dispersing into circular reserves is plotted in Figure 7 for a range of adult densities in the surrounding region: 100%, 50%, 25% and 10% of the density at carrying capacity. However, when $r < d < 2r$, use of equation (2) results in an underestimate because it evaluates the difference between the volumes of revolution of the sectors on each side of the y-axis rather than their sum, so values are not plotted for r in this range. To calculate N in this case, the separate volumes created by revolving the two

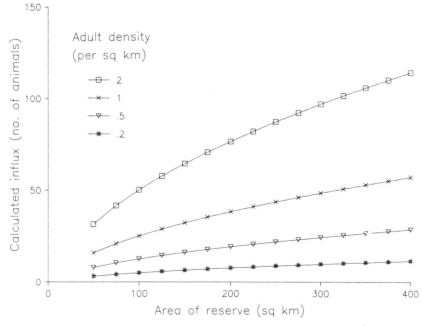

Fig. 7. Predicted influx of dispersing numbats into circular areas of 5000 ha (50 km^2) and above surrounded by continuous bushland containing numbat habitat, given the assumptions of the model. The four lines represent the number of young entering the area if the surrounding bushland contains densities of numbats at 100%, 50%, 25% and 10% of carrying capacity (2, 1, 0.5 or 0.2 adults.km^{-2} respectively).

triangles and one rectangle about the y-axis are calculated and summed, so

$$N = \int_{y_1}^{D_y/2} \pi x^2 \, dy + \int_0^{y_1} \pi x^2 \, dy + \pi y_1 (d-r)^2$$

where y_1 is the intercept of the line $x = \frac{y}{a} + b$ on the y-axis.

The case in very small reserves where $2r < d$, and many young are passing through the reserve, is not addressed. As the radius becomes smaller compared with the dispersal distance, the error due to the assumption of linearity becomes more significant, and this model will underestimate the influx of young.

Using equation (2), Table 1 shows the value of this variable for reserves of selected sizes.

Table 1. Carrying capacity for numbats of circular reserves of various sizes and expected annual influx of young from surrounding areas, occupied to various proportions of carrying capacity, predicted by immigration model.

	% of carrying capacity outside				
Area of reserve (ha)		5000	10,000	20,000	30,000
Carrying capacity (No. adults)		100	200	400	600
Predicted immigration (No. animals)	100%	32	50	77	97
	25%	8	13	19	24
	10%	3	5	8	10

DISCUSSION

In areas of up to 30,000 ha, even given that 50% of the surrounding countryside is fully occupied numbat habitat, the annual influx of dispersing animals according to this model is less than 100 animals. After a general decline across the region, the predicted influx of animals which will contribute to recovery within that area is quite small (i.e. using 25% or 10% occupancy of surrounding areas). Given that there are either surplus captive animals which can be released, or natural populations in other parts of the numbat's range which have not suffered a similar decline, then the transfer of small numbers of young to augment recovery in an isolated population is a practical possibility, which will approximate the missing natural influx of dispersing young.

In deciding on the size for translocation groups to aid recovery after decline of local populations, the approach used here is to artificially replace immigration of animals from surrounding areas which would have occurred if the population were not isolated. Certain assumptions have been made in order to produce a simple model, but as these are based on results of field studies it is felt that they are reasonable.

The rationale behind this approach is that it provides a non-arbitrary means of choosing the size of this pseudo-dispersal group, if some measure of the remaining population as a percentage of the carrying capacity can be made. In recommending the size of a translocation group required to establish a new population of the sea otter, *Enhydra lutris*, in California, Ralls et al. (1983) used as a starting point the arbitrary size of 25-30 animals each year for three to five years suggested by Jameson et al. (1982). This group size was based on previous experience of translocation attempts, and on practical considerations. Ralls et al. (1983) assumed that movement of a group of this size would result in the establishment of a population of 50 animals, which would then grow in a near-exponential fashion, as had occurred previously in a documented recovery. Using further assumptions based on field data they calculated that the new population would retain 77-86% of the genetic diversity present in the parent population.

The model proposed here allows decisions to be made regarding the desirable size of translocation groups without experience of prior attempts. The method can be applied to any species which is relatively evenly distributed within its habitat, by the use of equation (2); however, the model may need further development for colonial species. It does require certain basic information on the biology of the species, especially with regard to rates of juvenile survival, dispersal distance and adult density.

The fact that no real reserves are circular is not seen as a major problem, as this assumption merely provides a basis for relating influx of animals to the total area of habitat available. Use of a square or an even more realistic geometric figure would reduce the generality of the model as the chosen shape would have a strong influence, due to the strong effect of corners.

It should be emphasized that the intention is not to artificially enact annual immigration into the isolated population indefinitely, but to boost its numbers immediately after a population crash in order to minimize the deleterious effects of a prolonged bottleneck, which were outlined earlier. It may be necessary to transfer animals in the second year so that numbers more quickly approach the minimum viable population size. It has been pointed out, however, that genetic drift can be countered, in the absence of selection, by the exchange between populations of as few as one successfully breeding individual per generation, as the two subpopulations would then have the properties of a single breeding group (Franklin 1980). An exchange on one individual per generation is suggested by Allendorf (1983) because, in addition, this amount of exchange is not large enough to change allele frequencies in the presence of natural selection. In the case of the numbat, then, subsequent to the introduction of a relatively large pseudo-dispersal group, annual exchange of an individual between the two populations is recommended.

The numbat sighting data of Christensen *et al.* (1984) from Perup Forest (Fig. 2, this chapter) are uncontrolled, so their use as an index of population density should be approached with caution. During the period of the record, however, a reasonably consistent presence of observers was maintained in the study area (P. Christensen, pers. comm.). As the trends shown by these data are strong, the conclusion appears justified that the population declined to a small percentage of normal carrying capacity in 1976, but had recovered strongly by 1981.

The important point as regards loss of genetic variation through this period of decline is that the recovery of the population was rapid. While the Perup Forest is not an isolated vegetation remnant, surveys and compilation of sighting reports (Connell and Friend 1985) have indicated that there are probably no other numbat populations sufficiently close for frequent natural exchange of individuals. The increase between 1976 and 1981, therefore, was apparently due only to reproductive increase. Given that the recovery was probably rapid, following the findings of Nei *et al.* (1975) we may conclude that the loss of genetic variation subsequent to the population crash was most likely to have been small.

In this particular case of a decline and recovery, then, it appears in hindsight that the injection of a pseudo-dispersal group into the area was not necessary for a rapid return to previous densities. In any case, as the decline in numbat numbers was widespread, occurring also 140 km north at Dryandra Forest (see below), it is unlikely that a sufficient number of animals could have been found elsewhere, without seriously depleting other groups. It is probable, however, that the small size of the population surviving in 1975 meant that the subsequent level of variability was lower than that present before the decline. In order to help restore this level it is desirable that exchange of animals at the rate of one individual per generation be initiated between this and other populations. One of these should be the Dryandra population, where a simultaneous decline occurred (Turner and Borthwick 1980) followed by recovery (Fig. 3).

Frankel and Soulé (1981) suggest that all natural populations of less than 50 to 100 individuals require immediate genetic management if they are to escape rapid decline through inbreeding and loss of variability through genetic drift. Their conclusions are based on the experience of animal breeders as well as on theoretical considerations and therefore have particular relevance to the management of mammal populations. The implications of this statement are clear, in the context of areas such as the Western Australian wheatbelt where conservation of the mammal fauna relies on the survival of isolated populations in remnants of natural vegetation. In the absence of a concerted effort to implement programmes of genetic management, extinction of populations and genotypes of mammals is likely to continue.

ACKNOWLEDGEMENTS

I would like to express my thanks to all those, particularly Rod Burrows and Garry Connell, who have assisted in field studies at Dryandra Forest. Nick Caputi, Western Australian Fisheries Department, suggested and performed the test for randomness of dispersal direction. Andrew Bennett, Arthur Rylah Institute, and David Coates, Western Australian Wildlife Research Centre, made some helpful comments on the manuscript.

This work was supported by the Western Australian Department of Conservation and Land Management.

REFERENCES

Allendorf, F. W., 1983. Isolation, gene flow and genetic differentiation among populations. Pp. 51-65 *in* Genetics and Conservation: A Reference for Managing Wild Animal and Plant Populations ed by C. M. Schonewald-Cox, S. M. Chambers, B. MacBryde and W. L. Thomas. Benjamin/Cummins, Menlo Park, California.

Brown, A. M., 1983. Conservation genetics in Victoria. *Fish. Wildl., Victoria; Res. Plan. Br. Tech. Rep. Ser.* No. 1.

Calaby, J. H., 1960. Observations on the banded anteater *Myrmecobius f. fasciatus* Waterhouse (Marsupialia), with special reference to its food habits. *Proc. zool. Soc. (Lond.)* **135**: 183-207.

Carson, H. L., 1983. The genetics of the founder effect. Pp. 189-200 *in* Genetics and Conservation: A Reference for Managing Wild Animal and Plant Populations ed by C. M. Schonewald-Cox, S. M. Chambers, B. MacBryde and W. L. Thomas. Benjamin/Cummins, Menlo Park, California.

Chew, R. M., Butterworth, B. B. and Grechman, R., 1959. The effects of fire on the small mammal populations of chaparral. *J. Mammal.* **40**: 253.

Christensen, P. E. S., 1980a. A sad day for native fauna. *Forest Focus* **23**: 1-12.

Christensen, P. E. S., 1980b. The biology of *Bettongia penicillata*, Gray 1837, and *Macropus eugenii*, Desm. 1817, in relation to fire. *For. Dept. West. Aust. Bull.* **91**: 1-90.

Christensen, P. E. S. and Kimber, P., 1975. Effect of prescribed burning on the flora and fauna of south-west Australian forests. *Proc. Ecol. Soc. Aust.* **9**: 85-106.

Christensen, P., Maisey, K. and Perry, D. H., 1984. Radio-tracking the Numbat, *Myrmecobius fasciatus*, in the Perup Forest of Western Australia. *Aust. Wildl. Res.* **11**: 275-88.

Clout, M. N. and Efford, M. G., 1984. Sex differences in the dispersal and settlement of brushtail possums, *(Trichosurus vulpecula). J. Anim. Ecol.* **53**: 737-49.

Connell, G. W. and Friend, J. A., 1985. Searching for numbats. *Landscope* **1**: 21-6.

Cook, S. F. Jr., 1959. The effects of fire on a population of small rodents. *Ecology* 40: 102-8.

Frankel O. H. and Soulé, M. E., 1981. Conservation and Evolution. Cambridge University Press, New York.

Franklin, I. R., 1980. Evolutionary change in small populations. Pp. 135-49 *in* Conservation Biology: An Evolutionary-Ecological Perspective, ed by M. Soulé and B. Wilcox. Sinauer Associates, Sunderland, Massachusetts.

Friend, J. A., 1982. The Numbat: an endangered specialist. *Aust. Nat. Hist.* 20: 339-42.

Friend, J. A. and Burrows, R. G., 1983. Bringing up young numbats. *SWANS* 13(1): 3-9.

Friend, J. A. and Kinnear, J. E., 1983. Numbat (*Myrmecobius fasciatus*). Pp. 85 *in* The Complete Book of Australian Mammals ed by R. Strahan. Angus and Robertson, Sydney.

Gibbs, W. J. and Maher, J. V., 1967. Rainfall deciles as drought indicators. *Comm. Aust., Bur. Meteorol., Bull.* 48.

Jameson, R., Kenyon, K. W., Johnson, A. M. and Wright, H. M., 1982. The history and status of translocated sea otter populations in North America. *Wildlife Soc. Bull.* 10: 100-7.

Kimura, M. and Crow, J. F., 1963. The measurement of effective population number. *Evolution* 17: 279-88.

Kitchener, D. J., 1982. Predictors of vertebrate species richness in nature reserves in the Western Australian wheatbelt. *Aust. Wildl. Res.* 9: 1-7.

Kitchener, D. J., Chapman, A. and Barron, G., 1978. Mammals of the northern Swan Coastal Plain. Pp. 54-93 *in* Faunal studies of the northern Swan Coastal Plain, a consideration of past and future changes. Unpublished report, Western Australian Museum.

Kitchener, D. J., Chapman, A., Muir, B. J. and Palmer, M., 1980. The conservation value for mammals of reserves in the Western Australian wheatbelt. *Biol. Conserv.* 18: 179-207.

Leake, B. W., 1962. Eastern wheatbelt wildlife. B. W. Leake, Perth.

Lehmkuhl, J. F., 1984. Determining size and dispersion of minimum viable populations for land management planning and species conservation. *Environ. Manage.* 8: 167-76.

Main, A. M. and Yadav, M., 1971. Conservation of macropods in reserves in Western Australia. *Biol. Conserv.* 3: 123-33.

Nei, M., Maruyama, T. and Chakraborty, R., 1975. The bottleneck effect and genetic variability in populations. *Evolution* 29: 1-10.

Newsome, A., McIlroy, J. and Catling, P., 1975. The effects of an extensive wildfire on populations of twenty ground vertebrates in south-east Australia. *Proc. Ecol. Soc. Aust.* 9: 107-23.

Patton, J. L. and Yang, S. Y., 1977. Genetic variation in *Thomomys bottae* pocket gophers: macrogeographic patterns. *Evolution* 31: 679-720.

Ralls, K. and Ballou, J., 1983. Extinction: lessons from zoos. Pp. 164-84 *in* Genetics and Conservation: A Reference for Managing Wild Animal and Plant Populations ed by C. M. Schonewald-Cox, S. M. Chambers, B. MacBryde and W. L. Thomas. Benjamin/Cummins, Menlo Park, California.

Ralls, K., Ballou, J. and Brownell, R. L. Jr., 1983. Genetic diversity in California Sea Otters: theoretical considerations and management implications. *Biol. Conserv.* 25: 209-32.

Recher, H. F. and Christensen, P. E., 1981. Fire and the evolution of the Australian biota. Pp. 135-62 *in* Ecological Biogeography of Australia ed by A. Keast. Junk, The Hague.

Serventy, D. L., 1954. The recent increase of the rarer native mammals. I — Introduction. *West. Aust. Nat.* 4: 128-9.

Shortridge, G. C., 1910. An account of the geographical distribution of the marsupials and monotremes of south-west Australia having special reference to the specimens collected during the Balston Expedition of 1904-7. *Proc. zool. Soc. (Lond.)* 1909: 803-48.

Templeton, A. R., 1980. The theory of speciation via the founder principle. *Genetics* 92: 1011-38.

Turner, J. and Borthwick, J., 1980. Status of the Numbat, *Myrmecobius fasciatus*, population in Western Australia. Unpublished report, Western Australian Department of Fisheries and Wildlife.

Zar, J. H., 1984. Biostatistical Analysis. Prentice-Hall, Englewood Cliffs.

CHAPTER 6

Effects of Patch Area and Habitat on Bird Abundances, Species Numbers and Tree Health in Fragmented Victorian Forests

Richard H. Loyn[1]

Data were analysed on birds in 56 forest patches (0.1 to 1771 ha) that had been isolated by clearing for agriculture or pine plantations in the Latrobe Valley of southeastern Victoria.

Positive relationships were found for forest and total bird species and abundance with patch area or its logarithm, and several other variables made additional contributions, e.g. grazing intensity. Habitat indices were closely correlated and explained almost as much variation. Forest bird abundance did not increase with area above 10-30 ha. Numbers of forest species continued to increase with patch area but medium-sized patches (10-150 ha) supported a higher proportion of species than expected. The species present depended on habitats represented. Large reserves may be needed by some species but other forest species occurred only on medium-sized patches. More complex relationships were found for farmland birds and water birds.

Small heavily grazed patches (less than 10 ha) supported few forest birds and more farmland birds, including noisy miners *Manorina melanocephala* which aggressively excluded other species. Few birds fed on insects in the canopy in these patches, which showed signs of dieback due to insect damage. Protection of understorey habitat can have wide benefits in maintaining the health of the rural ecosystem.

INTRODUCTION

WHEN forests are cleared many plant and animal species become confined to remnant patches of native vegetation, especially when land is permanently altered for uses such as agriculture. These patches and their fauna can continue to play a role in the local ecology, protecting land from erosion, salinity and other degradation and providing habitat for wildlife including species that benefit from clearing (farmland species) (e.g. Breckwoldt 1983). Hence it is important to know how to design and manage systems of remnant vegetation, and how to conserve plant and animal species in these fragmented systems.

This chapter deals with a study of birds on 56 fragmented forest patches in the Latrobe Valley of southeastern Victoria, following a similar study of mammals on the same patches (Suckling 1980, 1982, 1984). Some preliminary data on birds have already been presented (Loyn 1984, 1985); the latter paper included a list of species with details of habitats occupied. Here more attention is paid to mathematical relationships and a subsequent paper is planned to analyse effects of fragmentation on individual species.

STUDY REGION

A large part of eastern Victoria is hilly or mountainous and remains as forested public land. The Latrobe Valley is an exception, as it has become a major centre for agriculture and industry.

Much of the study region has been cleared for agriculture and pine plantations over the last 100 years and in 1978 about 34% remained as fragmented native forests and woodlands, with 52% as agricultural land and 14% as pine plantations (Suckling 1982). By 1983 an additional 4% had been cleared (mainly for a new coal-fired power-station at

---

[1]Mountain Forest Research Station, Sherbrooke, Victoria 3789. Present Address: Arthur Rylah Institute for Environmental Research, 123 Brown Street, Heidelberg, Victoria, 3084 (Department of Conservation, Forests and Lands).
Pages 65-77 *in* NATURE CONSERVATION: THE ROLE OF REMNANTS OF NATIVE VEGETATION ed by Denis A. Saunders, Graham W. Arnold, Andrew A. Burbidge and Angas J. M. Hopkins. Surrey Beatty and Sons Pty Limited in association with CSIRO and CALM, 1987.

Loy Yang), including three of the 59 original fragments. The fragments studied ranged in size from 0.1 to 1771 ha and were scattered through the region mostly on private property. They had been isolated for periods from 7 to 100 years and three had been partly cleared for use as shire rubbish tips. The largest area has subsequently been converted to pine plantations. No fragment was further than 1.5 km from another patch of native forest. Two were isolated by as little as a main road; details are given in the Appendix. More extensive forests still exist on public land in the hills and coastal sand-plain south of the region, and continuous forest occurs on higher ground north of the Latrobe Valley.

The 56 fragments contained mixed-species dry sclerophyll forest including stands of narrow-leaf peppermint *Eucalyptus radiata,* manna gum *E. viminalis* and yertchuk *E. consideniana.* The understorey was dominated either by heathy shrublands, swards of austral bracken fern *Pteridium esculentum* or in heavily grazed patches by introduced grasses. Thickets of burgan *Leptospermum phylicoides* occurred frequently and some gullies contained tall gully shrubs, though generally the region was too low (60 to 360 m above sea level) and dry (about 750 mm mean annual rainfall at 180 m) for gully vegetation to be well developed. More information on individual study areas is given by Suckling (1980) and Loyn (1984).

METHODS

All 56 fragments were searched in each of three spring/summer seasons from November 1980 to March 1983. All birds observed were recorded, and numbers seen or heard on 20-minute counts were used as measures of abundance (Loyn 1986). On the smallest patches one or two counts were sufficient to cover the whole area on each visit, whereas on larger patches up to 19 counts were made on a visit with the aim of searching all habitats. Areas of about 3 ha could be covered in each count, so birds per count on smaller patches do not give a true reflection of bird density. At the end of the study, counts were continued until few extra species could be found. Basic observations were made on the use birds were making of their habitats.

ANALYSIS

Data were tabulated and statistics calculated for each patch (numbers of bird species, and bird abundances expressed as individual birds per count). Depending on their known use of habitats elsewhere, the 132 bird species observed in the region were classified as 78 forest species, 28 farmland (open-country) species and 26 water bird species. Most of the farmland species used trees for nesting or roosting, and some such as noisy miners *Manorina melanocephala* also fed among trees as well as pasture.

Values for bird species and abundance were regressed against characters of the individual patches, listed in Table 1. Variables 1 to 8 were described by Suckling (1982). The habitat indices 10 to 13 were derived from data in the Appendix; it was realised that they were coarse and that habitats of individual species depended on combinations of these features and more subtle features, which could often be described verbally but not numerically.

Initially regressions were run against A (area) or log A alone, and then against variables 2 to 8 together in a forward, stepwise, multiple linear regression package (SPSS, Nie *et al.* 1975). Then the variables 9 to 13 (altitude and habitat indices) were added as appropriate, and finally variable 14 (noisy miner abundance, considered in a separate section). Variables were only included in equations if they improved regressions significantly ($P < 0.05$).

Some of the characters were intercorrelated and hence hard to separate, especially in the enlarged package. Small patches tended to be heavily grazed, at low altitude, widely scattered, isolated for long periods, occupied by noisy miners and with few remaining forest habitats. Some other combinations were tested to allow for intercorrelation (see Nie *et al.* 1975), in case stepwise inclusion of one variable obscured effects of a correlated variable, or failed to maximize explained variance. When considering bird abundance, water birds were included with farmland birds as they were not abundant enough in the forest fragments to warrant separate analysis, and depended on farm dams or pasture for feeding.

RESULTS

Numbers of bird species and bird abundances on individual forest patches are shown in the Appendix. Bivariate correlations between these parameters and patch characters are shown in Table 2; stepwise contributions of patch variables to multiple regression equations are considered in Table 3, and the predictive equations for lines of best fit are shown in Table 4.

Bird Abundances

The patches fell clearly into two groups with high or low population densities of forest birds (Appendix). All but two patches larger than 10 ha supported reasonably high densities of forest birds (16.6 to 39.8 birds per count), making over 85% total populations. The exceptions were a long narrow patch, heavily grazed by sheep and cattle from surrounding pasture (F4) and a ring of remnant forest, much of it grazed, around the main rubbish tip for Traralgon (T29). In those patches densities of forest birds were low, making only 39 and 32% of total populations (11.5 and 12.7 birds per count). Two other disturbed patches retained a shrubby native understorey despite disturbance and they supported higher

Table 1. Measured characters of forest fragments.

Character	Brief description	Mean value
1. A:	its area in ha (0.1 to 1771)	69.3
2. log A:	the log to the base ten of A	0.93
3. Years isolated:	the time since isolation in years (7 to 100 years)	33.2
4. Dist:	the distance from the nearest other patch in km to (0.03 to 1.43)	0.24
5. LDist:	the distance from the nearest larger patch in km (0.03 to 4.43)	0.74
6. Graze:	an index of grazing history from 1 (very light) to 5 (very heavy)	2.98
7. Fire:	an index of fire history from 1 (no recent fires) to 5 (frequent or recent)	2.23
8. Timber:	an index of logging and other disturbance from 1 (light) to 5 (heavy)	1.86
9. Alt:	the mean altitude in metres above sea level (60 to 250)	146
10. Hab:	an index of the number of habitats on a patch, including eucalypt communities, types of understorey, special features and types of edge (0 to 17)	4.5
11. Forhab:	an index of the number of forest habitats on a patch, as above but taking rubbish tips, cattle-degraded understorey and pasture-edge as negative (-3 to 14)	1.7
12. Farmhab:	an index of the number of farmland habitats on a patch (0 to 6)	1.7
13. Waterhab:	an index of the number of water habitats on a patch (small dam, creek, or proximity to a large swamp, 0 to 3)	0.3
14. Noisy miners:	the measured abundance of aggressive noisy miners *Manorina melanocephala*, in birds per count (0 to 12)	2.8

Variable 14 was not used with abundances of farmland birds or total birds, as noisy miners were included in those categories. Variables 11-13 were only used with their respective groups of birds.

Table 2. Correlation coefficients (r) for regressions of bird numbers against patch characters.

Dependent variable: Patch character—independent variable	Forest bird species	Farmland bird species	Water bird species	Total bird species	Forest bird abundance	Farmland and water bird abundance	Total bird abundance	Noisy miner abundance
1. A	0.427***	0.053^{NS}	0.130^{NS}	0.428***	0.256^{NS}	−0.193^{NS}	0.116^{NS}	−0.168^{NS}
2. log A	0.861***	0.276*	0.237^{NS}	0.900***	0.740***	−0.402**	0.548***	−0.365**
3. Dist	0.284*	0.211^{NS}	0.142^{NS}	0.330*	0.238^{NS}	−0.078^{NS}	0.246^{NS}	−0.149^{NS}
4. LDist	0.191^{NS}	0.425***	0.202^{NS}	0.295*	0.097^{NS}	0.155^{NS}	0.356***	0.068^{NS}
5. Fire	0.294*	−0.103^{NS}	0.124^{NS}	0.268*	0.348**	−0.274*	0.142^{NS}	−0.214^{NS}
6. Timber	−0.056^{NS}	0.044^{NS}	0.095^{NS}	0.051^{NS}	−0.097^{NS}	0.155^{NS}	0.067^{NS}	0.042^{NS}
7. Graze	−0.700***	0.093^{NS}	−0.007^{NS}	−0.642***	−0.683***	0.639***	−0.140^{NS}	0.500***
8. Years	−0.500***	0.147^{NS}	0.101^{NS}	−0.430***	−0.516***	0.551***	−0.014^{NS}	0.398***
9. Alt	0.362**	−0.442***	−0.111^{NS}	−0.240^{NS}	0.382**	−0.501***	−0.118^{NS}	−0.364***
10. Hab	0.735***	0.234^{NS}	0.212^{NS}	0.756***	0.702***	−0.431***	0.452***	−0.416***
11. Forhab	0.821***	–	–	–	0.818***	–	–	–
12. Farmhab	–	0.428***	–	–	–	0.542***	–	0.427***
13. Waterhab	–	–	0.550***	–	–	–	–	−0.134^{NS}
14. Noisy miners	−0.659***	0.272*	0.265*	−0.557***	−0.710***	(0.709)***†	(−0.378)**†	–

– = not calculated. Correlations marked † were calcualted after subtracting noisy miner abundance from farmland or total bird abundance.
* = P <0.05 ** = P <0.01 *** = P< 0.001.

densities of forest birds (39.8 and 24.9 birds per count). One (F6) was the site of the smaller Gormandale rubbish tip, and the other (T18) was partly used as a trail-bike course; neither had been heavily grazed.

In contrast, many patches of 10 ha or less had been grazed or slashed for fire protection and they supported low densities of forest birds and high densities of farmland birds (Appendix). Of the 31 patches of this size class, forest birds made less than 20% of total populations in eleven, and more than 85% in only six, all of which had received a degree of protection from stock through fencing or being bounded by pine plantations rather than pasture. In those patches, densities of forest birds appeared

Table 3. Stepwise contribution of variables to multiple regression equations.

Dependent variable	Variables tested (see Analysis)	Independent variable added at each step (and significance of adding it)					Variance explained at each stem ($r^2 \times 100\%$)				
		Step 1	Step 2	Step 3	Step 4	Step 5	Step 1	Step 2	Step 3	Step 4	Step 5
Forest bird species	2-8, 2-10	log A***	–Graze**	–Years*	–	–	74.2	81.1	82.7		
	2-11	log A***	Forhab***	–Years**	–	–	74.2	82.5	85.2		
	2-14	log A***	–Noisy miners***	–Years**	–	–	74.2	87.9	90.0		
Farmland bird species	2-8	L Dist**	–	–	–	–	18.1				
	2-13, 2-14	–Alt***	log A***	Farmhab**	–	–	19.6	36.6	44.6		
Water bird species	2-8	–	–	–	–	–	no significant relationship				
	2-13	Waterhab***	–Hab*	–Alt*	Fire*	–	30.2	38.4	44.3	49.0	
	2-14	Waterhab***	Noisy miners**	–	–	–	30.2	41.9			
Total bird species	2-8, 2-13	log A***	–Years**	–	–	–	81.1	84.1			
	2-14	log A***	–Noisy miners*	–Years*	–	–	81.1	87.1	88.1		
Forest bird abundance	2-8, 2-10	log A***	–Graze***	–	–	–	54.7	65.3			
	2-11, 2-14	Forhab***	log A**	–Years*	–	–	66.8	71.5	74.5		
	2-8+14	log A***	–Noisy miners***	–Years**	–	–	54.7	77.0	79.9		
Farmland (+ water) bird abundance	2-8	Graze***	–	–	–	–	40.8				
	2-13, 2-14	Graze***	Farmhab**	–Hab**	LDist*	–	40.8	50.3	57.1	62.6	
	2-14 (not 5)	Graze***	Farmhab**	–Hab**	Timber*	–Alt*	40.8	50.3	57.1	60.7	63.9
Noisy miner abundance	2-8	Graze***	–	–	–	–	25.0				
	2-13	Graze***	Farmhab*	–Hab*	–	–	25.0	31.0	40.4		
	2-13 (not 6)	Farmhab***	–Hab***	–	–	–	18.2	39.6			
Total bird abundance	2-8	log A***	–	–	–	–	30.1				
	2-13	log A***	–Alt*	–	–	–	30.1	37.9			

* = P < 0.05, ** = P < 0.01, *** = P < 0.001. – = additional variables made no further significant contribution.

only slightly lower than in larger patches (Appendix). The apparent reduction in patches smaller than 3 ha may be partly an artefact of the study method which was designed to search areas of about 3 ha on each count.

Signs of eucalypt dieback associated with insect attack were evident in all patches lacking understorey, which contained low densities of insectivorous forest birds (Appendix).

Forest Bird Abundance — Because population densities of forest birds were low in the smallest patches, significant regressions were found for abundance against area for forest birds (Table 2). The relationship was improved greatly by considering log A not A, explaining 54.7% of variance, but was clearly not linear and for patches larger than 30 ha there was little difference in density with increasing patch size (Appendix). The relationship with log A was better mainly because large patches received less weighting in the equations.

Addition of the grazing index improved the relationship significantly, in a negative sense, explaining an additional 10.6% of variance (Table 3). No other variable in the first package (variables 2 to 8) made significant additional contributions, though several were strongly correlated (Table 2). In the enlarged package the variable most strongly correlated with forest bird abundance was the index of forest habitats, and its positive effect superseded the negative effect of the grazing index in the predictive equation. The years isolated also entered this equation (negatively), suggesting a progressive degradation of these patches as habitat for forest birds, apart from any effects of area reduction or changes in other measured variables. The equation explained 79.9% of variance (Table 3).

Farmland and Water Bird Abundance — Abundances of farmland and water birds were negatively correlated with log A (Table 2), but the regression only explained 16.0% of variance. The grazing index superceded the effect of log A, in a positive sense, explaining 40.8% of variance. Farmland birds were only abundant in heavily grazed patches, regardless of size but generally these patches were small. No other variable in the first package made additional contributions. In the enlarged package, 63.9% of variance was explained by including additional terms for farm habitats and the timber index (positively) and total habitats and altitude (negatively) (Table 3). This was the only equation where the timber index (effects of logging and other disturbance) made a significant contribution.

Total Bird Abundance — As abundances of forest and farmland birds responded in opposite ways to most measured variables (Table 2), the regressions for total birds were weaker, explaining only 37.9% of variance with the enlarged package, 30.1% from log A (positively, through the forest component) and the additional 7.8% from altitude (negatively, through the farmland component).

Numbers of Species

Numbers of forest species and total species increased with size of patch over the range of sizes studied, whereas numbers of farmland species showed little change (see Figure 2 in Loyn 1985).

Table 4. Regression equations for predictive lines of best fit.

Groups of birds	Variables tested	Bird species (numbers of species on patch) =	Bird abundance (individual birds per count) =
Forest birds	2, 2-8	16.04 log A + 12.78	9.78 log A + 10.40
	2-10	13.05 log A − 2.06 (Graze) − 0.09 (Years isolated) + 24.8	6.89 log A − 2.86 (Graze) + 21.61
	2-11	as above	4.33 log A + 2.30 (Forhab) − 0.081 (Years isolated) + 14.08
	2-8 + 14	12.83 log A − 1.29 (Noisy miners) − 0.09 (Years isolated) + 22.43	6.92 log A − 1.19 (Noisy miners) − 0.074 (Years isolated) + 18.88
	2-14	as above	as with 2-11
Farmland birds	2	1.06 log A + 6.34	−4.89 log A + 13.81
	2-8	1.45 LDist + 6.24	4.29 (Graze) − 3.55
	2-13	1.59 log A − 0.033 (Alt) + 1.19 (Farmhab) + 8.59	1.54 (Graze) + 5.79 (Farmhab) − 3.07 (Hab) + 2.76 (LDist) + 6.57
	2-13 (not 5)	as above	1.60 (Graze) + 5.16 (Farmhab) − 2.14 (Hab) + 2.22 (Timber) − 0.046 (Alt) + 12.32
	2-14	as above	not tested
Waterbirds	2-8	no significant relationship	not tested
	2-13	1.48 (Waterhab) − 0.23 (Hab) − 0.007 (Alt) + 0.21 (Fire) + 1.72	not tested
	2-14	0.98 (Waterhab) + 0.077 (Noisy miners) − 0.014	
Noisy miners	2	0 or 1	−1.80 log A + 4.49
	2-8	0 or 1	1.36 (Graze) − 1.26
	2-13	0 or 1	0.35 (Graze) + 2.14 (Farmhab) − 1.09 (Hab) + 3.05
	2-13 (not 6)	0 or 1	2.44 (Farmhab) − 1.29 (Hab) + 4.46
(Total mammals)	2	(6.09 log A + 2.11)	
Total birds	2	17.36 log A + 19.6	4.89 log A + 24.21
	2-13	16.34 log A − 0.11 (Years isolated) + 24.0	5.61 log A − 0.048 (Alt) + 30.61
	2-14	15.15 log A − 0.88 (Noisy miners) − 0.064 (Years isolated) + 26.25	not tested

Table 5. Comparison of three regression models of bird (and mammal) statistics against patch area.

Model:	Sign of correlation	Linear	Variance explained ($r^2 \times 100\%$) Exponential	Power function	Slope of power function (z)
Dependent variable:		Number of species	Number of species	log (Number of species)	
Independent variable:		A	log A	log A	
Forest bird species:	+	18.3***	74.2***	54.9***	.34
Farmland bird species:	+	2.8^{NS}	7.6*	5.2*	.06
Water bird species:	NS	0.17^{NS}	5.6^{NS}	6.6^{NS}	−
Total bird species:	+	18.3***	81.0***	69.1***	.24
(Total mammal species):	+	−	86.0**	−	.37

* = $P < 0.05$, ** = $P < 0.01$, *** = $P < 0.001$; NS = not significant; − = not calculated.

Hence for forest and total birds, area terms explained more variance for numbers of species than they had done for abundance (Table 3). Three regression models (linear, exponential and power function) of species numbers against area are compared in Table 5. All gave highly significant positive relationships for total bird species and forest bird species, but the linear model (untransformed species numbers against area) explained far less variance than the other two. The exponential model explained almost 20% more variance for forest birds than did the power function (74.2% for forest species, 81.1% for total species) (Table 5). All models overestimated numbers of species in the largest patches, but this effect was least in the exponential model, which gave least weighting to the large patches. To look at it another way, if it were proposed to reduce the size of a large patch of forest containing a given number of species, all models would overestimate the initial loss of species with decreasing patch size, but the exponential model would do so less than the others.

Forest Bird Species — The relationship of forest bird species with area was closer than for abundance and log A accounted for 74.2% of the 85.2% of variance explained in the enlarged package (Table 3). The grazing index made a significant additional contribution (in a negative sense) with the first package, but was superceded by the index of forest habitats (in a positive sense) in the enlarged package, suggesting that grazing acted on forest bird species by removing specific forest habitats. The years isolated contributed in a negative sense with both packages, and the multivariate equations (Table 4) suggest that about nine forest species would be lost from patches of constant size in the first century of isolation. Further extrapolation would be unwarranted without more evidence about whether this was happening at a steady linear rate.

No other variable contributed significantly. Altitude was positively correlated (Table 2), probably because the smallest, most degraded patches were lowest in the valley. Factors such as fire and logging are known to affect forest bird species and abundance elsewhere, but in this study fire and logging histories did not differ greatly between patches, and distances of isolation also covered a narrow range (Appendix). These variables were not strongly correlated with forest or total bird species (Table 2).

The index of forest habitats explained almost as much variance (67.4%) as the most strongly correlated variable log A (74.2%), but as these two variables were themselves intercorrelated (r = 0.723, r^2 = 52.3%, p < 0.01), the contribution of the habitat index was masked by the area effect. Large patches contained more habitats than small ones and the area of a patch (or its log) may be a better measure of habitat diversity than the coarse habitat indices measured.

Farmland Bird Species — Numbers of farmland species were positively correlated with log A (although their abundance was negatively correlated), but the relationship was weak (Table 2). Few species occurred in patches that lacked a farmland boundary. In the first package the closest correlation was with distance from the nearest larger forest patch, suggesting that isolation between patches (or the amount of surrounding farmland) helped determine numbers of farmland species. It was thought that a better measure could be the square or log of this distance term, but they gave even weaker relationships. A stronger regression was found with the enlarged package, explaining 44.6% of variance (still far less than for forest birds). The contributing variables were altitude (in a negative sense) and log A and the index of farmland habitats (in a positive sense) (Table 3). The negative effect of altitude reflects the greater degree of agricultural clearing in the valley, which has provided habitats for more farmland bird species.

Water Bird Species — Only a few water bird species made more than incidental use of the forest patches, although some small and large patches provided important nesting and roosting sites for them. No significant relationship was found with area terms or the first package of variables 2 to 8. Inclusion of terms for altitude and habitats gave a regression explaining 49.0% of variance (Table 3). The contributing variables were water habitats and the fire index (in a positive sense) and total habitats and altitude (in a negative sense).

The effect of altitude reflects the greater extent of natural and artificial wetlands in the valley, on a broader scale than considered in the habitat variables. The fire index made a significant contribution to no other equation and its positive effect may be a quirk of intercorrelation; the least fire-prone patches were the small heavily grazed patches in farmland containing dams and swamps.

Total Bird Species — As forest species, farmland species and water bird species all increased with log A, the relationship was even stronger for total species, explaining 81.1% of variance (Table 3). Forest bird species responded in different ways to most other variables (Table 2), and only years isolated made a significant additional contribution (in a negative sense), explaining a further 3.0% of variance. The multivariate equation (Table 4) suggests that about eleven species would be lost from patches of constant size in the first century of isolation.

Noisy Miners — A Special Factor

The most conspicuous bird in many of the small, grazed patches was an aggressive honeyeater, the noisy miner, which breeds communally and defends territories against all other bird species (Dow 1977). Noisy miners compounded the effects of understorey removal by physically expelling small birds that attempted to encroach on their territories.

The only other birds that were able to nest in noisy miner territories were larger farmland species (e.g. Australian magpie *Gymnorhina tibicen*), some large raptors and water birds (e.g. at T28 a pair of brown goshawks *Accipiter fasciatus* and a pair of yellow-billed spoonbills *Platalea flavipes*), a predatory forest passerine (grey butcherbird *Cracticus torquatus*) and a few hole-nesting species (e.g. eastern rosella *Platycercus eximius*, laughing kookaburra *Dacelo novaeguineae*, tree martin *Cecropis nigricans* and introduced common starling *Sturnus vulgaris*). Occasionally striated pardalotes *Pardalotus striatus* nested in tree hollows in small patches or isolated trees in paddocks, making raids into noisy miner colonies to snatch food before expulsion. Noisy miners fed in pasture and in the canopy (on larger invertebrates, exudates, etc.), and the other farmland birds fed primarily on insects and seeds in surrounding pasture. Apart from the pardalotes, few birds remained to feed on small invertebrates in the canopy of patches dominated by noisy miners, all of which suffered dieback and defoliation by insects. Noisy miners did not occur in those forest patches where intact understorey provided cover for competing birds. In large patches noisy miners occurred only locally, always on the edge of pasture and usually in grazed peninsulas of forest that required defence on only one edge.

Because noisy miners were locally dominant, regressions were run for their abundance against log A and other patch variables. Noisy miner abundance proved to be negatively correlated with log A (Table 2), but the regression only explained 13.3% of variance. The grazing index superseded the effect of log A, in a positive sense, explaining 25.0% of variance. In the enlarged package, 40.4% of variance was explained by including additional terms for farm habitats (in a positive sense) and total habitats (in a negative sense) (Table 3). These two terms superseded the effect of the grazing index with little loss in explained variance (Table 3), and a slight gain in significance of regression. This suggests that the substantial contribution of the grazing index was almost entirely reflected in the habitat indices. The other relationships were similar to those for farmland birds generally, except that less variance was explained and no extra contribution was made by altitude (despite a strong negative correlation, Table 2) or the timber index. A few farmland birds, but not noisy miners, entered larger forest patches after logging. Farmland birds and water birds generally favoured similar forest patches to noisy miners, and numbers were positively correlated despite aggression between them (Table 2).

When noisy miner abundance was included as a patch character in multiple regressions, it improved the best previous relationships for forest bird abundance, forest bird species and total bird species, but not for farmland or water bird species (Table 3). (Noisy miner abundance could not be regressed against farmland or total bird abundance as it formed a substantial part of each.) For forest birds, noisy miner abundance superseded the grazing index in the same negative sense, but was superseded by the forest habitat index in a positive sense (Table 3). These three characters were too closely intercorrelated to be separated easily (Table 2). It is likely that forest birds would be scarce wherever grazing had reduced the understorey, though not as scarce as when noisy miners compounded the effect through their evident aggression.

Observations on two patches suggest separate effects of these variables. One 1 ha patch (M8) retained dense swamp paperbark *Melaleuca ericifolia* thickets despite grazing, and it supported no noisy miners and a fairly high abundance of forest birds (Appendix), including species that fed in the eucalypt canopy as well as in the paperbark. On one 4 ha patch with heathy understorey (MF), noisy miners were initially resident on the edge and in nearby roadside trees. During a drought in late 1982 a few cattle were admitted to the patch and noisy miners encroached further, reducing populations of forest birds even where understorey remained intact. Thus a shifting balance was evident even in the short time-scale of the study.

DISCUSSION

Changes in Species with Clearing

Before European settlement the study region was almost entirely forested and present-day farmland birds would have been confined to a few open areas of woodland in the valley; it is certain that they are now more abundant and probable that more species are represented. It is also certain that forest birds are less numerous, as only 30% of forest habitat remains and population density was no higher in small patches than in more extensive forest (Appendix).

Historical records are not adequate to determine whether any forest species have been lost from the region but an assessment can be made from knowledge of current distribution. In a survey of a 30,000 ha block of forest nearby, only two additional bird species were found (Gilmore 1977); they were brown quail *Coturnix australis* and southern emu-wren *Stipiturus malachurus*, and both were confined to small areas of swampy heathland not represented in the study region. It is not known whether such heathland might have been represented before clearing. A similar situation held for mammals (Suckling 1982), with only one extra species found in the 30,000 ha block, also in a special localised habitat (Gilmore 1977).

A few extra forest bird species were found further away from the study region, but all were in different habitats such as coastal forest or wetter forest at higher altitude on either side of the Latrobe Valley, above the rainshadow of the study region (pers. obs.; Blakers *et al.* 1984; Loyn *et al.* 1980; Norris *et al.* 1979).

Those most likely to have had suitable habitat in the study region before clearing are superb lyrebird *Menura novaehollandiae* (wetter forests and gullies), flame robin *Petroica phoenicea* (now a passage migrant, breeding in slightly wetter forests at higher altitude), brown treecreeper *Climacteris picumnus* (dry forest including river red gums, but in SE Victoria mainly box woodland), bell miner *Manorina melanophrys*, yellow-tufted honeyeater *Lichenostomus melanops* and satin bowerbird *Ptilonorhynchus violaceus* (all in broad foothill gullies) and pied currawong *Strepera graculina* (slightly wetter forests, but rarely breeding south of the Latrobe Valley). In an American study, Bond (1957) found that birds of wet forest habitats were the most susceptible to fragmentation, but here the point is rather that the study region would only have contained small areas of suitable or marginal habitat (if any), while wetter forests nearby (especially north of the Valley) have not been greatly fragmented. Conversely, four forest species (and many farmland and water birds) were markedly more common in the study region than in continuous forest nearby. The forest species were emu *Dromaius novaehollandiae*, white-bellied cuckoo-shrike *Coracina papuensis*, leaden flycatcher *Myiagra rebecula* and noisy friarbird *Philemon corniculatus*. All are typically birds of dry forest, and were present because their habitats were represented not because they benefited from fragmentation. Indeed, the cuckoo-shrike was only found on the largest patches, and emus are absent from more heavily cleared parts of Victoria. As indicated by Lynch and Wigham (1984) from American studies, the regional distribution of habitat should be considered. In the Latrobe Valley, clearing has been concentrated in the more open forest and woodland habitats, so the more critical need locally is to conserve species associated with those habitats. This can now only be done by conservation of fragmented patches.

Most species whose habitats were represented in the study patches were recorded on them, and it seems that most if not all species have survived there despite the 70% reduction in area of forest habitat.

Distribution of Species Between Patches; Species not on Large Patches

Generally the distribution of species reflected their particular habitat requirements (Loyn 1985). No single patch supported all species, but the largest patch (F11, 1771 ha) supported the greatest number of forest species with 62 of the 79 recorded. This is a slightly lower proportion than for mammals where the same patch supported 18 of the 20 species recorded (Suckling 1982). Eight of the forest bird species missing from that patch were merely occasional or irregular visitors to other patches, and three others were observed on the second largest patch (M11, 973 ha). The remaining six apparently depended on smaller patches for survival in the study region. The six species and their main local habitats were emu (heathy woodland), rose robin *Petroica rosea* (silver wattles *Acacia dealbata*), spotted quail-thrush *Cinclosoma punctatum* (drier forest with open understorey), chestnut-rumped hylacola *Sericornis pyrrhopygius* (heathy woodland), noisy friarbird *Philemon corniculatus* (river red gums *E. camaldulensis*, forest red gums *E. tereticornis* and banksias) and white-winged chough *Corcorax melanorhamphos* (drier woodland with open understorey). Several other species occurred in the two largest patches only in very small numbers, and smaller patches were also important for their local survival; these species included birds of drier habitats such as leaden flycatcher. Elsewhere in the state all occurred in continuous forest, though little or no extensive habitat remains for some of them (notably noisy friarbird) in this region of southern Victoria. There appears to be no reason for these species to avoid larger patches of forest, but in this region their habitats were only represented in small patches.

Species only on Large Patches

No species was confined to the largest patch. At least one pair of powerful owls *Ninox strenua* was resident there and otherwise the species was not recorded except occasionally on the second largest patch (M11, 973 ha). Another forest bird, the beautiful firetail *Emblema bella*, was only observed once outside the largest study patch where small numbers were resident in heathy gullies. Just one other species, the white-bellied cuckoo-shrike, was recorded only in the three largest patches (F11, M11 and M2, 144 ha) where a few pairs occurred as regular summer migrants.

Just as some species (at least six in this study) had habitats that by chance were only represented on small forest patches, others would be expected to have habitats only represented on large forest patches. Therefore it would be wrong to conclude that the three birds mentioned above necessarily need large reserves. If the patches were reduced in size their particular habitats might be lost, but if the habitats remained the species might remain as well.

No obvious reason can be suggested for white-bellied cuckoo-shrikes to need large reserves. Their absence from extensive forest nearby suggests that they need particular habitats (undefined) that happened to be represented in this study region solely on large patches. However, possible reasons can be suggested for the other two species.

Powerful owls have large territories, sometimes in the order of 800 to 1000 ha (Fleay 1968; Seebeck 1976) though this would depend on prey availability. Resident pairs were studied by Tilley (1982) in fragmented forest patches of 142 ha and 165 ha, where their diet included common farmland birds as well as arboreal mammals. It is not known whether smaller areas could be used, or combinations close together, but in this study they did not appear to be.

Beautiful firetails inhabit flammable understoreys and may need large reserves to provide successions of suitable habitat after fires. They are slow to return after wildfires and have not been seen flying across open country, though they may do so.

Hence three reasons for species to be associated with large patches may be exemplified by these three species: white-bellied cuckoo-shrikes by chance, powerful owls needing large areas for hunting and beautiful firetails needing continuous regeneration of successional habitats.

Species/area Relationships

Many studies have attempted to explain variations in numbers of species on habitat islands in terms of the equilibrium theory of biogeography (MacArthur and Wilson 1967; Diamond 1975). This theory has provided a framework for discussing important conservation issues but suffers from a paucity of experimental evidence, and many of the observed relationships can be described well with other models (Connor and McCoy 1979). The theory has also been used wrongly to argue that single large reserves necessarily conserve more species than series of several small reserves (the SLOSS debate) whereas the theory itself has been shown not to answer the question (Simberloff and Abele 1976, 1982). A full analysis of the present data with respect to SLOSS is planned, but a comparison of five 50 ha combinations with varying degrees of fragmentation revealed very similar numbers of forest bird species (Loyn 1985), implying that the observed reduction in species numbers with reducing patch size was a basic consequence of patch area rather than an effect of fragmentation. This use of SLOSS to separate 'passive' area effects from effects of fragmentation *per se* has been overlooked and clouded by arguments about general answers to the SLOSS question.

All studies have shown that large habitat patches support more species than small patches, but the form of the relationship varies according to geographical factors and the taxonomic or ecological group concerned. The present data fitted an exponential model best, and this model has generally been associated with systems that are not fully isolated and where numbers of species are determined mainly by habitats represented. All birds observed on this study were capable of moving easily between patches and most were observed doing so, including the only flightless species (emu); it was observed more often in pine plantations and pasture than in natural habitat. However, the graphs sloped steeply with the power function model ($z=0.34$ for forest birds), which has been taken as typical of more isolated systems and the equilibrium model. High z values and good exponential fits have also been reported for mammals on the same patches (Suckling 1982) and lizards, birds and mammals in the wheatbelt of Western Australia (Kitchener *et al.* 1980a, b, 1982; Humphreys and Kitchener 1982). The explanation in all cases is probably that habitats decline rapidly with decreasing patch size.

In the Latrobe Valley, the three models overestimated numbers in larger forest blocks nearby (as they did for mammals, Suckling 1982). For example, extrapolation of the exponential equation predicted 65 diurnal forest species on the largest patch (1771 ha) and 85 forest species on the 30,000 ha forest block studied by Gilmore (1977), whereas numbers actually found were 59 and 52 respectively. If instead the graphs were fitted to extensive forest nearby, they would underestimate numbers of forest species currently surviving in medium or large forest patches.

All equations were improved significantly by including terms other than log A, and the entry of a term for years isolated suggests that numbers of species were declining at about ten species per century (mainly forest birds) and could be collapsing towards lower equilibrium levels fitting another model. The equations for farmland and water birds suggested their species numbers depended primarily on factors external to the patches, whereas their abundance on patches depended mainly on grazing intensity, reflecting their use of these habitats. Nevertheless, the patches provided essential nesting and roosting sites for many of these species. Farmland and water birds were most common at low altitude, probably because the greater extent of clearing in the valley has provided more diverse habitats for both groups. It should also be noted that many farmland species were originally confined to open habitats at low altitude or inland, and would have been even rarer in the forested ranges before clearing.

The grazing and habitat indices made major contributions despite their coarseness. Few bird species were absent from patches where their habitats were well represented, so that a more careful (but subjective) calculation of habitat indices might have yielded near-perfect correlations. Noisy miner abundance was correlated with these variables and may have been a more sensitive (indirect) measure of grazing effects on small patches than the other indices, though noisy miners had a special impact of their own (on small grazed patches), as described.

Hence in this study, habitat representation emerges as a prime factor contributing to species-area relationships, with grazing as a prime factor reducing habitats, specially on small patches. Noisy miners may reduce forest bird populations and species further on small heavily grazed patches. Fragmentation acts mainly by reducing area of habitat, though when patches are reduced below 10 ha they appear to become more susceptible to effects of grazing and noisy miners.

Ecological Interactions and Tree Health

It was noticed that eucalypt dieback from defoliating insects was more severe in small, heavily grazed patches than in forest patches with intact understorey and higher populations of forest birds. Similar observations were made elsewhere in eastern Australia (New England tablelands) by Ford and Bell (1981) and Davidson (1984), although in higher rainfall parts of Victoria there is less evidence of dieback in grazed forest patches.

Many factors can be involved in rural dieback (e.g. see Old *et al.* 1981; Marks and Idczak 1976) including climate, effects of stock on trees and soil compaction, increased exposure to weather and pollutants (the Latrobe Valley study region was downwind from a major industrial centre) and hydrologic changes.

Insects are often an important factor affecting the health of trees and birds can play a major role in controlling them. This was demonstrated recently when common forest birds controlled psyllid infestations after experimental removal of aggressively territorial bell miners, resulting in improved tree health (Loyn *et al.* 1983). In the Latrobe Valley noisy miners may exert a similar influence to bell miners, maintaining high populations of insect prey through territorial defence, despite other differences in the ecology of the two miners. Even without noisy miners, it seems that populations of forest birds would be low in small heavily grazed patches lacking understorey, and hence such patches would be relatively unprotected against insect attack.

Mammals such as sugar gliders *Petaurus breviceps* also consume substantial numbers of insects (Nagy and Suckling 1985) and benefit from understorey protection (Suckling 1980, 1984). The importance of understorey was discussed by Davidson (1984), as habitat for predatory and parasitic insects as well as birds, all of which can reduce pest insects. The solution is to protect small patches from grazing, improving both species diversity and tree health.

Implications for Conservation and Management

The message for conservation is that small and large patches of remnant forest have value, depending on the habitats represented. In a region such as the Latrobe Valley, some important habitats are represented on medium-sized patches (10-150 ha) and not on larger patches as well as *vice versa*. These patches also have value in maintaining regional populations of birds and other wildlife, including common forest and farmland species which can help maintain the health of the forest patches and probably pasture as well.

Good management of small and scattered forest patches presents logistic problems for a central agency (e.g. government), and can best be achieved by individual landholders with suitable incentives and encouragement. Conservationists and government should recognize the complementary value of scattered forest remnants as well as large reserves for wildlife conservation and other reasons.

Trees and patches of forest on farmland have value for many purposes apart from wildlife conservation. They provide shelter for stock, protection from erosion and salinity, and supplies of firewood and fencing material. When patches smaller than 10 ha are heavily grazed by stock, they do not regenerate and old trees senesce rapidly, at least in part through inadequate natural control of insect populations. The survival of these patches, and to some extent the whole landscape, may depend on greater attention to animal habitat including protection and encouragement of understorey as well as tree regeneration.

ACKNOWLEDGEMENTS

I would like to thank Dr Graeme C. Suckling (now of Geelong Region, Dept. of Conservation, Forests and Lands) who encouraged this work, and several people who helped with fieldwork or tabulation, especially Barry Traill and also David Eades, John Martindale, Mark Cavill, Steve Smith, Peta O'Donahue and Andrew Moore. Peter McHugh (Mountain Forest Research Station) helped greatly with advice on running the computer programmes, and Ruth Blackbourn typed the manuscript. Drs James F. Lynch and Harry F. Recher made valuable comments on an earlier draft.

REFERENCES

Blakers, M., Davies, S. J. J. F. and Reilly, P. N., 1984. The Atlas of Australian Birds. Melbourne University Press, Melbourne.

Bond, R. R., 1957. Ecological distribution of breeding birds in the upland forests of southern Wisconsin. *Ecol. Monogr.* 27: 351-84.

Breckwoldt, R., 1983. Wildlife in the Home Paddock. Angus and Robertson, Sydney.

Connor, E. F. and McCoy, E. D., 1979. The statistics and biology of the species — area relationship. *Am. Nat.* 113: 791-833.

Davidson, R. L., 1984. The unique value of natural areas to agriculture. Symposium on Small Natural Areas, their conservation and management. National Trust of Australia (NSW), University of Newcastle, 103-12.

Diamond J. M., 1975. The island dilemma: lessons of modern biogeographic studies for the design of natural reserves. *Biol. Conserv.* 7: 129-45.

Dow, D. D., 1977. Indiscriminate interspecific agression leading to almost sole occupancy of space by a single species of bird. *Emu* 77: 115-21.

Fleay, D., 1968. Nightwatchmen of Bush and Plain. Jacaranda Press, Brisbane.

Ford, H. A. and Bell, H., 1981. Density of birds in eucalypt woodland affected by varying degrees by dieback. *Emu* 81: 202-8.

Gilmore, A. M., 1977. A survey of vertebrate animals in the Stradbroke area of South Gippsland, Victoria. *Vic. Nat.* 94: 123-8.

Humphreys, W. F. and Kitchener, D. J., 1982. The effect of habitat utilization on species — area curves: implications for optimal reserve area. *J. Biogeogr.* 9: 391-6.

Kitchener, D. J., Chapman, A., Dell, J., Muir, B. G. and Palmer, M., 1980a. Lizard assemblage and reserve size and structure in the Western Australian wheatbelt — some implications for conservation. *Biol. Conserv.* 17: 25-62.

Kitchener, D. J., Chapman, A., Muir, B. G. and Palmer, M., 1980b. The conservation value for mammals of reserves in the Western Australian wheatbelt. *Biol. Conserv.* 18: 179-207.

Kitchener, D. J., Dell, J., Muir, B. G. and Palmer, M., 1982. Birds in Western Australian wheatbelt reserves — implications for conservation. *Biol. Conserv.* 22: 127-63.

Loyn, R. H., 1984. Bird diversity in isolates. Pp. 49-61. *in* Symposium on Small Natural Areas, their conservation and management. National Trust of Australia (NSW), University of Newcastle.

Loyn, R. H., 1985. Birds in fragmented forests in Gippsland, Victoria. Pp. 323-31 *in* Birds of eucalypt forests and woodlands — ecology, conservation, management ed by A. Keast, H. F. Recher, H. Ford and D. Saunders. Surrey Beatty and Sons, Sydney.

Loyn, R. H., 1986. The 20-minute search — a simple method for counting forest birds. *Corella* 10: 58-60.

Loyn, R. H., Macfarlane, M. A., Chesterfield, E. A. and Harris, J. A., 1980. Forest utilization and the flora and fauna in Boola Boola State Forest in southeastern Victoria. *Forests Commission Victoria, Bulletin* 28.

Loyn, R. H., Runnalls, R. G., Forward, G. Y. and Tyers, J., 1983. Territorial bell miners and other birds affecting populations of insect prey *Science* 221: 1411-3.

Lynch, J. P. and Whigham, D. F., 1984. Effects of forest fragmentation on breeding bird communities in Maryland, USA. *Biol. Conserv.* 28: 287-324.

MacArthur, R. H. and Wilson, E. O., 1967. The Theory of Island Biogeography. Princeton University Press, Princeton.

Marks, G. C. and Idczak, R. M., 1976. Eucalypt dieback in Australia. Report of a seminar held at Lakes Entrance, Victoria. Forests Commission, Victoria.

Nagy, K. A. and Suckling, G. C., 1985. Field energetics and water balance of sugar gliders, *Petaurus breviceps* (Marsupialia: Petauridae). *Aust. J. Zool.* 33: 683-91.

Nie, N. H., Hadlae Hull, C., Jenkins, J. G., Steinbrenner, K. and Bent, D. H., 1975. Statistical Package for the Social Sciences. 2nd ed. McGraw-Hill.

Norris, K. C., Gilmore, A. M. and Menkhorst, P. W., 1979. Vertebrate fauna of South Gippsland, Victoria. *Mem. Nat. Mus. Vic.* 40: 105-99.

Old, K. M., Kile, G. A. and Ohmart, C. P. (eds.), 1981. Eucalypt dieback in forests and woolands. C.S.I.R.O. Melbourne.

Seebeck, J. H., 1976. The diet of the Powerful owl (*Ninox strenua*) in Western Victoria. *Emu* 76: 167-70.

Simberloff, D. S. and Abele, L. G., 1976. Island biogeography theory and conservation practice. *Science* 191: 285-6.

Simberloff, D. and Abele, L. G., 1982. Refuge design and island biogeographic theory: effects of fragmentation. *Am. Nat.* 120: 41-50.

Suckling, G. C., 1980. The effects of fragmentation and disturbance of forest mammals in a region of Gippsland, Victoria. Ph.D. Thesis, Monash University.

Suckling, G. C., 1982. Value of reserved habitat for mammal conservation in plantations. *Aust. For.* 45: 19-27.

Suckling, G. C., 1984. Population ecology of the sugar glider, *Petaurus breviceps*, in a system of fragmented habitats. *Aust. Wildl. Res.* 11: 49-75.

Tilley, S., 1982. The diet of the powerful owl (*Ninox strenua*) in Victoria. *Aust. Wild. Res.* 9: 157-75.

Appendix: Habitats and birds on study patches.

Area	Size (ha)	Distance from nearest patch (m)	Distance from nearest larger patch (m)	Grazing index* (1-5)	Date of isolation	Altitude (m)	Main Vegetation †	Other features	Boundaries with #	Forest bird species	Farmland bird species	Water bird species	Total bird species (diurnal) P	Forest birds per count	Total birds per count	Forest birds as % of total birds	Noisy miners as % of total birds
M5	0.1	39	39	2	1968	160	N,o		O	8	4	0	12	5.3	10.7	50	34
M6	0.5	104	311	5	pre 1900	140	N,o		O	4	6	1	11	1.5	15.5	10	38
T27	0.5	234	234	5	pre 1920	60	N,o		O	3	7	0	10	2.0	10.6	19	21
M8	1	246	246	5	1890-1918	140	N,o,t		O	21	8	0	29	13.8	22.6	61	0
M10	1	234	285	3	1960-1965	130	N,o		O	22	10	0	32	14.1	36.3	39	10
M13	1	156	156	4	1960-1965	130	N,o		O	3	6	1	10	3.3	22.4	15	50
M21	1	311	571	5	-1900	130	N,o		O	4	9	1	14	3.0	23.8	13	1
ME	1	467	467	5	1955	90	N,o		O	6	11	0	17	3.8	25.6	15	27
T26	1	66	66	3	1966	160	N,o		P,O	21	3	0	24	17.1	18.1	94	0
T28	1	284	1829	5	1920	60	N,o		O	5	8	1	14	2.2	24.2	9	22
F10	2	402	402	1	1972	210	Y,h		P	27	3	0	30	14.6	15.5	95	0
F21	2	93	93	5	1890	160	Y,o		O	3	5	0	8	1.8	26.8	7	43
M7	2	130	623	5	pre 1900	145	Y,b,o	T	O	4	8	1	13	3.5	31.6	11	30
MB	2	259	259	5	pre 1900	120	N,o		O	10	7	0	17	8.0	27.8	29	17
T25	2	66	332	3	1926	210	N,b,o		P,O	23	4	0	27	27.6	29.8	93	0
M19	2	39	39	2	1973	210	Y,b,h		P,S	26	7	0	33	21.2	23.2	91	2
MD2	3	26	26	3	1962	110	Y,h,g,t		O	28	8	0	36	29.0	38.7	75	0
M15	3	26	26	3	1960-1965	140	N,o		O	10	8	0	18	2.9	26.9	11	41
F25	4	40	40	3	pre 1960	190	Y,h		O,R	32	5	0	37	24.3	25.9	94	0
M20	4	26	2205	5	pre 1923	120	N,o	E	O	7	10	3	20	5.0	39.6	13	38
M18	4	234	519	2	1973	220	Y,h,b	M	P,S	25	4	0	29	24.4	26.1	94	3
MF	4	259	259	2	1953	90	N,Y,h		O	25	10	0	35	19.4	32.0	61	17
M14	5	39	39	3	1960-65	130	N,b,o,t		O	20	10	0	30	15.8	32.8	48	25
TE	5	372	372	1	1967	100	Y,h		P,O	33	3	0	36	25.3	27.3	93	0
F20	6	93	226	3	1890	170	Y,h (sparse)		O,W	26	5	1	32	19.5	32.6	60	18
T11	6	159	1009	3	1960	150	N,b		P,S	31	2	0	33	20.9	21.2	99	0
M9	7	285	882	3	1960-65	130	N,o		O	29	10	0	39	16.3	37.9	43	23
TB	7	133	133	3	1960	210	Y,b		S	32	2	0	34	18.8	19.3	98	0
F8	8	266	266	4	1973	120	N,b,o		P,O	25	4	0	29	19.9	25.2	79	3
T31	9	350	1868	5	pre 1910	90	R,o		O	11	12	0	33	13.0	34.4	38	18
MD1	10	337	519	3 but slashed often	1962	110	N,o	D	O	13	10	4	27	6.8	36.0	18	29
F24	11	27	27	4	pre 1960	175	Y,h,g		O,R	35	3	0	38	21.7	22.7	96	0
F23	12	133	133	4	pre 1960	200	Y,h		O,S,W	33	8	1	42	26.8	30.4	88	1
F9	12	159	159	4	1963	110	N,b,g		P,O	40	7	0	47	27.8	28.5	98	0
M1	13	146	146	2	1968	190	N,b,o		O	24	5	0	29	16.6	18.9	88	1
T2	15	146	398	3	1966	120	Y,b,o		P,O	42	4	2	48	30.4	31.8	96	0
F4	17	389	2595	5	pre 1920	100	N,o,b		O	28	10	0	38	11.5	34.1	34	28
T19	18	40	3188	2	1966	190	N,b,h	recently reduced from 57 ha	P,O,W	31	6	0	37	26.0	28.5	90	1

Appendix — Continued.

Area	Size (ha)	Distance from nearest patch (m)	Distance from nearest larger patch (m)	Grazing index* (1-5)	Date of isolation	Altitude (m)	Main Vegetation †	Other features	Boundaries with #	Forest bird species	Farmland bird species	Water bird species	Total bird species (diurnal) P	Forest birds per count	Total birds per count	Forest birds as % of total birds	Noisy miners as % of total birds
F7	19	266	1302	4	1973	150	N,b	T major rubbish near Traralgon	P,O	34	7	0	41	25.7	28.8	89	3
T29	22	311	2672	5	pre 1918	90	N,b,o		O	29	13	1	43	12.7	40.1	32	9
F12	26	597	597	1	1963	140	Y,h		P,O	40	4	0	44	23.5	24.2	97	0
T18	29	545	770	3	1965	130	Y,h,b		P,O	42	8	0	50	24.9	27.3	91	1
M3	32	208	208	1	1974	170-210	Y,h		P,O	41	5	0	46	29.0	29.8	97	0
T10	34	66	345	1	1965	110	Y,h		P,O	33	9	1	43	28.3	30.5	93	1
M14A	40	285	178	2	1962	120	Y,N,h		O	45	13	1	59	28.3	33.5	85	5
F19	44	93	93	3	1960	190	N,h,g		P,O,W	37	5	1	43	32.0	33.9	95	3
M13A	48	259	1297	3	1928	120	N,G,b,g,t	C	O	47	12	1	60	37.7	41.1	92	1
F6	49	1427	1557	1	1972	180	Y,h	T,E	P	44	10	2	56	3.8	43.0	92	1
T1	50	398	1461	1	1967	100-180	Y,S,G,h	C	P,O	44	4	0	48	29.6	29.9	99	0
M23W	57	467	467	1	1963	120-210	Y,G,h	D	P,O	46	6	0	52	31.0	32.5	95	0
M4	96	39	519	2	1976	150-200	N,G,b,g	D	P,O	46	12	0	58	34.4	38.9	89	2
F1	108	402	4426	1	1964	70	N,R,G,b,t	D	P,O	46	12	1	59	27.3	32.1	85	0
M23E	134	467	2213	1	1964	120-200	Y,b,h	C	P,O	49	9	0	58	30.5	32.3	95	2
M2	144	146	319	1	pre 1940	160-220	Y,S,G,h	D,C	O	49	12	2	63	33.0	35.0	94	0
M11	973	337	1323	1	pre 1970	140-240	Y,S,G,h	D,C	P,O	52	12	3	67	29.5	31.3	94	0
F11	1771	332	332	1	1933	160-360	Y,S,G,h,t	C	P,O	59	5	0	64	30.9	31.6	98	0

Key to Appendix:
* Grazing pressure is indicated on a scale of 1 (little or no grazing by domestic stock) to 5 (heavy grazing pressure, little or no natural understorey remaining).
† The Main vegetation includes:
N — stands dominated by Narrow-leaf Peppermint *Eucalyptus radiata*, usually with Manna Gum *E. viminalis*, and But-but *E. bridgesiana*, and sometimes Swamp Gum *E. ovata* and other species.
Y — stands dominated by Yertchuk *E. consideniana*, with some of the species above and various stringybarks.
S — stands dominated by various stringybarks, specially Messmate *E. obliqua*, often with Narrow-leafed Peppermint and Mountain Grey Gum *E. cypellocarpa* in gullies and on wetter slopes, and an understorey containing Forest Wire-grass *Tetrarrhena juncea* and various shrubs.
R — stands dominated by River Red Gum *E. camaldulensis* or Forest Red Gum *E. tereticornis*.
G — some well developed gully vegetation.
b — understorey dominated by Bracken *Pteridium esculentum*.
t — thickets of tea-tree, specially Burgan *Leptospermum phylicoides*.
h — heathy understorey (e.g. with Common Heath *Epacris impressa* or *Banksia* spp), as well as Bracken, Burgan and various shrubs.
o — heavily grazed open understorey of introduced grasses.
g — naturally open understorey of native and introduced grasses.
Other features include rubbish tips (T), small dams (D), a small ephemeral swamp (E) or a permanently running creek (C). One area with unusually sparse trees and abundant Mistletoe also occurred in many patches).
Boundaries with pine plantations (P), open farmland (O) or scrub, e.g. regenerating farmland (S). Patches separated from other forest only by a road are marked R. Patches with extensive water nearby are marked W.
P Numbers of species exclude nocturnal birds, of which four species were resident with up to three on individual patches.

CHAPTER 7

The Incidence and Conservation of Animal and Plant Species in Remnants of Native Vegetation within New Zealand

C. C. Ogle[1]

New Zealand's native vegetation cover was modified during 1000 years or so of Maori settlement, and then largely reduced to remnants during 140 years of European presence. This chapter assesses the conservation values of the remaining native vegetation patches of various sizes. Both large and small patches of forests, wetlands and grasslands are shown to have values, and even advantages, for species conservation. Large patches appear necessary for some bird species such as kokako *Callaeas cinerea* and Australasian bittern *Botaurus stellaris poiciloptilus*, but small patches may be more suitable for other animals and plants, such as mistletoes and *Paryphanta* snails. Species able to survive in small patches might be conserved in a national pattern of representative reserves, but the survival of certain rare species which were once widespread (e.g. the skink, *Cyclodina whitakeri*) may depend upon habitat features which occur in very few patches.

Values of vegetation remnants, previously isolated by geological or other processes, should be recognized in any reserve system for the known or likely occurrence of local endemics.

INTRODUCTION

WHILE the main theme of this chapter is native vegetation patch size and its implications for conservation of animal and plant species, the range of examples chosen is broad. The chapter has been structured to show firstly the origins of the present pattern of large and small patches of native vegetation in New Zealand, and the extent to which these areas are reserved. Data are presented on the variety of bird and land snail species in patches of various sizes, and on the incidence of particular species of birds, native land snails, lizards, and plants which have relictual distributions. Sites with endemic taxa are considered briefly.

There has been extensive clearing of New Zealand's indigenous vegetation in the 1000 years or so of human presence. Maori-induced fires destroyed forests over about 25% of the land area, mostly to be replaced by native tussock grasses, scrub, and ferns (Molloy 1980). A further 30% of the land was cleared of forest following the arrival of Europeans (Wendelken, *in* Wards 1976), to make way for pastoral farming, urban development, and, increasingly, exotic afforestation, as well as for the native timber resource itself. In total, 14.7 million ha of indigenous forest had been removed by 1976 (ibid).

Today, much of the remaining native vegetation occurs as isolated patches and all are modified to some degree by man or his agents, including domestic stock, introduced wild mammals, exotic plants, and agricultural chemicals. Remaining native forests have often been selectively logged for choice timber trees, or are secondary, following clearance of the original forest cover.

Patches which survived 140 years of European impacts continue to be reduced in area or disappear completely. As an example, in Northland over the past five years, 14% of the area of freshwater wetlands disappeared (22.7% of 163 patches underwent some reduction in area), and forests lost 7.5% of their total area (55.4% of 271 sites were affected) (Anderson *et al.* 1984). Losses were from both Crown-owned and privately-owned lands.

---

[1] New Zealand Wildlife Service, Department of Internal Affairs, Private Bag, Wellington, New Zealand.
Pages 79-87 *in* NATURE CONSERVATION: THE ROLE OF REMNANTS OF NATIVE VEGETATION ed by Denis A. Saunders, Graham W. Arnold, Andrew A. Burbidge and Angas J. M. Hopkins. Surrey Beatty and Sons Pty Limited in association with CSIRO and CALM, 1987.

On the other hand, reservation of land is extensive; by 1983, 17% of New Zealand lay in formally protected natural areas, mostly under Crown ownership. The largest proportion is administered by the Department of Lands and Survey as national parks (7.66% of New Zealand's area) and other reserves (2.75%), followed by the New Zealand Forest Service which administers reserves in state forests (6.54%), and the New Zealand Wildlife Service's wildlife reserves (0.09%) (New Zealand Department of Lands and Survey 1984).

However, existing forested reserves are inadequate in several respects. Hackwell and Dawson (1980) and Dawson (1984) demonstrated that: lowland forest is greatly under-represented in the reserve system and frequently has been modified by past logging operations; there is a preponderance of small, isolated reserves; and intact, altitudinal sequences are seldom reserved.

Concerns have been expressed also at the negligible reserves of other types of natural areas such as tussock grasslands (Mark 1980; McSweeney 1983) and wetlands (Environmental Council Wetlands Task Group 1983). In general, mainland reserves of all types are dominated by subalpine and alpine areas and by forest. A disproportionately large amount of reserved forest is at high altitudes. Hackwell and Dawson (1980) stated that 43% of New Zealand lies below 300 m, yet only 16% of national parks and 7% of forest parks are below this altitude. Lowland reserves tend to be of small size, and are often on land such as coastal cliffs, river gorges, and other steeplands which had been uneconomic to clear, and which, by their topography, do not contain a representative selection of a region's biota.

A consequence of this lack of representativeness of existing reserves is that the survival of a number of species of indigenous animals and plants, some of them endemic to New Zealand, is inadequately assured. Some endemic species do not even occur in reserves.

There are a few, long-standing, special purpose reserves to protect individual species, mostly species and their habitats with some public appeal, such as mainland nesting sites for Australasian gannet *Sula bassana serrator* at Cape Kidnappers (Fig. 1), and royal albatross *Diomedea epomophora* near Dunedin. There have been recent additions to special purpose reserves to protect and manage the whole habitat of several rare or threatened animals such as North Island fernbirds *Bowdleria punctata vealeae* and kokako *Callaeas cinerea wilsoni* in the central North Island, a rare skink *Cyclodina whitakeri* and speargrass weevil *Lyperobius huttoni* near Wellington, and a chafer beetle *Prodontria lewisi* at Cromwell. A few rare or threatened plant species have received protection in specially created reserves. Examples include *Pittosporum turneri* (Erua State Forest Sanctuary), *P. obcordatum* (Hurumua Nature Reserve, near Wairoa), *Todea barbara* (Todea Barbara Ecological Area, Waitangi State Forest), and *Senecio turneri* (Mimi Scenic Reserve). In all cases, except for kokako, such special purpose reserves are small (less than 150 ha, and mostly much smaller). They cannot be regarded as representative reserves, and may not be viable in the long term.

In the past decade there have been moves to improve these aspects of the reserve system. A major stimulus has been biogeographic theory, originally developed outside New Zealand by MacArthur and Wilson (1967), Diamond (1975, 1976), and others, but tested since on data from local biological inventories (Hackwell and Dawson 1980). The general conclusions have been that patch size, degree of isolation, and altitudinal range within the patch are important parameters in predicting the variety of native bird species which can be expected in a given forest patch. Such studies have given us some empirical rules for reserve selection, which have been variously adapted, extended, and used for rating patches of wildlife habitat (Crook *et al.* 1977; Imboden 1978; Ogle 1981) and remnant forests (Park and Walls 1978).

The conservation of threatened species also requires that general principles of reserve selection be integrated with the results of autecological studies. A recent major study of the endemic, threatened, North Island kokako *Callaeas cinerea wilsoni* by Hay (1981) involved the simultaneous study of vegetation and phenology in its habitat (Leathwick 1981), and comparisons of the diets of kokako and potential mammalian competitors (Leathwick *et al.* 1983). These complemented other findings on the incidence of kokako nationally, in forest patches of various sizes (e.g. Hackwell and Dawson 1980; Ogle 1982; O'Donnell 1984), and provide the basis for a national recovery plan for kokako now being prepared (C. R. Veitch, unpublished data).

BIRDS AND PATCH SIZE

Forests: Almost all extant species of native landbirds on the mainland were widespread in historic times, with only Cook Strait acting as a major barrier to the dispersal of some species between the North and South Islands. Some now have much reduced ranges, and a few have become confined to only one area (e.g. the endemic rail, takahe *Notornis mantelli*), or a very small number of areas (e.g. kokako, black stilt *Himantopus novaezelandiae*, and the endemic parrot, kakapo *Strigops habroptilus*).

Studies based on the large isolated tracts of forest on the whole of the New Zealand mainland, and on all forest patches over 10 ha in size in the Northland

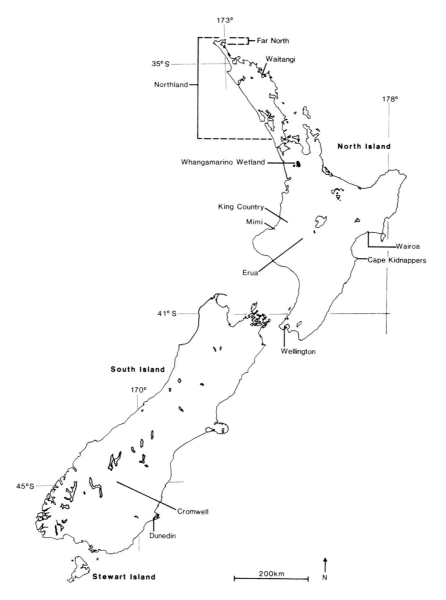

Fig. 1. Map of New Zealand, showing locations of places named in text.

region, have shown that larger patches tend to hold more native bird species (Hackwell and Dawson 1980; Ogle 1981, 1982; Dawson 1984). Forest species with a high level of endemism seem to be particularly vulnerable to local extinctions when patch size is reduced. As examples, in Northland, the two rarest forest birds are kokako (the sole mainland survivor of an endemic New Zealand family) which occurs in only three patches of forest, three of the four patches in the region which exceed 10,000 ha, and kaka *Nestor meridionalis*, a representative of an endemic genus, which is also mostly in larger patches (Ogle 1982). Similar trends have been found for these birds and other species over the whole of New Zealand (Hackwell and Dawson 1980).

The forest bird species which are most widespread and occur in forest patches of all sizes tend to have low levels of endemism, e.g. fantail *Rhipidura fuliginosa*, grey warbler *Gerygone igata*, and silver-eye *Zosterops lateralis*.

One of the major concentrations of kokako remaining in New Zealand is in King Country of central and western North Island. In 1982-83, the New Zealand Wildlife Service's survey of all forest remnants of King Country revealed kokako in just nine patches. Within the past 30 years kokako were known in 12 further patches (Lavers 1978; O'Donnell 1984), and, in two of these sites, kokako have gone as recently as 1979 and 1980.

Figure 2 shows that not only do large forest patches in King Country tend to contain more native bird species than small patches, but also that it is the largest patches which have kokako, or have had them until quite recently. There is no obvious reason for patch size *per se* to be the factor controlling the presence or absence of kokako. In fact, a

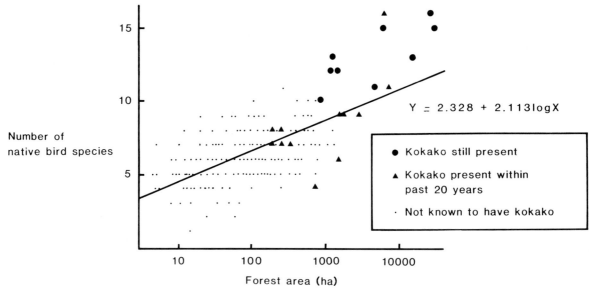

Fig. 2. Number of native forest bird species in different-sized forest patches in King Country, central North Island. Patches with kokako *Callaeas cinerea wilsoni,* and those which had kokako within the past 20 years are highlighted.

recent study (Hay 1981) showed that kokako occur at densities of up to 0.23 birds per ha. If Franklin's (1980) proposition of a minimum of 500 individuals for long term population maintenance is applied to kokako at the maximum density found by Hay (1981), a kokako population requires a little over 2000 ha of suitable habitat. That Hay (1981) found territories occupied by either unpaired birds or paired birds reinforces the cautions expressed by Frankel and Soulé (1981) in accepting Franklin's (1980) figure of 500 individuals, and an *effective* population size of 500 means that a much larger number of kokako would be needed, and hence an area much greater than 2000 ha. Figure 2 shows that at least three patches greater than 2000 ha in King Country have lost kokako within the last 20 years, although they still remain in four smaller patches.

Habitat which supports kokako at lower densities would also need to be more extensive to contain an effective population. At the worst, if it is assumed that 1000 birds are needed for an effective population of 500, and the habitat is such that it supported birds at the lowest densities found by Hay (1981) of 0.16 birds per ha, then possibly there is no patch of forest which is large enough for long-term survival of kokako in King Country.

Whether large patches have a greater variety of vegetation types, and can hence supply a greater range of food items the year round, or are less affected by disturbance by introduced predators, browsers, and man, or have some other intrinsic quality favouring the kokako's existence, is unknown.

Northland is a hilly region without high mountains (maximum 776 m above sea level). The highest hills are mostly in the large patches of native forest, and almost all large patches of forest contain land over 300 m. For this reason, large patch size can mask the effects of a real limiting factor, such as altitude. This is demonstrated by the distribution of pied tits *Petroica macrocephala toitoi* in Northland. Ogle and Anderson (1979) showed that pied tits occurred more frequently in the large forest patches than in small patches. A detailed wildlife survey of the five largest forest patches in 1979 included the counting of birds in seven blocks (Moynihan 1980). Three of these surveyed blocks were in separate patches of forest, while the remaining four blocks were in two other patches, two blocks in each. Figure 3 results from a recent, more detailed analysis of Moynihan's (1980) data, and shows the altitudinal distribution of pied tits within these seven blocks. Tits were found at most stations in blocks at high altitudes (Raetea, Mataraua and Warawara), and at significantly higher altitudes in three blocks which covered a wide range of altitudes (Omahuta, Waipoua and Russell).

These findings allow tit presence in smaller patches to be more readily interpreted. In at least five of the small, isolated, forest patches in Northland with pied tits (Mareretu 2640 ha, Moirs Hill 350 ha, Kaikanui 850 ha, Pukearenga 300 ha and Conical Peak 320 ha), pied tits occur only around the highest peak or ridge in the patch (354 m, 358 m, 479 m, 274 m and 386 m respectively; Fig. 3). High peaks have more cloud cover and hence more mist or rain than surrounding land. This results in increased epiphytic plant growth, such as filmy ferns (Family Hymenophyllaceae), mosses, liverworts, and lichens, which, in turn, is likely to support a richer and more abundant invertebrate fauna, the dietary items of pied tits (Falla *et al.* 1979).

This hypothesis may explain the absence of a clear altitudinal zonation of pied tits in Puketi Forest (Fig. 3), where the surveyed block extended from the

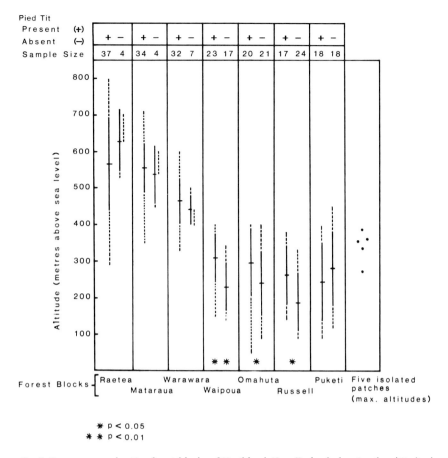

Fig. 3. Seven surveyed native forest blocks of Northland, New Zealand, showing the altitudinal distribution of pied tits, Petroica macrocephala toitoi. For each block, the mean, standard deviation, and range of altitudes of bird-count stations with and without pied tits is shown. The highest altitudes of a further five, isolated and much smaller forest patches in Northland (named in the text), where pied tits occur only at or close to the highest peak in each patch, are shown in the extreme right-hand column.

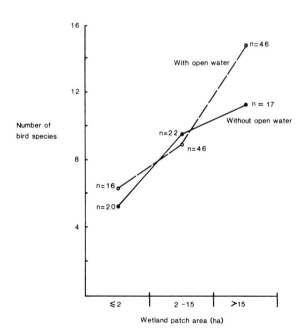

Fig. 4. Mean numbers of species of freshwater-inhabiting birds, in different-sized freshwater wetland patches with and without open water in Northland, New Zealand (data from Ogle 1982).

crest of an exposed ridge with previously logged forest, down a south-facing slope, and into a valley with dense, wet forest; the forest of the higher altitude parts surveyed is not obviously wetter than the lower altitude forest.

Wetlands: Comparatively little zoogeographic work has been done in New Zealand in vegetation other than forest. An analysis of wetland bird data from 167 freshwater wetlands in Northland (Ogle 1982) indicated that the number of wetland bird species also increases with increasing size of a wetland patch. Furthermore, the presence of areas of open water within a wetland allows more bird species to use the site (Fig. 4) since it increases habitat diversity.

A single example is presented of a species' incidence in relation to wetland size. The Australasian bittern *Botaurus stellaris poiciloptilus* may number as few as 1000 individuals in New Zealand. Its incidence in specific wetlands may be obscured by both its mobility and practical difficulties in locating birds, but in a survey of Northland, bitterns were found or were reliably reported in 61 of 226 freshwater wetlands. In these, bitterns occurred in 14% of small (<2 ha) wetlands, 40% of larger (2-15 ha) wetlands, and in 56% of wetlands larger than 15 ha (Ogle 1982). An estimate of the densities of bitterns in one of New Zealand's largest wetlands, the 7100 ha Whangamarino Wetland, gave a maximum of one bittern per 8.3 ha in a 100 ha block of apparently ideal habitat, and one per 145 ha over the entire wetland (Ogle and Cheyne 1981).

These data suggest that large wetlands, or perhaps a number of smaller wetlands close together, are needed to sustain bittern populations, but data on the incidence and the variety of bird species in wetlands throughout New Zealand require more rigorous analysis to confirm and extend these preliminary findings. This was also one of the major recommendations of a one-day workshop called by the Royal Society of New Zealand in March 1983, on the theme 'Biological reserve design in mainland New Zealand' (King 1984).

OTHER FAUNA AND FLORA

Up to this time, regional or national studies of species incidence in remnants of native vegetation have been based on birds rather than other fauna, since birds are the only animals for which data are available from a large number of patches. Birds pose special problems for such studies, since there are relatively few species of native land-birds in New Zealand, and birds generally move between patches, e.g. those isolated swamps of less than two ha where Australasian bitterns were recorded are probably only temporary habitat for these birds, although several together could be the permanent range.

It is necessary to distinguish two groups of species when examining the incidence of plants and slow-dispersing animal species; those which have relictual distributions, which reflect man-induced changes in their habitats (including the effects of predators), and those which exhibit local endemism, the result of periods of isolation.

(i) *Relictual Distributions*

(a) *Mistletoes*. A national decline in the range and abundance of New Zealand's six species of leafy mistletoes (Family Loranthaceae) has been attributed to the selective browsing of these plants by the Australian brushtailed possum *Trichosurus vulpecula* (Wilson 1984; Ogle and Wilson 1985). These authors documented a number of instances where mistletoes have been eliminated from the large tracts of forest, but record their occurrence nearby in isolated host trees, or in small groves of trees, in developed farmland, orchards, and urban areas. Possibly there is a wide range of alternative preferred foods for possums in agricultural land, and high levels of possum trapping, particularly in fruit orchards (Ogle and Wilson 1985)[1].

(b) *Land snails*. A similar example to that of mistletoes is found in Northland, where the large (up to 8 cm diameter), carnivorous land snail, *Paryphanta b. busbyi*, occurs quite widely in large and small forest remnants. Paradoxically, its survival seems more assured in small rather than large remnants because of the high levels of predation by wild pigs which occur in almost all of the large forest patches, but in few small patches (Ogle 1982).

Paryphanta snails can disperse up to 800 m in 15-18 years (Penniket, *in* Ogle 1982), which has one unfortunate consequence for their survival in very small patches of forest; individual animals have been found desiccated beyond the forest edge, having unwittingly wandered into surrounding pasture at night and not returned to the forest by daybreak.

New Zealand has probably more than 1000 species of endemic terrestrial snails (F. Climo, pers. comm.) and many of these are widespread in suitable habitat. Up to 57 species have been recorded from a single four ha patch of forest surrounded by farmland near Auckland (Solem *et al.* 1981). It is likely that small land snails do not require large areas of habitat in order to persist, so their conservation, and that of many other invertebrates, could be achieved by the retention of small remnants of native vegetation across the whole of New Zealand.

(c) *Cyclodina skinks*. Some large skinks in the endemic genus *Cyclodina* are particularly vulnerable to predation by rats, and these lizards are now largely on rat-free islands (Whitaker 1978; Towns and Robb 1986). One species, Whitaker's skink *C. whitakeri*, is known only from two small, rat-free islands off the north-east coast of the North Island, and from one mainland site near Wellington. The

---

[1] During the post-workshop tour and subsequent touring in the wheatbelt of the south-west of Western Australia, in September-October 1985, I noted that some mistletoes there have similar patterns of distribution and abundance to leafy mistletoes in New Zealand. *Amyema miquelii* and *A. preissii* sometimes occur on most of the available host trees, *Eucalyptus wandoo* and *Acacia* spp. respectively, where these hosts grow beside roads in agricultural land. In places, the quantity of mistletoe foliage on a tree exceeds that of its host. The same mistletoes are in large, isolated patches of forest such as the 28,000 ha Dryandra Forest, but on scattered host trees only, and never as heavy infestations.

Can these differences in mistletoe occurrence be related to possum browsing? Literature on habitat use of brushtailed possums in southwestern Australia has been reviewed by Winter (1979) and Kerle (1984). Sampson (*in* Kerle 1984) in 1971 found a clear preference by possums for valley woodland associations of *Eucalyptus wandoo* and *Casuarina huegeliana*. Winter (1979) cited unpublished comments that possums had declined in numbers or that their range had contracted in drier, more sparsely wooded areas.

In Dryandra Forest, south-east of Perth, Serventy (1970) observed that brushtailed possums were becoming more numerous after a period of decline through hunting and, possibly, disease. He cited older residents' views that mistletoes had both increased in abundance and were browsed by possums. Freeland and Winter (1975) recorded that consumption of flowers and fruit of *Amyema miquelii* by brushtailed possums was frequent and seasonal.

Since possums browse mistletoe, have declined in numbers, at least until recently, and use forest patches rather than trees in agricultural land, it does appear that they may be a significant factor in producing the current patterns of mistletoe distribution and abundance in Western Australia's wheatbelt.

Although mistletoes are widely regarded as pests in Western Australia, because they damage their hosts, it appears that if, or when, their conservation becomes a matter for concern that small forest patches and isolated trees will make the best mistletoe reserves, just as they do in New Zealand.

latter site is a steep, scrub-covered, north-facing, coastal hillside built up of loosely-fitting, but relatively stable, rocks and boulders. Despite grazing and other disturbances Whitaker's skink seems to survive here because it can evade predators by withdrawing into crevices between the rocks. Of 22 known localities with one or more of the four largest species of *Cyclodina*, only three are in areas with rats, and in all three the lizards are confined to boulder banks (Towns and Robb 1986).

(d) *Peat bog biota.* Species can become confined to a very few sites because their habitat has been destroyed or heavily modified elsewhere. Lowland peat bogs were once very extensive in New Zealand, but unmodified examples, and much of their characteristic flora and fauna, are now extremely rare. The endemic black mudfish *Neochanna diversus*, a native clubmoss *Lycopodium serpentinum*, a bladderwort *Utricularia delicatula*, which may be identical to the Australian *U. lateriflora*, and a tall restiad rush in the monotypic, endemic genus *Sporadanthus*, are some threatened species of this habitat-type. Peat bogs exemplify a situation where a suite of species with relict distributions may be protected by direct management of the habitat. Low fertility sites are extremely sensitive to inputs of extra nutrients, and resulting changes in such habitats may be avoided, or at least retarded, by controlling sources of nutrients, such as artificial fertilisers, farm wastes and soil erosion, in the catchment.

By way of contrast, high fertility sites, such as once occurred widely on river flood plains, limestone areas and recently weathered volcanic soils, are particularly vulnerable to invasion by exotic plants. Many species requiring unmodified, high fertility sites also have relict distributions in New Zealand.

Examples could be cited from other habitat types which are much reduced or highly modified from their primitive state, such as sand dunes, non-seral native shrublands, and inland salt flats.

It does appear that the presence of a rare, but previously widespread, species in a site is seldom a random happening, but is the result of special conditions there which favour that species' survival in that site. It is possible to define the precise nature of these special conditions in some sites, such as peat bogs, but in others it may not be easy to recognize or quantify these factors. Nevertheless, the recognition of these sites as habitats for once widespread species makes them vital for reservation, irrespective of their size.

(ii) *Local Endemism*

Although many species of small, native land snails are widespread, a number exhibit local endemism, in areas isolated for long periods by geological processes. The Far North of Northland (Fig. 1) was a cluster of volcanic islands in the lower Pliocene (6 million years B.P.) but became linked by a sand isthmus to the mainland during the Pleistocene. The area comprises more than 16,000 ha of *Leptospermum* scrub, mostly resulting from repeated fires, and some remnants of intact forest, most of it secondary. When 25 litter samples were taken from a range of sites here in March 1985, 18 contained native land-snails, comprising 51 species, 23 of which are endemic to the Far North.

Local endemics occur in the native vascular flora of the Far North, particularly on a small outcrop of serpentine. At least 13 endemic taxa have been recorded from an area of 110 ha. These include unnamed species of *Hebe* and *Coprosma*, and distinctive, local subspecies of *Phyllocladus trichomanoides* and a range of flowering plants (Druce *et al.* 1979). Elsewhere in New Zealand there are local endemics on other ultramafic outcrops, and on other rock types such as limestone, marble and ancient volcanic rocks, and on other land which has been physically separated from adjoining country at some time in the past. All such areas require to be identified and then protected within the national reserve system.

CONCLUSIONS

Biogeographic principles have been useful in New Zealand since they provided some empirical rules for species habitat protection, particularly in forests, when our knowledge of individual species requirements was (and frequently still is) inadequate for their conservation. Large patches of habitat tend to contain more bird species than do small patches. This is a result, at least in part, of the greater habitat diversity which tends to occur in larger patches; e.g. larger forests tend to have greater altitudinal ranges and hence more vegetation types, and larger wetlands are more likely to contain interspersed vegetation and water. Large patches of habitat appear to be essential for the survival of certain birds, such as kokako and Australasian bittern, in forests and wetlands, respectively. Many other species of flora and fauna persist in quite small patches of habitat, and there are some, including *Paryphanta* land snails and mistletoes, which, for a variety of reasons, are more likely to survive in small rather than large patches.

Every reserve system should reflect differences in individual species' requirements. While fully supporting the need for large reserves, this chapter stresses the conservation values of suitably chosen and well-managed small reserves, since these are valuable for common biota, for those species with relict distributions or which are endemic to small parts of New Zealand, and for those species whose survival is enhanced by being in small reserves.

ACKNOWLEDGEMENTS

I wish to thank the following members of the New Zealand Wildlife Service for their contributions to this paper: members of the Fauna Survey Unit for data on distributions of small land snails in the Far North and on birds nationally, Dr D. Towns for advice on *Cyclodina whitakeri*, Messrs K. Moynihan, G. Elliott and J. MacDonald for data analysis and drafting of figures, Drs M. Crawley and M. Williams for their valuable suggestions and criticisms of drafts of this paper. Dr F. Climo (National Museum, Wellington) is thanked for his identification of land snail specimens.

REFERENCES

Anderson, R., Hogarth, I., Pickard, R. and Ogle, C., 1984. Loss of wildlife habitat in Northland, 1978-83, with notes on recently identified wildlife values. Technical Report No. 6 (unpublished). New Zealand Wildlife Service, Department of Internal Affairs, Wellington, New Zealand. 46 p.

Crook, I. G., Best, H. A. and Harrison, M., 1977. A survey of the native bird fauna of forests in the proposed beech project area of North Westland. *Proc. N.Z. Ecol. Soc.* 24: 113-27.

Dawson, D. G., 1984. Principles of ecological biogeography and criteria for reserve design (extended abstract). *In* Proceedings of a workshop on biological reserve design in New Zealand. *J. Roy. Soc. N.Z.* 14: 11-5.

Diamond, J. D., 1975. The island dilemma: lessons of modern biogeographic studies for the design of natural reserves. *Biol. Conserv.* 7: 129-45.

Diamond, J. D., 1976. Relaxation and differential extinction on landbridge islands: applications to natural preserves. *Proc. Int. Ornith. Cong.* 16: 616-28.

Druce, A. P., Bartlett, J. K. and Gardner R. O., 1979. Indigenous vascular plants of the serpentine area of Surville Cliffs and adjacent cliff tops, north-west of North Cape, New Zealand. *Tane* 25: 187-206.

Environmental Council Wetlands Task Group, 1983. Wetlands: a diminishing resource: a report for the Environmental Council (G. K. Stephenson, convenor). Water and Soil Miscellaneous Publication No. 58. Water and Soil Division, Ministry of Works and Development, Wellington, New Zealand.

Falla, R. A., Sibson, R. B. and Turbott, E. G., 1979. The new guide to the birds of New Zealand and outlying islands. Collins, Auckland and London. 247 p.

Frankel, O. H. and Soulé, M. E., 1981. Conservation and evolution. Cambridge University Press, Cambridge. 327 p.

Franklin, I. R., 1980. Evolutionary change in small populations. Pp. 135-49 *in* Conservation biology: an evolutionary-ecological perspective ed by M. E. Soulé and B. A. Wilcox. Sinauer Associates, Sunderland, Massachusets.

Freeland, W. J. and Winter, J. W., 1975. Evolutionary consequences of eating: *Trichosurus vulpecula* (Marsupialia) and the genus *Eucalyptus. J. Chem. Ecol.* 1: 439-55.

Hackwell, K. R. and Dawson, D. G., 1980. Designing forest reserves. *Forest and Bird* 13: 8-15.

Hay, J. R., 1981. The kokako. Forest Bird Research Group Report. Jointly published by Royal Forest and Bird Protection Society, New Zealand Wildlife Service, New Zealand Forest Service, Wellington.

Imboden, C., 1978. The valuation of wildlife habitats. *Wildlife — a review* 9: 54-8. New Zealand Wildlife Service, Department of Internal Affairs, Wellington, New Zealand.

Kerle, J. A., 1984. Variation in the ecology of *Trichosurus*: its adaptive significance. Pp. 115-28 *in* Possums and gliders ed by A. Smith and I. Hume. Surrey Beatty & Sons Pty Ltd, Sydney.

King, C. M., 1984. Open discussion. Proceedings of a workshop on biological reserve design in New Zealand. *J. Roy. Soc. N. Z.* 14: 39-44.

Lavers, R. B., 1978. Distribution of the North Island kokako *(Callaeas cinerea wilsoni)*. A review. *Notornis* 25: 165-85.

Leathwick, J. R., 1981. The vegetation of kokako and general bird study areas in some central North Island indigenous forests. Forest Bird Research Group Report. Jointly published by New Zealand Forest Service, New Zealand Wildlife Service, Royal Forest and Bird Protection Society, Wellington.

Leathwick, J. R., Hay, J. R. and Fitzgerald, A. E., 1983. The influence of browsing by introduced mammals on the decline of North Island kokako. *N.Z. J. Ecol.* 6: 55-70.

MacArthur, R. H. and Wilson, E. O., 1967. The theory of island biogeography. Princeton University Press, Princeton.

Mark, A. F., 1980. A disappearing heritage — tussock grasslands of the South Island rain-shadow region. *Forest and Bird* 13: 18-24.

McSweeney, G., 1983. South Island tussock grasslands — a forgotten habitat. *Forest and Bird* 14: 50-4.

Molloy, L. F. (Compiler), 1980. Land alone endures: land use and the role of research. New Zealand Department of Scientific and Industrial Research Discussion paper No. 3. Department of Scientific and Industrial Research, Wellington. 286 p.

Moynihan, K. T., 1980. Birdlife in seven State Forests surveyed in Northland. Fauna Survey Unit Report No. 22. New Zealand Wildlife Service, Department of Internal Affairs, Wellington. 16p.

New Zealand Department of Lands and Survey, 1984. Register of protected natural areas in New Zealand ed by S. Timmins and K. King. Department of Lands and Survey, Wellington. 468p.

O'Donnell, C. F. J., 1984. The North Island kokako *(Callaeas cinerea wilsoni)* in the western King Country and Taranaki. *Notornis* 31: 131-44.

Ogle, C. C., 1981. The ranking of wildlife habitats. *N.Z. J. Ecol.* 4: 115-23.

Ogle, C. C., 1982. Wildlife and wildlife values of Northland. Fauna Survey Unit Report No. 30. New Zealand Wildlife Service, Department of Internal Affairs, Wellington, New Zealand. 272p.

Ogle, C. C. and Anderson, P., 1979. Northland habitat survey. *Wildlife — a review* 10: 30-4. New Zealand Wildlife Service, Department of Internal Affairs, Wellington.

Ogle, C. C. and Cheyne, J., 1981. The wildlife and wildlife values of the Whangamarino wetlands. Fauna Survey Unit Report No. 28. New Zealand Wildlife Service, Department of Internal Affairs, Wellington. 94p.

Ogle, C. C. and Wilson, P. R., 1985. Where have all the mistletoes gone? *Forest and Bird* 16: 10-3.

Park, G. N. and Walls, G. Y., 1978. Inventory of tall forest stands on lowland plains and terraces in Nelson and Marlborough Land Districts, New Zealand. Unpublished report, Botany Division, Department of Scientific and Industrial Research, Nelson, New Zealand. 127p.

Serventy, V., 1970. Dryandra. The story of an Australian forest. A. H. and A. W. Reed, Sydney. 205p.

Solem, A., Climo, F. M. and Roscoe, D. G., 1981. Sympatric species diversity of New Zealand land snails. *N.Z. J. Zool.* 8: 453-85.

Towns, D. R. and Robb, J., 1986. The importance of northern offshore islands as refugia for endangered lizard and frog species. Pp. 197-210 *in* The offshore islands of northern New Zealand ed by A. E. Wright and R. E. Beever. New Zealand Department of Lands and Survey Information Series No. 16. Wildlife Service publication No. 273. Department of Internal Affairs, Wellington.

Wards, I. (Ed.), 1976. New Zealand atlas. Government Printer, Wellington. 292p.

Whitaker, A. H., 1978. The effects of rodents on reptiles and amphibians. Pp. 75-88 *in* The ecology and control of rodents in New Zealand nature reserves ed by P. R. Dingwall, I. A. E. Atkinson and C. Hay. New Zealand Department of Lands and Survey Information Series No. 4.

Wilson, P. R., 1984. The effects of possums on mistletoe on Mt Misery, Nelson Lakes National Park. *In* Proceedings of Section A4e, 15th Pacific Science Congress, Dunedin, February 1983 ed by P. R. Dingwall. New Zealand Department of Lands and Survey, Wellington.

Winter, J. W., 1979. The status of endangered Australian Phalangeridae, Petauridae, Burramyidae, Tarsipedidae and the koala. Pp. 45-59 *in* The status of endangered Australasian wildlife ed by M. Tyler. Proceedings of the Royal Zoological Society, South Australia.

CHAPTER 8

Assessing the Conservation Value of Remnant Habitat 'Islands': Mallee Patches on the Western Eyre Peninsula, South Australia

C. R. Margules[1] and A. O. Nicholls[1]

The problem addressed in this chapter is how to compare the conservation values of remnant vegetation patches in a given region to select a set of patches for a nature reserve network. The conservation value of a patch is defined as the degree to which it represents the range of regional vegetation variation.

A survey and numerical classification of mallee vegetation in remnant habitat patches on part of the Eyre Peninsula, South Australia, revealed six floristic groups. The survey sampled only a subset of the patches, so that the probability of finding each of the floristic groups in each of the remnant patches was estimated from statistical models. These models related the groups to three mapped environmental variables, age of calcium carbonate layer, distance from the coast and latitude. Thus, a uniform set of information about the floristic composition of each patch was generated so patches could be compared. The problem of defining adequate representation is discussed, and a determination is made of the minimum set of patches required to represent all communities according to different definitions of adequate representation.

INTRODUCTION

THE kinds of research necessary for selecting reserves for conservation form a continuum from manipulative field experiments, through observational and correlational analyses, so-called natural experiments, to the stage of identifying suitable reserves, which is actually a planning procedure. This final stage should be based on information from experiments and correlational analyses but such information is rarely available, especially in the detail desired. If the problem is ignored until sufficient experimental results are available, conservation options will be severely reduced. The urgency for decisions makes it necessary to generate some minimum set of information for all options, with known reliability.

The term conservation value is used in a similar sense to capability or suitability when these terms are used in relation to agriculture or forestry. Essentially, conservation value is the extent to which a given area of land, or a patch of habitat, contributes to representing the biological diversity of a region. A variety of conservation objectives is reflected in a plethora of criteria used to assess conservation value, which have been identified and discussed previously (Margules and Usher 1981). Most refer either directly or indirectly to a common underlying theme; to represent the greatest possible diversity of living organisms in the reserve system and to ensure their survival in perpetuity. The criteria of diversity and representativeness both most obviously address this goal. However, diversity as a criterion of conservation value has most often been used in a very parochial sense to describe local, usually atypical, species-rich sites (Margules 1986). The concept of representativeness (Austin and Margules 1984) seems more adequately to reflect the purpose of a reserve network; to represent the range of biological variation.

This chapter describes a method for generating a minimum common set of information across all habitat patches in a defined region, from a field

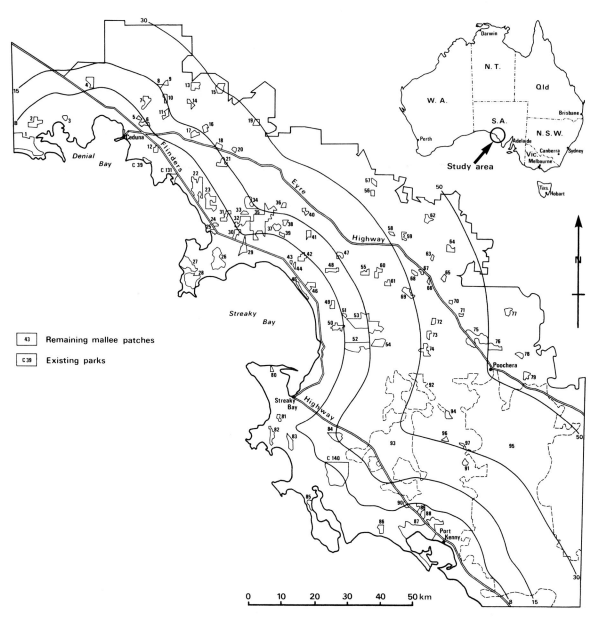

Fig. 1. Location of the study area showing the mallee patches and contours of distance from the coast. The north eastern boundary of the study area is the approximate limit of the croplands. The 101 patchs are numbered 1 to 98 and C39, C131 and C140, which are State National Parks and Wildlife Service codes for the three existing mallee conservation parks in the region. The patches were mapped from photo mosaics dating from 1978 and 1983.

survey typically restricted in scope by the resources of time, personnel and funds, and shows how that information can be used to compare the representativeness of habitat patches in that region.

THE STUDY AREA

The region is approximately 15,500 km² on the western side of the Eyre Peninsula in South Australia, from Poochera south to near Port Kenny, west to longitude 133° 20′ just west of Ceduna, and north to an arbitrary boundary just outside the margin of the agricultural lands (Fig. 1). Lithologically, the area consists of successive deposits of calcareous sand in various stages of calcrete formation, overlying granite bedrock which outcrops occasionally to form inselbergs, notably between Port Kenny and Calca in the south of the region. The inselbergs have not been included in this study. Along the coast, recent calcareous dunes occur, and, towards the inland boundary of the study area there are dunes of mixed calcareous sand and siliceous sand originating further inland, which overlie the calcrete deposits. The coastal dunes are in the order of 80% calcium carbonate, whereas the inland dunes are about 30-40% calcium carbonate (K. Wetherby pers. comm.).

In general, the area is cleared and the dominant land use is rotation cereal cropping and wool sheep grazing (Laut *et al.* 1977). Where the oldest, and therefore hardest, carbonate layer outcrops, mainly in the south, there has been less clearing, though the mallee scrubs and woodlands are still used for grazing. Elsewhere, remnant patches of mallee remain; as woodlots for firewood and general farm timber or as backblocks undeveloped through lack of capital or the resources to manage them efficiently. Altogether, there are 101 patches in the study area, ranging in size from about 35 ha to over 10,000 ha. Three are conservation parks; Wittelbee Conservation Park, Laura Bay Conservation Park and Calpatanna Waterhole Conservation Park, designated C39, C130 and C140 respectively.

The northern half of the study area is characterized by Crocker's (1946) *Eucalyptus oleosa — E. gracilis — E. dumosa* edaphic complex, and the southern half by his *E. diversifolia — Allocasuarina verticillata — Melaleuca lanceolata* edaphic complex (Lange and Lang 1985).

Like many parts of Australia the region is remote from large population centres and has received little attention from biologists or ecologists. Yet vegetation clearance is proceeding apace, so that habitat patches which remain are under threat (Anon. 1976). In addition, development of lands on the edges of the siliceous dune fields to the north, which are marginal for agriculture has increased in recent years and the technology has been developed to prepare the areas of hard calcrete in the south for cropping. There is an urgent need for a detailed survey and inventory of the biological resources of the region if proper planning and management of land use change is to occur. Unfortunately, there are too few biologists, too little time and not enough commitment from the responsible authorities for such a survey. Almost certainly, there will never be a detailed inventory. Thus, a rapid survey utilizing available resources is required, the results of which can be extrapolated throughout the region with a known level of confidence. The survey described below was of vegetation only, assuming a correlation with other biota, largely for logistic reasons. The survey is the first of two steps necessary for a conservation evaluation. The second involves a selection procedure based on comparative conservation value.

THE SURVEY AND DATA ANALYSIS

Survey Methods

A preliminary reconnaissance indicated the range of floristic variation across the region and that plant distributions were especially influenced by the degree of development of the calcium carbonate layer from soft to hard (Wetherby and Oades 1975), proximity to the coast, and the presence of the more siliceous inland dunes. Thus, areas where different stages of calcrete formation occurred close together were singled out because a number of sites likely to contain different flora could be sampled together, increasing efficiency. Because a distance from the coast factor includes effects of both coastal proximity and the occurrence of the inland dunes, the remaining sites were chosen to provide a relatively even (though constrained by access) geographic coverage of the area to include the entire range of distances from the coast. This meant that two sample strategies were combined. One selected steep local gradients of environmental change for intensive sampling (Gillison and Brewer 1985) and the other approximated a stratified random sample. For various logistic reasons, the survey was limited in time to three weeks. In this time it was planned to collect data from at least 100 sites.

The samples consisted of 0.1 ha plots on which the presence of every vascular plant species was recorded. The plots were located so as to sample only one landform element i.e. on a dune flank, a dune swale, a stony rise, etc. Altogether, 106 plots were recorded. One plot was on a granite inselberg and one on a coastal cliff to better define the floristic limits of the study area. These two were discarded for the purpose of classifying the mallee communities, leaving 104. Only 21 of the 101 mapped mallee patches (Fig. 1), were sampled, some of the larger ones more than once, while some of the 104 plots came from continuous habitat on the inland boundary of the study area and a few from roadside vegetation.

Vegetation Classification

The plot data were classified by an hierarchical agglomerative numerical method. Association between plots was measured with the Jaccard coefficient (Jaccard 1908; Clifford and Stephenson 1975) and the sorting strategy was an algorithm called homogeneity clustering (Belbin *et al.* 1984). The classification was stopped at the 6 group level for two reasons. Below this level small changes in the coefficient of association produced new groups, and the six resulting communities could be placed easily into the broader edaphic complexes of Crocker (1946).

Statistical Models

The probabilities of the six communities from the vegetation classification occurring in each remnant patch were estimated with statistical models. The models used three environmental variables, called factors in the models as predictors of the plant communities: calcium carbonate layer, distance from the coast and latitude. Carbonate layer was divided into three levels corresponding to Wetherby and Oades' (1975) classes 2, 3 and 4. Their class 1 does not occur on the Eyre Peninsula. These classes more or less represent a sheet calcrete carbonate layer, a nodular carbonate layer, and a soft layer just forming in dunes, respectively. Distance from the coast was divided into five levels; 0-8 km, 8.1-15 km, 15.1-30 km, 30.1-50 km and more than 50 km. Latitude was divided into two levels, north and south. The boundary between the two corresponds roughly to a geological change from a formation characterized by carbonate nodules and lumps, to one characterized by platy carbonate accretions and calcrete outcrops. These are related in broad terms to the levels in the carbonate layer factor, but any carbonate layer can occur in either geological formation.

Logistic regression analyses, in which the error structure is assumed to be binomial and the data are logit transformed, were carried out using the computer package GLIM (Baker and Nelder 1978). The statistical models are mathematical statements of the relationships between the plant communities and the three concomitant environmental factors. The fitted model provides a summary of this relationship and can also be used to predict the relationship in unsampled areas. The logistic regression model is a particular form of generalized linear models, and is used when the response is binary (Dobson 1983; McCullagh and Nelder 1983). In ecology, logistic regression analysis has been used successfully to determine the distributions of plant species in relation to major environmental variables (Austin *et al.* 1984).

The procedure was to identify the single factor accounting for the greatest change in deviance, then to test the other two, one at a time, in combination with the first. If neither were significant, the procedure stopped. If one was significant, it was added to the model and the third factor was tested again in combination with the other two. If it was significant it was added to the model. If both were significant at the second step, the factor with the most significant change in deviance was added as the second factor, and the third was tested in a combined model as above.

RESULTS

The vegetation classification is summarized in Table 1. The first four communities fall into Crocker's (1946) *E. oleosa* — *E. gracilis* — *E. dumosa* edaphic complex and the sixth into his *E. diversifolia* — *Allocasuarina verticillata* — *Melaleuca lanceolata* edaphic complex. The fifth is characteristic of the more siliceous inland dunes, but is probably still a marginal community of the *E. oleosa* — *E. gracilis* — *E. dumosa* complex. Figure 2 shows histograms of the proportion of plots from each community on each carbonate layer and Figure 3 shows their frequency in different distance from the coast classes.

Table 2 gives the results of fitting each factor to each community separately, in the absence of the other factors. Using these results, the multi-factor models summarized in Table 3 were built up.

Table 1. Results of an agglomerative hierarchical classification of the vegetation plots using the Jaccard association measure and the homogeneity clustering sorting strategy. Nomenclature follows Jessop (1983).

Community	Number of plots	Common species	Frequency
1	18	Eucalyptus dumosa	0.89
		Westringia rigida	0.89
		Melaleuca acuminata	0.67
		Eucalyptus oleosa	0.61
		Olearia muelleri	0.61
2	17	Stipa eremophila	0.88
		Enchylaena tomentosa	0.71
		Melaleuca lanceolata	0.65
		Threlkeldia diffusa	0.59
		Anagallis arvensis	0.59
3	33	Melaleuca lanceolata	0.76
		Eucalyptus oleosa	0.73
		Rhagodia crassifolia	0.61
		Zygophyllum aurantiacum	0.58
		Sclerolaena diacantha	0.55
4	11	Maireana trichoptera	0.82
		Eucalyptus oleosa	0.73
		Stipa setacea	0.73
		Zygophyllum ovatum	0.73
5	9	Eucalyptus socialis	1.00
		Melaleuca adnata	0.78
		Triodia irritans	0.56
		Rhagodia preissii	0.56
6	16	Acrotriche patula	0.81
		Gahnia lanigera	0.75
		Eutaxia microphylla	0.69
		Eucalyptus diversifolia	0.69

For all communities except community two, two or more factors were included in the final model. Note that for community four, carbonate layer is not a significant factor alone, whereas distance is significant alone (Table 2). However, in the multi-factor model, carbonate layer replaces distance, which is not significant in combination with latitude and carbonate layer. This proved to be unrealistic when estimating probabilities from the model coefficients, as community 4 does not occur in distance classes 1

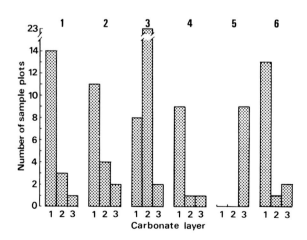

Fig. 2. The number of plots from each of the six plant communities that were recorded on each of the three carbonate layers. The community numbers are along the top.

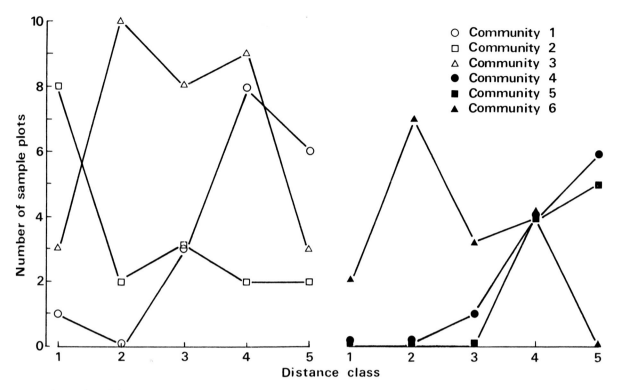

Fig. 3. The number of plots from each community in distance from the coast classes. The classes are, from 1 to 5, <8 km, 8.1-15 km, 15.1-30 km, 30.1-50 km and >50 km.

Table 2. Single factor models of the relationships between each of the three factors and the plant communities, from logistic regression analysis. Statistical significance is indicated by the following: n.s. not significant; * $0.05 \geq p > 0.01$; ** $0.01 \geq p > 0.001$; *** $p \leq 0.001$. d.f. stands for degrees of freedom and △ deviance for change in deviance.

Model	Deviance	d.f.	△ deviance	d.f.	Significance
Community 1					
Null	46.96	22			
Carbonate	41.05	20	5.91	2	n.s.
Distance	35.49	18	11.47	4	*
Latitude	30.67	21	16.29	1	***
Community 2					
Null	34.80	22			
Carbonate	33.64	20	1.16	2	n.s.
Distance	18.75	18	16.05	4	**
Latitude	33.69	21	1.11	1	n.s.
Community 3					
Null	63.64	22			
Carbonate	29.64	20	34.00	2	***
Distance	53.12	18	10.52	4	*
Latitude	48.74	21	14.90	1	***
Community 4					
Null	44.90	22			
Carbonate	36.21	20	4.69	2	n.s.
Distance	27.87	18	13.03	4	*
Latitude	28.26	21	12.64	1	***
Community 5					
Null	48.20	22			
Carbonate	10.46	20	37.71	2	***
Distance	34.10	18	14.10	4	**
Latitude	38.00	21	10.20	1	***
Community 6					
Null	50.74	22			
Carbonate	42.81	20	7.93	2	*
Distance	38.07	18	12.67	4	*
Latitude	33.31	21	17.43	1	***

Table 3. Multi-factor models of the relationships between the three factors, carbonate layer, distance from the coast and latitude, in combination, and the plant communities from logistic regression analysis. Statistical significance is indicated by the following: n.s. not significant; * $0.05 \geq p > 0.01$; ** $0.01 \geq p > 0.001$; *** $p \leq 0.001$. d.f. stands for degrees of freedom and Δ deviance for change in deviance.

Model	Deviance	d.f.	Δ deviance	d.f.	Significance
Community 1					
Null	46.96	22			
+ Latitude	30.67	21	16.29	1	***
+ Distance	17.54	17	13.13	4	*
−Distance + Carbonate	30.54	19	0.13	2	n.s.
Latitude + Distance	17.54	17			
+ Carbonate	17.34	15	0.20	2	n.s.
Final Model: Latitude and Distance					
Community 2					
Null	34.80	22			
+ Distance	18.75	18	16.05	4	**
+ Carbonate	14.11	16	4.64	2	n.s.
−Carbonate + Latitude	15.67	17	3.08	1	n.s.
Final Model: Distance					
Community 3					
Null	63.64	22			
+ Carbonate	29.64	20	34.00	2	***
+ Latitude	21.79	19	7.85	1	**
−Latitude + Distance	25.93	16	3.71	4	n.s.
Carbonate + Latitude	21.79	19			
+ Distance	16.68	15	5.11	4	n.s.
Final Model: Carbonate + Latitude					
Community 4					
Null	40.90	22			
+ Latitude	28.26	21	12.64	1	***
+ Distance	14.51	17	13.75	4	**
−Distance + Carbonate	12.93	19	15.33	2	***
Latitude + Carbonate	12.92	19			
+ Distance	5.01	15	7.92	4	n.s.
Final Model: Latitude + Carbonate					
Community 5					
Null	48.20	22			
+ Carbonate	10.46	20	37.71	2	***
+ Distance	0.0019	16	10.46	4	*
−Distance + Latitude	8.887	19	1.57	1	n.s.
Carbonate + Distance	0.0988	16			
+ Latitude	0.0019	15	0.0969	1	n.s.
Final Model: Carbonate + Distance					
Community 6					
Null	50.74	22			
+ Latitude	33.31	21	17.43	1	***
+ Distance	20.28	17	13.03	4	*
−Distance + Carbonate	29.86	19	3.45	2	n.s.
Latitude + Distance	20.28	17			
+ Carbonate	14.08	15	6.2	2	*
Final Model: Latitude + Distance + Carbonate					

or 2 at all, but because distance was not in the model, probabilities as high as 0.475 were being predicted in these distance classes. Therefore distance, a statistically non-significant factor, was used to estimate the probabilities of occurrence of community 4 because inspection of the data showed it to be an important factor. Its lack of statistical significance in the regression model is probably the result of an uneven distribution of this community over the environmental space and the small number of occurrences recorded during the survey (Table 1). In addition, for the estimation of model coefficients, latitude 1, where community 4 does not occur at all, was set to 0.

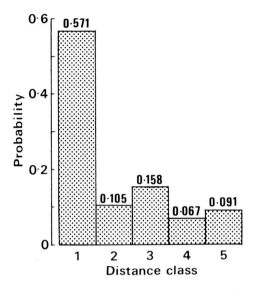

Fig. 4. Histogram of the probabilities of community 2 in distance from the coast classes, derived from the model in Table 3.

The coefficients from each model were back-transformed to probabilities. Figure 4 is a histogram of the probabilities of community 2 in different distance classes because distance was the only significant factor. Table 4 gives the probability estimates for each cell created by combinations of the factors used in the models for the other five communities.

Expressing the model for community 1 in the more usual regression equation form:

$$y = a + lat(i) + dist(j) \qquad (1)$$

where y = logit (probability of community 1 occurring)

a = -1.368
$lat(1)$ = 0
$lat(2)$ = -2.588
$dist(1)$ = 0
$dist(2)$ = -8.994
$dist(3)$ = 0.644
$dist(4)$ = 1.341
$dist(5)$ = 1.238.

Table 4. Estimated probabilities for each plant community (except community 2; see Fig. 4) in relation to the categories of the environmental factors used in the final models from Table 3 (though note the exception of community 4 discussed in the text). Probabilities of less .001 are recorded as 0.

Community 1

			Distance		
	1	2	3	4	5
Latitude 1	.203	0	.326	.492	.583
Latitude 2	.019	0	.035	.067	.095

Community 3

	Carbonate		
	1	2	3
Latitude 1	.070	.410	.025
Latitude 2	.289	.790	.124

Community 4

			Distance		
	1	2	3	4	5
Carbonate 1	0	0	.272	.515	.653
Carbonate 2	0	0	.054	.141	.240
Carbonate 3	0	0	.021	.058	.106

Community 5

			Distance		
	1	2	3	4	5
Carbonate 1	0	0	0	0	0
Carbonate 2	0	0	0	0	0
Carbonate 3	0	0	0	.667	.833

Community 6
Latitude 1

			Distance		
	1	2	3	4	5
Carbonate 1	.437	.764	.422	.272	0
Carbonate 2	.089	.291	.085	.045	0
Carbonate 3	.715	.913	.703	.547	0

Latitude 2

			Distance		
	1	2	3	4	5
Carbonate 1	.053	.189	.085	.026	0
Carbonate 2	.007	.029	.007	.003	0
Carbonate 3	.153	.431	.146	.080	0

Thus, from equation (1) the estimate for latitude (1), distance (1) is,

$$-1.368 + 0 + 0 = -1.368,$$

and for latitude (2), distance (4) is,

$$-1.368 - 2.588 + 1.341 = -2.635.$$

Probabilities are calculated from these estimates by,

$$\frac{e^{(\text{estimate})}}{e^{(\text{estimate})} + 1} \quad (2)$$

because logit $(p) = \log_e(p/_{1-p})$. Thus, for latitude (1), distance (1) the probability is,

$$\frac{e^{-1.368}}{e^{-1.368} + 1} = 0.203$$

For latitude (2), distance (4) the probability is,

$$\frac{e^{-2.635}}{e^{-2.635} + 1} = 0.067$$

Using these estimates, it is possible to assign to each patch of remnant vegetation a probability of occurrence of each community. Distance from the coast and latitude for each patch are easily measured from a map. Carbonate layer is not so straightforward. The three carbonate layers represent different stages in a hardening process that finishes with the calcrete of layer 1. The layers reflect successive calcareous sand deposits, and degree of hardening can vary considerably within each layer leading to gradational changes. The northern half of the study area is dominated by the nodular carbonates of layer 2, but the harder layer 1, which underlies layer 2, outcrops regularly. Thus, in latitude 2, carbonate layers 1 and 2 were assumed to occur on all patches unless direct field observation showed otherwise, and carbonate layer 3 occurs wherever there are dunes. The southern half of the study area is dominated by the hard calcrete outcrops of layer 1 and there is little evidence of softer, more recent deposits. Thus, the nodular carbonates of layer 2 were assumed, conservatively, to occur only in the largest patches or where field observation showed otherwise. Again, carbonate layer 3 occurs wherever there are dunes. Table 5 shows the estimated probabilities of each community occurring on each remnant patch. Note that in this table, estimates of less than one in a thousand are recorded as 0. This table is the end point of the survey and data analysis stage. It represents a uniform set of information at a common scale for each patch. Using this information it is now possible to proceed and compare the patches in terms of their conservation value.

CONSERVATION EVALUATION

The conservation value of a habitat patch has been defined as the contribution that patch makes to representing the biota of the region. In this example, plant communities represent the biota, so the conservation value of any given patch depends on the plant communities present.

Table 5. Probabilities of each plant community occurring in each patch as determined from Table 4 and Figure 4.

Patch	1	2	3	4	5	6
1	.019	.571	.79	0	0	.053
2	.019	.571	.79	0	0	.053
3	.019	.571	.79	0	0	.007
4	0	.105	.79	0	0	.189
5	.019	.571	.79	0	0	.053
6	.019	.571	.79	0	0	.053
7	0	.105	.79	0	0	.189
8	.035	.158	.79	.272	0	.007
9	.035	.158	.79	.272	0	.007
10	.035	.158	.79	.272	0	.050
11	0	.105	.79	0	0	.187
12	.019	.571	.79	0	0	.053
13	.035	.158	.79	.272	0	.007
14	.035	.158	.79	.272	0	.050
15	.068	.067	.289	.515	.667	.026
16	.035	.158	.79	.272	0	.007
17	0	.105	.79	0	0	.007
18	.035	.158	.79	.272	0	.007
19	.035	.158	.79	.272	0	.007
20	.035	.158	.79	.272	0	.007
21	0	.105	.79	0	0	.029
22	.019	.571	.79	0	0	.053
23	.019	.571	.79	0	0	.053
24	.019	.571	.79	0	0	.053
25	.019	.571	.79	0	0	.053
26	.019	.571	.79	0	0	.153
27	.019	.571	.79	0	0	.053
28	.019	.571	.79	0	0	.153
29	.019	.571	.79	0	0	.053
30	.019	.571	.79	0	0	.053
31	.019	.571	.79	0	0	.053
32	0	.105	.79	0	0	.189
33	0	.105	.79	0	0	.187
34	0	.105	.79	0	0	.189
35	.203	.158	.79	0	0	.189
36	.035	.158	.79	.054	0	.050
37	0	.105	.79	0	0	.189
38	0	.105	.79	0	0	.187
39	0	.105	.79	0	0	.189
40	.035	.158	.79	.054	0	.050
41	0	.105	.79	0	0	.187
42	.019	.571	.79	0	0	.189
43	.019	.571	.79	0	0	.189
44	.109	.571	.79	0	0	.053
45	.019	.571	.79	0	0	.053
46	.019	.571	.79	0	0	.053
47	.035	.158	.79	.054	0	.050
48	0	.105	.79	0	0	.189
49	.019	.571	.79	0	0	.053
50	.019	.571	.79	0	0	.053
51	0	.105	.79	0	0	.189
52	0	.105	.79	0	0	.187
53	.035	.158	.79	.054	0	.050
54	.035	.158	.79	.054	0	.189
55	.035	.158	.79	.054	0	.050
56	.068	.067	.79	.515	.667	.080
57	.068	.067	.79	.515	.667	.080
58	.068	.067	.79	.515	0	.026
59	.068	.067	.79	.515	0	.026
60	.035	.158	.79	.054	0	.050
61	.035	.158	.79	.054	0	.050
62	.068	.067	.79	.515	0	.026
63	.068	.067	.79	.515	0	.026
64	.068	.067	.79	.515	0	.026
65	.068	.067	.79	.515	0	.026
66	.068	.067	.79	.515	0	.026
67	.068	.067	.79	.515	0	.026
68	.068	.067	.79	.515	0	.026
69	.035	.158	.79	.272	0	.050
70	.068	.067	.79	.515	0	.026

Table 5. Continued.

Patch	Community					
	1	2	3	4	5	6
71	.068	.067	.79	.515	0	.026
72	.068	.067	.79	.515	0	.026
73	.068	.067	.79	.515	0	.026
74	.493	.067	.79	0	0	.026
75	.068	.067	.79	.515	0	.026
76	.095	.091	.79	.653	0	0
77	.095	.091	.79	.653	0	0
78	.583	.091	.41	0	0	0
79	.583	.091	.41	0	0	0
80	.203	.571	.07	0	0	.437
81	.203	.571	.07	0	0	.437
82	.203	.571	.025	0	0	.715
83	.203	.571	.025	0	0	.715
84	.203	.571	.07	0	0	.764
85	.203	.572	.07	0	0	.437
86	.203	.571	.07	0	0	.437
87	.203	.571	.07	0	0	.715
88	.203	.571	.07	0	0	.437
89	.203	.571	.07	0	0	.437
90	.203	.105	.07	0	0	.715
91	.326	.158	.07	0	0	.422
92	.493	.068	.410	0	0	.272
93	.493	.158	.41	0	0	.764
94	.493	.068	.07	0	0	.272
95	.583	.158	.41	0	0	.764
96	.326	.158	.07	0	0	.272
97	.326	.158	.07	0	0	.272
98	.203	.571	.025	0	0	.437
C140	.203	.571	.025	0	0	.913
C131	.019	.571	.79	0	0	.153
C39	.019	.571	.79	0	0	.153

It was accepted initially that a community would be said to be present in a group of patches if there was a 95% chance of finding it. The number of patches necessary to have a 95% chance of finding at least one representative of any community can be calculated by:

$$n = \frac{\ln(.05)}{\ln(1-p)} \quad (3)$$

where p is the probability of success. Thus, for a probability of success of 0.571 (as for community 2 in patches 1, 2 and 3 of Table 5), the number of patches required to have a 95% chance of the community being present is:

$$n = \frac{\ln(.05)}{\ln(1-.571)}$$

$$= \frac{-3.0}{-0.8463} = 3.545 = 4.$$

Four patches are necessary to be confident the community is represented at least once if its probability of occurrence in an individual patch is 0.571.

An heuristic search for the minimum set of patches which, by the 95% criterion has every community at least once, was made in the way described below and set out in Table 6.

For each community, the patch was found with the highest probability of success, also listing the probabilities of the other communities in those patches. Thus, initially, a maximum of six patches was selected. The next step was to calculate the probability of failure, that is, of not finding the community in each of the selected patches, and to find their product for each community. Then the procedure was repeated with the first set of patches excluded, i.e. the second best patch was found for each community. The probabilities of failure were continued to be multiplied within each community, and the third best, fourth best patches were continued to be found, and so on, until the product of the probability of failure was less than 0.05 for all communities. The result was the minimum set of patches with a 95% probability of including each community at least once.

For every community except community six there was more than one patch with equal highest probability. One approach to deciding which one to

Table 6. The procedure for selecting the minimum set of patches necessary to represent each community at least once. When patches are otherwise equal, the largest patch is chosen. All predicted probabilities from Table 4 of less than 10% have been set to 0. p is the probability of finding the community and q is the probability of not finding it.

Patch	Community											
	1		2		3		4		5		6	
	p	q	p	q	p	q	p	q	p	q	p	q
95	.583	.417	.158	.852	.410	.590	0	1	0	1	.764	.236
C140	.203	.797	.571	.429	0	1	0	1	0	1	.913	.087
1	0	1	.571	.429	.790	.210	0	1	0	1	0	1
76	0	1	0	1	.790	.210	.653	.347	0	1	0	1
15	0	1	0	1	.289	.711	.515	.485	.667	.333	0	1
product (q)	.332		.152		.019		.168		.333		.021	
79	.583	.417	0	1	.410	.590	0	1	0	1	0	1
87	.203	.797	.571	.429	0	1	0	1	0	1	.715	.285
77	0	1	0	1	.790	.210	.653	.347	0	1	0	1
57	0	1	0	1	.790	.210	.515	.485	.667	.333	0	1
Cumulative product (q)	.110		.065		0		.028		.111		.006	
78	.583	.417	0	1	.410	.590	0	1	0	1	0	1
22	0	1	.571	.429	.790	.210	0	1	0	1	0	1
56	0	1	0	1	.790	.210	.515	.485	.667	.333	0	1
Cumulative product (q)	.046		.028		0		.014		.037		.006	

choose was to take the first one encountered. However, the result would then be order dependent and since there are a number of possible combinations of patches which could all have the minimum number required to have each community at least once, there has to be a way of choosing between them. In Table 6, the procedure was to choose the largest patch with the highest probability for each community and then, if the probabilities were still equal, the second largest patch and so on. One further decision was necessary. As there is no single agreed criterion for testing the adequacy of the statistical models used to estimate the probabilities, and the standard errors associated with many of the estimates were so large that confidence in the estimate could not be high, small probabilities are likely to be especially unreliable. Therefore, all probabilities in Table 5 less than 10% were set to 0 for the purpose of patch selection.

Looking then at Table 6, patch 95 is the largest patch with the highest probability of community 1 (p), 0.583. The probability of not finding community 1 (q) therefore is 0.417. Calpatanna Waterholes Conservation Park, C140, is the largest patch with the highest probability of community 2, 0.571. The probability of not finding community 2 in patch C140 is 0.429. Patch 1 is the largest of many patches with a probability of 0.790 for the ubiquitous community 3. The probability of not finding community 3 in patch 1 is 0.210. Patch 76 is the largest with the highest probability of community 4. The probability of not finding community 4 in patch 76 is 0.347. Patch 15 is the largest with the highest probability of community 5, 0.667. The probability of not finding community 5 in patch 15 therefore is 0.333. Patch C140, already selected for community 2, is the largest patch with the highest probability of community 6, 0.913. The probability of not finding community 6 in patch C140 is a remote 0.087. Thus, only five patches were selected at the first step. The products of the probabilities of not finding each community in the first five patches then are 0.332, 0.152, 0.019, 0.168, 0.333 and 0.021 respectively. For communities 3 and 6 this figure already is below the criterion of 0.05, but more patches are needed for communities 1, 2, 4 and 5. Thus the process was continued until the product of the failures for all communities fell below 0.05.

The result (Table 6 and Fig. 5) was that 12 patches were required to ensure, with 95% confidence and choosing the largest patch where there was a choice, that each community is represented at least once. Table 7 shows the five extra patches that would be necessary to be 99% confident that each community is represented at least once. Note that there is no change for community 5. Only three of the mapped patches have a prediction for community 5, 0.667 in each case. At that probability level all three patches are required for a 95% chance or better of the community being present (Equation 3). In terms of planning a conservation reserve network in this region, it might be logical to ignore these three patches and set aside a substantial area of the still uncleared adjacent dune fields, as community 5 is characteristic of the more siliceous inland dunes. In fact, it is possible that Yumbarra Conservation Park, on the northwestern edge of the study area, already acts as a sufficient reserve for community 5. More survey work beyond the boundaries of the study area would be necessary before any relevant decisions could be made because changes in species complements are likely to be substantial in that area, a reflection of the rapid reduction in rainfall with distance inland. For the purpose of this study, only those three patches have any likelihood of containing community 5 so there are no more patches that could be selected to improve the chances of representing it.

Adequate Representation

Minimum viable population sizes may suggest an answer to the question 'what is adequate representation?' However, minimum population sizes can be estimated accurately only for a few species throughout the world which have been studied in sufficient detail, for example mice (Berry 1964) and grizzly bears (Shaffer 1981). There are no Australian examples. In addition, the kinds of survey data available for decisions on reserve selection are often inadequate to map species distributions let alone abundances. Further, Boecklen and Simberloff (in press), in reviewing evidence for minimum viable population sizes, conclude that whilst there is a (usually unknown) lower limit, other factors such as a species' biology or its habitat requirements are as likely to cause extinction if reserves are not selected with such requirements in mind. Since there is usually as little knowledge of biological or habitat requirements as there is of population sizes, any definition of adequate representation will be arbitrary and unreliable until an enormous amount of research on the autecology of species is completed.

However, adequate representation is unlikely to be achieved simply by ensuring that each community is represented at least once. Natural catastrophes such as fire, or political pressure for land use change, could mean that the only occurrence of some community was lost from the reserve system if each were represented only once. The question remains: is adequate representation twice, five times, ten times or more? Whilst it is arbitrary, that decision is constrained by the number of patches available and the number of objects to be represented: in this case 101 patches (including the three existing conservation parks) and six communities. Beginning with the minimum set necessary for single representation (Table 6) it is possible to explore options for increasing representation given various constraints.

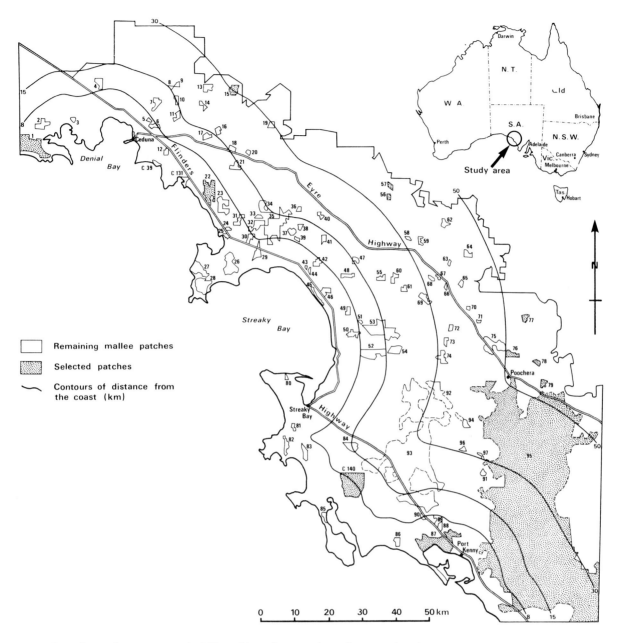

Fig. 5. The mallee patches necessary to be 95% confident of representing each community at least once.

Table 7. The extra patches which would have to be added to Table 6 to be 99% confident of representing all communities (except 5) at least once. The probability of failure has to drop below 0.01. p is the probability of finding the community and q is the probability of not finding it.

Patch	Community											
	1		2		3		4		5		6	
	p	q	p	q	p	q	p	q	p	q	p	q
35	.203	.797	.158	.852	.790	.210	0	1	0	1	.189	.811
93	.493	.507	.158	.852	.410	.590	0	1	0	1	.764	.236
75	0	1	0	1	.790	.210	.515	.485	0	1	0	1
92	.493	.507	0	1	.410	.590	0	1	0	1	.272	.785
26	0	1	.571	.429	.790	.210	0	1	0	1	.153	.847
product (q)	.205		.311		0		.485		1		.127	
product (q) from Table 6	.046		.028		0		.014		.037		.006	
Final product (q)	.009		.009		0		.007		.037		.001	

Table 8. Minimum set of patches necessary to have a 95% chance of representing each community at least once, including the three mainland reserves already in the region at the very beginning of the procedure shown in Table 6.

Patches in order of admission
C140
C131
C39
95
1
76
15
79
78
77
57
93
56

Table 6 includes Calpatanna Waterholes Conservation Park, C140, which has the highest probability for community 6. Table 8 lists the minimum set of patches if the other two mainland conservation parks in the region, Wittelbee, C39, and Laura Bay, C131, are included before any other patches are added. In other words, the extra patches, on top of the existing reserves, necessary for the representation of every community at least once. There are some differences between Tables 6 and 8. In 8, there is an extra patch, patch 93, which is the largest with the second highest probability of community 1. It is not needed in Table 6 because community 1 has a high enough probability in patch 87 to give it a low enough cumulative probability of failure. In addition, two large patches representing community 2, patches 22 and 87, have been replaced by the two much smaller conservation parks.

Table 9 lists the additional patches required for representation twice, four times, five times and six times. Three notable features emerged from the analysis to produce Table 9. First, neither of the two small Conservation Parks in Table 8 are included, even for representation six times. There are many more large patches capable of representing the same communities as those two parks represent. Second, there are not enough patches in the whole region with a high probability of community 1 to be 95% confident of representing community 1 at least five times. If adequate representation is defined as representation at least five times, then it is no longer possible to represent community 1 adequately. Third, to represent communities 2, 3, 4 and 6 at least five times, community 1 four times and community 5 once (see above), it would be necessary to preserve 45 patches. For representation of communities 2, 3, 4 and 6 six times, 48 patches are required, that is almost half of all the patches remaining in the study area, yet representation five times might be considered by many to be the absolute minimum necessary to secure conservation.

CONCLUSIONS

The assessment of conservation value has at least two aspects; a definition of conservation value, and a way of comparing areas of land and their associated plants and animals in terms of that definition. Conservation value was defined here as the contribution a given area of land makes to representing the range of biological variation in a region, and a procedure for comparing the representativeness of remnant habitat patches was described.

The results depend critically on four features of the procedure. In the first place there is the initial collection of data and the choice of attributes. The more comprehensive the survey, in terms of both sample size and attributes measured, the more reliable will be the results. More data will always be an improvement. However, many land use planning decisions, of which conservation decisions are a subset, are made on very few field data. There must be a minimum acceptable amount of data, but given the inevitability of continued changes in land use, it is desirable to use methods which produce a result from limited data. The use of plants alone, the recording of presence rather than abundance, the relatively small sample size, and the use of logistic regression models to make predictions to unsampled areas are all limitations on the reliability of results, but are all the results of realistic limitations imposed on the data gathering stage.

The second aspect is the classification. The choice of attributes, the measure of association and the sorting strategy are all decisions which affect the

Table 9. Minimum sets of patches required to represent each community twice, four times, five times and six times, with 95% confidence. Again, all predicted probabilities from Table 5 of 10% or less have been set to 0. As the number of representatives required increases the probability of failure for community 5 increases rapidly to 1.0 as there are only 3 patches with community 5. Note that the probabilities of failure for the other communities (except 3) also increase towards the end, and that the probability of failure for community 1 exceeds the 0.05 threshold for representation more than four times.

Number of times	Patches	Probability of failure Communities					
		1	2	3	4	5	6
2	As for 1 representative + 35, 75, 92, 93, 26, 74 and 28	.042	.036	0	.033	.259	.013
4	As for 2 + 54, 84, 62, 90, 94, 64, 83, 91, 27, 59, 96, 23, 73, 97, 63, 98, 88, 82, 86	.047	.003	0	.042	1.000	.001
5	As for 4 + 65, 89, 68, 85, 72, 81, 80	.074	.001	0	.034	1.000	.001
6	As for 5 + 2, 58, 67	.172	.003	0	.042	1.000	.003

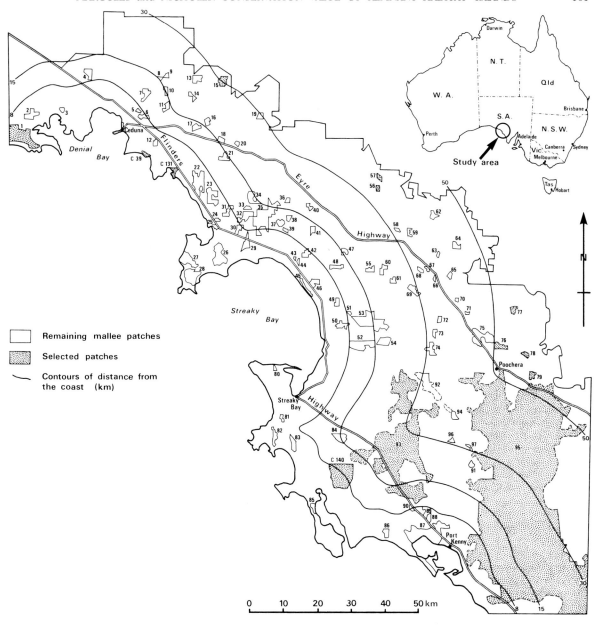

Fig. 6. The patches necessary to represent each community at least once (with 95% confidence) if the three existing reserves in the study area, C39, C131 and C140, are included initially.

outcome (Austin and Margules 1984). The decision to stop the classification at six groups, whilst based on a knowledge of local plant ecology and the structure of the classificatory dendrogram, is very much a subjective one. By recognizing fewer communities, the chances of representing each one a given number of times in a minimum set of patches would be enhanced. On the other hand, fewer communities might need to be represented more often because they are more heterogeneous entities.

Thirdly, there are the logistic regression models. Like all regressions, they attempt to minimize some measure of the deviance from the mean over the whole set of observations, so local, possibly large, but infrequent departures from the mean might get hidden. Their reliability also suffers from the small data set and an uneven distribution of observations over the independent variables. Specific to logistic regression analysis is the lack of a widely accepted adequate procedure for testing the models. However, they do provide a means of predicting to unsampled areas and thereby generating a uniform set of information about all the patches; a necessary prerequisite to making comparisons. Implicit in this step is the need to establish some ground verification of the predictions upon which the selection is based. Follow up work to test the selection procedure will involve two parts. The first will be to check the predicted probabilities of community occurrence in a subset of patches. Given that the predictions prove reliable, the second part of the verification will be to check that, at the stated level of probability, all communities are represented in the minimum set of patches.

Finally, there is the selection procedure itself. The patches in Tables 6, 7 and 9 were chosen with the constraint that if probabilities were equal, the largest patch would be chosen each time. Table 8 includes an additional constraint; that the three existing reserves be included in the minimum set. In a similar way it would be possible to incorporate other constraints. For example, patches known to contain rare species or patches with atypical species richness could be included initially in the same way. Alternatively, patches known to be severely degraded might be left out of the procedure, as might other patches for reasons of policy such as commitments to land use change.

An alternative, or even additional, constraint to patch size in Tables 6, 7, 8 and 9, might be distance to the next nearest patch, with the objective of minimizing inter-reserve distances.

Previous conservation evaluation procedures have been wholly subjective and not repeatable, or based on hypothetical immigration and extinction rates which ignore species interactions and environmental variability. In spite of the sorts of problems and possible refinements listed above, the evaluation procedure suggested here has resulted in a partly objective, wholly explicit assessment of the relative conservation values of the remnant mallee patches in the study area.

ACKNOWLEDGEMENTS

Darryl Kraehenbuehl identified most of the plant species, except the grasses, which were identified by Mike Lazarides. Alex Gibson and Sue Ward helped collect the data and Elaine Cork helped analyse them. Ken Wetherby advised on the carbonate layers in the soil. Norm McKenzie and Colin Ogle provided valuable criticism of the manuscript.

REFERENCES

Anon, 1976. Vegetation Clearance in South Australia. Department for the Environment, Adelaide.

Austin, M. P., Cunningham, R. B. and Fleming, P. M., 1984. New approaches to direct gradient analysis using environmental scalars and statistical curve-fitting procedures. *Vegetatio* **55**: 11-27.

Austin, M. P. and Margules, C. R., 1984. The concept of representativeness in conservation evaluation with paticular reference to Australia. CSIRO Aust. Div. Water Land Resour. Tech. Memo. 84/11.

Baker, R. J. and Nelder, J. A., 1978. The GLIM System: Release 3. Royal Statistical Society, Oxford.

Belbin, L., Faith, D. P. and Minchin, P. R., 1984. Some algorithms contained in the numerical taxonomy package NTP. CSIRO Aust. Div. Water Land Resour. Tech. Memo. 84/23.

Berry, R. J., 1964. The evolution of an island population of the house mouse. *Evolution* **18**: 468-83.

Boecklen, W. J. and Simberloff, D. (in press). Area-based extinction models in conservation. *In* Dynamics of Extinction ed by D. Elliot. Wiley, New York.

Clifford, H. T. and Stephenson, W., 1975. An Introduction to Numerical Classification. Academic Press, New York.

Crocker, R. L., 1946. An introduction to the soils and vegetation of Eyre Peninsula, South Australia. *Trans. R. Soc. S. Aust.* **70**: 83-107.

Dobson, A. J., 1983. An Introduction to Statistical Modelling. Chapman and Hall, London.

Gillison, A. N. and Brewer, K. R. W., 1985. The use of gradient directed transects or gradsects in natural resource survey. *J. Environ. Manage.* **20**: 103-27.

Jaccard, P., 1908. Nouvelles réchercises sur la distribution florale. *Bull. Soc. Vaudoise Sci. Nat.* **44**: 223-70.

Jessop, J. P. (ed.), 1983. A list of the vascular plants of South Australia. Adelaide Botanic Gardens and State Herbarium and Department of Environment and Planning, Adelaide.

Lange, R. T. and Lang, P. J., 1985. Vegetation. *In* Natural History of Eyre Peninsula ed by C. R. Twidale, M. J. Tyler and M. Davies. Royal Society of South Australia Inc., Adelaide.

Laut, P., Heyligers, P. C., Keig, G., Löffler, E., Margules, C., Scott, R. M. and Sullivan, M. E., 1977. Environments of South Australia. 8 Province Reports and Handbook. CSIRO, Melbourne.

Margules, C. R., 1986. Conservation evaluation in practice. Pp. 297-314 *in* Wildlife Conservation Evaluation ed by M. B. Usher. Chapman and Hall, London.

Margules, C. R. and Usher, M. B., 1981. Criteria used in assessing wildlife conservation potential: a review. *Biol. Conserv.* **21**: 79-109.

McCullagh, P. and Nelder, J. A., 1983. Generalized Linear Modelling. Chapman and Hall, London.

Shaffer, M. L., 1981. Minimum population sizes for species conservation. *Bio. Science* **31**: 131-4.

Wetherby, K. G. and Oades, J. M., 1975. Classification of carbonate layers in highland soils of the Murray Mallee, South Australia, and their use in stratigraphic and land use studies. *Aust. J. Soil Res.* **13**: 119-32.

CHAPTER 9

Effects Of Fragmentation on Communities and Populations: A Review with Applications to Wildlife Conservation

Michael B. Usher[1]

Conservationists have been obsessed with surveys. These give 'snapshots' of the community, and hence many techniques used by conservationists could be labelled 'static' since they describe the community at one instant in time. Repeat surveys allow for an assessment of changes in measurable features of the community, recognising its 'dynamic' nature. As an example, the results of two surveys (in 1973/4 and in 1985) of the limestone pavement flora of Ingleborough, northern England, are discussed. The results clearly demonstrate the dynamic nature of these communities on naturally fragmented habitats, though the summary, represented by the species-area relationship, changed little.

The community is one level in the ecological hierarchy, but it is argued that the conservationist needs to concentrate more on the species or population level of that hierarchy. The review picks out four features of population dynamics that have application in the management of habitat fragments for wildlife conservation. How are populations initiated by immigration and establishment processes? How are populations maintained, paying due regard to their regeneration and to any essential spatial processes? What are the genetic effects, particularly in the long term, of small population size and isolation? What causes populations to become extinct?

The literature is full of contradictions, due partly to confusion about levels in the ecological hierarchy, partly to inherent biotic variability, and partly to the historical development of the present communities. Should conservationists accept the dire warnings of species extinctions when population sizes become small and of unfavourable genetic consequences, or should they ask for evidence of such effects in habitats which have been fragmented for a long period of time (such as the semi-natural woodlands and lowland grasslands in Britain)?

INTRODUCTION

FRAGMENTATION reduces the size of a population. If one considers an area A, which is subsequently reduced to area a and fragmented into p fragments, each of area a_i ($i = 1, 2, \ldots, p$) and with total area

$$a = \sum_{i=1}^{p} a_i \quad (a < A),$$

then the critical question is 'How is the population of species k influenced when the area is reduced from A to a?'. Two of the important preliminary questions that need to be answered relate to the distribution of species k within the overall area A and to the way in which this area was fragmented.

First, for simplicity, assume that the fragmentation was a random process. If species k was distributed regularly over A with a uniform population density, then fragmentation would reduce the population from N to aN/A. The population density, which is N/A per unit of area remains constant in the fragments. However, regular distributions are rare in nature. Random distributions, whilst comparatively frequent in some species over small areas (cf. Usher 1976 for distribution of the soil fauna), are infrequent over large areas that might approximate to the size of a nature reserve. Aggregated or patchy distributions, leading to pattern in plants (cf. Greig-Smith 1983), are by far the most common. Simple simulation with artificial populations undergoing fragmentation indicates that the spatial distribution

[1] Department of Biology, University of York, York YO1 5DD, United Kingdom.

of species k affects the number that are likely to be contained within the fragments. The mean number remains constant at aN/A, but the variance increases with increasing aggregation, and so the probability of having no individuals of species k in a fragment also increases.

Secondly, fragmentation is unlikely to be a random process. In most parts of the world the primary cause of fragmentation has been the change of a natural or semi-natural ecosystem into one subject to more intense agricultural (or forest) production. This tends to mean that the fragments that remain are on poorer soils, are more difficult to drain, or are less accessible, etc. This negative process of selection of the fragments implies that the species that tend to be aggregated on the poorer soils, etc., will be well represented in the fragments, whereas those that tend to aggregate in the areas best suited to agriculture will be under-represented in the fragments. This process again contributes to an increase in the variance of the number of individuals of species k as A is fragmented to a. However, occasionally conservationists are able to select the fragments that are to remain. This is perhaps best exemplified by the 1% criterion used by ornithologists (Fuller and Langslow 1986): sites are selected as important if they contain at least 1% of the world's total breeding population of the bird species. Similarly, Terborgh and Winter (1983) recognized that the forests of Ecuador and Colombia would become fragmented, and hence they suggested siting reserves in areas where particularly large numbers of species of endemic birds are concentrated.

Such considerations about the effects of fragmentation are essentially instantaneous, addressing the question 'Given that there are N individuals of species k, how many will there be after fragmentation?'. This, however, ignores the more subtle and long-term effects of fragmentation. Once the population of N individuals is reduced to an expected aN/A individuals, dispersed into p separate populations each with an expected $a_i N/A$ individuals, will it still be effectively a single population, or will each of the fragments have a separate population? It is the longer term implications of fragmentation, related to the dynamics of populations and communities, that forms the focus for this review. However, before this can be undertaken one must describe in detail what happens during the fragmentation process. Such description is essentially 'static', since the majority of ecological survey work provides a 'snapshot' of an ecosystem at one time. It is the repeat surveys, so that the states of the ecosystem can be compared at two or three times, that lead to data that can be used to evaluate the 'dynamics' of fragmented populations and communities. Monitoring also relates to 'dynamics' since a population, or some component of a community such as the concentration of SO_2, will be recorded repeatedly.

The arrangement of this review is to start by investigating some features of communities, notably the species-area relationship. This summarises what is happening to a large number of populations and, in a sense, provides an overview of the effects of fragmentation on communities. Data from the limestone pavement habitats in northern England will be used to demonstrate some of the concepts reviewed. The review will then move to a more detailed level in the ecological hierarchy, investigating how fragmentation affects populations, and how individuals in those populations may respond to their changed circumstances.

THE COMMUNITY: SURVEYS AND THE SPECIES-AREA RELATIONSHIP

In many parts of the world, the last two decades have seen a considerable effort, on the part of conservation organizations and agencies, in ecological survey work. In the United Kingdom, the *Nature Conservation Review* (Ratcliffe 1977) demonstrated a major effort to enumerate the sites of greatest importance for wildlife conservation. In Ireland the remit was wider so that sites of geological or geographical importance were also included in their list (An Foras Forbartha 1981). In areas of the world which are biologically less well known, the survey work has often taken the form of defining biogeographical regions and vegetational associations so that there was a scientific basis for the criterion of 'representativeness' (cf. Austin and Margules 1986): examples of this would be the characterization of the South Australian environments by Laut *et al.* (1977) and of Western Australia by Beard (1980).

When working with fragmented ecosystems it has often been tempting to treat them as islands, and to ask questions about how the number of species is related to the size of the fragment. Usher (1973) used the data which had been compiled on the higher plants on nature reserves in Yorkshire, England, up to 1970, and found that these gave a statistically significant fit to the model

$$S = CA^z, \qquad (1)$$

where S is the number of species, A is the area (ha), and the constants C and z took the values 69.0 and 0.207 respectively. During the 1970s better lists were compiled, and more of the reserves were surveyed. This resulted in a change of the relationships (Usher 1979) to

$$S = 59.3 A^{0.290}.$$

Numerous studies have now calculated species-area relationships for habitat fragments, and these were reviewed by Rafe (1983). However, a few studies that discuss this relationship, investigating its stability, application to various groups of organisms, and changes with different components of a community, will be reviewed here.

After a survey of arctic-alpine plants on the tops of the Adirondack Mountains of New York, Riebesell (1982) found that 70% of the variance in the number of species could be accounted for by area (z in equation (1) was 0.19), and that 59% of the remaining variance was accounted for by an immigration model that assumed that plant propagules were carried between mountains by mammals or birds. Another study, in the southern Appalachians of northeastern North America (White, Miller and Ramseur 1984), confirmed that the number of vascular plants on mountaintops was related to area, but the number of species was also related to the number of peaks forming a top, the maximum elevation of the mountain, and the number of community types present. The rare species had less scatter about the species-area relationship than the common species, and this led the authors, after a review of the rare species distribution elsewhere in USA, to hypothesize that local extinctions are more important influences on species richness than immigration.

Not all botanical data from naturally fragmented communities show convincing species-area relationships. An analysis of the plants in the hanging gardens of the Zion National Park, USA (Malanson 1982), suggested that these small isolates, between 1 and 100 m² at water seepage points on a vertical canyon wall, contain random assortments of individuals. Statistical analysis of the plant species lists tended to indicate that dispersal of propagules was a causal factor in the assembly of these hanging garden communities. Also, even when a species-area relationship is shown to hold, some small sites may be unusually species rich [as discussed by White, Miller and Ramseur (1984) for two of their southern Appalachian mountaintops, and by Helliwell (1983) for vascular plants in ponds in Worcestershire, England]. There seems to be no explanation yet either as to why some small sites have this unusually great species richness, or as to whether it is just a transitory occurrence. Järvinen (1982) was unable to use a species-area relationship for the plants in deciduous woodland fragments on the Åland Islands to decide whether it was better to have one large or several small nature reserves. He found that a collection of small areas tended to have more endangered plant species than a single large site of equal area, and concluded that 'conservation ought to be based on solid autecology and ecological genetics instead of debatable biogeographical theory'.

Species-area relationships have also been investigated for some groups of animals, notably birds (see, for example, Usher 1985a). Rafe et al. (1985) have shown that both the number of breeding species (S_b) and the number of wintering species (S_w) was related to the area of the nature reserve being surveyed (see Fig. 1), with the species-area relationships having equations

$$S_b = 14.5 A^{0.176} \qquad (2)$$

and

$$S_w = 14.8 A^{0.240} \qquad (3)$$

The amount of variance accounted for was increased from 34 to 58% and from 31 to 47% respectively when a term related to the habitat heterogeneity of the reserves was included in the regression model. Woolhouse (1983) analysed the breeding bird data for 30 woodlands, varying in size from 3.6 to 40.1 ha. The data had been collected in the five consecutive years from 1976 to 1980. There was no statistical difference between the five annual regression lines, and hence he pooled them to give

$$S_b = 13.9 A^{0.227}. \qquad (4)$$

Further analysis of his data indicated that surveyors spent more time, proportionally, in smaller woodlands than in larger woodlands, and hence that the species-area relationship for these woodland birds would tend to be depressed, i.e. that the z value of 0.227 in equation (4) was underestimated. The efficiency of surveyors during a single census has been analysed by Haila (1983), who indicated that between 70 and 95% of the breeding pairs were recorded. Both Woolhouse (1983) and Haila (1983) highlight an important aspect of data collection during survey work: is the efficiency of the survey known, and is the effort put into all areas surveyed approximately equal?

In Illinois, Blake's (1983) study of the breeding birds of isolated forests, varying from 1.8 to 600 ha, divided the birds into five analysable trophic groups — omnivores, and four guilds of insectivores (bark, ground, foliage and aerial). Although all five groups showed a positive species-area relationship, omnivores dominated the small forests and frequently foraged in the surrounding farmland (farmland was also found to be an important component for larger avian diversity by Rafe et al. 1985). Foliage insectivores were uncommon in the small forests, but formed the largest component of the avifauna of the large forests. Another possible division of bird communities is into groups of species with similar behaviour: permanent residents, short-distance migrants and long-distance migrants were the groups used by Lynch and Whigham (1984). They found that, although the number of bird species, irrespective of behavioural group, was related to area, isolation, forest structure and floristics, the highly migratory species tended to be most abundant in the largest forests, whilst permanent residents and short-distance migrants were less affected by area. However, Lynch and Whigham (1984) conclude that 'the impacts of forest fragmentation on bird populations are complex and species-specific'.

The highly fragmented remnants of heathland vegetation in the Poole Basin of Dorset, England, were sampled for spiders and beetles by Webb and

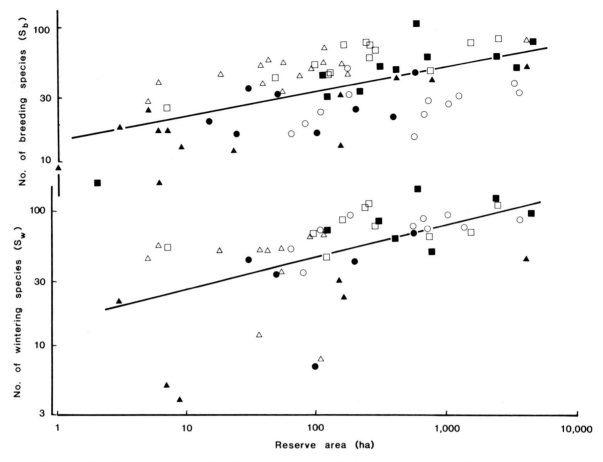

Fig. 1. Two examples of species-area relationships. The data relate to breeding bird species (upper graph) and wintering bird species (lower graph) on reserves managed by the Royal Society for the Protection of Birds in the British Isles. The following symbols have been used for woodlands (△), offshore islands (▲), low coasts (○), cliffs (●), freshwater wetlands (□), and either other habitats or mosaics of different habitats (■). The lines represent equations (2) and (3) in the text. The data are redrawn from Rafe *et al.* (1985).

Hopkins (1984) and Webb *et al.* (1984). The species-area relationships were negative and, often, statistically significant, unlike the majority of other relationships reviewed above. These results are particularly interesting since they show that conservationists should not be solely interested in the numbers of species, but that they should be interested in what could loosely be called the 'quality' of those species. Webb *et al.* (1984) demonstrated that the number of species of spiders and beetles was positively related to the number of different habitats adjacent to these heathland remnants. However, using the set of those species that are true heathland species (Webb *et al.* 1984; Hopkins and Webb 1984), a positive species-area relationship was found. Thus for all spiders the correlation coefficient between the number of species and the logarithm of the area was -0.06, whilst for the 'quality species', the heathland spiders, it was +0.38.

Whether one reviews studies of birds, plants or arthropods, the species-area relationship is found to be lacking since it only accounts for a proportion of the observed variation. Criticisms of it abound:

Gilbert (1981) essentially argued that the ecological sciences could never have the rigour of the simple models that are associated with the physical sciences, whilst Boecklen and Gotelli (1984) discounted such models since they rarely account for more than half of the variation in the number of species. Haila (1983), however, gave a more reasoned criticism. On the one hand the underlying species-abundance distribution cannot be regarded as a satisfactory explanation of the species-area relationship if the processes leading to the distribution are unclear; on the other hand, a good statistical fit of a set of data to a species-area relationship is no proof for the equilibrium hypothesis. This hypothesis stems from the original work on island biogeography by MacArthur and Wilson (1967), and is shown diagramatically in Figure 2. The equilibrium hypothesis, often referred to as 'the theory of island biogeography', is beguilingly simple: the number of species on an island or on a habitat fragment will increase if the number is initially below the equilibrium level, or it will decrease if the number is above that level. These changes will occur until the rate of immigration of species not currently present on the area equals the rate of extinction of

▷

PLATE 1
CHAPTER 2

Wandoo *Eucalyptus wandoo* woodland in the Mt Lesueur area. (E. G. Griffin).

◁

PLATE 2
CHAPTER 2

Ecotone between woodland and heath in the Mt Lesueur area. (E. G. Griffin).

▷

PLATE 3
CHAPTER 9

One of the largest of the limestone pavements that surround Ingleborough, North Yorkshire, England. The mountain in the distance is Pen-y-Ghent. (M. B. Usher).

◁

PLATE 4

CHAPTER 9

Rigid buckler-fern *Dryopteris villarii montana,* a species that is nationally rare and virtually confined to the limestone pavement habitat. (M. B. Usher).

▷

PLATE 5

CHAPTER 9

Primrose *Primula vulgaris,* a common species that is usually associated with woodlands, but which frequently occurs in deep grykes on limestone pavements. (M. B. Usher).

◁

PLATE 6

CHAPTER 12

Flowers of *Eucalyptus caesia* subsp. *magn* Each is up to 6 cm across. Their striking colouration is associated with pollination by birds. (S. Hopper).

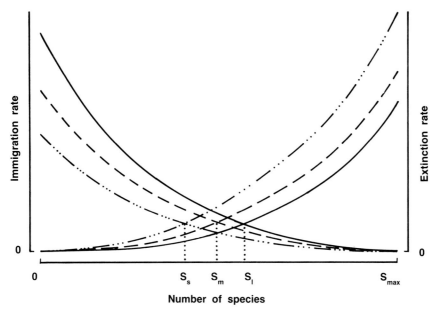

Fig. 2. A diagrammatic representation of the equilibrium theory of island biogeography. The continuous lines represent large islands or fragments, dashed lines the medium-sized islands or fragments, and dashed and dotted lines the small islands or fragments. The number of species ranges from 0, when the extinction rates will be zero, to S_{max}, which is the number of species in the source pool, when the immigration rates will be zero. The equilibrium number of species is indicated where the immigration and extinction rate lines cross, and are S_l, S_m and S_s for large, medium and small islands or fragments respectively.

the species present. As the immigration curves are area dependent (small islands provide a smaller target for propagules), as are the extinction curves (small islands have smaller populations which are more extinction prone), the equilibrium number of species reduces from S_l to S_m to S_s as one moves from a large to a medium-sized to a small island (see Fig. 2). Note that this relates just to the number of species, and does not predict which particular species will, or will not, be present. Criticisms of this theory and its application to nature reserve design can be found in McCoy (1982) and Margules et al. (1982). In choosing reserves, both Reed (1983) and Usher (1985b) conclude that several factors, and not just species-area relationships or equilibrium theory, need to be taken into consideration.

Before investigating these other factors, many of which are related to the dynamic aspects of populations, one should ask about the relationships between islands, fragments and archipelagoes. Kobayashi (1983) has investigated the archipelago phenomenon both theoretically and using data on the mammals of the Ryukyu Islands, beetles of the Izu Islands, and birds of four archipelagoes. He predicted that, in the relationship between ln$S$ and ln$A$, the value of z in equation (1) would decrease monotonically towards 0 as the area increased. The precise nature of the species-area relationship would be determined by the inter-island distances in the archipelago (and hence by migration between islands) as well as to the distance between the archipelago and the mainland. He showed that archipelagoes with migration between islands may have numbers of species on the islands more reminiscent of samples from a mainland where the species are themselves aggregated.

This section of the review can, however, be concluded by making one general observation that applies to islands, habitat fragments or archipelagoes. As the area of habitat (either an island or fragment) increases, the number of species, that are ecologically dependent upon that type of habitat, will also increase. Such a statement does not imply any mechanism for the increase, nor does it assume any mathematical model, but it does imply that the correlation implicit in the species-area relationship need not apply to the number of all species, but only to that group of species which are ecologically dependent upon that type of habitat [cf. the heathland spiders of Hopkins and Webb 1984]. The species-area relationship is a *summary* of what is happening to a community of species, and hence its use is limited. In managing a nature reserve, a reduction in the number of species may indicate that one or more unwanted processes are occurring: only detailed studies, asking which species are disappearing, will reveal what these processes are and how they may either be stopped or moderated. Criticisms of the species-area relationship have usually focussed on the species themselves, rather than on the community as a whole, and hence it is important to understand how fragmentation affects both the populations of species that are summed to form the species-area relationship and the various measures of diversity that have been used by conservationists.

Fig. 3. A view to the summit of Ingleborough (723 m), North Yorkshire, England, from the west side. A limestone pavement, at approximately 420 m, occupies the foreground of the photograph.

AN EXAMPLE: THE LIMESTONE PAVEMENTS OF INGLEBOROUGH

Limestone pavements are a feature of horizontally bedded limestone rocks in a number of areas of Britain. A survey in the early 1970s listed 537 pavements in the British Isles (Ward and Evans 1976), the great majority of them occurring on the Carboniferous Limestone of the northern Pennines, in the counties of North Yorkshire, Lancashire and Cumbria. Ingleborough (723 m) is one of the highest peaks in this limestone area (see Fig. 3) and at the time of the original survey in 1973/4 it was surrounded by 77 limestone pavements. These pavements are scattered through an area of some 64 km^2, generally at altitudes of about either 330 m on 420 m, and hence form series of naturally fragmented habitats with areas of intact pavement between 0.1 and 51 ha.

The pavements probably resulted from weathering of the limestone rock whilst it was covered by glacial till. The flat expanses of exposed rock, known as 'clints', are more or less devoid of higher plants, except where there are shallow cups that retain some water and mineral or organic material, and

Fig. 4. A close up view of a limestone pavement showing the flat, unvegetated limestone surfaces (the 'clints') and the deep fissures (the 'grykes') which contain characteristic plant communities.

these are usually colonized by spring flowering annual plants, the most abundant being *Saxifraga tridactylites* [nomenclature follows Clapham *et al.* (1981)]. The fissures, called 'grykes', can sometimes be deep, to 5 or 6 m, though they are often not deeper than 1 or 2 m (see Fig. 4). The microclimate in the grykes is damp and shady, and they contain a rich flora that consists of three separate elements. One element is the flora of the sheep-grazed grasslands that surround the pavements, another element is a woodland flora, whilst the third element, by far the smallest in numbers of species, is of plants that are more or less confined to the limestone pavement habitat. Arbitrarily, Ward and Evans (1975a) defined grykes as being at least twice as deep as wide: this excluded peripheral fissures that were sufficiently wide to be grazed by the grassland herbivores (sheep, rabbits and hares). The 1973/4 survey only recorded species that occurred in grykes defined in this manner.

Ward and Evans (1975b) recorded a total of 183 higher plant species in their 1973/4 survey of 77 pavements, and they classified these species into four groups (Ward and Evans 1975a). Group D were species that occurred in the surrounding grasslands, and hence did not rely on the limestone pavement habitat for their continued existence. Groups A, B and C were, however, relying on the pavement for their survival, since these species did not occur on the grazed grasslands of Ingleborough or other areas. The difference between the three groups is that group A species are nationally rare, many of them being limestone pavement specialists (see Fig. 5a and 5b for one of these, *Actaea spicata*, and Fig. 5c for a map of its distribution in the British Isles). Group B species are nationally uncommon (one of these, *Cirsium helenioides*, is shown in Fig. 6a, and its distribution in Fig. 6b). Group C species are nationally common or abundant, as, for example, *Phyllitis scolopendrium*, in Figure 7. Usher (1985b) fitted the species-area relationship in equation (1) to the data collected by Ward and Evans (1975b): group A species are too infrequent for statistical analysis, but groups A and B combined have a relationship

$$S = 8.0A^{0.167} \tag{5}$$

that is statistically significant ($P < 0.001$). Groups A, B and C pooled ($z = 0.100$) and all species ($z = 0.084$) also have statistically significant species-area relationships with $0.05 > P > 0.01$. Both Haila (1983) and Usher (1985b) were more interested in the outlying points in species-area plots, asking 'would those above the regression equation tend to loose species, and would those below the line tend to gain species?'.

Fig. 5. Actaea spicata, a group A species of the limestone pavements (see text for definition). (a) The species in flower in a gryke, showing the group D species, *Sesleria albicans,* in the upper right. (b) A close-up of the flower, and (c) the species' distribution in the British Isles, indicating 10 km national grid squares where it has been recorded from 1960 onwards (●) and only prior to 1960 (○).

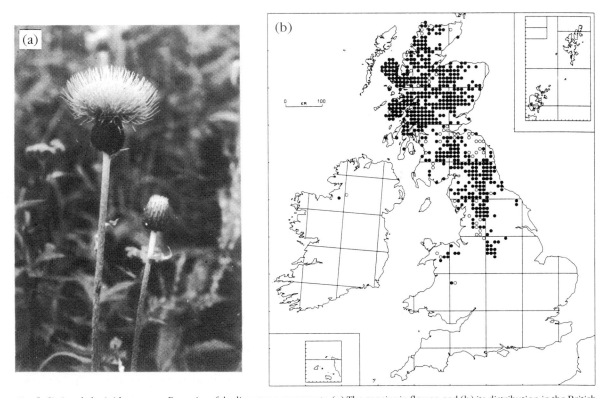

Fig. 6. Cirsium helenioides, a group B species of the limestone pavements. (a) The species in flower, and (b) its distribution in the British Isles, using the same symbols as Figure 5c.

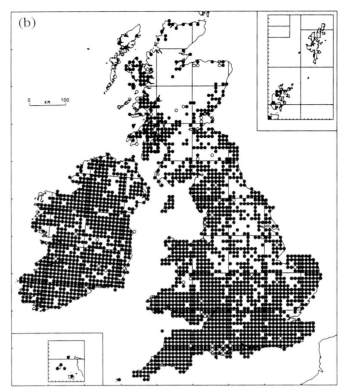

Fig. 7. Phyllitis scolopendrium, a group C species of the limestone pavements. (a) The species growing from the side of a gryke, with *Geranium robertianum* and *Asplenium viride* (at right) and *Mercurialis perennis* beneath the fern. (b) The species' distribution in the British Isles, using the same symbols as Figure 5c.

In order to investigate such questions, the limestone pavements of Ingleborough were re-surveyed in July 1985 using the same methods as the original 1973/4 survey. Two of the 77 pavements surveyed in 1974 had been destroyed, and access to another of the pavements was not possible. Hence, the 1985 data relate to only 74 pavements: the 1973/4 data have been re-analysed so that the same set of 74 pavements can be compared in the two surveys. Table 1 lists some of the statistics of the species-area relationship. As rarity is a major concern of many conservationists (see Margules and Usher 1981; Usher 1986a), the data in Table 1 relate only to the rare and uncommon species (groups A and B).

One result of the analyses is that the constant terms in the 1985 equations are smaller, indicating that fewer species were recorded in the second survey. It is possible that two factors have contributed to this drop in records of group A and B species (721 records, or a mean of 9.7 per pavement in 1974; 641 records, or a mean of 8.7 per pavement, in 1985). First, the survey was carried out slightly later in the year, and hence some spring flowering species, such as *Arum maculatum* and *Primula vulgaris*, were recorded less frequently (the original survey was in September 1973, but it was supplemented in May and June 1974). Secondly, the abnormally wet summer of 1985 resulted in lush growth of the taller plants with the consequence that many of the shorter-growing species were recorded less frequently.

Another result from Table 1 is that both a linear model and a log-log model as in equation (1) give a statistically reasonable fit to the data, the former being slightly better in 1974 whilst the latter was better in 1985. The log-log model has been used for subsequent analysis.

Table 1. Species-area relationships for higher plant species (groups A and B only; see text for definition) on 74 limestone pavements around Ingleborough, North Yorkshire, England. In this table S is the number of species, A is the area (ha), r is the product moment correlation coefficient (with 72 degrees of freedom), and F is the variance ratio (with 1, 72 degrees of freedom) in the analysis of variance of the regression coefficient.

	Year	
Statistic	1974	1985
Linear Model		
Equation	$S = 8.21 + 0.500A$	$S = 7.73 + 0.303A$
Variance ratio, F	15.53	10.94
Correlation coefficient, r	0.42	0.36
Log-log Model		
Equation	$S = 7.94A^{0.170*}$	$S = 7.19^{0.179}$
Variance ratio, F	12.60	16.64
Correlation coefficient, r	0.39	0.43

* This equation differs from equation (5) due to the exclusion of three pavements, as discussed in the text.

(a)

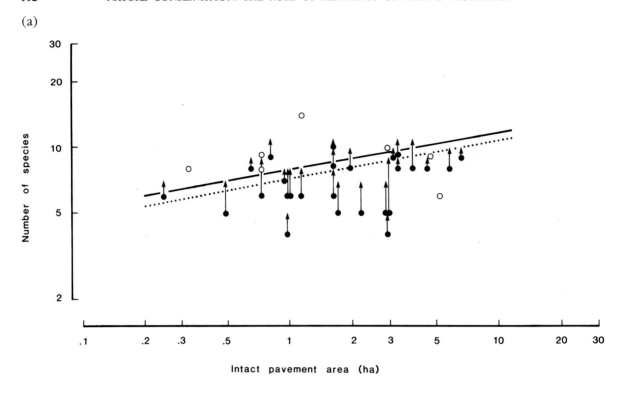

(b)

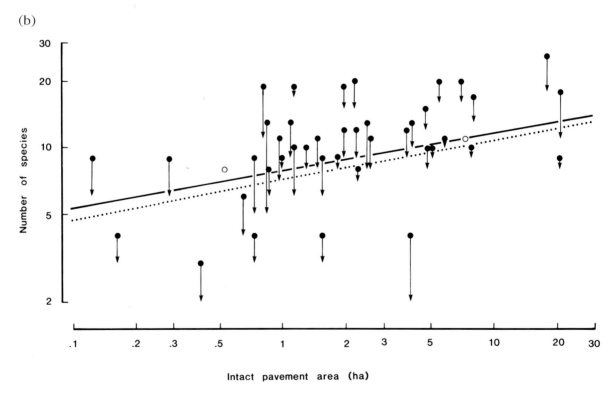

Fig. 8. The numbers of group A and B species on 74 limestone pavements around Ingleborough, England. (a) The pavements which had the same number of species (○), or more species (●), in 1985 than in 1974. (b) The pavements which had fewer species in 1985 than in 1974 (●), or which had been destroyed (○). In both illustrations the 1974 number of species is represented by an open or filled circle, and the 1985 number by the tip of the arrow showing the direction of movement. The continuous lines indicate the 1974 species-area relationship, $S = 7.94A^{0.170}$, and the dotted lines the 1985 relationship, $S = 7.19A^{0.179}$.

The aim of the 1985 survey was to understand how the individual pavements had changed in the 11 years between the two surveys. In Figure 8a the 33 pavements which have either retained or increased their number of group A and B species are plotted. Only three of the pavements that have increased numbers of species lie above the 1974 species-area regression line, whilst 23 pavements lie below this line. Taking the median pavement size of 1.8 ha gives 14 of these pavements smaller than the median, and 12 larger than the median. Pavements showing an increase in the number of species are, therefore, scattered throughout much of the range of pavement areas, but tend to be those that had fewer than the expected number of species in 1974. With the increases recorded in 1985, 13 of the pavements in Figure 8a moved either to or above the 1974 species-area regression line.

The 41 pavements for which the number of species in groups A and B had declined between 1974 and 1985 are shown in Figure 8b. Again, the area of the pavement seems to have had no effect on the decrease in number of species: 19 of the pavements have areas smaller than the median, whereas 22 of the pavements are larger. However, the species richness in 1974 seems to have been important since 30 of these pavements lie above the 1974 species-area regression line, whilst only 11 pavements were below the line. With the decreases recorded in 1985, 28 of the pavements moved to or below the 1974 regression line in 1985.

The overall result of the survey and resurvey is that limestone pavements with more than the expected number of species in 1974 (on the basis of the species-area relationship) have tended to loose species during the subsequent 11 years (i.e. relaxation has occurred). Pavements with fewer, but not greatly fewer, species have tended to gain species (i.e. it is possible that there has been some colonisation of species-poor pavements). However, the most unexpected result is that the extremely species-poor pavements, those that deviated most in a negative direction from the 1974 species-area relationship, have continued to loose species (Fig. 8b), and hence were poorer in 1985 than in 1974. It is interesting to note that 11 of the pavements that were above the 1974 species-area line are below the 1985 line, and that 15 pavements below the 1974 line are above the 1985 line. Surveys such as these provide 'snapshots', indicating which fragments are species-rich or species-poor. In this one example, 26 of the 74 pavements (38%) had 'changed sides' over the 11-year period between the two surveys. Unquestionably, fragments are dynamic entities, and conservationists need to be aware of these dynamic properties of the habitats that they are preserving. However, the dynamics of the community is merely an integration of the dynamics of the various species that together form that community.

THE POPULATION: DETAILED EFFECTS OF FRAGMENTATION

The classical approach to studying populations has been to write an equation such as

$$N_{t+1} = N_t + B - D + I - E, \qquad (6)$$

where N is the number of individuals in the population, B and D are the numbers of births and deaths between times t and $t+1$, and I and E are the numbers of immigrants and emigrants, also between times t and $t+1$. Clearly, for the conservation management of any species, it is important to understand the fecundity and mortality of the species, as well as to understand movements which contribute to the I and E terms of equation (6). However, equation (6) describes a relatively short term situation (changes over one period of time), whereas the manager of a biological resource will also be interested in the longer-term effects of management.

In discussing the effects of fragmentation on populations, four related features of populations will be considered. First, although fragments will have existing populations on them, they may be open to colonization by species not present when the fragment was created. It was implicit in the equilibrium theory shown in Figure 2 that immigration was a continuous process, with the fragment being bombarded with propagules: why is it that only a few of these manage to estsblish themselves? Secondly, populations must be able to maintain themselves: referring to equation (6) this implies that, averaged over a period of time, $B+I$ must approximately equal $D+E$. The long-term maintenance of the populations of a few of the species in the community is probably one of the most important functions of the conservation manager. Thirdly, over a long period of time there may be genetic changes in populations that have become isolated, at least to some extent, from other populations. Frequently this long-term effect is forgotten in the management of biological reserves, but it can have far-reaching implications for the species. Finally, extinction is perhaps what the conservationist fears most, and there are all sorts of questions which can be framed, but rarely answered, like 'is there an area for a reserve below which the populations will go extinct?', or 'is there a minimum size for a viable population?', or 'do extinction-prone populations have certain characteristics?'. The initiation, maintenance, genetics and extinction of populations are reviewed in the following sections.

Population Initiation: Immigration, Colonization and Establishment

Colonization of newly-formed oceanic islands, although not directly the subject of this review, does demonstrate what biotic potential there is. Thornton (1984) summarized the re-colonization of Krakatau by many groups of species during the century following the total destruction of its fauna and flora by the volcanic eruption in 1883. The number of butterfly species has levelled off at rather less than 40, though the total number of species recorded is about 60: at least a third of the species that have arrived on the island have subsequently become extinct (but one cannot guess how many species arrived and became extinct, but were never recorded). The number of species of ants, on the other hand, is apparently still increasing, with the 68 species recorded in 1982 being nearly double the number recorded in the 1920s and 1930s. A habitat devoid of species will gain species during time and, as evidenced by the butterflies but not yet by the ants, there may be a saturation level for the number of species which is appreciably less than the total number of potential colonists. A wide variety of mechanisms is available to colonizing species. Gandawijaja and Arditti (1983) suggested that long distance aerial dispersal enabled the orchid seeds to reach Krakatau rapidly, and such an explanation may account for the unexpected appearance of *Orchis simia* on a well-recorded coastal site in Yorkshire, England, nearly 300 km beyond its previously known distribution. The small size and lightness of orchid seeds means they are adapted for long-distance dispersal by wind. However, many seeds of grassland forbs have rather limited dispersal, 0.3 to 3.5 m being quoted by Verkaar *et al.* (1983). These are average distances, and occasionally a seed may be distributed much further. Peterken's (1981) data for woodland forbs do, however, suggest that their speed of spread along hedgerows, and colonization of secondary woodlands, is extremely slow, also possibly indicating limited seed dispersal and small probabilities of establishment.

Slow dispersal of seed and a small probability of establishment would imply that corridors between habitat fragments are of no use for conservation management. Diamond's (1975) diagram, which has been copied into *The World Conservation Strategy* (IUCN 1980), is beguilingly simple, indicating in the fifth pair of better/worse situations that fragments with corridors are better than an equal area of isolated fragments. The diagram has been criticised as being without empirical evidence (Margules *et al.* 1982), although these authors also discussed some data that could support the case for corridors. Further support for corridors comes from a study of chipmunks, *Tamias striatus*, in Ottawa, Canada (Henderson *et al.* 1985). Extinction of chipmunks in woodland fragments did not last for long periods of time: the woods were recolonized by animals, coming from other fragments, travelling along hedges and fences. Järvinen (1982), on the other hand, definitely concluded that such corridors between deciduous woods in the Åland Islands were of no apparent advantage in maintaining plant species richness.

The data are clearly equivocal. In the conservation of plant species, it would seem that corridors between habitat fragments are of little or no value in conservation management. Average seed dispersal is for far too short a distance, and the probability of establishment is far too low, for plant species to move along corridors. However, in the conservation of animal species, there appear to be far more indications of the value of corridors (or 'stepping stones' for migrating birds). Animal species are able to travel the length of the corridor and breed at the other end, while plants would need to stop and breed repeatedly along the length of the corridor. However, besides wind (cf. the Krakatau orchids discussed above), many plants rely on animal vectors for their dispersal, and, for these species, corridors may be indirectly beneficial. Riebesell (1982) considered that immigration to mountaintops by plant propagules carried by birds and other animals was important for the maintenance of species richness. Corridors and stepping stones may therefore prove to be more important in maintaining immigration into habitat fragments than the data available at the present time suggest.

Population Maintenance: Pattern, Process and Regeneration

Watt (1947) started his Presidential Address to the British Ecological Society by drawing attention to the fact that the community is a working mechanism. Despite the community bias of his address, Watt was often concerned with the analysis of populations. For example, he described in detail the rhizome structure of a clump of bracken, demonstrating its advance into grass heath and the eventual degeneration phase: and the building, death and erosion cycle associated with the grass *Festuca ovina*. In studying a single spatial point in a community, the dynamic processes which affect populations will result in the population at that point changing from year to year, often in a cyclic manner over a period of many years. There are two important considerations for the conservationist that follow from studies such as Watt's on the dynamic properties of populations and ecosystems.

First, how does one decide whether a population is stable or not? The terms 'stable' and 'stationary' are often confused, and hence it is important to distinguish them. Stable implies that, under constant environmental conditions, the population size will remain within some limits, not necessarily at an equilibrium, and will return to these limits if perturbed slightly. Stationary means that the mean, variance and time structure of the population size are not time dependent. Williamson (in press) shows that a stationary population is stable, and that a stable population is stationary in a constant environment. In the real world the environment is never constant, and hence stable and unstable movements in the population would be extremely difficult to distinguish. Because of the difficulties of interpretation, these concepts are probably best forgotten by the manager: they should, however, be borne in mind by the surveyor since they do indicate that repeat surveys, even at reasonably short time intervals, are unlikely to yield closely similar results even if the populations were stable.

Second, since a population may occur in a number of different phases — pioneer, building, mature and degenerate — space will be needed for all of these phases to be represented on the habitat fragment. It is also possible that vegetation patches may themselves move around in space: Benedict (1984) showed that tree-islands of *Picea engelmannii* and *Abies lasiocarpa* moved downwind across the forest/tundra ecotone in Colorado. The leeward edge of these islands expanded at rates of 1.5 to 2.6 cm yr^{-1}, whilst the windward side contracted at 0.9 to 1.9 cm yr^{-1}. Although these rates are slow, they can be considerable during the life of an individual tree. Space is also needed for regeneration: sites for the regeneration of the species being conserved may be a particular bottleneck in maintaining that population. Runkle (1981) has highlighted the need for gaps (c. 200m^2) for the regeneration of trees in mixed forests in eastern USA, and Grubb (1977) has discussed the importance of what he termed 'the regeneration niche'. If the processes of regeneration are viewed more widely than just the germination of the seeds, also including the establishment of the young plants, then the studies of Connell *et al.* (1984) on the subtropical rainforest in Queensland are important. Commoner species had lower rates of recruitment than rarer species, implying that there are compensatory trends among the many populations that form a complex community. Such trends were thought to be essential in maintaining the diversity of this sub-tropical forest.

Conservationists should not look for a non-changing population or community in a habitat fragment. All of the evidence suggests that cycles of change can be expected as the structural plant species of that community regenerate, mature and finally degenerate. The important aspect of management is that all of the stages of the cycle are represented in the fragment so that species do not become extinct for the want of the correct micro-site to regenerate, etc. This may well imply that there are minimum area requirements for the fragments.

Population Genetics: Evolution, Variability and Viability

Studies of island populations have often focussed on the speciation that has occurred [see, for example, Williamson (1984) for a discussion of Sir Joseph Hooker's pioneering contribution to the study both of evolution on islands and of relic distributions]. Habitat fragments have not yet evolved their own endemic species, though Mader (1984) provide numerous observations on Carabid beetles and small mammals to suggest that even a relatively narrow road will drastically reduce the gene flow between the populations on each side of the road. It is, therefore, pertinent to ask what genetic effects fragmentation is likely to have.

Some of the genetical consequences of small population size have been reviewed by Miller (1979) and Harris *et al.* (1984), and the subject is more fully discussed in the book by Frankel and Soulé (1981). Essentially there are six main effects, all to some degree interrelated, that should be of concern to the conservationist, namely

(a) inbreeding,
(b) heterozygosity,
(c) bottleneck, or founder, effects,
(d) effective population size,
(e) genetic drift, and
(f) mutation rates.

Inbreeding, the breeding of genetically related individuals, results in an increased proportion of homozygous pairs of alleles, especially recessive alleles. There are two main problems for the conservation of a species. First, with the reduced proportion of heterozygous genes there will, in the long term, be a reduction in genetic variability, and, second, since many maladaptive traits are recessive, inbreeding may result in their more frequent expression in the population. The documented problems of inbreeding generally concern the conservation of large mammal species: the white tiger, *Panthera tigris*, okapi, *Okapia johnstoni*, dorcas gazelle, *Gazella dorcas*, and Przewalski's wild horse, *Equus przewalskii*, have all been quoted.

Heterozygosity is generally recognized as being desirable, since the fitness of a population increases with heterozygosity. Soulé (1979), for example, analysed the relationship between heterozygosity and morphological asymmetry in fifteen populations of lizards. He considered that asymmetry represented accidents during development, and hence that genetically superior lizards would be less asymmetrical than genetically inferior ones. The strongly negative relationship found implied, at least in this lizard, that populations rich in genetic variation were more fit than genetically depauperate populations. Other studies reviewed by Frankel and Soulé (1981) indicated that such a result may be generally applicable.

The bottleneck or founder effect results from a population going through a 'bottleneck', i.e. being very small for one or more generations before expanding again. During such a period inbreeding is likely to occur, with its consequent loss of genetic variation, as well as the relic population only containing a sample of the genetical variation in the population that entered the bottleneck. Most reviews mention the northern elephant seal, *Mirounga angustirostris*, that now has almost no genetic variation: prior to this century the population had been reduced to less than 30 individuals. There are, however, a number of studies of small mammals (Berry 1973, 1977) that indicate that the seriousness of the bottleneck effect is likely to vary both with species and with habitats, the genetic variation in some populations that have passed through bottlenecks apparently not being drastically reduced. Large mammals may be the most vulnerable to this genetical effect.

However, it is not the total population size that is important, but the genetically active part of the population that should be considered at times of low population density. This part of the population consists of the breeding adults, omitting the post-reproductive adults and the non-breeding juveniles. Since many species show behavioural patterns that prevent the whole of this adult population breeding, an effective population size, N_e has been defined as

$$N_e = 4 N_f N_m / (N_f + N_m),$$

where N_f and N_m are the numbers of breeding females and males respectively. It can be seen that N_e equals the number of males and females in a monogamous species, but is less than this total in polygamous species.

Genetic drift is caused by the sampling of parental genes for each successive generation. In large populations little or no change in gene frequency would be expected from such a random process, but in small populations there can be detectable changes from generation to generation. It may be that the conservation manager, particularly if managing small relic populations of large mammals, will wish to select which individuals breed with each other in order to reduce as far as possible the effects of genetic drift that can, in time, result in alleles being fixed in or lost from the population.

The five genetic effects already reviewed are all related to a loss in genetic variability, and hence to a loss of within-species diversity. Evolution can only proceed adaptively if there is sufficient genetic variation within the species. One need not only think of evolution of new species, but the evolutionary process is also part of the adaptation of the species to an ever-changing environment. Lamb (1969), for example, has shown that there are long-term climatic fluctuations, of the order of centuries, in the northern hemisphere and Bowler (1982) has discussed aridity in the southern hemisphere. Species that are short-lived in comparison to these fluctuations will require some genetic variability in order to adapt: species that are long-lived may be able to reproduce and then survive until the climate is once again favourable for reproduction to occur again. Mutation, although a comparatively rare event, is about the only process that can compensate for the loss of genetic variability. Data for mutation rates in species of concern to conservationists seem to be lacking, since most estimates of mutation rates involve laboratory animals and insects, as well as man (Berry 1977). However, it is important to realize that this small and relatively unknown rate of the creation of new genetic variation has to balance the loss of genetic variation due to inbreeding, bottlenecks, reduced gene flow, etc., all of which increase as a result of population fragmentation.

Population Extinction: Minimum Area, Minimum Viable Size and Extinction Rate

There is an intuitive feeling amongst many conservationists that as the area of a nature reserve is reduced there comes some point when its value for conservation is lost. This was explicitly stated in Diamond's (1975) 'better/worse' diagrams, and is implicit in many surveys of nature resources. For example, Ratcliffe (1977) stated 'The nature conservation value of a woodland tends to increase with size. Only in a large wood may it be possible to encompass the local range of diversity or to achieve diversity by means of different silvicultural systems or management practices.... Some organisms, e.g. large birds, require a fairly large minimum area, depending on territory size...'. Although no minimum area was stated, it can be inferred from a study of those woods that have been selected. Using the woods selected in England (since those in Scotland are less fragmented and hence are larger by an order of magnitude), the range of sizes is shown in Figure 9. Approximately 30 ha would seem to be the implicit minimum size, though there are a number of smaller woods in the series that are either parts of larger sets of neighbouring woods or have been selected for very particular reasons. Similar studies of sites selected in other habitat groupings usually suggested operational minimum areas of the order of 50 ha.

However, although nature conservation value increases with size, what happens to the value as size decreases? Usher (1986b) has suggested three models (Fig. 10). If there genuinely is a minimum area, then one would expect model I in which the value is suddenly lost as the reserve size drops below A_{min}. Alternatively, there may be a small area range within which the value declines steeply, as in model II, so that above A_u there is a considerable value whereas below A_l the value is negligible. Both models I and II imply that there are minimum areas for a large number of the characteristic species of the habitat either at A_{min} (model I) or between A_l and A_u (model II). Although intuitively plausible, there are no data that would indicate that either model I or model II is realistic. Perhaps more realistically one could postulate model III where the conservation value decreases monotonically with decreasing area. This model would imply that the minimum areas for the species characteristic of the habitat are distributed across the whole area spectrum, though the convex nature of the curve indicates that more

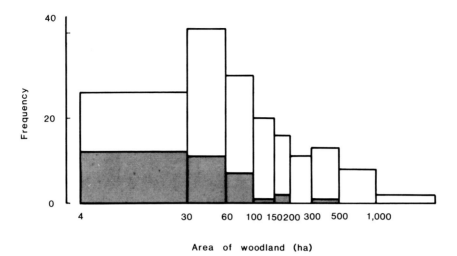

Fig. 9. The areas of woodland selected in England and included in the *Nature Conservation Review* (Ratcliffe 1977). The horizontal axis is logarithmic, and the vertical axis records the frequency of the 167 woodlands in the appropriate size classes. Although many woodlands are recognized as distinct sites (unshaded bars), others are grouped together to form single sites in the *Nature Conservation Review* (these woodlands are shown individually as shaded bars).

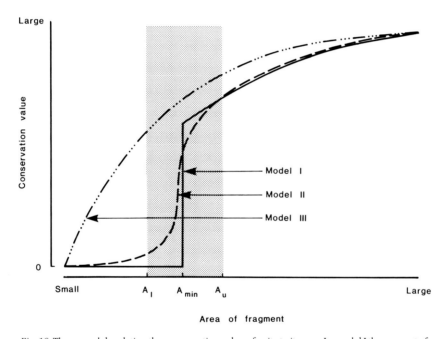

Fig. 10. Three models relating the conservation value of a site to its area. In model I the concept of a minimum area is valid since sites with area greater than A_{min} have a reasonably large value whereas sites less than A_{min} have no value. Model II is a more generalized form of model I since the value decreases steeply between A_u, above which there is considerble value, and A_l, below which there is little or no value. In model III the value declines with decreasing area, but, as there is no sharp decline in value, no minimum area can be defined. The illustration is redrawn from Usher (1986b).

species are lost at smaller areas than at larger areas. The fact that most woodlands in the English series of key sites (Ratcliffe 1977) have an area of 30 ha or more, whereas a number of smaller areas are included because they have particular species, may imply a model III situation. The important considerations are how much individual species are valued, and what are the minimum area and population size requirements of these species. McCoy's (1983) methods of estimating minimum areas suggest that this topic can only be approached through a study of the individual species, and not at the community level in the hierarchy.

Frankel and Soulé (1981) have suggested, on genetic grounds, that the minimum viable effective population size is of the order of 500. If, in the long term, this figure is correct, then many of the large herbivores on game reserves in Africa will become extinct. East (1983) has reviewed the literature on population sizes of such mammals in 13 African states, and many of the species have individual population sizes less than 500. Of the data included in East's (1983) Table 1, only 37% of the individual populations were larger than 500, whilst 21% of the populations had fewer than 50 individuals.

Wright and Hubbell (1983) have discussed minimum size from the point of view of demographic stochasticity rather than from genetic considerations, and so as to reduce the probability of chance extinction they conclude that the larger the reserve the better for rare species where the possibility of immigration to the population is virtually non-existent. Goodman (in press) has taken such an analysis further by using a model to predict the persistence time of a small population in a randomly changing environment. His results are particularly interesting since they compare the within-species variation with the effects of a stochastic environment. The predicted persistence time when altering within-species variation increased as a power of the ceiling on the population size, whereas with the stochastic environment the persistence time increased approximately linearly with the ceiling on the population size. The study did not, therefore, indicate that there was a minimum viable population size, but that the probability of chance extinction was reduced as the population size increased.

Extinctions undoubtedly do occur, but it may be as well to distinguish two types of extinction. First, when a habitat is fragmented, some species may become extinct because of the minimum area or minimum population size effects. Karr (1982a, b) has, for example, documented the extinctions of bird species on Barro Colorado Island, a fragment of forest created during the construction of the Panama Canal. This aspect of extinction can be related to the phenomenon of saturation [cf. Terborgh and Faaborg (1980) for a discussion in relation to West Indian bird communities]. Secondly, in any habitat there are species that 'come-and-go', being present during some years and absent in others. The example of the butterflies on Krakatau, quoted above, indicated that one-third of the species known to have arrived on the island have subsequently become extinct. Williamson (1983) listed the numbers of breeding pairs of each species of land and freshwater bird on Skokholm for each year between 1928 and 1979: of the 29 species recorded on the island, only six bred every year, whilst eight species were either initially absent and then, after arrival, continuously bred, or continuously bred and then became extinct. The remaining 15 species showed the 'come-and-go' behaviour, breeding some years but not others, although eight of these species were recorded as breeding only once or twice. This latter form of extinction is often referred to as 'turnover'.

The conclusions that can be drawn from this review are (i) that there probably are no such things as definitive minimum areas and minimum population sizes, and (ii) that the extinctions of concern to the conservationist are those of the permanently breeding species rather than those of the 'come-and-go' species. Perhaps the best maxim is that the larger the area, or the larger the population size, the better, since the chance of extinction is likely to be reduced. However, in relation to such a discussion one of the unexplained results relates to East and Williams' (1984) study of the indigenous forest birds of New Zealand. They concluded that area requirements increased with the degree of endemism: will it prove generally to be the case that endemic species, often the focus of greatest conservation attention, need larger reserves than non-endemic species?

DISCUSSION

The most difficult feature of preparing a review that attempts to answer the question 'What are the effects of fragmentation?' is the bewildering series of contradictions in the literature. Almost always when one has found a few studies that indicate that there may be some 'rule', one then finds a study that points in exactly the opposite direction. In reviewing the results of research one must always bear in mind a historical perspective (cf. Grime 1984) since the majority of the world's ecosystems have already been affected, either directly or indirectly, by man. Given that research reports will originate from a spectrum of ecosystems, from the little altered to the drastically altered, are there any general principles that emerge? There are perhaps four general concepts that can be enumerated that cut across the hierarchy from community to species that has already been reviewed.

First, conservationists must recognize the existence of succession, spatial processes and ecological gradients, and reserves must be large enough to encompass the range of such processes and gradients. Watt's (1947) data, already reviewed, highlighted the spatial processes occurring in many plant species that provide the structure to the ecosystem, and van der Maarel (1980) urged managers to consider the gradient structure of the systems that they are conserving. Successional processes are all too frequently forgotten by conservationists (Jefferson and Usher 1986) both in North America and Western Europe. In Australia, with the great awareness of fire ecology, successional processes are perhaps taken into greater consideration. Studies such as that of Fox (1982) on the small mammals invading burnt areas of the Myall Lakes National Park in New South Wales are clearly important when management is aiming to retain the whole of the community. In Watt's terminology, 'pattern and process' should be as important to the conservationist as to the ecologist.

Secondly, the community level (considering either plants or animals) in the hierarchy has little to offer when considering the design of nature reserves or national parks. The majority of authors are critical, either because evidence can be found both in favour of or against any design (Margules *et al.* 1982), or because the more theoretical approaches explain so little of the variation found in the real world (Boecklen and Gotelli 1984) or are meaningless [Abbott (1983) criticized equation (1) since z has no unique biological meaning]. The arguments in favour of several small as opposed to one large reserve (e.g. Higgs and Usher 1980) have tended to concentrate on the number of species on reserves, whereas those in favour of a single large reserve tend to concentrate on the stochastic extinction of species when the population size is small (e.g. Wright and Hubbell 1983). Other studies (e.g. Blake and Karr 1984; White, Miller and Bratton 1984) have tended to come down on neither side, indicating that single large areas are better for some species, whilst several small areas are better for other species. Amidst such a welter of contradictions, perhaps the only conclusion is that the best policy is to have a mixed collection of both large and small areas, such as in Figure 9 (though for the English woodlands it could be argued that there are insufficient large areas).

Thirdly, many studies point to the importance of the species level in the hierarchy. Studies that have reviewed the possibilities of island biogeographical theory (e.g. Abbott 1983; Lynch and Whigham 1984) have concluded that counting the numbers of species is insufficient or that estimation under field conditions of immigration and extinction rates is extremely difficult. One of the problems with any survey procedure is that the sighting, trapping or collection of an individual of species k proves that that species is present, whereas a lack of observations of species k does not prove that it is absent. After a critical review, Abbott (1983) concludes 'A more fruitful approach would involve studying the biology of the individual species, establishing their requirements for minimum area and habitat and determining which factors restrict their ... dispersal'. Such an approach is similar to Grubb's (1976) emphasis on the strategies of plants on chalk grassland reserves, or to Kitchener's (1982) division of vertebrate species in the Western Australian wheatbelt into u-species (those of natural vegetation) and d-species (those of disturbed habitats). Conservationists are often not interested in *all* species that occur on fragment, but only in those that are *characteristic* of that habitat [as, for example, the set of all spiders as opposed to the set of heathland spiders studied on the fragmented Dorset heathlands by Webb *et al.* (1984)].

Fourthly, the majority of data has been assembled as a result of surveys rather than as a result of experimentation. Lovejoy *et al.*'s (1983) experiments in Amazonia, which aimed to monitor the changes in isolated blocks of tropical forest of varying size, provided an example of the kind of experiment that would be useful in other habitat types. Evidence from the literature tends to indicate that conservationists are excellent surveyors but reluctant experimenters!

Finally, does fragmentation of habitats lead to a loss of species? Intuitively, an affirmative answer would be expected. Ratcliffe (1984) documented the rates of fragmentation of some habitats in the British Isles: 95% of natural lowland grasslands have been converted to other forms of land use, and semi-natural woodland only covers about 1.3% of Britain's land surface. However, very few species associated with either of these habitats are known to have become extinct. Ratcliffe lists as losses from the flora seven species of flowering plants prior to 1930, and 12 species after 1930. The lists tend to indicate that these were species of wet habitats, or weeds of arable fields, rather than of species characteristic of grasslands and woodlands. In the USA Whitney and Somerlot (1985) indicated that during woodland fragmentation the common species of vascular plants have been proportionally largely reduced in abundance, but that extinctions are minimal. Despite the fears of adverse genetic consequences, and the problems of chance extinction in small isolated populations, are many species robust enough to avoid final extinction?

ACKNOWLEDGEMENTS

I should like to thank the British Ecological Society for funding the 1985 survey of the limestone pavements on Ingleborough, and Dr S. D. Ward for making available data from the 1973/4 survey and for help in 1985. I should also like to acknowledge the financial contributions from the Organizing Committee of the Workshop on the role of remnants (Australia) and from both the Royal Society and the Natural Environment Research Council (U.K.) towards the cost of my visit to Australia.

REFERENCES

Abbott, I., 1983. The meaning of z in species-area regressions and the study of species turnover in island biogeography. *Oikos* 41: 385-90.

An Foras Forbartha, 1981. Areas of Scientific Interest in Ireland. An Foras Forbartha, Dublin.

Austin, M. P. and Margules, C. R., 1986. Assessing representativeness. Pp. 45-67 *in* Wildlife Conservation Evaluation ed by M. B. Usher. Chapman and Hall, London and New York.

Beard, J. S., 1980. A new phytogeographic map of Western Australia. *West. Aust. Herb. Res. Notes* 3: 37-58.

Benedict, J. B., 1984. Rates of tree-island migration, Colorado Rocky Mountains, USA. *Ecology* 65: 820-3.

Berry, R. J., 1973. Chance and change in British long-tailed field mice *(Apodemus sylvaticus)*. *J. Zool., Lond.* 170: 351-66.

Berry, R. J., 1977. Inheritance and Natural History. Collins, London.

Blake, J. G., 1983. Trophic structure of bird communities in forest patches in east-central Illinois. *Wilson Bull.* 95: 416-30.

Blake, J. G. and Karr, J. R., 1984. Species composition of bird communities and the conservation benefit of large versus small forests. *Biol. Conserv.* 30: 173-87.

Boecklen, W. J. and Gotelli, N. J., 1984. Island biogeographic theory and conservation practice: species-area or speciousarea relationships? *Biol. Conserv.* 29: 63-80.

Bowler, J. M., 1982. Aridity in the late Tertiary and Quaternary. Pp. 35-45 *in* Evolution of the Flora and Fauna of Arid Australia ed by W. H. Barker and P. J. M. Greenslade. Peacock Publications, Frewville, South Australia.

Clapham, A. R., Tutin, T. G. and Warburg, E. F., 1981. Excursion Flora of the British Isles, 3rd edition. Cambridge University Press, Cambridge.

Connell, J. H., Tracey, J. G. and Webb, L. J., 1984. Compensatory recruitment, growth, and mortality as factors maintaining rainforest tree diversity. *Ecol. Monogr.* 54: 141-64.

Diamond, J. M., 1975. The island dilemma: lessons of modern biogeographic studies for the design of nature reserves. *Biol. Conserv.* 7: 129-46.

East, R., 1983. Application of species-area curves to African savannah reserves. *Afr. J. Ecol.* 21: 123-8.

East, R. and Williams, G. R., 1984. Island biogeography and the conservation of New Zealand's indigenous forest-dwelling avifauna. *N.Z. J. Ecol.* 7: 27-35.

Fox, B. J., 1982. Fire and mammalian secondary succession in an Australian coastal heath. *Ecoloy* 63: 1332-41.

Frankel, O. H. and Soulé, M. E., 1981. Conservation and Evolution. Cambridge University Press, Cambridge.

Fuller, R. J. and Langslow, D. R., 1986. Ornithological evaluation for wildlife conservation. Pp. 274-69 *in* Wildlife Conservation Evaluation ed by M. B. Usher. Chapman and Hall, London and New York.

Gandawijaja, D. and Arditti, J., 1983. The orchids of Krakatau: evidence for a mode of transport. *Ann. Bot.* 52: 127-30.

Gilbert, F. S., 1981. What use, island biogeography? *Ecos* 2: 18-20.

Goodman, D. (in press). The minimum viable population problem. I. The demography of chance extinction. *In* Viable populations for Conservation ed by M. Soulé. Cambridge University Press, Cambridge.

Greig-Smith, P., 1983. Quantitative Plant Ecology, 3rd edition. Blackwell, Oxford.

Grime, J. P., 1984. The ecology of species, families and communities of the contemporary British flora. *New Phytol.* 98: 15-33.

Grubb, P. J., 1976. A theoretical background to the conservation of ecologically distinct groups of annuals and biennials in the chalk grassland ecosystem. *Biol. Conserv.* 10: 53-76.

Grubb, P. J., 1977. The maintenance of species-richness in plant communities: the importance of the regeneration niche. *Biol. Rev.* 52: 107-45.

Haila, Y., 1983. Ecology of island colonization by northern land birds: a quantitative approach. Doctoral Thesis, University of Helsinki.

Harris, L. D., McGlothlen, M. E. and Manlove, M. N., 1984. Genetic resources and biotic diversity. Pp. 93-107 *in* The Fragmental Forest ed by L. D. Harris. University of Chicago Press, Chicago and London.

Helliwell, D. R., 1983. The conservation value of areas of different size: Worcestershire Ponds. *J. Environ. Man.* 17: 179-84.

Henderson, M. T., Merriam, G. and Wagner, J., 1985. Patchy environments and species survival: chipmunks in an agricultural mosaic. *Biol. Conserv.* 31: 95-105.

Higgs, A. J. and Usher, M. B., 1980. Should nature reserves be large or small? *Nature, Lond.* 285: 568-9.

Hopkins, P. J. and Webb, N. R., 1984. The composition of the beetle and spider faunas on fragmented heathlands. *J. Appl. Ecol.* 21: 935-46.

IUCN, 1980. World Conservation Strategy. International Union for the Conservation of Nature and Natural Resources, Gland, Switzerland.

Järvinen, O., 1982. Conservation of endangered plant populations: single large or several small reserves? *Oikos* 38: 301-7.

Jefferson, R. G. and Usher, M. B., 1986. Ecological succession and the evaluation of non-climax communities. Pp. 69-91 *in* Wildlife Conservation Evaluation ed by M. B. Usher. Chapman and Hall, London and New York.

Karr, J. R., 1982a. Avian extinction on Barro Colorado Island, Panama: a reassessment. *Am. Nat.* 119: 220-39.

Karr, J. R., 1982b. Population variability and extinction in the avifauna of a tropical land bridge island. *Ecology* 63: 1975-8.

Kitchener, D. J., 1982. Predictors of vertebrate species richness in nature reserves in the Western Australian wheatbelt. *Aust. Wildl. Res.* 9: 1-7.

Kobayashi, S., 1983. The species-area relation for archipelago biotas: islands as samples from a species pool. *Res. Popul. Ecol.* 25: 221-37.

Lamb, H. H., 1969. The new look of climatology. *Nature, Lond.* 223: 1209-15.

Laut, P., Heyligers, P. C., Keig, G., Loffler, E., Margules, C., Scott, R. M. and Sullivan, M. E., 1977. Environments of South Australia, 8 province reports and handbook. CSIRO, Melbourne.

Lovejoy, T. E., Bierregaard, R. O., Rankin, J. M. and Schubart, H. O. R., 1983. Ecological dynamics of tropical forest fragments. Pp. 377-84 *in* Tropical Rain Forest: Ecology and Management ed by S. L. Sutton, T. C. Whitmore and A. C. Chadwick. Blackwell, Oxford.

Lynch, J. F. and Whigham, D. F., 1984. Effects of forest fragmentation on breeding bird communities in Maryland, USA. *Biol. Conserv.* 28: 287-324.

MacArthur, R. H. and Wilson, E. O., 1967. Island Biogeography. Princeton University Press, Princeton.

McCoy, E. D., 1982. The application of island biogeographic theory to forest tracts: problems in the determination of turnover rates. *Biol. Conserv.* 22: 217-27.

McCoy, E. D., 1983. The application of island-biogeographic theory to patches of habitat: how much land is enough? *Biol. Conserv.* 25: 53-61.

Mader, H. J., 1984. Animal habitat isolation by roads and agricultural fields. *Biol. Conserv.* 29: 81-96.

Malanson, G. P., 1982. The assembly of hanging gardens: effects of age, area and location. *Am. Nat.* 119: 145-50.

Margules, C. R., Higgs, A. J. and Rafe, R. W., 1982. Modern biogeographic theory: are there any lessons for nature reserve design? *Biol. Conserv.* 24: 115-28.

Margules, C. R. and Usher, M. B., 1981. Criteria used in assessing wildlife conservation potential: a review. *Biol. Conserv.* 21: 79-109.

Miller, R. I., 1979. Conserving the genetic integrity of faunal populations and communities. *Environ. Conserv.* 6: 297-304.

Peterken, G. F., 1981. Woodland Conservation and Management. Chapman and Hall, London and New York.

Rafe, R. W., 1983. Species-area relationships in conservation. D. Phil. Thesis, University of York.

Rafe, R. W., Usher, M. B. and Jefferson, R. G., 1985. Birds on reserves: the influence of area and habitat on species richness. *J. Appl. Ecol.* 22: 327-35.

Ratcliffe, D. R. (ed.), 1977. A Nature Conservation Review, vols. 1 and 2. Cambridge University Press, Cambridge.

Ratcliffe, D. R., 1984. Post-medieval and recent changes in British vegetation: the culmination of human influence. *New Phytol.* 98: 73-100.

Reed, T. M., 1983. The role of species-area relationships in reserve choice: a British example. *Biol. Conserv.* 25: 263-71.

Riebesell, J. F., 1982. Arctic-alpine plants on mountaintops: agreement with island biogeography theory. *Am. Nat.* 119: 657-74.

Runkle, J. R., 1981. Gap regeneration in some old-growth forests of the eastern United States. *Ecology* 62: 1041-51.

Soulé, M. E., 1979. Heterozygosity and developmental stability: another look. *Evolution* 33: 396-401.

Terborgh, J. and Winter, B., 1983. A method for siting parks and reserves with special reference to Colombia and Ecuador. *Biol. Conserv.* 27: 45-58.

Terborgh, J. W. and Faaborg, J., 1980. Saturation of bird communities in the West Indies. *Am. Nat.* 116: 178-95.

Thornton, I. W. B., 1984. Krakatau — the development and repair of a tropical ecosystem. *Ambio* 13: 216-25.

Usher, M. B., 1973. Biological Management and Conservation: Ecological Theory, Application and Planning. Chapman and Hall, London.

Usher, M. B., 1976. Aggregation responses of soil arthropods in relation to the soil environment. Pp. 61-94 *in* The Role of Terrestrial and Aquatic Organisms in Decomposition Processes ed by J. M. Anderson and A. Macfayden. Blackwell, Oxford.

Usher, M. B., 1979. Changes in the species-area relations of higher plants on nature reserves. *J. Appl. Ecol.* 16: 213-5.

Usher, M. B., 1985a. An assessment of species-area relationships using ornithological data. Pp. 159-70 *in* Statistics in Ornithology ed by B. J. T. Morgan and P. M. North. Springer-Verlag, Berlin.

Usher, M. B., 1985b. Implications of species-area relationships for wildlife conservation. *J. Environ. Man.* 21: 181-91.

Usher, M. B. (ed.), 1986a. Wildlife Conservation Evaluation. Chapman and Hall, London and New York.

Usher, M. B., 1986b. Wildlife conservation evaluation: attributes, criteria and values. Pp. 3-44 *in* Wildlife Conservation Evaluation ed by M. B. Usher. Chapman and Hall, London and New York.

van der Maarel, E., 1980. Towards an ecological theory of nature conservation. *Verh. Ges. Oekol.* 8: 13-24.

Verkaar, H. J., Schenkeveld, A. J. and van de Klashorst, M. P., 1983. The ecology of short-lived forbs in chalk grasslands: dispersal of seeds. *New Phytol.* 95: 335-44.

Ward, S. D. and Evans, D. F., 1975a. A botanical survey and conservation assessment of limestone pavements based on botanical criteria: collection and treatment of the data. Duplicated report of the Institute of Terrestrial Ecology, Bangor, Gwynedd.

Ward, S. D. and Evans, D. F., 1975b. The limestone pavements of Ingleborough: a botanical survey and conservation assessment based on botanical criteria. Duplicated Report of the Institute of Terrestrial Ecology, Bangor, Gwynedd.

Ward, S. D. and Evans, D. F., 1976. Conservation assessment of British limestone pavements based on floristic criteria. *Biol. Conserv.* 9: 217-33.

Watt, A. S., 1947. Pattern and process in the plant community. *J. Ecol.* 35: 1-22.

Webb, N. R., Clarke, R. T. and Nicholas, J. T., 1984. Invertebrate diversity on fragmented *Calluna*-heathland: effects of surrounding vegetation. *J. Biogeogr.* 11: 41-6.

Webb, N. R. and Hopkins, P. J., 1984. Invertebrate diversity on fragmented *Calluna* heathland. *J. Appl. Ecol.* 21: 921-33.

White, P. S., Miller, R. I. and Bratton, S. P., 1984. Island biogeography and preserve design: preserving the vascular plants of Great Smoky Mountains National Park. *Natural Areas J.* 3: 4-13.

White, P. S., Miller, R. I. and Ramseur, G. S., 1984. The species-area relationship of the southern Appalachian high peaks: vascular plant richness and rare plant distributions. *Castanea* 49: 47-61.

Whitney, G. G. and Somerlot, W. J., 1985. A case study of woodland continuity and change in the American Midwest. *Biol. Conserv.* 31: 265-87.

Williamson, M., 1983. The land-bird community of Skokholm: ordination and turnover. *Oikos* 41: 378-84.

Williamson, M., 1984. Sir Joseph Hooker's lecture on insular floras. *Biol. J. Linn. Soc.* 22: 55-77.

Williamson, M. (in press). Are communities ever stable? *In* Colonization, Succession and Stability ed by A. J. Gray, P. J. Edwards and M. J Crawley. Blackwell, Oxford.

Woolhouse, M. E. J., 1983. The theory and practice of the species-area effect, applied to the breeding birds of British woodlands. *Biol. Conserv.* 27: 315-32.

Wright, S. J. and Hubbell, S. P., 1983. Stochastic extinction and reserve size: a focal species approach. *Oikos* 41: 466-76.

CHAPTER 10

Responses of Breeding Bird Communities to Forest Fragmentation

James F. Lynch[1]

Field studies in eastern North America indicate that local densities of most forest-dwelling bird species are directly or indirectly influenced by forest insularization. Relevant site variables among those measured include patch area and isolation, tree stature and density, and development of herbaceous and shrub understorey. In general, highly migratory species that specialize on forest-interior habitat are adversely affected by forest fragmentation, whereas forest-edge species, particularly year-round inhabitants (here termed 'residents'), tend to achieve higher local densities in fragmented forests. In eastern North America, rates of nest parasitism and nest predation are correlated with patch insularity. Several life history characteristics appear to make highly migratory species especially sensitive to these direct agents of reproductive failure.

The regional integrity of eastern North America's avifauna is maintained by frequent exchange of propagules among forest patches, few of which are sufficiently large to maintain a stable avifauna *in vacuo*. This pattern of frequent re-invasion of small forested tracts may be less common at lower latitudes, where many bird species are sedentary habitat specialists.

In attempting to determine the optimal size and spatial arrangement of forest reserves for bird conservation, the absolute geometric scale of potential reserves, the functional 'grain' of the regional habitat mosaic, the degree of ecological specialization of the bird species to be conserved, and their dispersal all must be considered.

INTRODUCTION

OVER the past two decades, theoretical ecologists and conservationists have attempted to assess the effects of habitat insularity on bird populations and communities. MacArthur and Wilson's (1963, 1967) equilibrium theory of island biogeography has provided a common theoretical framework for many of these studies (for a review, see Abbott 1980). The original focus on actual islands has expanded to include disjunct patches of terrestrial habitat. The latter are often (if somewhat euphemistically) depicted as 'islands' surrounded by a 'sea' of disturbed or otherwise unsuitable terrain. The extension of the insular metaphor from real to virtual islands was anticipated by MacArthur and Wilson (1967: 114), who also pointed out two important distinctions: (1) immigration rates are generally much lower on true islands because of their high degree of effective isolation, and (2) competition and other negative interactions with species intruding from neighboring habitats occur less frequently on true islands than in mainland habitat patches. Typically, the surroundings of mainland habitat 'islands' are not mere passive barriers to dispersal, but are staging areas for potential predators, parasites, and competitors.

Because the analogy between true islands and mainland habitat patches is imperfect, and because the validity of the equilibrium theory itself has increasingly been called into question (e.g., Simberloff 1976; Abbott 1980; Gilbert 1980; Higgs 1981; Margules *et al.* 1982; Simberloff and Abele 1982), attempts to apply the specific tenets of MacArthur and Wilson's equilibrium theory to habitat patches are controversial. Nevertheless, certain responses of birds to habitat fragmentation have been documented, whatever their underlying causal mechanisms ultimately prove to be.

The aim of this chapter is to summarize some of the major empirical findings concerning the effects of habitat insularization on forest-dwelling birds. I

[1]Smithsonian Environmental Research Centre, P.O. Box 28, Edgewater, Maryland 21037, USA.

will emphasize my own research and that of other investigators who have worked in the Middle Atlantic region of the USA, an area that has been better studied in this regard than has any other section of North America. I will describe the patterns that have emerged, discuss possible causal mechanisms, note some conservation applications, and identify important questions which invite additional research. I will briefly compare the results of work done in North America with studies done elsewhere, particularly Britain and Australia. Throughout, I will emphasize small birds (mainly passerines) that inhabit forested tracts.

EFFECTS OF FRAGMENTATION ON THE AVIFAUNA OF EASTERN NORTH AMERICA

The Process and Pattern of Habitat Insularization

The arrival of European settlers in eastern North America in the early 17th century heralded two centuries of relentless forest-cutting. What must have been an essentially continuous expanse of forest was reduced to patches and strips of woodland in most of what is now the eastern United States. The farms cleared by settlers in the 17th and early 18th centuries were clustered along the Atlantic seaboard, but the most complete deforestation occurred in the midwestern states following the crossing of the Appalachian Mountain barrier in the late 18th and early 19th centuries. Thus, the Atlantic seaboard has remained a complex mosaic of farms, abandoned fields, woodlands, and riparian strips, and the percentage of forested land there has actually increased over the past century (Aldrich and Robbins 1970; Morse 1980). Although no sizeable tracts of virgin forest have survived at lower elevations anywhere in the eastern USA, second growth forest covers 30-50% of many of the states east of the Appalachian crest. In contrast, deforestation in much of the American midwest has been inexorable (e.g., Curtis 1956; Whitney and Somerlot 1985). Patterns of land use in the latter region resemble those in such well-studied regions as Britain and the wheatbelt of Western Australia. In all of these areas, forest survives mainly as small, disjunct patches.

Changes in the Avifauna and their Causes

What have been the net effects of post-settlement deforestation on the resident avifauna of the eastern USA? We can say with certainty that few species have become extinct. A continental avifauna that included perhaps 660 species at the time of the arrival of Europeans has suffered fewer than ten global extinctions. Within the eastern deciduous forest region, only the passenger pigeon, *Ectopistes migratorius* and Carolina parakeet, *Conuropsis carolinensis* are certainly extinct. Two additional species, the ivory-billed woodpecker, *Campephilis principalis* and Bachman's warbler, *Vermivora bachmanii*, have probably been extirpated, but only a few others are sufficiently rare to be in serious danger of global extinction.

At the local and regional levels, however, the future survival of the avifauna associated with the eastern deciduous forest is problematical (Whitcomb *et al*. 1981). Analysis of long-term census data for isolated parks and preserves in the eastern U.S. (Lynch and Whitcomb 1978) has revealed two major patterns. Firstly, the bird communities inhabiting individual forest patches are highly unstable. Mean annual species turnover rates are so high (often 10-25%) that only a minority of the local pool of bird species is present every year within any given forest tract. Secondly, a systematic decline in the abundance of a number of forest-dwelling species began in the 1950's and 1960's (Fig. 1). The explanation for these reductions, which have resulted in the total disappearance of many previously common species from wooded tracts, remains obscure. Among the causal mechanisms that have been suggested (Whitcomb *et al*. 1976, 1981; Lynch and Whitcomb 1978; Terborgh 1980; Morse 1980; Lynch and Whigham 1984) are: (1) increasing isolation of reserves and parks due to urbanization of their surroundings, (2) encroachment by predators, competitors, and nest-parasites from adjacent disturbed habitats, (3) mortality due to widespread use of pesticides in agriculture and forestry, and (4) global reduction in the abundance of many migratory species due to massive habitat destruction on their tropical wintering grounds. This last factor is potentially of overwhelming importance in eastern North America, where 50-75% of the individual breeding pairs of forest-dwelling birds are neotropical migrants. Moreover, these long-distance migrants are the very species that have shown the most severe declines in parks and woodlots of eastern North America (Lynch and Whitcomb 1978; Whitcomb *et al*. 1981). Most resident species, on the other hand, have either held their own, or actually have increased in abundance within isolated forest patches.

Effects on Breeding Birds of Forest Fragmentation in the Maryland Coastal Plain

With the foregoing historical perspective in mind, Dennis Whigham and I undertook a 2-year study of breeding birds associated with forested tracts in the Atlantic Coastal Plain province of Maryland, east of Washington, D.C. (Fig. 2). Although the 6-county study area lies within the heavily populated urban-suburban corridor of the eastern United States, considerable forest remains. Most water courses are bordered by strips of riparian woodlands, and upland forest covers from 35-65% of the land area in each of the six counties. Most of this forest has regenerated following the widespread abandonment of farming in the late 19th and early 20th centuries. The high proportion of forest extant in the study

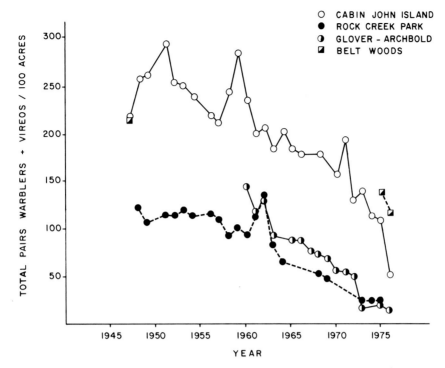

Fig. 1. Historical decline in the combined abundance of migratory wood warblers (Parulinae) and vireos (Vireonidae) in four isolated forest tracts in Maryland and the District of Columbia (USA). The patches themselves have undergone only minor disturbance over the past several decades, but the surrounding region has become increasingly urbanized. Figure redrawn from Lynch and Whitcomb (1978).

area, together with the high degree of connectivity imparted by riparian woodlands has resulted in a relatively low degree of functional insularization for the eastern Maryland landscape.

We surveyed breeding bird communities in 270 forest patches. We measured area, isolation, and various floristic and physiognomic features of the vegetation in 183 of these patches. Using stepwise linear multiple regression, we tested 13 habitat variables as predictors of (a) the local occurrence of each of the 31 most common forest-dwelling bird species, and (b) the magnitudes of several community variables (species richness, total number of pairs, number of species and pairs of migrants, number of species and pairs of residents). Our analysis (Lynch and Whigham 1984) revealed several major patterns (Tables 1 and 2).

1. Local abundance of all but one of the 31 bird species showed statistically significant associations with one or more of the habitat variables. The single exception was the pileated woodpecker *Dryocopus pileatus*; occurrence data for this species were too few for meaningful statistical analysis.

2. Vegetation descriptors and patch isolation indices were more efficient predictors of local abundance than was patch area *per se*. Although patch area (or its logarithm) made statistically significant contributions to the regressions for several species, it was the most important predictor of abundance for only one, the Kentucky warbler, *Oporornis formosus*.

3. As a group, long-distance migrants tended to respond negatively to what might loosely be termed 'habitat degradation' factors (e.g., reduction in patch area, increase in patch isolation, decrease in tree stature, reduction in plant diversity within patches). Residents tended to be less responsive to these factors, or to react in the opposite direction to long-distance migrants. Short-distance migrants were intermediate in their response to habitat insularization.

The density responses of highly migratory species (most of which require closed-canopy forest as breeding habitat) and residents (which, as a group, are more tolerant of 'edge' and other nonforest conditions) tend to be complementary. As a result, overall species diversity and total abundance of individuals remain fairly constant over a wide range of patch configurations and floristic types. If anything, bird abundance and species diversity tend to be *higher* in small forest patches in Maryland, due to an influx of common 'edge' species. This result highlights the pitfall of a simplistic numbers game in conservation planning. If one chooses to consider all species to be equal in conservation value, habitat fragmentation may not appear to be very harmful.

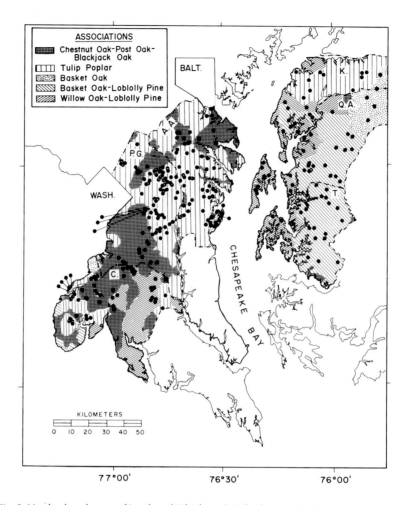

Fig. 2. Maryland study area of Lynch and Whigham (1984), showing the locations of the forest patches that were studied in relation to major vegetation types. Point surveys of breeding birds were conducted in 270 patches.

Table 1. Statistically significant predictors of breeding bird community properties in 270 forest patches located on the Atlantic coastal plain of Maryland (USA). Western Shore localities (n = 196) are west of Chesapeake Bay; Eastern Shore localities (n = 74) are east of the Bay. Table is based on results of stepwise linear multiple regression analysis (Lynch and Whigham 1984). Significance levels: * = P<.05; ** = P<.01; *** = P<.001. 'Neg' indicates that regression coefficient has negative sign.

Community Characteristic	Significant Predictors	
	Western Shore	Eastern Shore
Total bird species	Abundance of pines (neg)* Isolation (neg)*	Patch area (neg)*** Canopy height (neg)* Herbaceous cover (neg)*
Forest-interior species	Herb diversity** Isolation (neg)**	None
Edge-field species	Patch area (neg)*** Canopy height (neg)*	Patch area (neg)* Herbaceous cover* Canopy height*
Total pairs of birds	Isolation (neg)** Isolation (neg)*** Shrub density*	Patch area (neg)**
Forest-interior pairs	Herbaceous cover*** Isolation (neg)*** Total basal area of trees***	Isolation (neg)**
Edge-field pairs	Patch area (neg)*** Canopy closure (neg)* Total basal area of trees (neg)*	Patch area (neg)*** Number of plant species* Canopy height*

Table 2. Statistical significant predictors of the point density of individual breeding bird species in 270 forest patches on the Atlantic coastal plain of Maryland (data combined for eastern and western sides of Chesapeake Bay). Significance levels and source of data as in Table 1. Data from Lynch and Whigham (1984).

Species	Significant Predictors
NEOTROPICAL MIGRANTS	
Cuculidae	
Coccyzus americanus	Total plant species (neg)***, Isolation***
Trochilidae	
Archilochus colubris	Shrub diversity*
Tyrannidae	
Empidonax virescens	Total plant species***, Isolation (neg)***
Contopus virens	Canopy height***, Patch area*
Myiarchus crinatus	Isolation*, Herbaceous cover (neg)*, Abundance of pines**
Turdinae	
Hylocichla mustelina	Isolation (neg)***, Canopy height**, Total basal area of trees**
Parulinae	
Mniotilta varia	Total basal area of trees (neg)***, Isolation (neg)*, Patch Area*
Wilsonia citrina	Total plant species**, Herbaceous cover**, Shrub diversity (neg)*, Tree density (neg)*
Oporornis formosus	Patch area***, Herbaceous cover**, Shrub diversity (neg)*, Tree density (neg)*
Parula americana	Total plant species***, Patch area**
Seiurus auricapillus	Isolation (neg)***, Tree density***, Shrub diversity***, Herbaceous cover*
Helmitheros vermivorus	Abundance of pines***, Isolation (neg)**
Vireonidae	
Vireo olivaceus	Total plant species***, Abundance of pines (neg)**
Vireo griseus	Herbaceous cover**, total basal area of trees (neg)**
Vireo flavifrons	Shrub diversity**, Isolation (neg)**
Thraupinae	
Piranga olivacea	Shrub diversity**, Abundance of pines (neg)*
SHORT-DISTANCE MIGRANTS	
Picidae	
Colaptes auratus	Isolation***, Total plant species**
Mimidae	
Dumatella carolinensis	Patch area (neg)***, Shrub diversity (neg)**, Abundance of pines (neg)*
Corvidae	
Cyanocitta cristata	Total plant species (neg)***, Isolation**, Shrub density**, Canopy height*
Parulinae	
Dendroica pinus	Abundance of pines***, Canopy closure (neg)*, Shrub diversity*
Sylviidae	
Polioptila caerulea	Canopy closure (neg)**, Isolation (neg)**
Cardinalinae	
Pipilo erythrophthalmus	Tree density**, Tree diversity (neg)**
RESIDENTS	
Picidae	
Picoides pubescens	Patch area (neg)**, Abundance of pines**, Isolation (neg)**, Canopy height*
Picoides villosus	Isolation**
Dryocopus pileatus	None
Melanerpes carolinus	Total basal area of trees***, Tree density (neg)***
Troglodytidae	
Thryothorus ludovicianus	Isolation**
Sittidae	
Sitta carolinensis	Canopy height***
Paridae	
Parus carolinensis	Shrub diversity (neg)**
Parus bicolor	Canopy closure (neg)*
Cardinalinae	
Cardinalis cardinalis	Herbaceous cover***, Shrub density (neg)**

Other Surveys in Maryland

General Background. A very extensive study of birds in forest patches has recently been completed by Robbins et al. (ms.), who studied 469 forested tracts in Maryland and adjacent portions of the states of Pennsylvania, Virginia, and West Virginia. Some of the results of this important study were discussed in earlier publications by Robbins and his co-workers (Anderson and Robbins 1981; Robbins 1980; Whitcomb et al. 1981). Robbins et al. used the same point census technique as did Lynch and Whigham (1984), and their study area encompassed most of Maryland, except for the six Coastal Plain counties surveyed by Lynch and Whigham. Robbins *et al.* quantified vegetation characteristics, as well as patch area and isolation, and used stepwise curvilinear regression to relate these habitat descriptors to bird occurrence in forest patches. Because of the broader geographic scope and greater sampling intensity of their study, Robbins *et al.* obtained sufficient data to analyse the distributions of 76 species (in comparison to the 31 considered by Lynch and Whigham).

Although the study of Robbins *et al.* is very similar in orientation to that of Lynch and Whigham, there are also some important differences between the two studies that must be kept in mind if they are to be compared. First, Robbins *et al.* included a wider range of major physiographic regions (coastal plain, piedmont, and montane formations were studied) than did Lynch and Whigham, whose study was restricted to the coastal plain. In addition, Robbins *et al.* surveyed both riparian and upland forests, whereas Lynch and Whigham considered only upland plots.

Robbins *et al.*'s study included a very broad range of patch sizes (0.1 to 3300 ha). Lynch and Whigham intentionally rejected patches smaller than 10 ha in order to minimize 'contamination' of their point surveys by species living outside the forest patches, and their largest tracts were only about 1000 ha. The greater range in size of the tracts surveyed by Robbins *et al.*, together with their use of curvilinear regression models (as opposed to the linear models employed by Lynch and Whigham), would be expected to increase the magnitude of any underlying correlations between forest area and bird occurrences, as indeed proved to be the case. On the other hand, Robbins *et al.* surveyed only mature stands of forest, whereas Lynch and Whigham also considered successional woodlands, so the latter authors placed more emphasis on local (as opposed to regional) habitat gradients. Finally, some of the specific habitat descriptors used by Robbins *et al.* differ from those employed by Lynch and Whigham, so the absolute values of correlation or regression coefficients cannot always be directly compared in the two studies.

Results of the Statewide Survey of Maryland. Robbins *et al.* (ms.) found patch isolation to be the single best predictor of bird occurrence. Isolation made a statistically significant contribution to the predictive multiple regression equation in 43 of 76 (57%) species. Patch area was a statistically significant predictor variable for 28 of 76 (37%) species. In at least nine of these 28 regressions the significance level (hence, partial correlation coefficient) for patch area was substantially lower than for one or more other predictor variables. Thus, patch area played a major role in predicting the occurrence of no more than 25% of the species studied. By comparison, isolation (estimated by Robbins *et al.* as the percentage of regional forest cover) played a substantial role in 45% of the species regressions, and various aspects of forest physiognomy were highly significant predictors for all but a handful of the 76 species. In discussing the results of their study Robbins *et al.* emphasize the utility of forest area as a predictor of bird occurrence, but their data suggest that within-patch habitat characteristics and regional forest cover together play a more important role than patch area *per se* in determining the suitability of individual forest tracts for particular bird species. This result is in agreement with the findings of Lynch and Whigham (1984). Both studies also agree in indicating that there is no absolute minimum critical area for the occurrence of the smaller bird species in the middle Atlantic region, although many species tend to be encountered more frequently in larger tracts. Given the existing mixture of large and small forest patches in Maryland, together with the relatively low degree of interpatch isolation that now characterizes much of the state, even small tracts are at least occasionally inhabited by almost any of the forest-associated bird species occurring in the state.

In summary, the Maryland studies indicate that complex of ecological variables, some of which are intercorrelated with patch area, act together to determine the suitability of a particular tract for any given bird species. The identities and the relative importance of these determining factors vary from species to species and from one habitat type to another. Nevertheless, if one restricts attention to bird species whose abundance is significantly correlated with patch area, and if patch area is viewed as a convenient index to a plethora of more immediate ecological factors, area-specific 'incidence functions' (*sensu* Diamond 1975) can be helpful in envisioning the impacts of insularization on different bird species (Figs 3 and 4). It is evident that some species, particularly highly migratory forms associated with the interior of forests, are considerably more likely to be encountered in a given sampling period within large tracts than in smaller ones (Fig. 3). The opposite is true for other species (Fig. 4), particularly 'edge' species, whose territories are organized along forest margins. Forest-interior species, on the other hand, either tend to avoid forest-edge, or to utilize the entire forest area for territories (Fig. 5). Finally, some species show no consistent positive or negative relationship with forest area or any of its ecological correlates (Fig. 4).

Limitations of the Point Survey Method

Although the Maryland studies just discussed constitute an extensive data base (more than 700 forest patches were sampled), they cannot be directly

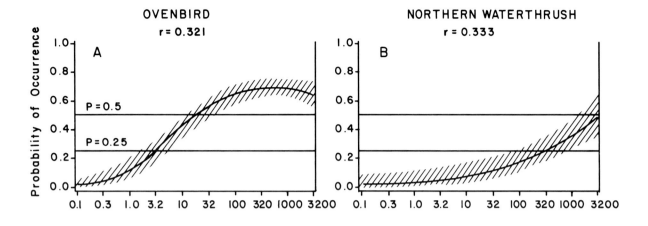

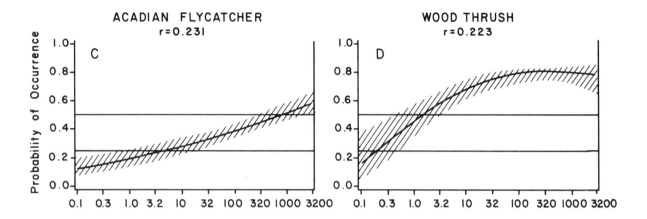

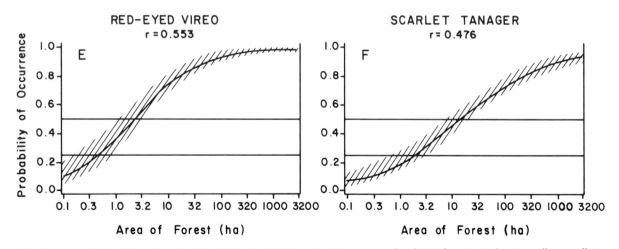

Fig. 3. Area-specific 'incidence functions' for six Maryland bird species that are more abundant in large tracts than in small ones. All six species are long-distance migrants: ovenbird *Seiurus aurocapillus*, Northern waterthrush *S. noveboracensis*, acadian fly-catcher *Empidonax virescens*, wood thrush *Hylocichla mustelina*, red-eyed vireo *Vireo olivaceus*, and scarlet tanager *Piranga olivacea*. Figure modified from Robbins *et al.* (ms). Hatched areas represent 95% confidence limits of the regressions.

compared with studies that employed different methods. The point survey technique calls for equal sampling intensity (three visits of 20 min each) at a single station within each patch. This differs substantially from traditional spot-mapping methods, which attempt to enumerate all species and individuals within a large census plot.

Point surveys are quite efficient for detecting the presence of species within a small patch of forest. Because the detection distance for most Maryland bird species is in the range of 100-200 m (Whitcomb *et al.* 1981), the effective area censused in a point survey is in the order of 3-10 ha. This limitation unavoidably results in undersampling of larger tracts,

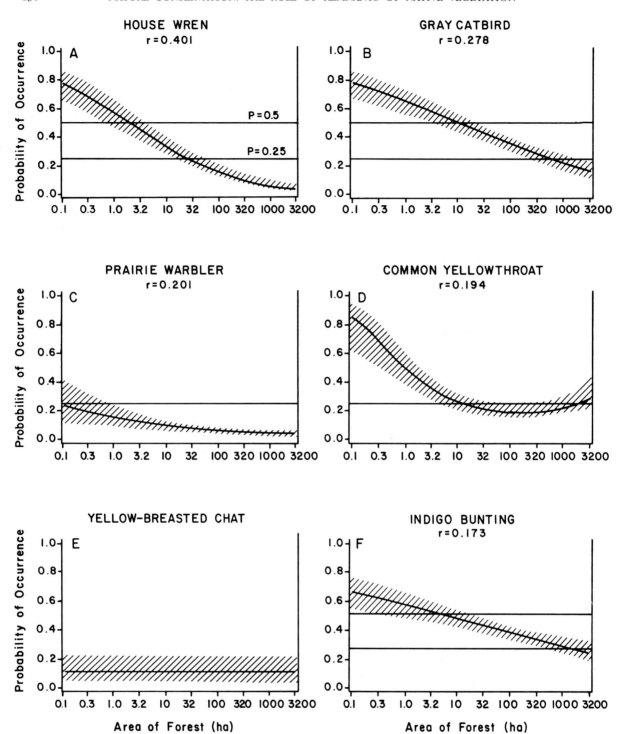

Fig. 4. Area-specific 'incidence functions' for five Maryland bird species that are more abundant in small forest tracts than in large ones, and a sixth species that shows no significant density response to patch area. All six species are more common in forest edge and scrub than in forest-interior. House wren *Troglodytes aedon*, gray catbird *Dumatella carolinensis*, and common yellowthroat *Geothlypis trichas* are short distance migrants, although some individuals of the latter two species migrate as far south as northern Central America. Prairie warbler *Dendroica discolor*, yellow-breasted chat *Icteria galbula*, and indigo bunting *Passerina cyanea* are neotropical migrants. Figure modified from Robbins *et al.* (ms).

and causes one to miss at least some of the uncommon species that may be scattered through an extensive area of forest. To give a concrete example, consider a species such as the Kentucky warbler which in Maryland has a 25% probability of being detected in a set of point surveys conducted within a 100 ha tract (Robbins *et al.* ms). If the area adequately covered by a point survey is 10 ha, then the probability of encountering at least one Kentucky warbler in an exhaustive census of a uniform 100 ha tract can be estimated as $(1-.75^{10}) = .96$. Thus, even if the point density (i.e., abundance per unit

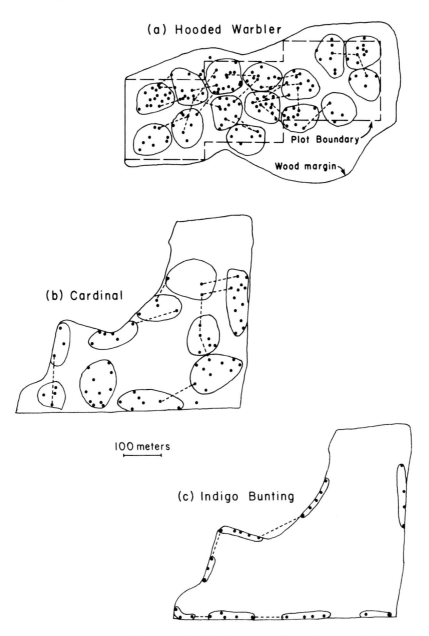

Fig. 5. Mapped territories of three Maryland bird species in forest fragments. Simultaneous registrations of territorial males are connected by dashed lines. (a) Hooded warbler *Wilsonia citrina*, a highly migratory forest-interior specialist that avoids forest edge. One additional unmapped territory was located outside the boundary of the study plot, in the lower right portion of the patch. (b) Cardinal *Cardinalis cardinalis*, an interior-edge species that is most common near the forest edge, but occurs throughout the forest. (c) Indigo bunting *Passerina cyanea*, a species that establishes its territories along the forest margins. Figure modified from Whitcomb *et al.* (1981).

area) of Kentucky warblers is similar in a 100 ha tract as in a 10 ha tract, there is a much higher probability that the species is present *somewhere* within the larger tract.

Thus, point survey data underestimate the frequency of species occurrence per tract, as opposed to the occurrence of individuals per unit area. As a general rule, however, adjusting point survey results to account for this bias will only *increase* the strength of any positive relationship between species incidence and forest area.

Differential Responses of Birds to Forest Fragmentation

In recent years a severely reductionist stance has emerged in zoogeography and conservation (e.g., Simberloff and Abele 1976a, 1976b; Higgs and Usher 1980; Simberloff and Gotelli 1984; Zimmerman and Bierregaard 1986). This position disparages previous attempts at community analysis, and invokes detailed autecological studies as the only valid basis for conservation action. A reductionist programme is indeed warranted if one is concerned with conserving individual species that are of sufficient concern to justify the expenditure of large amounts of time and money. Among North American birds, examples of such 'flagship' species would include the California condor, *Gymnogyps californianus*, with fewer than 20 surviving individuals in the wild (Snyder and Hamber 1985), and a few others. However, conservationists also must consider the long-term preservation of entire assemblages of species, most of which are relatively obscure, and which have no public constituency. Moreover, if the reductionist programme is taken literally, every local situation must be individually evaluated before recommending conservation action. Detailed studies are of course laudable, but given inevitable limitations of time and financial resources, there is not the slightest possibility of conducting in-depth autecological studies of every bird species in every local community, and one must therefore seek ways to generalize about groups of species. In addition to this purely practical consideration, there is the desire shared by many biologists to discern and to explain patterns in nature, as opposed to focusing exclusively on the idiosyncracies of individual species. To dismiss pattern-seeking on the grounds that every species is different is as misguided as to suppose that any two species are alike in every detail of their ecology. The trick, of course, is to define and analyze groups of species in ways that are simultaneously practical, realistic, and revealing. A judicious mix of in-depth autecological studies and broader, more inclusive community-level approaches would appear to offer the best basis for rational conservation action.

As was noted above, one useful way to categorize North American bird species is according to migratory strategy. Whitcomb *et al*. (1981) showed that, as a group, highly migratory species in eastern North America share a suite of ecological and behavioural traits (Table 3). For example, in comparison to sedentary species, long-distance migrants show a greater tendency to nest on or near the ground, to be insectivorous, and to produce smaller clutches. They also tend to react more strongly than do short-distance migrants or resident species to forest insularization. In a similar vein, Greenberg (1980) has drawn attention to a number of demographic correlates of long-distance migration in North American passerines.

Another useful grouping for North American forest birds differentiates 'forest-interior' species (which tend to avoid forest margins) from 'interior-edge' species (which occur throughout the forest) and 'edge-field' species (which occur only at the forest margin and adjacent open habitats). Examples of these three categories are shown in Figure 5. These three broad habitat categories can be combined with the three migratory categories to form nine habitat/migratory strategy classes (Table 3). The distribution of species among these categories has a strong phylogenetic element (Whitcomb *et al*. 1981). In the Maryland study area, for example, all Picidae, Sittidae, Paridae and Corvidae (12 species) are residents, showing, at most, local movements during the non-breeding season. Most of these 12 species are feeding generalists, eating both animal and plant material. On the other hand, 13 of the 15

Table 3. A 3X3 classification of 63 Maryland breeding bird species according to habitat and migratory pattern, showing mean values of key ecological variables. n = number of species in category; CS = mean number of eggs in first clutch; TF = tolerance to forest fragmentation, measured as the point density in small (6-14 ha) forest patches divided by the point density in large (>70 ha) patches; NH = mean nest height (m). Data from Whitcomb *et al*. (1981).

	Habitat Association			
Migratory Pattern	Forest Interior	Interior and Edge	Edge and Scrub	Totals
Resident	n = 3 spp CS = 4.1 TF = 0.5 NH = 8.4	n = 7 spp CS = 6.2 TF = 1.2 NH = 4.8	n = 4 spp CS = 9.9 TF = 2.2 NH = 4.0	n = 14 spp CS = 6.8 TF = 1.2 NH = 5.3
Short-Distance Migrant	n = 1 spp CS = 3.8 TF = 0.5 NH = 10.4	n = 6 spp CS = 7.7 TF = 1.5 NH = 2.0	n = 11 spp CS = 8.2 TF = 1.4 NH = 2.5	n = 18 spp CS = 7.8 TF = 1.4 NH = 2.8
Neotropical Migrant	n = 11 spp CS = 4.6 TF = 0.2 NH = 2.5	n = 11 spp CS = 4.7 TF = 0.7 NH = 4.2	n = 9 spp CS = 4.7 TF (NA) NH = 3.1	n = 31 spp CS = 4.7 TF = 0.4 NH = 3.3
Totals	n = 15 spp CS = 4.4 TF = 0.3 NH = 4.2	n = 24 spp CS = 6.1 TF = 1.0 NH = 3.8	n = 24 spp CS = 7.2 TF = 1.6 NH = 3.0	n = 63 spp CS = 6.0 TF = 0.9 NH = 3.6

local species belonging to the subfamily Parulinae (New World warblers) are insectivorous neotropical migrants, the only two exceptions being short-distance migrants. Of the nine habitat/migration groups, the forest interior/neotropical migrant category shows the strongest negative reaction to forest fragmentation in Maryland (Table 3) and several species in this group have disappeared from entire regions in Maryland and nearby states as urbanization has increased (Robbins 1980; Whitcomb et al. 1981). It appears, then, that evolutionary constraints, migratory strategy and habitat use combine to determine the sensitivity of Maryland bird species to forest reduction. Interestingly, observations of North American migrants on their neotropical wintering grounds reveal that species which as breeders are restricted to forest-interior are often substantially more tolerant of disturbed vegetation and 'edge' conditions during the non-breeding season. Such observations indicate considerable latent flexibility in the behavioural-ecological repertoires of even the more highly specialized forest-interior species, and underscore the need for additional ecological data covering the non-breeding season of highly migratory species.

BIRDS IN WESTERN AUSTRALIAN WHEATBELT RESERVES

A classification of bird species based on their taxonomic affiliation, habitat specificity, and migratory pattern was used by Kitchener and his co-workers (Kitchener et al. 1982; Humphreys and Kitchener 1982) to define 10 categories of birds that occur in forest reserves within the wheatbelt of southwestern Western Australia. In comparison to the avifauna of eastern North America, Western Australia's avifauna includes a higher proportion of non-passerines and more species with irregular and rainfall-triggered temporal and spatial patterns of breeding and migration. This irregularity, which does not pertain to all parts of Australia (R. Loyn, *in litt.*), contrasts sharply with the highly predictable seasonality of breeding and latitudinal migration in temperate North America. Another notable difference in the two avifaunas is the much greater number of nectarivorous species (10) in the wheatbelt reserves than in Maryland forest patches, where only a single specialized nectarivore occurs. Such differences are potentially important in a conservation context, because nectarivores tend to be more specialized on particular plant species than are other birds. For example, flowers of *Dryandra* provide the key food resource for the entire honeyeater community in some of the areas studied by Kitchener et al. (1982). Such an extreme of food specialization, which is virtually non-existant in temperate North America bird communities, raises the possibility that Australian nectarivores might be especially sensitive to the absence of particular plant species from reserves, and hence to forest fragmentation.

Kitchener et al. (1982) showed that the total number of passerine species in the wheatbelt reserves was strongly influenced by the floristic composition of the vegetation, and by the number of major physiognomic types of vegetation present. Reserve area played a lesser role in predicting the total number of passerine species. However, the abundance of the species with the most specialized habitat requirements (so-called 'P_5' species) was highly correlated with reserve area. Kitchener et al. (1982) concluded that larger reserves are more likely to include the particular habitats required by P_5 species. This view was supported by the observation that the correlation exists between reserve area and habitat complexity only for reserves larger than about 600 ha. In the wheatbelt reserves, forest was found to be the vegetation type with the highest total number of resident and migratory bird species, the highest density of birds, and the highest degree of endemism.

One can envision a continuum of forest insularization ranging from the non-insularized forest of pre-settlement eastern North America to highly insularized patches of forest on offshore islands. The Maryland forest patches studied by Lynch and Whigham (1984) represent a relatively low degree of insularization, given the large total area of existing forest and the short average distance between patches. For even the most generalized forest-dwelling bird species of Western Australia, the wheatbelt reserves are functionally more insular than are Maryland's forests, although these generalists are capable of using a reasonably broad range of habitats both within and outside the reserve system; from the point of view of such species, the wheatbelt reserves are not completely isolated, either from one another or from other patches of woody vegetation outside the reserve system. By contrast, the P_5 species of Western Australia show greater habitat specialization, and tend to be restricted to particular sub-formations of native vegetation within the wheatbelt reserves (Kitchener et al. 1982). For such species, even the pristine landscape must have been patchy, hence insularized, and the existing archipelago of forest remnants is even more so. Indeed, Kitchener et al. (1982) showed that the community of P_5 species in the wheatbelt reserves resembles the avifauna of southern Australia's offshore islands in the slope (if not the absolute number of species) of its diversity response to habitat fragmentation.

The result of extreme insularization of a continental avifauna is exemplified by places such as Britain, where centuries of deforestation and other human disturbance may have winnowed out all bird species except those that are tolerant of extreme forest fragmentation. We lack reliable information on the pre-settlement avifauna of Britain, but none of the 56 bird species recorded by Moore and Hooper (1975) as breeding in British forest patches are restricted to forest, although some species attain their highest local density there. According to Arnold (1982), all of the British forest-dwelling species in question also breed in gardens, hedgerows, and other non-forest habitats. Thus, in the terminology of Kitchener et al. (1982), there are no P_5 species in these British woods. Compared with species found in Australia or North America, extant British birds are relatively tolerant of forest fragmentation, but this seems to reflect the fact that they are in fact rather generalized in their habitat requirements. Whether the depauperate nature of the British avifauna reflects the aftermath of glacial extinctions, human-related habitat disturbance, the insular character of Britain, or (as seems most likely) a combination of all of these factors, forest-interior specialist birds are markedly under-represented compared with what is seen in more pristine regions of the world. A similarly depauperate avifauna has come to be associated with small forest patches in extensively deforested regions of the eastern and midwestern regions of the USA (e.g., Galli et al. 1976; Whitcomb et al. 1976, 1981; Blake and Karr 1984), and is characteristic of much of present-day New Zealand (Diamond 1984).

MECHANISMS BY WHICH INSULARIZATION INFLUENCES BIRDS

The most obvious means by which forest fragmentation diminishes the diversity and abundance of forest-dwelling birds is through simple diminution of habitat. No bird species can persist if even one critical element of its habitat is completely eliminated. However, even if all critical elements are retained in a given patch of forest, the areal extent of habitat may be insufficient to maintain a viable population over the long term. Minimum population size is an especially serious constraint in the attempt to conserve larger species (e.g., birds of prey), whose feeding territories may encompass tens or hundreds of hectares.

As long as the total extent of forested land in a local landscape remains high, the area of individual forest patches above some critical minimum may not be a crucial factor in determining their use by birds, particularly if riparian strips, road verges, or other corridors of wooded habitat serve to interconnect the patches (e.g., Saunders 1980, 1982). However, many sedentary forest-dwelling birds, particularly tropical species, appear to have strong psychological aversions to crossing even narrow unforested barriers (Willis 1974; Diamond 1984). The failure of some 13 species of forest-adapted birds to persist in seemingly favourable habitat on Barro Colorado Island, which was isolated about 80 years ago by the rising waters of the Panama Canal, has been attributed to their moderate initial population size, followed by local population extinctions, and the lack of a 'rescue-effect' (*sensu* Brown and Kodric-Brown 1977) by immigrants from flourishing populations that exist less than a kilometre away on the mainland of Panama (Willis 1974).

As regional deforestation proceeds, inter-patch dispersal first lessens, then ceases altogether for habitat-specialized sedentary birds. In fully insularized reserves, bird populations will persist only if they are large enough to survive normal demographic fluctuations without periodic subsidization. Even very large populations may be driven to extinction by environmental changes (Leigh 1981), and an extensive and diverse mosaic or gradient of habitats is the best hedge against such extinctions. In practical terms, this means that truly isolated avifaunal reserves usually must be quite large, and that they may require active management interventions to prevent or recoup local extinctions. Again, it should be emphasized that the conservation problems posed by truly isolated reserves are fundamentally different from those encountered in a regional system of reserves that are capable of exchanging propagules. Species turnover events tend to dominate the avifaunal make-up of small forest patches in the eastern USA (Lynch and Whitcomb 1978); this reflects the fact that this avifauna is organized at the level of the regional landscape, not of the individual habit patch (Smith 1975). It is also important to recognize that species of birds with different habitat requirements and dispersal abilities may perceive a given configuration of reserves in totally different ways. Species-specific information is indeed needed on the ability of birds to cross barriers, utilize corridors, and occupy the available habitat gradient.

Non-sedentary birds, particularly long-distance migrants, present conceptual difficulties in the context of forest isolation. Thus, it seems unlikely that inter-patch isolation could directly control the distributions of birds whose annual migrations are several orders of magnitude greater than the normal inter-patch distances within the breeding range. Nevertheless, both major Maryland studies demonstrated that neotropical migrants have shown greater negative responses than other birds to forest fragmentation. How can this seeming anomaly be explained?

Several independent lines of evidence indicate that birds living in small forest patches are subject to a variety of ecological pressures that are correlated with the process and pattern of forest fragmentation. 'Edge' conditions, both physical and biotic, extend much further into a woodlot than is commonly realized (Levenson 1981; Ranney et al. 1981). Differences in plant species composition, canopy height and density, temperature, shading, and development of the understorey all can be detected far into the forest beyond the apparent edge boundary.

More importantly, many predators and nest-parasites concentrate their activities along forest margins, and range hundreds of metres into the forest proper (Gates and Gysel 1978; Kroodsma 1982; Brittingham and Temple 1983).

The activities of a single species of nest-parasite, the brown-headed cowbird *Molothrus ater*, has been implicated as a devastating cause of mortality for many North American forest passerines, including the globally endangered Kirtland's warbler (Mayfield 1977). Prior to the era of widespread forest-clearing, cowbirds were virtually confined to the prairie and prairie-forest ecotone of central North America, but over the past century and a half deforestation has allowed this species to spread throughout the entire eastern half of the continent, and has placed cowbirds into direct contact with an assemblage of forest-adapted birds that has not evolved the necessary behavioural mechanisms for dealing with nest parasitism. In contrast, species that have evolved in the same communities as cowbirds frequently thwart nest parasitism by ejecting cowbird eggs, or covering them with new nest material, or by renesting. Although cowbirds retain a strong association with open and 'edge' habitats, they are capable of deep penetration of forest patches from their margins. Cowbird nest parasitism declines with increasing distance from the forest margin, and is virtually non-existent in the interior of very large forested tracts, such as those in the southern Appalachian Mountains (D. Wilcove, pers. comm.).

That the brown-headed cowbird is not unique in its adverse impact on nesting birds is indicated by Wiley's (1985) report of a recent unaided range expansion of a tropical nest parasite, the shiny cowbird *M. bonariensis*, from South America into the West Indies. The effect of this species on the nesting success of evolutionary naive host passerines is every bit as dramatic as that of the brown-headed cowbird on North America species: 75-100% of the nests of some of the shiny cowbird's host species are parasitized in Puerto Rico. In the Old World, the common cuckoo, *Cuculus canorus* has been implicated as a causal agent in the extirpation of the sedge warbler *Acrocephalus schoenobaenus* in one local population (*fide* Wiley 1985), despite the long-term association of these two species elsewhere. The effects of the Cuculidae on reproductive success of host species in Old World forest reserves should be studied in relation to reserve area and isolation.

Other deleterious effects that emanate from edge habitat bordering on forest patches can be as pervasive as they are non-obvious. As an example, Kroodsma (1982) showed that many forest-dwelling birds are actually attracted to edge habitat for nesting. The original selective basis (if any) for this attraction is not known, but may involve better feeding conditions, denser concealing foliage, or other factors associated with treefall gaps and other small natural clearings. In any event, birds that nest in small forest patches, or along the margins of larger patches, suffer much higher rates of nest predation than do conspecifics that place their nests farther from the edge or in larger patches (Gates and Gysel 1978; Kroodsma 1982; Wilcove 1985).

OPTIMAL SIZES FOR RESERVES

Many recent studies have suggested that ecological attributes of reserves outweigh area *per se* as determinants of species occurrence (e.g., Lynch and Whigham 1984; Zimmerman and Bierregaard 1986). Nevertheless, reserve area is a useful 'shorthand' predictor of species diversity for many animal groups, including birds, and the question of optimal allocation of patch sizes for conservation purposes continues to generate heated controversy. Given the well-established and universally-accepted generalization that larger reserves will hold more species than smaller but otherwise similar ones, the conventional wisdom has been that 'bigger is better' when one plans reserves. Disagreement arises, however, when one seeks the optimum strategy for allocating a given total area of reserved land among individual parcels. Early attempts to apply the theoretical precepts of MacArthur and Wilson (1963, 1967) to the problem (e.g., Diamond 1975; May 1975; Terborgh 1975; Wilson and Willis 1975) were faulted by Simberloff and Abele (1976a, 1976b); the latter authors in turn were criticized by conservation-minded ecologists (Diamond 1976; Terborgh 1976; Whitcomb *et al.* 1976) who feared that Simberloff and Abele's largely theoretical critique would be seized upon by developers as a biologically defensible rationale for carving up large reserves into small pieces.

Since 1976, numerous theoretical and empirical studies have addressed the question of the optimal allocation of reserve areas (e.g., Gilpin and Diamond 1980; Higgs and Usher 1980; Simberloff and Abele 1982; Kindlman 1984; Simberloff and Gotelli 1984; Wilcox and Murphy 1985; McClellan *et al.* 1986). The results of theoretical simulations depend on the parameter values and functional forms chosen to represent species colonization probabilities, as summarized by Kindlman (1984). If one makes the simplifying (but patently false) assumption that all species have equal colonization probabilities, then the number of species preserved clearly increases with the degree of reserve subdivision (Fig. 6a). However, if species differ significantly in their area requirements and colonization potential, the advantages of subdivision are greatly reduced (6b), and may be eliminated or even reversed (Fig. 6c).

Empirical support for the conservation strategy of sequestering numerous small reserves has come from studies of plant diversity in small tracts (e.g.,

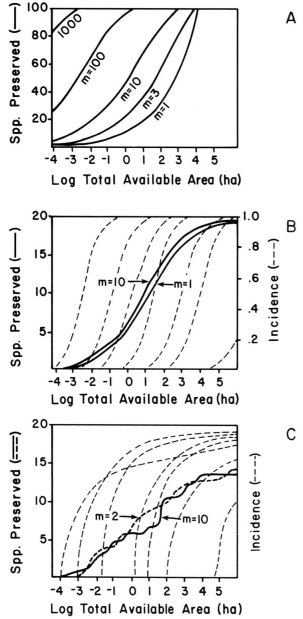

Fig. 6. Theoretical effects of reserve subdivision on the total number of species preserved, under varying assumptions of colonization potential. Details concerning construction of incidence functions and species richness curves are given in Kindlman (1984). Solid lines represent the number of species predicted to be preserved by subdividing a given total land area into m fragments. (a) All species have equal probabilities of colonizing reserve fragments of a given size. In this case, increasing reserve subdivision increases the total number of species contained in the reserve system as a whole. (b) Species differ in their area-specific colonization potentials and individual incidence functions (dashed lines) are sigmoidal. In this case, there is only a slight advantage gained by subdividing a single large reserve into 10 segments. (c) Species differ in their area-specific colonization potential, but incidence functions are hyperboloid. In this case, which Kindlman (1984) argues is more realistic than either (a) or (b), there is no consistent advantage to either reserve subdivision strategy.

Higgs and Usher 1980; Game and Peterken 1984; Simberloff and Gotelli 1984). This result reflects the fact that plants, as contrasted with vertebrate animals, are notably unresponsive to reserve area (e.g., Levenson 1981; Weaver and Kellman 1981). Given the empirical observation that few (if any) plant species are demonstrably 'area-sensitive,' the best strategy for preserving the maximum total species diversity of plants within a given total area of reserved land may be to place a number of relatively small reserves so as to include all important plant habitats and microhabitats. However, such an arrangement might fail to preserve important functional characteristics of large intact ecosystems (e.g., mineral cycling, population structure, co-evolved interactions, the normal mosaic of successional habitats).

As metabolically active and wide-ranging organisms, birds require much more habitat space than do individual plants of comparable (or even much greater) biomass. Accordingly, the concept of critical minimum habitat size is highly relevant for the survival of bird species populations, and even more so for the maintenance of intact avian communities. Nevertheless, over a considerable range of patch sizes, several fairly small reserves may be found to hold more species than does a single large reserve of the same total area. This can be illustrated by computing the number of species that would be encountered by combining survey data for various subsets of patch sizes for the Maryland data (McClellan et al. 1986; Fig. 7). However, this numerical exercise does not address the crucial problem of long-term survival probabilities of bird populations in truly isolated reserves (Wilcox and Murphy 1985). The list of birds found in a given tract of Maryland forest reflects not only the characteristics of that particular patch, but also the entire regional configuration of habitat.

Typically, theoretical studies of the optimal size allocation of reserves treat all species as having equal conservation significance, which of course is never the case. Area-sensitive species (e.g., birds with large body size, raptors, other feeding specialists) tend to be less common, and therefore, more in need of active preservation, than are small-sized generalist species, which tend to be relatively abundant. Aesthetic or economic consideration also may cause some species to have higher priority for preservation than others (Gilpin and Diamond 1980).

Humphreys and Kitchener (1982) demonstrated empirically that Western Australian bird species with specialized habitat requirements respond differently to reserve subdivision than do generalist species, which occupy a wider range of natural and disturbed vegetation types (Fig. 8). Within the latter group, scattered small reserves often will preserve more species than a single large reserve; for habitat specialists, on the other hand, a higher number of species will often be contained in a single large

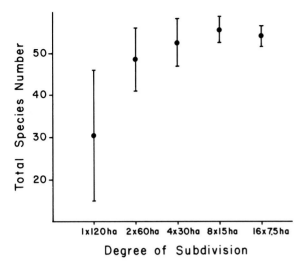

Fig. 7. Effect of patch subdivision on the number of bird species contained in Maryland forest tracts, based on point survey data in Lynch and Whigham (1984). The mean expected number of species (± one standard error) is plotted for varying numbers of small forest patches that add up to a total area of 120 ha. The figure does not take into account the undersampling of large tracts (< 60 ha) by the point survey method (see text), nor are forest-interior species distinguished from interior-edge and field-edge species. Nevertheless, the data suggest that forest subdivision does not increase total species richness for individual plots smaller than about 30 ha. Figure redrawn from McClellan et al. (1986).

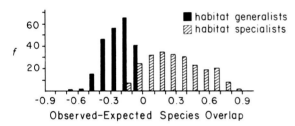

Fig. 8. Effect of reserve subdivision on bird species diversity in Western Australian wheatbelt reserves. The abcissa plots the difference between the observed and expected overlap in species composition between a single large reserve and two small reserves of the same total area, based on a random sampling model. Positive overlap values indicate that a single large reserve contains more species than two smaller ones combined. Negative values indicate the opposite. Note the very different responses of habitat specialists ('P_5' species) and habitat generalists. Figure redrawn from Kitchener et al. (1982).

reserve than in two smaller ones of equivalent aggregate area. A preliminary analysis of birds in Victorian forest patches revealed more farmland bird species in groups of small forest patches than in single large patches, but numbers of forest-associated species were almost identical in the two patch-size configurations (Loyn, this volume). In a study of birds in Illinois (USA) forest patches, Blake and Karr (1984) concluded that aggregations of small reserves contained, on average, more sedentary, edge-associated species than did single, large reserves of the same total area, but the opposite was true for highly migratory forest-interior bird species.

It is important to recognize that the optimal strategy for reserve subdivision depends not only upon the taxa in question, but also on the absolute scale of the geographic region of interest. If one considers an entire continent, for example, several moderately large reserves that together encompass all major climatic-vegetational zones would be expected to contain more species than would any one reserve, no matter how large, that represents only a single ecological region. A very different situation arises, however, when one contemplates subdividing relatively small reserves within a single ecological association. In the extreme case, even a very large number of small preserves will be useless for the conservation of a species if each is smaller than its minimum territorial requirements, unless (as in the case of some raptors, parrots, and other vagile species) the species is capable of integrating a number of small patches into a single territory or home range. A continental strategy of establishing reserves could come close to maximizing all three of the main components of species diversity (Whittaker 1970) — purely local ('alpha'), gradiential ('beta'), and regional ('gamma') by judiciously placing moderately large reserves in all major biomes. A purely local strategy of reserve sub-division, on the other hand, would by definition forego the regional component of species diversity, and would not necessarily augment the two remaining components significantly.

If the bird species to be conserved are sufficiently sedentary that isolated reserves function as true islands, then extinction, not colonization, is the process that must dominate reserve planning and maintenance (Pickett and Thompson 1978). In the absence of a natural 'rescue effect' (Brown and Kodric-Brown 1977), by which the extinction of local populations is prevented by frequent recolonization by propagules from outside source areas, the number of individuals per population must be sufficiently large to reduce the risk of stochastic extinction to an acceptably low level. Maintenance of a diverse range of habitats, which almost by definition entails large areas, also provides buffering against potentially devastating environmental changes (Leigh 1981). A major challenge to conservationists is to determine the appropriate geographic scale for individual reserves and systems of reserves, based on the properties of real species or communities. Again, the obvious fact that area requirements and other ecological characteristics vary from species to species (and region to region), does not preclude useful generalizations based on body size, feeding category, migratory habits, or other salient characteristics.

THE NEED FOR A REGIONAL PERSPECTIVE

Fortunately, most continental bird species are not yet confined to a single park or reserve. Instead, we are normally dealing with *systems* of reserves, together with other, often much more extensive, areas of habitat that lie outside formal reserves. The most sensible way to approach avian conservation in such a context is on a regional basis, such that the entire assemblage of parks and unprotected lands is viewed as a system. From this broad perspective it may make more sense to devote scarce resources to assuring the survival of threatened species in the region as a whole, rather than becoming unduly distracted by the status of inherently non-viable subpopulations of birds in small, isolated forest patches.

The current resurgence of interest in landscape ecology (e.g., Noss 1983) reflects a growing awareness that ecosystems do not exist as isolated entities. For example, few, if any, of the forested tracts we studied in Maryland were large enough in themselves to support viable populations of any but the most abundant forest-interior passerines. The long-term survival of Maryland's avifauna at the regional level depends on the persistence of large numbers of forest patches, and on free movement of colonizing individuals among them. In Maryland, as in many parts of the world, most forest is not part of a formally designated reserve network, but protected and unprotected forests are functionally linked through their exchange of avian propagules. As has been pointed out by earlier investigators (e.g., Terborgh 1975; Wilson and Willis 1975), singular island preserves, for which recolonization is by definition precluded, must be large (often thousands or tens of thousands of hectares) if they are to maintain even short-term stability in the species composition of their vertebrate fauna. On the other hand, fairly small reserves (say 10-100 ha) can serve valuable conservation functions for birds, provided they are functionally connected, either to one another or to sources of colonists outside the reserve system. In extreme cases, as where formerly extensive species distributions have been reduced to a few scattered relicts, intensive human intervention may be required to maintain the viability of individual populations and to increase the functional connectivity of reserves.

In order to counter some of the negative side-effects of forest insularization, managers can work to increase effective reserve size. This can be done by increasing contiguous acreage of reserves, by promoting compatible land-use practices in adjacent non-reserved lands, or by allowing secondary succession to increase the proportion of forested land within a region. The isolation of individual forest patches can be reduced by planting (or by not removing) roadside forest strips, hedge rows, riparian corridors, etc. This technique has been advocated in Britain (Arnold 1982) and Australia (Suckling 1982). In Western Australia, Saunders (1980, 1982) has suggested that such corridors appear to be crucial as an incentive to inter-patch movement by Carnaby's black cockatoo *Calyptorhynchus funereus latirostris*. For some species corridors may provide important nesting or foraging habitat, as well as passageways between forest patches. However, the desirability, as well as the practicality, of increasing reserve connectivity must be judged on a case-by-case basis. In some situations, a high degree of reserve isolation actually serves a *positive* conservation function. In Australia and New Zealand, vermin-free offshore islands now harbour the sole (or main) populations of several previously widespread forest-dwelling species (Merton 1975; Williams 1977). Some of these island populations are natural remnants; others are the result of successful translocations by conservation officials, who have introduced endangered species onto islands in order to isolate them from adverse mainland conditions. Maintenance of several carefully monitored, highly isolated populations of such species may be better insurance against the ravages of disease, predation, or other catastrophic disturbance than would be the sequestering of a single poorly-isolated population of the same (or even larger) aggregate size.

SUMMARY AND CONCLUSIONS

The detailed effects of forest fragmentation on birds are complex and species-specific, yet some generalizations emerge. Firstly, reduction in habitat area is causally related to an entire syndrome of physical and biotic changes that may have a more direct influence than does area *per se* on populations. Edge effects, including changes in microclimate, floristic composition, vegetation structure, predation rates, and risk of nest parasitism, often extend far into a forest from its margins. Functionally, small forest patches up to a few hectares in an area may consist of nothing but 'edge' habitat (Levenson 1981). Breeding success of forest-dwelling birds is often depressed in the vicinity of forest margins, and it is this reproductive failure, rather than simply the amount of habitat available that may account for the scarcity or absence of many highly mobile forest-interior birds from small forest patches in eastern North America.

Where extensive mainland habitat exists nearby, and where predation/parasitism rates are not excessive, even tiny near-shore islands may support breeding populations (more accurately, sub-populations) of forest-interior neotropical migrants (Morse 1977; Whitcomb *et al.* 1981). However, such islands do not constitute self-contained ecosystems, relying as they do on continuous recruitment from mainland populations. Small, subsidized islets cannot be viewed as analogues of truly isolated preserves that are established for the long-term conservation of sedentary bird populations.

Our North American studies indicate that the responses of bird species to forest fragmentation are correlated with migratory habits, degree of habitat specialization, feeding ecology and other general biological characteristics. The North American data demonstrate that highly migratory bird species, particularly small insectivores, are more sensitive to the forest fragmentation syndrome than are year-round residents. The observed sensitivity to forest fragmentation appears to reflect facultative and numerical responses by migrants to immediate ecological factors, rather than being a direct result of the migration process *per se*. Different patterns may exist at low latitudes, where sedentary habitat specialists are more prevalent, and where regular large-scale latitudinal migration is uncommon.

In eastern North America, non-migratory species, particularly edge-loving forms, tend to profit from forest fragmentation, such that total species diversity and bird density may hold steady, or even increase, in fragmented forests. Such a pattern of compensation underscores the importance of differentiating between forest-interior specialists, many of which are regionally endangered by insularization of forest, and ecologically generalized edge-tolerant species, which are of little concern in a conservation context.

Finally, it is increasingly evident that long-term maintenance of some semblance of balanced natural bird communities in a fragmented landscape will require a regional perspective. In most (but not all) situations, efforts should be made to maintain or increase the functional connectivity of forest remnants. The optimal configuration of reserves needed to maintain regional diversity under real-world political and budgetary constraints will depend on the biologies of the particular species that are of concern, the aggregate area of available habitat, the functional connectivity of habitat patches, and the regional diversity and geographic scale of the habitat mosaic. When considering conservation of bird species, all of these factors must be assessed from a 'bird's eye' point of view.

ACKNOWLEDGEMENTS

C. S. Robbins, R. F. Whitcomb, and D. F. Whigham collaborated with me in much of the research described in this paper, although they do not agree with me in all details of interpretation. C. S. Robbins kindly permitted me to cite data from his important manuscript on the distribution of Maryland birds. The manuscript benefited from critical readings by G. Arnold, R. Loyn and D. Saunders. Financial support for field research came from the Maryland Power Plant Siting Commission, The World Wildlife Fund-U.S., and the Smithsonian Institution. Margaret McWethy prepared the figures; Eve Huntington, Margaret McKim, and Jeanine Cheek typed the numerous drafts of the manuscript.

REFERENCES

Abbott, I., 1980. Theories dealing with the ecology of land birds on islands. *Adv. Ecol. Res.* 11: 329-71.

Aldrich, J. W. and Robbins, C. S., 1970. Changing abundance of migratory birds in North America. Pp. 17-25 *in* The Avifauna of Northern Latin America ed by H. K. Buechner and J. H. Buechner. *Smithsonian Contrib. Zool.* 26.

Anderson, S. H. and Robbins, C. S., 1981. Habitat size and bird community management. *Trans. 46th North Amer. Wildl. and Natur. Res. Conf.* 511-20.

Arnold, G. W., 1982. The influence of ditch and hedgerow structure, length of hedgerows, and area of woodland and garden on bird numbers on farmland. *J. Appl. Ecol.* 20: 731-50.

Blake, J. G. and Karr, J. R., 1984. Species composition of bird communities and the conservation benefit of large vs. small forests. *Biol. Conserv.* 30: 173-87.

Brittingham, M. C. and Temple, S. A., 1983. Have cowbirds caused forest songbirds to decline? *Bioscience* 33: 31-5.

Brown, J. H. and Kodric-Brown, A., 1977. Turnover rates in insular biogeography: effect of immigration on extinction. *Ecology* 58: 445-9.

Curtis, J. T., 1956. The modification of mid-latitude grasslands and forests by man. Pp. 21-736 *in* Man's Role in Changing the Face of the Earth ed by W. L. Thomas. University of Chicago Press, Chicago.

Diamond, J. M., 1975. The island dilemma: lessons of modern biogeographic studies for the design of natural reserve. *Biol. Conserv.* 7: 129-46.

Diamond, J. M., 1976. Island biogeography and conservation: strategy and limitations. *Science* 193: 1027-9.

Diamond, J. M., 1984. Distribution of New Zealand birds on real and virtual islands. *New Zealand J. Ecol.* 7: 37-55.

Galli, A. E., Leack, C. F. and Forman, R. T. T., 1976. Avian distribution patterns within different sized forest islands. *Auk* 93: 356-65.

Game, M. and Peterken, G. F., 1984. Nature reserve selection strategies in the woodlands of central Lincolnshire, England. *Biol. Conserv.* 29: 157-81.

Gates, J. E. and Gysel, L. W., 1978. Avian nest dispersion and fledging success in field-forest ecotones. *Ecology* 59: 871-83.

Gilbert, F. S., 1980. The equilibrium theory of island biogeography: fact or fiction? *J. Biogeogr.* 7: 209-35.

Gilpin, M. E. and Diamond, J. M., 1980. Subdivision of nature reserves and the maintenance of species diversity. *Nature, Lond.* 285: 567-8.

Greenberg, R., 1980. Demographic aspects of long distance migration. Pp. 493-504 *in* Migrant Birds in the Neotropics: Ecology, Behaviour, Distribution and Conservation ed by A. Keast and E. S. Morton. Smithsonian Institution Press, Washington.

Higgs, A. J., 1981. Island biogeography theory and nature reserve design. *J. Biogeogr.* 8: 117-24.

Higgs, A. J. and Usher, M. B., 1980. Should nature reserves be large or small? *Nature, Lond.* 285: 568-9.

Humphreys, W. F. and Kitchener, D. J., 1982. The effect of habitat utilization on species-area curves: implications for optimal reserve area. *J. Biogeogr.* 9: 391-6.

Kindlman, P., 1984. Do archipelagoes really preserve fewer species than one island of the same area? *Oecologia* 59: 141-4.

Kitchener, D. J., Dell, J., Muir, B. G. and Palmer, M., 1982. Birds in Western Australian wheatbelt reserves: implications for conservation. *Biol. Conserv.* 22: 127-63.

Kroodsma, R. L., 1982. Edge effect on breeding forest birds along a powerline corridor. *J. Appl. Ecol.* **19**: 361-70.

Leigh, E. G., 1981. The average lifetime of a population in a varying environment. *J. Theor. Biol.* **90**: 213-39.

Levenson, J. B., 1981. Woodlots as biogeographic islands in southeastern Wisconsin. Pp. 13-39 *in* Forest Island Dynamics in Man-dominated Landscapes ed by R. L. Burgess and D. M. Sharpe. Springer-Verlag, New York.

Lynch, J. F. and Whigham, D. F., 1984. Effects of forest fragmentation on breeding bird communities in Maryland, USA. *Biol. Conserv.* **28**: 287-324.

Lynch, J. F. and Whitcomb, R. F., 1978. Effects of the insularization of the eastern deciduous forest on avifaunal diversity and turnover. Pp. 461-89 *in* Classification, Inventory, and Evaluation of Fish and Wildlife Habitat ed by A. Marmelstein. Washington, U.S. Fish and Wildlife Service Publ. OBS-78716.

MacArthur, R. H. and Wilson, E. O., 1963. An equilibrium theory of insular zoogeography. *Evolution* **17**: 373-87.

MacArthur, R. H. and Wilson, E. O., 1967. The Theory of Island Biogeography. Princeton University Press, Princeton, N.J.

Margules, C., Higgs, A. J. and Rafe, R. W., 1982. Modern biogeographic theory: are there any lessons for nature reserve design? *Biol. Conserv.* **24**: 115-28.

May, R. M., 1975. Island biogeography and the design of wildlife preserves. *Nature, Lond.* **254**: 177-8.

Mayfield, H., 1977. Brown-headed cowbird: agent of extermination? *Amer. Birds* **31**: 107-13.

McClellan, C. H., Dobson, A. P., Wilcove, D. S. and Lynch, J. F., 1986. Effects of forest fragmentation on new and old world bird communities: Empirical observations and theoretical implications. *In* Modeling Habitat Relationships of Terrestrial Vertebrates ed by J. Verner, M. L. Morrison and C. J. Ralph. University of Wisconsin Press, Madison (in press).

Merton, D. V., 1975. Success in re-establishing a threatened species: the Saddleback — its status and conservation. *Bull. Int. Counc. Bird Preserv.* **12**: 150-8.

Moore, N. W. and Hooper, M. D., 1975. On the number of bird species in British woods. *Biol. Conserv.* **8**: 239-50.

Morse, D. H., 1977. The occupation of small islands by passerine birds. *Condor* **79**: 399-412.

Morse, D. H., 1980. Population limitation: breeding or wintering grounds? Pp. 505-16 *in* Migrant Birds in the Neotropics: Ecology, Behaviour, Distribution and Conservation ed by A. Keast and E. S. Morton. Smithsonian Institution Press, Washington.

Noss, R. F., 1983. A regional landscape approach to maintain diversity. *Bioscience* **33**: 700-6.

Pickett, S. T. A. and Thompson, J. N., 1978. Patch dynamics and the design of nature reserves. *Biol. Conserv.* **13**: 27-37.

Ranney, J. W., Bruner, M. C. and Levenson, J. B., 1981. The importance of edge in the structure and dynamics of forest islands. Pp. 67-95 *in* Forest Island Dynamics in Man-dominated Landscapes ed by R. L. Burgess and D. M. Sharpe. Springer-Verlag, New York.

Robbins, C. S., 1980. Effects of forest fragmentation on breeding bird populations in the piedmont of the Middle-Atlantic region. *Atl. Natur.* **33**: 31-6.

Saunders, D. A., 1980. Food and movements of the short-billed form of the White-tailed Black Cockatoo. *Aust. Wildl. Res.* **7**: 257-69.

Saunders, D. A., 1982. The breeding behaviour and biology of the short-billed form of the White-tailed Black Cockatoo *(Calyptorhynchus funereus)*. *Ibis* **124**: 422-55.

Simberloff, D. S., 1976. Species turnover and equilibrium island biogeography. *Science* **194**: 572-8.

Simberloff, D. S. and Abele, L. G., 1976a. Island biogeography theory and conservation practice. *Science* **191**: 285-6.

Simberloff, D. S. and Abele, L. G., 1976b. Island biogeography theory and conservation: strategy and limitations. *Science* **193**: 1032.

Simberloff, D. S. and Abele, L. G., 1982. Refuge design and island biogeography: effects of fragmentation. *Amer. Natur.* **120**: 41-50.

Simberloff, D. and Gotelli, N., 1984. Effects of insularization on plant species richness in the prairie-forest ecotone. *Biol. Conserv.* **29**: 27-46.

Smith, F. E., 1975. Ecosystems and evolution. *Bull. Ecol. Soc. Amer.* **56**: 2-6.

Snyder, N. F. R. and Hamber, J. A., 1985. Replacement clutching and annual nesting of California Condors. *Condor* **87**: 374-8.

Suckling, G. C., 1982. The value of reserved habitat for mammal conservation in plantations. *Aust. For.* **45**: 19-27.

Terborgh, J., 1975. Faunal equilibria and the design of wildlife preserves. Pp. 369-80 *in* Tropical Ecological Systems: Trends in Terrestrial and Aquatic Research ed by F. B. Golley and E. Medina. Springer-Verlag, New York.

Terborgh, J., 1976. Island biogeography and conservation: strategy and limitations. *Science* **193**: 1029-30.

Terborgh, H. W., 1980. The conservation status of neotropical migrants: present and future. Pp. 21-30 *in* Migrant Birds in the Neotropics: Ecology, Behaviour, Distribution and Conservation ed by A. Keast and E. S. Morton. Smithsonian Institution Press, Washington.

Weaver, M. and Kellman, M., 1981. The effects of forest fragmentation on woodlot tree biotas in southern Ontario. *J. Biogeogr.* **8**: 199-210.

Whitcomb, R. F., Lynch, J. F., Opler, P. L. and Robbins, C. S., 1976. Island biogeography and conservation: strategy and limitations. *Science* **193**: 1030-2.

Whitcomb, R. F., Robbins, C. S., Lynch, J. F., Whitcomb, B. L., Klimkiewicz, M. K. and Bystrak, D., 1981. Effects of forest fragmentation on avifauna of the eastern deciduous forest. Pp. 125-205 *in* Forest Island Dynamics in Man-dominated Landscapes ed by R. L. Burgess and D. M. Sharpe. Springer-Verlag, New York.

Whitney, G. G. and Somerlot, W. J., 1985. A case study of woodland continuity and change in the American midwest. *Biol. Conserv.* **31**: 265-87.

Whittaker, R. H., 1970. Communities and Ecosystems. MacMillan Company, London.

Wilcove, D. S., 1985. Nest predation in forest tracts and the decline of migratory songbirds. *Ecology* **66**: 1211-4.

Wilcox, B. A. and Murphy, D. D., 1985. Conservation strategy: the effect of fragmentation and extinction. *Am. Nat.* **125**: 879-87.

Wiley, J. W., 1985. Shiny cowbird parasitism in two avian communities. *Condor* **87**: 165-76.

Williams, G. R., 1977. Marooning — a technique for saving threatened species from extinction. *Int. Zoo. Yearbook* **17**: 102-6.

Willis, E. O., 1974. Populations and local extinctions of birds on Barro Colorado Island, Panama. *Ecol. Monogr.* **44**: 153-69.

Wilson, E. O. and Willis, E. O., 1975. Applied biogeography. Pp. 522-34 *in* Ecology and Evolution of Communities ed by M. C. Cody and J. M. Diamond. Belknap Press of Harvard University, Cambridge, Massachusetts.

Zimmerman, B. L. and Bierregaard, R. O., 1986. Relevance of the equilibrium theory of island biogeography and species-area relations to conservation with a case from Amazonia. *J. Biogeogr.* **13**: 133-43.

CHAPTER 11

Consequences of Faunal Collapse and Genetic Drift for the Design of Nature Reserves

William J. Boecklen[1] and Graydon W. Bell[2]

We examine the mathematical properties of faunal collapse models and demonstrate that these models are neutral with respect to single large reserves versus several small reserves as the better design strategy. Depending on the slope of the extinction coefficient-area relationship and the degree of faunal similarity among small reserves, an archipelago of refuges may actually preserve more species longer than would a single large reserve of equal total area. We also compare through computer simulations rates of genetic drift in populations occurring in single large reserves to those in populations subdivided among several small refuges. Intact populations typically preserve more heterozygotes than do subdivided populations of equal total numbers of individuals. However, subdivided populations preserve more alleles and exhibit lower probabilities of allele fixation than do corresponding intact populations.

INTRODUCTION

THERE is much recent controversy among ecologists and conservation biologists regarding the optimal design of nature reserves. Particularly contentious is the proposition that single large reserves will preserve more species than would several small reserves of equal total area (Simberloff and Abele 1976, 1982, 1984; Jarvinen 1982, 1984; Willis 1984; Wilcox and Murphy 1985). This proposition is often referred to by the acronym SLOSS: single large or several small (Simberloff and Abele 1982).

The history of SLOSS begins with the application of island biogeographic theory, especially the dynamic equilibrium model (MacArthur and Wilson 1967), to conservation practice. The dynamic equilibrium model states that the number of species on an island is the result of a dynamic balance of immigrations and extinctions such that species compositions change, but total species numbers remain constant (Fig. 1A). The model assumes that extinction rates are related inversely to island area (Fig. 1B) and that immigration rates are related inversely to island isolation (Fig. 1C). The dynamic equilibrium model makes two predictions that are relevant to SLOSS: equilibrium species numbers increase with island area but decrease with island isolation.

These predictions have motivated general recommendations regarding reserve design and management policy. Wilson and Willis (1975) present several rules of design of natural preserves, based on current biogeographic theory that are often repeated (e.g. Terborgh 1974, 1975; Diamond 1975; Diamond and May 1976) and that have become firmly entrenched in conservation literature. These include recommendations regarding reserve shape, spatial arrangement of reserves, and the importance of interconnecting corridors. The most frequently cited contention is that species-area relationships dictate that single large reserves will preserve more species than would several small refuges of equal total area. Subsequent related approaches include faunal collapse models (e.g. Terborgh 1974; Soulé et al. 1979) and calculations of relaxation times (e.g. Diamond 1972). These models are predicated on the prediction that equilibrium species numbers decrease with increasing island isolation and have been used to forecast species extinctions in nature reserves as reserves become insular. Both species-area relationships and faunal collapse models have been adduced in support of single large reserves as the optimal design strategy.

Equilibrium theory and associated rules of refuge design have become widely accepted by conservation planners and biologists. The International

---

[1]Department of Mathematics, Department of Biological Science, Northern Arizona University, Flagstaff, AZ, 86011, USA.
[2]Department of Mathematics, Northern Arizona University, Flagstaff, AZ, 86011, USA.

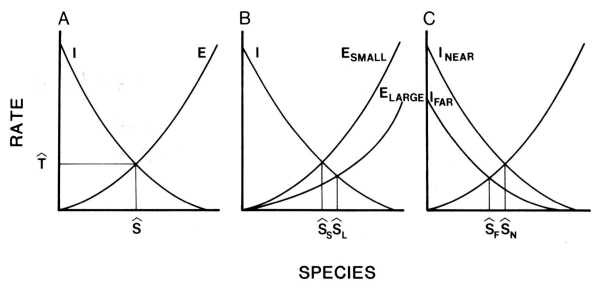

Fig. 1. Dynamic equilibrium model (A). Extinction rate (E) is balanced by immigration rate (I) to produce an equilibrium number of species, $\hat{S}$, that has a turnover rate, $\hat{T}$. The model predicts an area effect (B) and an isolation effect (C).

Union for Conservation of Nature in its summary statement of worldwide conservation strategy (IUCN 1980) states that reserve design and management practices should embrace the tenets of island biogeographic theory, as outlined in Wilson and Willis (1975). Simberloff (1985) documents the adoption of the IUCN recommendations for policy statements by other conservation organizations. The influence of equilibrium theory is not restricted to general policy statements; it has been invoked to justify specific management practices. For example, Temple (1981) advocates inter-island transfers of endangered species of birds within the Mascarene archipelago to minimize risk of extinction. Temple justifies this scheme partly on biological grounds, partly on equilibrium theory and the expectation of faunal collapse.

The application of island biogeographic theory to conservation practice is not without detractors. Critics argue that equilibrium theory itself is not well established (Connor and McCoy 1979; Gilbert 1980) and does not warrant legal principles regarding refuge design (Simberloff and Abele 1984; Jarvinen 1984; Boecklen and Simberloff 1986). Boecklen and Gotelli (1984) contend that species-area relationships and faunal collapse models have limited use for conservationists because these models have low explanatory power, are sensitive to influential cases, and give unreliable estimates. Boecklen and Simberloff (1986) review the development and application of faunal collapse and relaxation models and conclude that these models have low predictive power and are predicated on untenable assumptions. Simberloff and Abele (1976) demonstrate that the species-area relationship is actually neutral with respect to SLOSS; depending on the slope of a species-area relationship and the degree of faunal similarity among islands, an archipelago may contain more species than would a single island of equal total area. This result has been affirmed mathematically (Higgs and Usher 1980; Higgs 1981; Margules *et al.* 1982; Kindlmann 1983) and empirically (Jarvinen 1982; Simberloff and Abele 1982; Simberloff and Gotelli 1984). Despite these criticisms, some ecologists (e.g. Willis 1984; Wilcox and Murphy 1985) argue that equilibrium theory is still relevant to reserve design and SLOSS remains contentious.

Concurrent with the SLOSS debate, there has been much interest in the genetic consequences of refuge design. The optimal design will preserve sufficient genetic variability for long-term evolutionary change (Franklin 1980; Soulé 1980) and avoidance of inbreeding depression (Shaffer 1981; Allendorf 1983; Chesser 1983; Soulé 1983). Inbreeding depression is a general loss of fitness in highly inbred populations that results from accumulation of deleterious recessive alleles as population sizes decrease and, more importantly, loss of heterozygosity through genetic drift. Severe inbreeding depression can significantly reduce reproductive success, perhaps facilitating local extinctions of inbred populations. For example, Ballou and Ralls (1982) document increased juvenile mortality and delayed sexual maturity in inbred populations of 12 ungulate species. Shaffer (1981) and Soulé (1983) consider inbreeding depression to be the major long-term threat to small populations.

Schemes to maximize genetic variation and minimize genetic drift and inbreeding depression closely parallel SLOSS. Early suggestions stressed the need for large effective population sizes to maintain sufficient within-population variation to slow the rate of genetic drift. Estimates of minimum effective population sizes have been quite varied. Berry (1971) suggests a minimum effective population of

50, while Franklin (1980) recommends 500. Recent interest has focused on population subdivision as a means of maintaining genetic diversity (Chesser et al. 1980; Franklin 1980; Allendorf 1983; Chesser 1983; Lekmkuhl 1984; Boecklen 1986). By subdividing populations, between-population variation may be maximized. Thus, SLOSS could be rephrased within the context of conservation genetics to single intact or several subdivided populations.

Here, we examine the consequences of faunal collapse and genetic drift for reserve design. We evaluate SLOSS with respect to species loss following refuge insularization, preservation of heterozygotes and alleles, and probabilities of allele fixation. We compare single large reserves and single intact populations to systems of five small reserves and five subpopulations of equal magnitude.

FAUNAL COLLAPSE MODELS

Faunal collapse models attempt to forecast species extinctions in nature reserves as reserves become insular. The expectation of species loss following refuge insularization is based on the inverse relationship between equilibrium species numbers and island isolation, as predicted by the dynamic equilibrium model, and on comparisons of continental and insular species-area relationships. Preston (1960) observed that when progressively larger areas are sampled from continuous regions like continents, species-area relationships typically have larger intercepts and lower slopes than when isolated areas like islands are sampled (Fig. 2). Preston suggests that these species-area characteristics are intrinsic to non-isolated and isolated areas since non-isolated areas have an underlying truncated log-normal distribution of individuals into species that has a larger species-to-individual ratio than does the complete log-normal distribution characteristic of isolated areas. Consequently, small samples from continuous areas have disproportionately more species, depressing the slope of the species-area relationship. MacArthur and Wilson (1967) suggest that transients inflate species counts in non-isolated areas.

Faunal collapse models predict species loss as reserves change in status from non-isolated to isolated areas (Fig. 2). Most refuges currently are contiguous with areas of similar habitat and should have species richness characteristic of non-isolated areas. However, as more and more of the surrounding areas become uninhabitable, reserves become insular and species richness should 'relax' to new equilibria appropriate for isolated areas. This insularization of the landscape is occurring in natural habitats of all sorts (Burgess and Sharpe 1981) but is particularly acute in the Amazon Basin (Lovejoy et al. 1984) and in the Western Australian wheatbelt (Kitchener et al. 1980).

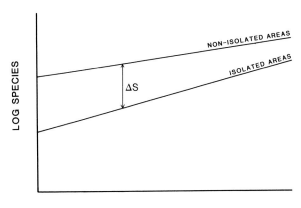

Fig. 2. Species-area relationships characteristic of non-isolated and isolated areas. Species loss, ΔS, is assumed to occur as reserves change in status from non-isolated to isolated areas.

Various assumptions regarding the rate of species loss following reserve insularization have led to a plethora of faunal collapse models (Table 1). Little is known about extinction rates in natural systems (but see, Simberloff 1976; Rey 1981) and the choice of an extinction rate for these models is arbitrary. Nevertheless, five distinct extinction rates have been proposed, all with the general form

$$dS/dt = -KS^\alpha \quad (1)$$

where K, the extinction coefficient, is a constant and α takes integer values from 0 to 4. The extinction coefficient is assumed to be taxon specific (Wilcox 1980), invariant over time, and inversely related to reserve area.

Table 1. Extinction rates and faunal collapse models.

Author	α	Extinction Rate	Faunal Collapse Model
Terborgh 1975	0	$dS/dt = -K$	$S(t) = S_0 - Kt$
Terborgh 1975; Soulé et al. 1979; Wilcox 1980	1	$dS/dt = -KS$	$S(t) = S_0 e^{-Kt}$
Terborgh 1974; Soulé et al. 1979; Wilcox 1980	2	$dS/dt = -KS^2$	$S(t) = S_0/(S_0 Kt+1)$
Soulé et al. 1979; Wilcox 1980	3	$dS/dt = -KS^3$	$S(t) = S_0/(2S_0 Kt+1)^{1/2}$
Soulé et al. 1979; Wilcox 1980	4	$dS/dt = -KS^4$	$S(t) = S_0/(3S_0 Kt+1)^{1/3}$

Faunal collapse models start with these extinction rates and describe species loss through time. The models represent two general cases:

$$S(t) = S_0 e^{-Kt} \text{ for } \alpha = 1; \text{ and} \quad (2)$$

$$S(t) = S_0 [1 - Kt(1-\alpha)/S_0^{1-\alpha}] \text{ for } \alpha \neq 1 \quad (3)$$

where, S(t) is the number of species remaining at time t and S_0 is the initial number of species in the refuge. The fraction of species that have become extinct at time t is given by the lifetime distribution

$$F(t) = 1 - S(t)/S_0 \quad (4)$$

and the faunal collapse for a system of r reserves of equal area is given by

$$S(t) = \sum_{i=1}^{r} n_i [1 - F(t)^i] \qquad (5)$$

where n_i is the number of species that occur in exactly i reserves.

Consequences of Faunal Collapse Models to SLOSS

We use data from Soulé *et al.*'s (1979) forecast of impending relaxation of the large mammal fauna in East African nature reserves. We compare species loss in the Samburu-Isiolo Game Reserve, as predicted by Soulé *et al.*, to that expected in a system of five small reserves, each a fifth the area of Samburu-Isiolo. Samburu-Isiolo is 298 km^2 and contains 56 species of large mammals. Logarithms of extinction coefficients for Samburu-Isiolo and for a reserve of 59.6 km^2 for each of the four models used by Soulé *et al.* are given in Table 2. We assume the following distribution of species within the system of five reserves: 22 species occur in exactly one reserve, 11 species occur in two reserves, eight species occur in three reserves, eight species occur in four reserves, and seven species occur in all five reserves.

Rates of species loss in the system of five small reserves are uniformly lower than those in Samburu-Isiolo (Fig. 3). For example, Model 1

Table 2. Logarithms of extinction coefficients for Samburu-Isiolo Game Reserve (298 km^2) and for a reserve one-fifth as large (59.6 km^2). (Data from Soulé *et al.* 1979).

Extinction Rate	Log Extinction Coefficient	
	Samburu-Isiolo	Small Refuge
dS/dt = -KS	-3.85	-3.75
dS/dt = -KS^2	-4.87	-4.65
dS/dt = -KS^3	-5.71	-5.37
dS/dt = -KS^4	-6.58	-6.11

predicts that 42 species will persist in Samburu-Isiolo 2000 years after refuge insularization, while 48 species will persist in the archipelago of refuges. After 5000 years of relaxation, Model 1 predicts that 10 more species will persist in the system of small reserves than will persist in Samburu-Isiolo. Model 2 predicts a five species difference throughout the relaxation process, Models 3 and 4 a three species difference.

Rates of species loss in the system of refuges for the 22 species found in only one of the small reserves are slightly higher than those for Samburu-Isiolo (Fig. 4). Model 2 predicts that 13 of the 22 species will persist in Samburu-Isiolo and 10 species will persist in the system of small reserves. The other models predict a 1-2 species difference throughout the relaxation process.

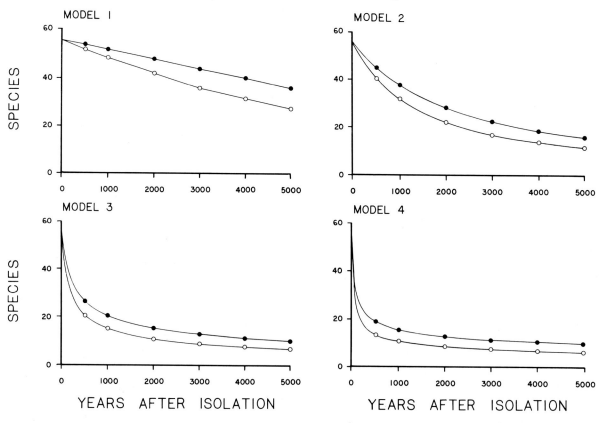

Fig. 3. Faunal collapse of Samburu-Isiolo Game Reserve (○) as predicted by Soulé *et al.* (1979) and of a system of five refuges (●) as predicted by equation 5.

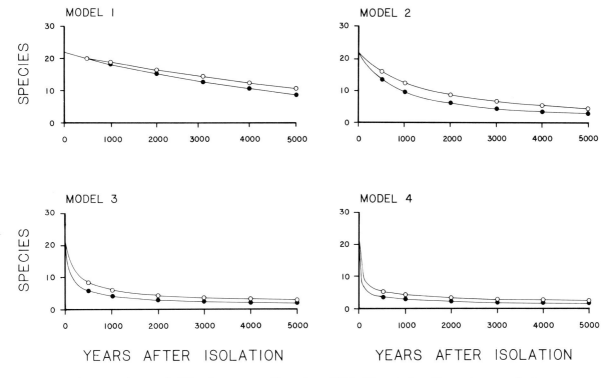

Fig. 4. Species loss in Samburu-Isiolo (○) and in a system of five refuges (●) for 22 species found in exactly one small refuge.

Systems of small reserves will not always preserve more species than will single large reserves of equal total area. The outcome of this comparison depends on two factors: the slope of the extinction coefficient-area relationship and the degree of faunal similarity among small reserves. Large negative slopes and low faunal similarity favour single large reserves, shallow slopes and high faunal similarity favour several small ones. Since faunal similarity and the extinction coefficient-area relationship (even if it could be calculated) will vary according to the taxa considered and the particular set of reserves examined, no generalization regarding optimal reserve design is possible based on faunal collapse models.

GENETIC DRIFT

The optimal reserve design will maximize the number of alleles preserved and minimize the risk of inbreeding depression by preserving heterozygosity.

We use Monte Carlo simulations to compare rates of genetic drift in intact populations to those in subdivided populations with equal total numbers of individuals. In particular, we compare persistence of heterozygotes, persistence of alleles, and probabilities of allele fixation in intact populations of 50, 250, and 500 individuals to those in five subpopulations of equal total size as a function of inter-subpopulation migration rate. We examine four migration rates: no migration, a migration every five generations, every 10 generations, and every 50 generations.

The genetic model is a single locus, six allele system with initial allele frequencies set at 0.16. The model assumes constant population size, constant sex ratio (1:1), no mutation, no natural selection, and random mating. The model spans 1000 generations and results are based on 500 replicates.

The simulation protocol is as follows. A unit line is partitioned based on allele frequencies into segments that represent different genotypes. The lengths of the segments correspond to the relative frequencies of the genotypes. For example, let A_1, A_2, A_3, A_4, A_5, and A_6 represent the six alleles and let p_1, p_2, p_3, p_4, p_5, and p_6 be the relative frequencies of those alleles. The initial condition (generation 0) is $p_1=p_2=p_3=p_4=p_5=p_6=1/6$. The probability of producing an offspring that is homozygous for the first allele (A_1A_1) is $p_1^2 = 1/36$. This genotype is represented on the unit line by the interval $(0, 1/36)$. The probability of producing an offspring that is homozygous for the second allele (A_2A_2) is $p_2^2 = 1/36$. This genotype is represented on the unit line by the interval $(1/36, 2/36)$. The unit line is partitioned further until all 21 genotypes are represented. Pseudorandom, uniform $(0, 1)$ deviates are compared to the partitioned line. If a uniform deviate fell in the interval $(0, 1/36)$ an individual with A_1A_1 genotype would be represented in the next generation. Indicator variables record numbers of each genotype, numbers of heterozygotes, and numbers of alleles represented. From these, new allele frequencies are calculated and a new partitioning of the unit line constructed.

Migrations occur with the first pseudorandom deviate. These numbers are compared to the partitioned line of the donor population but the appropriate indicator variables of the recipient population are incremented. There is no guarantee that a migrant will reproduce, the migrant only affects the next partitioning of the unit line for the recipient populations. A migration event consists of an exchange of one individual among subpopulations. This exchange occurs in a circular fashion; an individual from subpopulation 1 migrates to subpopulation 2, one from subpopulation 2 to subpopulation 3, etc.

Preservation of Heterozygosity

Persistence of heterozygotes is related positively to population size for both intact and subdivided populations (Fig. 5). For example, intact populations

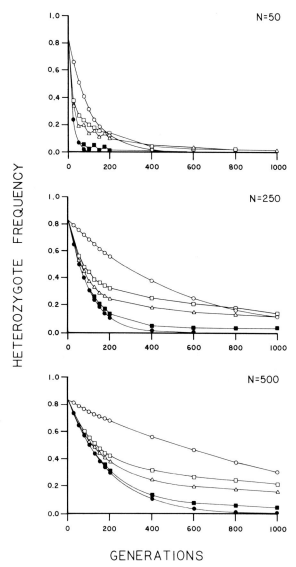

Fig. 5. Heterozygote frequencies for intact populations (○) of 50, 250, and 500 individuals and for subdivided populations with no migration (●), migrations every five generations (□), migrations every 10 generations (△), and migrations every 50 generations (■).

of 50 individuals contain, on average, 12% heterozygotes after 200 generations, while intact populations of 500 average 68% heterozygotes. Subdivided populations of 50 individuals with migrations every five generations average 13% heterozygotes after 200 generations; subdivided populations of 500 individuals average 43% heterozygotes.

Subdivided populations typically lose heterozygotes faster than do corresponding intact populations. Differences are largest over the first 200 generations; thereafter, heterozygote frequencies in subdivided populations approach those in intact populations and may actually exceed them. For example, intact populations of 50 individuals are indistinguishable, after 200 generations, from subdivided populations with migrations every five generations and every 10 generations; but after 500 generations, the intact populations are fixed, while the subdivided populations still contain heterozygotes. For populations of 250 individuals, subdivided populations with migrations every five generations contain more heterozygotes than do intact populations after 800 generations, subdivided populations with migrations every 10 generations after 1000 generations.

Persistence of heterozygotes in subdivided populations is related positively to inter-subpopulation migration rates. Subdivided populations of 50 individuals average 20% heterozygotes after 100 generations when migrations occur every five generations, 14% for migrations every 10 generations, and 2% for migrations every 50 generations. Subdivided populations with no migration exhibit the lowest heterozygote frequencies for all three population sizes.

Preservation of Alleles

Subdivided populations typically preserve more alleles than do intact populations with equal total numbers of individuals (Fig. 6). For example, subdivided populations of 250 individuals with no migration average 3.6 alleles after 400 generations; corresponding intact populations average 2.4 alleles.

Persistence of alleles in subdivided populations is related inversely to inter-subpopulation migration rates. Subdivided populations of 250 individuals with no migration contain, on average, 3.5 alleles after 600 generations; those with migrations every 50 generations average 3.3 alleles and those with migrations every five generations 2.5 alleles.

The number of alleles preserved in a system of refuges not only depends on inter-subpopulation migration rates, it also depends on the number of refuges. The more reserves, the more alleles are expected to be represented. Population size has no effect on the ultimate number of alleles preserved but it does affect the rate of allele loss. For example,

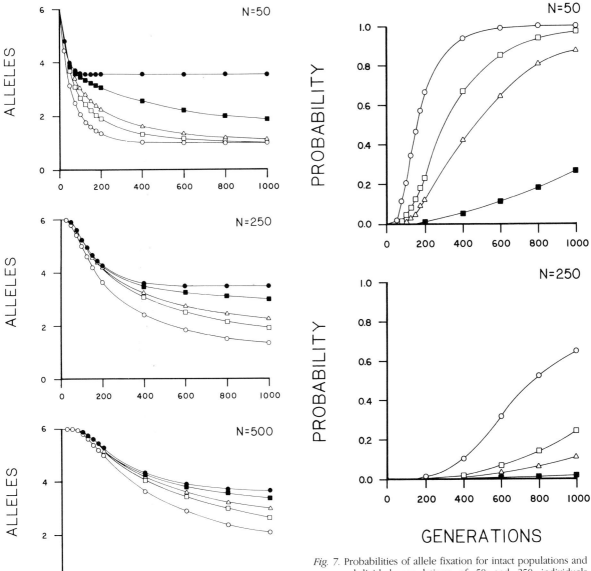

Fig. 6. Numbers of alleles represented in intact populations and subdivided populations of 50, 250, and 500 individuals (symbols as in Fig. 5).

Fig. 7. Probabilities of allele fixation for intact populations and subdivided populations of 50 and 250 individuals (symbols as in Fig. 5).

the number of alleles represented in subdivided populations of 50 individuals with no migration reaches a lower limit of 3.5 alleles after approximately 150 generations; populations of 250 and 500 individuals reach the same limit, but after approximately 600 and 1000 generations, respectively.

Probability of Allele Fixation

Probabilities of allele fixation in intact populations are uniformly higher than those in corresponding subdivided populations (Fig. 7). For example, the probability that an intact population of 50 individuals will become fixed after 200 generations is approximately 0.66, while corresponding probabilities for subdivided populations with migrations every five and every 10 generations are 0.23 and 0.12, respectively. Probabilities of allele fixation in subdivided populations of 50 individuals with migrations every 50 generations are lower than those for intact populations of 250 individuals.

CONCLUSIONS

Preservation and management of natural communities requires a thorough understanding of the biology and ecology of constituent species. However, much of the discussion regarding SLOSS has taken place in a theoretical arena that ignores species idiosyncratic responses to biotic and abiotic perturbations. This alone renders equilibrium theory a doubtful representation of natural processes that are often complex and species specific. Even within the relatively simple world of theoretical abstraction, it is clear that equilibrium theory, either in the form of species-area relationships or, as we have shown here, faunal collapse models, says little about reserve design. Like species-area relationships, faunal collapse models are actually neutral with respect to SLOSS.

Equilibrium theory cannot resolve SLOSS; nevertheless, SLOSS remains a valid issue for conservation planners and biologists. Shaffer (1981) and Soulé (1983) cite five factors that place small populations at risk: inbreeding depression, demographic stochasticity, environmental stochasticity, dysfunction of social behaviour, and natural catastrophies. Resolution of SLOSS for any specific application requires empirical evidence for each of these factors. Unfortunately, relevant empirical data are few. Systems of reserves that range over several orders of magnitude in area, such as those in the Western Australian wheatbelt, provide an ideal backdrop to gather such data.

We demonstrate some potential advantages of several small reserves over single large reserves: lower overall rates of extinction, larger numbers of alleles preserved, and lower probabilities of allele fixation. On the other hand, single large reserves typically preserve more heterozygotes. This suggests that the two stated aims of conservation genetics, preservation of heterozygotes and preservation of alleles, may be antagonistic from a management standpoint. However, Boecklen (1986) demonstrates for a two allele system that two sub-populations may simultaneously preserve more alleles and more heterozygotes than do intact populations over a wide range of migration rates. Of the two aims, preservation of alleles probably should have higher priority since heterozygosity can be reconstituted through gene flow among populations. However, in the absence of migration, small populations become fixed in relatively short time and may suffer inbreeding depression. Occasional exchanges of individuals among small isolated populations may represent a suitable compromise. Of course, these results are based on simplistic genetic models and need empirical validation.

Reserve design and management practice should be based primarily on the autecologies of targeted species. Small reserves are probably adequate for some animal and plant species (Simberloff and Gotelli 1984), but are clearly inadequate for species with extensive home ranges such as grizzly bears *Ursus horribilis* (Shaffer and Sampson 1985). Systems of small and large reserves probably represent the optimal design strategy.

REFERENCES

Allendorf, F. W., 1983. Isolation, gene flow, and genetic differentiation among populations. Pp. 51-65 *in* Genetics and Conservation: a Reference for Managing Wild Animal and Plant Populations ed by C. M. Schoenwald-Cox, S. M. Chambers, B. MacBryde and L. Thomas. Benjamin/Cummings, Menlo Park, California.

Ballou, J. and Ralls, K., 1982. Inbreeding and juvenile mortality in small populations of ungulates: a detailed analysis. *Biol. Conserv.* 24: 239-72.

Berry, R. J., 1971. Conservation aspects of the genetic constitution of populations. Pp. 177-206 *in* The Scientific Management of Animal and Plant Communities for Conservation ed by E. Duffy and A. S. Watt. Blackwell, Oxford.

Boecklen, W. J., 1986. Optimal design of nature reserves: consequences of genetic drift. *Biol. Conserv.* (in press).

Boecklen, W. J. and Gotelli, N. J., 1984. Island biogeographic theory and conservation practice: species-area or specious-area relationships? *Biol. Conserv.* 29: 63-80.

Boecklen, W. J. and Simberloff, D., 1986. Area-based extinction models in conservation. Pp. 247-76 *in* Dynamics of Extinctions ed by D. K. Elliott. Wiley, New York.

Burgess, R. L. and Sharpe, D. M., 1981. Introduction. Pp. 1-5 *in* Forest Island Dynamics in Man-dominated Landscapes ed by R. L. Burgess and D. M. Sharpe. Springer-Verlag, New York.

Chesser, R. K., 1983. Isolation by distance: relationship to the management of genetic resources. Pp. 66-77 *in* Genetics and Conservation: a Reference for Managing Wild Animal and Plant Populations ed by C. M. Schoenwald-Cox, S. M. Chambers, B. MacBryde and L. Thomas. Benjamin/Cummings, Menlo Park, California.

Chesser, R. K., Smith, M. H. and Brisbin, I. L., Jr., 1980. Management and maintainance of genetic variability in endangered species. *International Zoo Yearbook* 20: 146-54.

Connor, E. F. and McCoy, E. D., 1979. The statistics and biology of the species-area relationship. *Am. Nat.* 113: 791-833.

Diamond, J. M., 1972. Biogeographic kinetics: estimation of the relaxation times for avifaunas of the southwest Pacific islands. *Proc. natn. Acad. Sci. USA* 69: 3199-203.

Diamond, J. M., 1975. The island dilemma: lessons of modern biogeographic studies for the design of nature reserves. *Biol. Conserv.* 7: 129-46.

Diamond, J. M. and May, R. M., 1976. Island biogeography and the design of nature reserves. Pp. 163-86 *in* Theoretical Ecology ed by R. M. May. Saunders, Philadelphia.

Franklin, I. R., 1980. Evolutionary change in small populations. Pp. 135-49 *in* Conservation Biology: an Evolutionary-Ecological Perspective ed by M. E. Soulé and B. A. Wilcox. Sinauer, Sunderland, Massachusetts.

Gilbert, F. S., 1980. The equilibrium theory of island biogeography: fact or fiction? *J. Biogeogr.* 7: 209-35.

Higgs, A. J., 1981. Island biogeography theory and nature reserve design. *J. Biogeogr.* 8: 117-24.

Higgs, A. J. and Usher, M. B., 1980. Should nature reserves be large or small? *Nature* 285: 568-9.

I.U.C.N., 1980. World conservation strategy. International Union for Conservation of Nature and Natural Resources, United Nations Environmental Programme, World Wildlife Fund. Gland, Switzerland.

Lehmkuhl, J. F., 1984. Determining size and dispersion of minimum viable populations for land management planning and species conservation. *J. Environ. Mgmt.* 8: 167-76.

Lovejoy, T. E., Rankin, J. M., Bierregaard, R. O., Jr., Brown, K. S., Jr., Emmons, L. H. and Van der Voort, M. E., 1984. Ecosystem decay of Amazon forest fragments. Pp. 296-325 *in* Extinctions ed by M. H. Nitecki. University of Chicago Press, Chicago.

Jarvinen, O., 1982. Conservation of endangered plant populations: single large or several small reserves? *Oikos* 38: 301-7.

Jarvinen, O., 1984. Dismemberment of facts: a reply to Willis on subdivision of reserves. *Oikos* 42: 402-3.

Kindlmann, P., 1983. Do archipelagoes really preserve fewer species than one island of the same total area. *Oecologia* **59**: 141-4.

Kitchener, D. J., Chapman, A., Dell, J., Muir, B. G. and Palmer, M., 1980. Lizard assemblage and reserve size and structure in the Western Australian wheatbelt: some implications for conservation. *Biol. Conserv.* **17**: 25-62.

MacArthur, R. H. and Wilson, E. O., 1967. The Theory of Island Biogeography. Princeton University Press, Princeton.

Margules, C., Higgs, A. J. and Rafe, R. W., 1982. Modern biogeographic theory: are there any lessons for nature reserve design? *Biol. Conserv.* **24**: 115-28.

Preston, F. W., 1960. Time and space and the variation of species. *Ecology* **41**: 612-27.

Rey, J. R., 1981. Ecological biogeography of arthropods on Spartina islands in northwest Florida. *Ecol. Monogr.* **51**: 237-65.

Shaffer, M. L., 1981. Minimum population sizes for species conservation. *Bio. Sci.* **31**: 131-4.

Shaffer, M. L. and Sampson, F. B., 1985. Population size and extinction: a note on determining critical population size. *Am. Nat.* **125**: 144-52.

Simberloff, D. S., 1976. Experimental zoogeography of islands: effects of island size. *Ecology* **57**: 629-48.

Simberloff, D., 1985. Design of nature reserves. *In* Wildlife Conservation Evaluation ed by M. B. Usher. Chapman and Hall, London (in press).

Simberloff, D. S. and Abele, L. G., 1976. Island biogeography theory and conservation practice. *Science* **191**: 285-6.

Simberloff, D. and Abele, L. G., 1982. Refuge design and island biogeographic theory: effects of fragmentation. *Am. Nat.* **120**: 41-50.

Simberloff, D. and Abele, L. G., 1984. Conservation and obfuscation: subdivision of reserves. *Oikos* **42**: 399-401.

Simberloff, D. and Gotelli, N., 1984. Effects of insularization on plant species richness in the prairie-forest ecotone. *Biol. Conserv.* **29**: 27-46.

Soulé, M. E., 1980. Thresholds for survival: maintaining fitness and evolutionary potential. Pp. 95-118 *in* Conservation Biology: an Evolutionary-ecological Perspective ed by M. E. Soulé and B. A. Wilcox. Sinauer, Sunderland, Massachusetts.

Soulé, M. E., 1983. What do we really know about extinctions? Pp. 111-24 *in* Genetics and Conservation: a Reference for Managing Wild Animal and Plant Populations ed by C. M. Schoenwald-Cox, S. M. Chambers, B. MacBryde and L. Thomas. Benjamin/Cummings, Menlo Park, California.

Soulé, M. E., Wilcox, B. A. and Holtby, C., 1979. Benign neglect: a model of faunal collapse in the game reserves of East Africa. *Biol. Conserv.* **15**: 259-72.

Temple, S. A., 1981. Applied island biogeography and the conservation of endangered island birds in the Pacific Ocean. *Biol. Conserv.* **20**: 147-61.

Terborgh, J., 1974. Preservation of natural diversity: the problem of extinction prone species. *Bio. Sci.* **24**: 715-22.

Terborgh, J., 1975. Faunal equilibria and the design of wildlife preserves. Pp. 369-80 *in* Tropical Ecological Systems: Trends in Terrestrial and Aquatic Research ed by F. B. Golly and E. Madina. Springer-Verlag, New York.

Wilcox, B. A., 1980. Insular ecology and conservation. Pp. 95-117 *in* Conservation Biology: an Evolutionary-ecological Perspective ed by M. E. Soulé and B. A. Wilcox. Sinauer, Sunderland, Massachusetts.

Wilcox, B. A. amd Murphy, D. D., 1985. Conservation strategy: the effects of fragmentation on extinction. *Am. Nat.* **125**: 879-87.

Willis, E. O., 1984. Conservation, subdivision of reserves, and the anti-dismemberment hypothesis. *Oikos* **42**: 396-8.

Wilson, E. O. and Willis, E. O., 1975. Applied biogeography. Pp. 522-34 *in* Ecology and Evolution of Communities ed by M. L. Cody and J. M. Diamond. Belknap Press, Cambridge, Massachusetts.

CHAPTER 12

Conservation of the Genetic Resources of Rare and Widespread Eucalypts in Remnant Vegetation

G. F. Moran[1] and S. D. Hopper[2]

> Knowledge of the amount and geographic distribution of genetic diversity in a species is a major prerequisite to any management decisions about the role of remnant vegetation in conserving genetic resources. The number of species needing this kind of investigation exceeds current abilities to undertake the task. Hence, the pursuit of indicator species whose variation patterns may be of some general applicability is advocated.
>
> This chapter reviews the genetic structure of populations of trees with an emphasis on isozyme variation in eucalypts, and presents new data on rare localized mallees from Western Australia. Eucalypts, like other trees, have high levels of genetic diversity compared to other organisms. The distribution of genetic resources varies with population structure in eucalypts. Most variation is within populations in widespread eucalypts, but the reverse may apply in regionally distributed species with disjunct populations such as *E. caesia*. This result has clear implications for reserve design: a few large reserves across the geographic range would suffice for widespread species, whereas both large and small reserves over a finer geographic scale are needed for species like *E. caesia*.
>
> Many small populations of localized mallees in Western Australia have considerable genetic diversity. The often-quoted minimum population sizes of the order of 500 seem inappropriate for these species. If this finding is of general validity, remnants of native vegetation may play a more significant role in the conservation of genetic resources than has been appreciated in the past.

INTRODUCTION

A PRIME concern in the conservation of species is the maintenance of the maximum amount of their genetic resources. These resources include all inherited variation occurring within all individuals of a species. Genetic variation provides insurance against extinction, enabling populations to persist through environmental change. Innumerable examples of this exist from the fields of population genetics, domestication of stock, cultivation of crop plants, etc.

The optimal procedure for maintaining the genetic resources of individual species would be to conserve all types of biological communities in very large national parks or reserves. Obviously not all communities and their constituent species can be conserved in this way. It is here that remnant vegetation has a major role to play. Reserves of remnant vegetation can be used to conserve rare associations of native plants but perhaps their main function could be in maintaining the genetic resources of particular species. The latter role may be necessary either because a species is not represented at all in large reserves, or because the proportion of the genetic resources of such species in national parks is inadequate.

The use of remnant vegetation to conserve genetic resources of selected species will be dependant on their minimum viable population size. A breeding population needs to be big enough to retain its viability through successive generations. Although a central tenet of current conservation theory, this concept has been addressed rigorously in few empirical studies.

Theoretical calculations suggest that five hundred is an appropriate minimum viable population size for sexual outbreeding vertebrates (Frankel and Soulé 1981). However, the assumptions on which these calculations are based are unlikely to be met in most organisms. The question needs to be addressed through assays of genetic variation in populations of varying size in a broad spectrum of animal and plant groups before reliable generalizations (if any) can be made. Therefore, knowledge of the amount of genetic diversity over the geographic range of the species is a major prerequisite to any management decisions about the role of remnant vegetation in conserving genetic resources. This knowledge would form a basis for the determination and subsequent evaluation of optimal procedures for gene conservation.

It is impossible to survey thousands of species to obtain such genetic data, so it is important to establish whether general conservation strategies can be formulated for particular biological groups.

Several reviews have surveyed the available data on genetic diversity in plant species and looked for generalized relationships between ecological and life history traits and patterns of genetic diversity (Hamrick *et al.* 1979; Loveless and Hamrick 1984). These studies, taken together with reviews of the conservation of plant genetic resources (Brown 1978; Brown and Moran 1981), suggest that plants with similar mating systems, longevity, population structure, etc. may have similar patterns of genetic diversity. Hence by selecting a small number of indicator species in a large group with common attributes, general strategies for *in situ* conservation may be forthcoming.

For example, trees are characterized by their large size, longevity, high fecundity and high outcrossing rates. Of these factors, the predominantly outcrossing breeding system appears to be the major cause of the similarity in the patterns of genetic variability of trees. However, major differences in geographic population structure is a subject warranting further consideration in this field as well. For instance, do rare localized species have similar patterns of genetic diversity? Do they differ from widespread continuously distributed species?

In Australia, eucalypt and acacia trees dominate most vegetation associations. Since both these genera are large and have numerous species that are either economically important or rare and endangered (Rye and Hopper 1981; Boland *et al.* 1984), any successful generalizations concerning conservation strategies may be of widespread and immediate benefit. Furthermore the area occupied by a breeding population of a species of tree is generally much larger than for the majority of understorey species. Hence the conservation of a population of trees leads to the protection of large numbers of other organisms as well. For these reasons, eucalypts have considerable merit as high priority organisms for study in remnant reserves.

This chapter reviews the genetic structure of populations of trees and presents new data on rare localized Western Australian mallees. In particular, the distribution of genetic diversity within and between populations of eucalypts is examined and the implications these patterns have for the conservation of their genetic resources is explained. Emphasis is placed on species that are rare and localized and the extent to which these can provide insights into the long term effects on the genetics of species isolated in remnant reserves. The distribution and levels of genetic diversity under different geographic population structures is also evaluated, especially in relation to population size.

GEOGRAPHIC POPULATION STRUCTURE AND THE SPECIES UNDER REVIEW

The number, size and distributional range of the populations of a species collectively define its geographic population structure. The geographic ranges of trees may be divided into three main types (Fig. 1):

1. widespread. These are species with a range of 600 km in at least one direction;
2. regional. A range between 150-600 km;
3. localized. A small number of populations usually of limited size and endemic to a narrow geographic area of less than 100 km.

Within each of these three types the distribution of populations may vary from continuous to discontinuous and disjunct.

There are many eucalypt species that fit each of these categories of geographic range (Pryor and Johnson 1971; Chippendale and Wolf 1981). For the purposes of this review, we will consider further only those eucalypts (Fig. 1) for which isozyme data are available (Table 1). While the widespread Australian trees in Table 1 are well known and illustrated (e.g. in Boland *et al.* 1984), the Western Australian species with regional and localized distributions are not. They are briefly described and illustrated below.

Eucalyptus caesia

E. caesia is a widely cultivated and distinctive member of a group of Western Australian mallees that are confined to large granite outcrops. The species is rare in the wild, known from about 2000 adult plants distributed across a geographic range of 270 km (Fig. 1) in 15 disjunct populations of 1-580 individuals (Fig. 1; Hopper *et al.* 1982).

Eucalyptus lateritica (Fig. 4c)

Discovered and named only recently (Brooker and Hopper 1986), *E. lateritica* is a small mallee confined to a range of 30 km in an area of lateritic hills to the north of Perth (Fig. 1). It is known from about

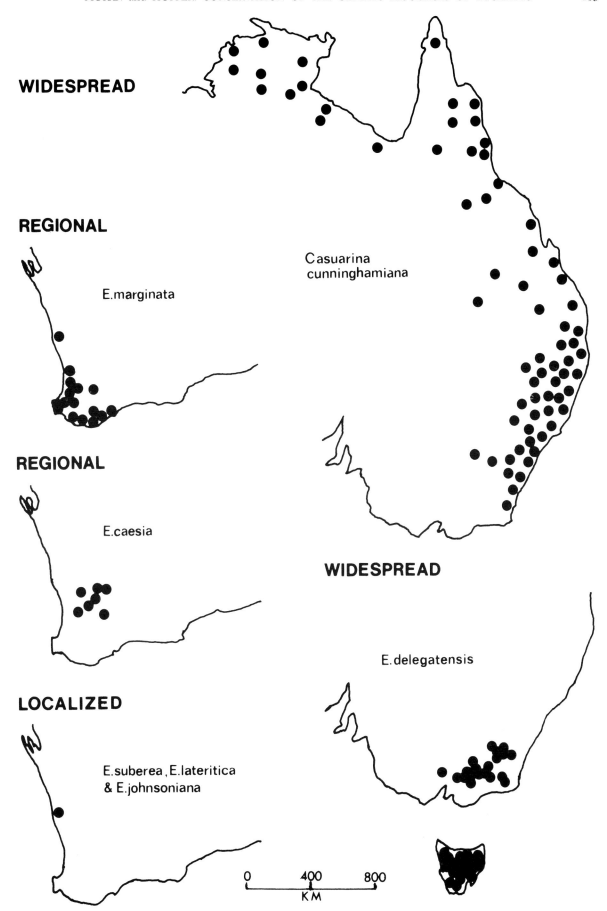

Fig. 1. Examples of tree or mallee species with widespread, regional and localized distributions as defined in the text.

Table 1. The average number of isozyme alleles and the percentage of alleles with various kinds of distribution in trees of widespread, regional and localized geographic distribution.

Species	No. Species/ Populations	Loci	A[a]	Distribution of Alleles[b]			
				Common		Rare	
				W	L	W	L
WIDESPREAD							
gymnosperms[c]	12	91	3.00	49	7	27	17
eucalypts[d]	4	52	3.17	53	6	28	13
Casuarina cunninghamiana	20	19	4.32	41	27	18	14
REGIONAL							
E. caesia	13	18	1.94	41	12	29	18
LOCALIZED							
E. lateritica	2	16	2.56	24	32	8	36
E. pendens	7	16	2.88	24	24	14	38
E. johnsoniana	3	16	1.88	0	50	0	50
E. suberea	3	16	2.44	17	35	13	35

a A = mean number of alleles per locus
b W = widespread, L = localized — see text
c Data from Brown and Moran (1981)
d eucalypts are *E. cloeziana*, *E. delegatensis*, *E. grandis* and *E. saligna* (data sources listed in Table 2).

10 populations of 1-100 individuals, usually on the upper scree slopes and edges of breakaways of flat-topped mesas.

Eucalyptus pendens (Fig. 3)

E. pendens, a striking whipstick mallee, has a range of 60 km in the same area occupied by *E. lateritica*, *E. johnsoniana* and *E. suberea*. Originally considered to be very rare (Brooker 1972) *E. pendens* is now known to be more abundant than these other mallees occurring in 14 populations of 1-3000 individuals (Fig. 2). A total of about 5500 individuals of the species are known in the wild.

Eucalyptus johnsoniana (Fig. 4a)

One of Western Australia's least fecund and most taxonomically isolated mallees, this species was named from two nearby populations by Brooker and Blaxell (1978). Further survey has established that *E. johnsoniana* occurs in about 10 populations (Fig. 1), most consisting of a few individuals. *E. johnsoniana* grows in sand or sand over laterite, usually emerging up to 3 m from low heath.

Eucalyptus suberea (Fig. 4b)

Like *E. johnsoniana*, *E. suberea* appears to be a very rare relic species with no close relatives (Brooker and Hopper 1986). It has an identical range and population structure to *E. lateritica*, growing in heath on lateritic breakaways in small disjunct populations across a 30 km geographic range to the north of Perth (Fig. 1).

MEASURING GENETIC DIVERSITY

To assess the genetic resources of a species, details of its population genetic structure are required. This involves the measurement of the genetic diversity within and between populations. Currently, the isozyme technique represents the best means of direct evaluation of the genetic variation in a population (Brown and Moran 1981). Isozyme data are used, therefore, in this review. Technical details of the electrophoretic procedures for eucalypts are given by Moran and Bell (1983) and Moran and Hopper (1983).

Single locus measures of genetic variation are widely used to characterize genetic diversity in populations. The resultant statistics are based on allelic frequencies at individual loci and summary measures are obtained by taking averages across loci. These statistics are:

A, the mean number of alleles per locus;
P, the proportion of loci which are polymorphic; and
H, the expected panmictic heterozygosity.

The first two statistics are very dependent on sample size whereas H is relatively unaffected by sample size (Brown and Weir 1983). For this reason most emphasis in this review is placed on estimates of the expected panmictic heterozygosity. Observed heterozygosity (H_o) at a single locus is the other widely used statistic. It can also be averaged over loci to give a population estimate of heterozygosity.

Genetic diversity can be thought of as having two components, allelic richness and allelic evenness (Brown and Weir 1983). Both should be considered in assessment of the genetic resources of species.

GENETIC DIVERSITY — ALLELIC RICHNESS

Allelic richness can be defined as the number of alleles in a population sample. It is most directly measured by A, the number of alleles per locus. Widespread tree species typically have a high value

△ Plate 7

△ Plate 8

Plate 9 ▽

PLATE 7
CHAPTER 12

Eucalyptus caesia subsp. *caesia*
(grey-white foliage)
in a soil pocket on a massive granite
outcrop in the wheatbelt of Western
Australia. Many such remnants
of native vegetation owe
their persistance to the rugged
nature of the terrain on which they
grow. (S. Hopper).

PLATE 8
CHAPTER 12

Buds, flowers and fruit
of *Eucalyptus pendens*.
(S. Hopper).

PLATE 9
CHAPTER 12

Eucalyptus suberea with characteristic,
yellow-brown corky bark.
(S. Hopper).

△ Plate 10   ▽ Plate 11

△ Plate 12

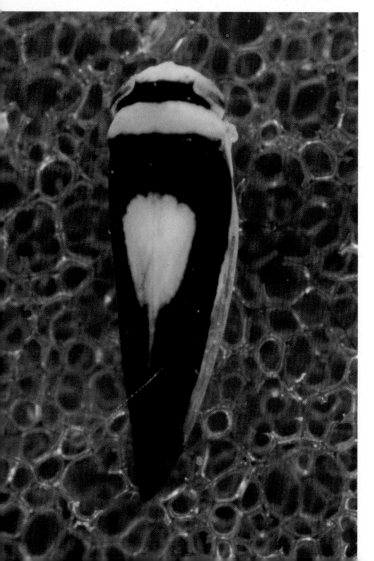

PLATE 10
CHAPTER 13

Iowa tall-grass prairie. A lush growth of tall grasses is interspersed with forbs such as purple coneflower *Echinacea angustifolia*. (R. Whitcomb).

PLATE 11
CHAPTER 13

Colladonus clitellarius, a savanna species: this is one of many insects that alternates between woody and herbaceous hosts. Considered uncommon, it is nonetheless a fairly general feeder. Cicadellids that alternate hosts and that also have narrow host preferences are apt to appear even rarer. (R. Whitcomb).

PLATE 12
CHAPTER 13

Wheatgrass *Agropyron smithii* prairie, in an austere Montana setting. This land, located in the rain shadow of the Rocky Mountains, receives scant rainfall, and, as the abandoned buildings attest, could not be farmed economically. Most of the National Grasslands consist of such abandoned land that has been reclaimed and enjoys more gentle management than private landholders could afford. (R. Whitcomb).

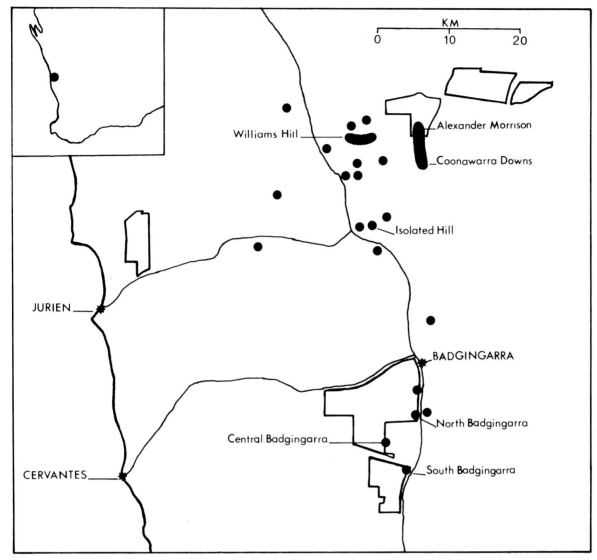

Fig. 2. Distribution of *Eucalyptus pendens* to the north of Perth, Western Australia, with place names used in Table 3.

of A (Table 1) compared to other plant groups (Hamrick *et al.* 1979). In sharp contrast, the regional species *E. caesia* with several disjunct populations has a low value of A. On average the localized species have estimates intermediate between those of species of regional and widespread distribution. Despite the fact that localized species have very small populations, they still exhibit appreciable allelic richness.

The percentage of allelic variants in four classes of distributional occurrence are listed in Table 1. These classes have been slightly modified from Marshall and Brown (1975) and are as follows:

1. common: occurring in at least one population with a frequency greater than 10%; and

2. rare: not occurring in any population with a frequency greater than 10%.

Within these two categories, alleles can either be widespread (W) when they occur in two or more populations or localized (L) when they are in only one population. The percentages were calculated by standardizing for any differences in the number of invariant loci reported (Brown 1978). These were computed by subtracting the number of loci studied from the number of alleles in the common and widespread class.

The distribution of alleles among populations of widespread species is similar for gymnosperms and eucalypts. In both, the proportion of locally common alleles is less than 10%. Alleles in this class are potentially important to conserve. Only a limited number of populations would be required to achieve this. Widespread eucalypts and gymnosperms both have a significant fraction of alleles in the rare but localized category. However, it will be difficult to devise realistic *in situ* conservation strategies to maintain a large proportion of these alleles.

In *Casuarina cunninghamiana*, a widespread species with a discontinuous distribution, 27% of its alleles are locally common. This high value appears

Fig. 3. Photographs of isolated stands of *E. pendens* in Badgingarra National Park (a, b), and plants in the largest continuous population of the species (Williams Hill c).

to be due in part to the large genetic discontinuities between geographic areas (i.e. races). This suggests that more populations of this species should be conserved. The distribution of alleles in *E. caesia* is similar to that for widespread eucalypts.

The localized species have markedly different patterns of allelic distribution. In fact the locally common and locally rare classes predominate, rather than the widespread classes. Some caution should be exercised in interpretation of these data as a limited number of populations have been surveyed. Nevertheless, over half the known populations of *E. pendens* were surveyed and the same trend was evident. To maintain a significant proportion of the allelic richness component of the genetic diversity in such species, a number of populations would need to be conserved.

GENETIC DIVERSITY — ALLELIC EVENNESS

The distribution of the genetic diversity across the geographic range of a species is a vital element in determining the optimal strategy for the conservation of genetic resources. The expected panmictic heterozygosity primarily measures the evenness component of genetic diversity. This diversity can be apportioned between and within populations (Nei 1975, see Table 2). In most cases widespread tree species have high levels of genetic diversity overall. A number of North American conifers have similar estimates to the *Pinus ponderosa* data shown in Table 2. Estimates for eucalyptus are of the same magnitude. There are exceptions as *P. resinosa* has almost no variation (Fowler and Morris 1977) and *P. radiata* with a low value of H_T (Table 2).

The proportion of the genetic diversity between populations of gymnosperms is less than 5% as shown by the example of *P. ponderosa* (Table 2). The highest level of interpopulation diversity for gymnosperms is 16.2% for *P. radiata*. In widespread eucalypts the interpopulation genetic diversity is about twice that of gymnosperms. One possible explanation is that because eucalypts are pollinated by animals the gene flow will be less between populations and hence interpopulation genetic diversity greater. Another possible factor is that edaphic fidelity may be much more precise for eucalypts compared to gymnosperms. Thus, even in widespread eucalypts the patchiness in occurrence associated with this edaphic specialization is much greater than in gymnosperms.

The other notable feature of Table 2 is that the wind pollinated but dioecious *Casuarina cunninghamiana* has a relatively high level (28.7%) of the genetic diversity between populations when compared with other widespread trees. This species has a distinctive riverine distribution over its very large geographic range. A significant fraction of the interpopulation component of the diversity for both *E. delegatensis* and *C. cunninghamiana* can be apportioned to geographic areas of the species.

With regard to allelic evenness, the conservation of genetic resources in widespread species will be influenced by the fact that most of the genetic diversity is within populations. *In situ* conservation of a few large populations in different geographic areas would be the prime requirement.

In *E. caesia*, a species with a geographic range of 270 km but very small disjunct populations, 61% of the genetic diversity is among populations (Table 2).

Table 2. Total genetic diversity and distribution of diversity within and between populations in trees of widespread, regional and localized geographic distribution.

Species	No. Populations	H_T[a]	H_S	G_{ST}(%)		[Source[b]]
WIDESPREAD						
Pinus ponderosa	11	.289	.284	1.5		(1)
P. radiata	5	.117	.098	16.2		(2)
Eucalyptus grandis	12	.190	.167	12.0		(3)
E. saligna	7	.260	.239	8.0		(3)
E. cloeziana	17	.270	.240	11.0		(4)
E. delegatensis	23	.272	.237	12.9	(8.8)[c]	(4)
Casuarina cunninghamiana	20	.292	.208	28.7	(13.7)	(2)
REGIONAL						
E. caesia	13	.176	.068	61.4		(5)
LOCALIZED						
E. lateritica	2	.318	.278	12.6		(2)
E. pendens	7	.170	.156	8.2		(2)
E. johnsoniana	3	.139	.084	39.6		(2)
E. suberea	3	.197	.170	13.7		(2)

a H_T = total genetic diversity, H_S = mean genetic diversity within populations, G_{ST} = $(H_T - H_S)/H_T$ x 100 (percentage of diversity between populations).
b Sources of data: (1) Hamrick (1983), (2) Moran and associates, unpublished, (3) Burgess and Bell (1983), (4) Turnbull (1980), (5) Moran and Hopper (1983).
c Component of diversity between geographic races.

Fig. 4. Photographs of plants of three localized mallees included in the study *E. johnsoniana* (a), *E. suberea* (b), and *E. lateritica* (c). All stems illustrated in (c) are identical in their isozyme profiles and presumably constitute one clonal individual 10 m across.

Moreover, the isozyme data indicate that differentiation between populations is almost twice in subspecies *caesia* compared to that in subspecies *magna* (Moran and Hopper 1983). The extensive interpopulation genetic differentiation suggests that a considerable number of the populations should be set aside in reserves for optimal *in situ* conservation, especially the southern populations of subsp. *caesia* (Hopper *et al.* 1982). Other eucalypts with regional distributions need to be studied to establish whether these species as a group have similar high levels of interpopulation genetic diversity.

In localized eucalypts, most of the diversity is within populations rather than between populations (Table 2). Localized species tend to have less total genetic diversity than widespread species. This is partly due to the smaller number of variable loci contributing to the diversity. The interpopulation diversity in *E. johnsoniana* is substantial and it is partly due to the low total diversity in the species. Associated with this interpopulation diversity there was significant multilocus organization within populations. For such species both components of genetic diversity, richness and evenness should be considered in formulating conservation strategies.

POPULATION SIZE AND GENETIC DIVERSITY

The range in size of populations of a tree species can be very large. Therefore it is important to know whether there is a relationship between population size and genetic diversity. In particular is there a minimum population size below which significant reductions in genetic diversity can occur? Minimum viable population size can be defined as that which will maintain genetic diversity, be capable of adaptive evolution and also long-term survival. This concept of minimum viable population size is crucial to the conservation of remnant vegetation.

The theory of population genetics suggests that fragmentation of species into populations with low numbers may result in a reduction in genetic variability. With a transient reduction in population size, the loss of alleles (especially the rare alleles) could be substantial whereas the loss in heterozygosity would not be severe (Frankel and Soulé 1981). The minimum population size required to counteract the effect of inbreeding has been proposed as 50 (for sexual outbreeding vertebrates) and 500 as a long term viable population size (Franklin 1981; Frankel and Soulé 1981). It may be that the success or failure in formulating conservation strategies in relation to geographic structure may ultimately depend on the relative roles of selection, genetic drift, gene flow and migration in determining genetic diversity patterns. The lack of information on the evolutionary history of populations is a problem. For instance the extent of fluctuations in the effective population size (N_e) will mostly be unknown. With few studies done on genetic diversity in small plant populations, especially of trees, the relevance of the often-quoted minimum viable population sizes to the real world is a matter for further research.

Genetic diversity measures for seven populations of *E. pendens* show no obvious relationship between population size and genetic diversity either in the richness or evenness components (Table 3). Similarly in *E. caesia* there was not a significant correlation between A and H_e and population size (Moran and Hopper 1983). Among the four localized species listed in Table 4 there was no relation between average population size and genetic diversity either for the allelic richness or evenness component. It seems that species may have intrinsic levels of diversity determined more by long term evolutionary history than recent phenomena such as current population size.

Clearly for all these species the generalization that small populations have less genetic variability than large populations is not true. However, all that is really known about these small populations is that in the current generation their size could be an effective bottleneck. These species probably have generation times in the hundreds of years. If the time scale of concern is a thousand years then it would seem practical to maintain their genetic resources using remnants of native vegetation. The minimum population size proposed in the literature are dependent on assumptions of random mating and a lack of population structure. For many plants these assumptions clearly do not hold. Remnants of native vegetation containing small populations of eucalypts do have an important role to play in the conservation of genetic resources.

SUBPOPULATION GENETIC STRUCTURE

The spatial array of individuals and of genotypes in a population may or may not be random. If the genotypic arrays are non-random then there can be neighbourhoods in a population. Often concomitant with these neighbourhoods is some level of inbreeding. The extent of this genetic differentiation will be an important determinant of whether large or small populations or parts thereof should be conserved.

Table 3. Population sizes and single-locus diversity measures for *Eucalyptus pendens*.

Population (see Fig. 2)	N	Genetic Diversity[a]			
		A	P	H_o	H_e
South Badgingarra	83	1.73	50	.145	.126
Central Badgingarra	33	1.63	44	.204	.191
North Badgingarra	27	1.69	63	.183	.208
Isolated Hill	35	1.75	56	.156	.148
Alexander Morrison	56	1.88	75	.113	.131
Williams Hill	3000	1.75	69	.163	.163
Coonawarra Downs	1000	1.44	44	.105	.123

a A = mean no. of alleles per locus, P = av. percentage of polymorphic loci, H_o = observed heterozygosity, H_e = expected panmictic heterozygosity.

Table 4. Mean sizes and genetic diversity measures per population for localized Western Australia eucalypts. Standard errors are given in brackets.

Species	N	A	P	H_e	H_o
E. johnsoniana	52	1.29 (.30)	27 (20)	.084 (.018)	.147 (.085)
E. lateritica	34	2.16 (.13)	75 (0)	.278 (.040)	.385 (.067)
E. suberea	87	1.85 (.20)	60 (14)	.170 (.016)	.200 (.030)
E. pendens	605	1.80 (.14)	57 (12)	.156 (.033)	.152 (.036)

In some inbreeding plants significant genetic differentiation between stands within populations has been demonstrated (Allard et al. 1972). In contrast there is no published evidence of extensive differentiation between stands within populations of outcrossing trees.

Some limited data on subpopulation structure in eucalypts is available to consider this problem. For example, in E. caesia there is distinct differentiation over distances of about 300 m between the small stands in the two largest populations of E. caesia. This differentiation would suggest that the whole populations should be maintained for in situ conservation of these genetic resources.

In Table 5 the genetic diversity in stands of E. pendens are presented as an example of one possible type of subpopulation structure in localized eucalypts. For the large populations of E. pendens there is little difference between overall diversity levels in the populations compared to that in stands within these populations. This suggests the absence of any significant subpopulation structure. In the small populations on the other hand, genetic diversity is considerably less within stands compared to that in the overall populations. The extensive differentiation between the disjunct stands within the small populations of E. pendens is a reflection of the occurrence of large clumps (up to 15 m across) of mallees having identical genotypes at least as detected by isozyme assays. In some cases, therefore, a stand may in fact be a single clonal individual.

Small populations of the other localized species (as in Table 2) also have similar subpopulation structure. Given this spatial genetic organization, the genetic resources in such small populations will be conserved by maintaining most of the stands whereas in larger continuous populations this may not be necessary.

These kinds of data are useful in deciding on the conservation value of remnant reserves in the wheatbelt of southwest Western Australia. However, it would be premature to make generalizations for eucalypts with similar distributions throughout Australia. Whether there is any subpopulation structure in small continuous populations of trees generally needs to be established.

OUTLIERS OF WIDESPREAD SPECIES

In establishing remnant reserves of plant communities disjunct or continuous outliers of widespread species must be considered. The nature and extent of the genetic resources in such outliers will determine the usefulness of them for in situ conservation. In E. delegatensis, outlying populations were found to have 96.2% of the genetic diversity present in populations of the main range. Similarly no significant differences in quantitative characters such as growth in provenance trials have been demonstrated for these outliers. Even so a cluster analysis based on genetic distance gave a distinct separation of the outliers from the main geographic groups of the main range. This result appears to be due to small shifts in allelic frequencies especially of rare alleles rather than differences in allelic richness per se. For this species remnant reserves of the outliers would maintain a reasonable fraction of the genetic diversity in the genome but it may not be necessary if the main range is adequately conserved in reserves.

From a much more limited sample of populations of E. marginata the outliers have 87% of the genetic diversity present in one population from the main range. The limited data suggest that outliers of widespread species are certainly not depauperate in genetic diversity and such populations should be considered for remnant reserves. The justification for such reserves may be not so much genetic, however, as other factors such as isolation from the mainstands in case of catastrophes, etc.

CONCLUSIONS

This review advocates a major role for the population genetics approach to the in situ conservation of biological organisms. Conservation at the species level is ultimately concerned with maintaining the genetic resources of the species. The central problem in the conservation of gene pools is the assessment of genetic resources. Two aspects of this assessment are of concern. The first is the amount of

Table 5. Genetic diversity estimates for stands within small and large populations of E. pendens.

Population Type	Level[a]	Genetic Diversity		
		A	P	H_e
Small disjunct	P	1.67	55	.185
	S	1.24	22	.094
Large continuous	P	1.60	56	.143
	S	1.48	46	.133

a P = average across estimates for populations, S = average across estimates for stands within populations.

the genome that can be characterized directly at the population level. Isozyme techniques provide the best available tool for such characterization, although they have their limitations (Brown and Moran 1981).

The second problem is that probably much less than 1% of plant species in Australia could be studied genetically in the foreseeable future. Hence the possibility of using key indicator species, particularly for remnant vegetation should be fully explored. There is also a strong need for generalized conservation strategies for groups of organisms and eucalypts were examined for this purpose. We would caution that for eucalypts more data are required for species of all classes of geographic population structure, expecially for those with regional distributions. Despite this lack of data some preliminary conclusions may be made.

Eucalypts, like other trees, have high levels of genetic diversity compared to other organisms. The distribution of these genetic resources is primarily within populations rather than between, although eucalypts with regional distributions (such as *E. caesia*) may possibly be an exception to this generalization. We conclude that, for widespread species, a limited number of large populations covering the geographic range would be a minimum requirement for effective conservation. However, it needs to be established whether there is significant subpopulation structure in such species before such a prescription is fully accepted. For eucalypts with localized distributions, the types and portions of populations needed to be conserved will depend on their subpopulation structure. With the apparent differences in structure between small and large populations found in *E. pendens* (Table 5), an initial management prescription would be to conserve intact a few small and large populations of each species. However, it should be emphasized that there is no evidence as yet of significant subpopulation structure in small populations with a continuous distribution of individuals.

The populations of localized species are often similar in size to those isolated by land clearance and now restricted to natural remnants of vegetation. Since many of the small populations of localized species have considerable genetic diversity the suggestions of minimum population sizes of the order of 500 seem inappropriate. In the ancient landscape of southwest Western Australia it may be that eucalypts have evolved to cope with small population sizes. Many species of the southwest flora have localized distributions. It seems likely that over a period of a 1000 years these species can be conserved in remnant reserves and at the same time maintain a reasonable fraction of their genetic resources. Nevertheless it may be that on an evolutionary time scale many of these species will become extinct.

The considerable loss of vegetation that has already occurred in the southwest of Western Australia, especially in the wheatbelt, suggest that the genetic resources of species occupying this zone and having regional distributions should be a high priority for study. Only then will we know whether the geographic pattern seen in the genetic resources of naturally rare and localized species is relevant to that of regionally distributed species that have been divided into similar sized remnant populations by rapid land clearance for agriculture.

ACKNOWLEDGEMENTS

We are grateful to the Division of Forest Research, CSIRO, for a visiting scientist's allowance awarded to S.D.H. for a two month period in 1984 in Canberra during which most of the collaborative work for this chapter was completed. Dr A. H. D. Brown, Dr D. J. Coates, Dr A. R. Griffin, an anonymous referee and the editors provided useful criticisms on an earlier draft. A. P. Brown prepared Figures 1 and 2. C. Bell and R. E. Sokolowski assisted with isozyme assays and collection of material in the field respectively.

REFERENCES

Allard, R. W., Babbel, G. R., Clegg, M. T. and Kahler, A. L., 1972. Evidence for co-adaptation in *Avena barbata*. *Proc. Natl. Acad. Sci.* 69: 3043-48.

Boland, D. J., Brooker, M. I. H., Chippendale, G. M., Hall, N., Hyland, B., Johnston, R. D., Kleinig, D. A. and Turner, J., 1984. Forest Trees of Australia. Nelson, Melbourne.

Brooker, M. I. H., 1972. Four new taxa of *Eucalyptus* from Western Australia. *Nutysia* 1: 242-53.

Brooker, M. I. H. and Blaxell, D. F., 1978. Five new taxa of *Eucalyptus* from Western Australia. *Nutysia* 2: 220-31.

Brooker, M. I. H. and Hopper, S. D., 1986. Notes on the informal subgenus 'Monocalyptus (Myrtaceae)' and the description of three new upland species from southwest Western Australia. *Nutysia* 5: 341-56.

Brown, A. H. D., 1978. Isozymes, plant population genetic structure and genetic conservation. *Theor. Appl. Genet.* 52: 145-57.

Brown, A. H. D. and Moran, G. F., 1981. Isozymes and the genetic resources of forest trees. Pp. 1-10 *in* Isozymes of North American Forest Trees and Forest Insects ed by M. T. Conkle. U.S. Department Agriculture. Berkeley, California.

Brown, A. H. D. and Weir, B. S., 1983. Measuring genetic variability in plant populations. Pp. 219-39 *in* Isozymes in Plant Genetics and Breeding ed by S. D. Tanksley and T. J. Orton. Elsevier, Amsterdam.

Burgess, I. P. and Bell, J. C., 1983. Comparative Morphology and allozyme frequencies of *Eucalyptus grandis* Hill ex Maiden and *E. saligna* Sm. *Aust. For. Res.* 13: 133-49.

Chippendale, G. M. and Wolf, L., 1981. The natural distribution of *Eucalyptus* in Australia. Special Publication 6, Australian National Parks and Wildlife Service, Canberra.

Fowler, D. P. and Morris, R. W., 1977. Genetic diversity in red pine: evidence of low heterozygosity. *Can. J. For. Res.* 7: 341-7.

Frankel, O. H. and Soulé, M. E., 1981. Conservation and Evolution. Cambridge University Press, Cambridge.

Franklin, I. R., 1981. Evolutionary change in small populations. Pp. 135-49 *in* Conservation biology ed by M. E. Soulé and B. A. Wilcox. Sinauer Associates, Massachusetts.

Hamrick, J. L., 1983. The distribution of genetic variation within and among natural plant populations. Pp. 335-48 *in* Genetics and Conservation ed by C. M. Schonewald-Cox, S. M. Chambers, B. MacBryde and L. Thomas. Benjamin-Cummings, London.

Hamrick, J. L., Linhart. Y. B. and Mitton, J. B., 1979. Relationships between life history characteristics and electrophoretically detectable genetic variation in plants. *Ann. Rev. Ecol. Syst.* 10: 173-200.

Hopper, S. D., Campbell, N. A. and Moran, G. F., 1982. *Eucalyptus caesia*, a rare mallee of granite rocks from southwestern Australia. Pp. 46-61 *in* Species at risk: Research in Australia ed by R. H. Groves and W. D. L. Ride. Australian Academy Science, Canberra.

Loveless, M. D. and Hamrick, J. L., 1984. Ecological determinants of genetic structure in plant populations. *Ann. Rev. Ecol. Syst.* 15: 65-95.

Marshall, D. R. and Brown, A. H. D., 1975. Optimum sampling strategies in genetic conservation. Pp. 53-80 *in* Genetic Resources for today and tomorrow ed by O. H. Frankel and J. G. Hawkes. Cambridge University Press, Cambridge.

Moran, G. F. and Bell, J. C., 1983. Eucalyptus. Pp. 423-41 *in* Isozymes in Plant Genetics and Breeding ed by S. D. Tanksley and T. J. Orton. Elsevier, Amsterdam.

Moran, G. F. and Hopper, S. D., 1983. Genetic diversity and the insular population structure of the rare granite rock species, *Eucalyptus caesia* Benth. *Aust. J. Bot.* 31: 161-72.

Nei, M., 1975. Molecular Population Genetics and Evolution. North Holland, Amsterdam.

Pryor, L. D. and Johnson, L. A. S., 1971. A Classification of the Eucalypts. Australian National University Press, Canberra.

Rye, B. L. and Hopper, S. D., 1981. A guide to the gazetted rare flora of Western Australia. Western Australian Department of Fisheries and Wildlife Report No. 42. Department of Fisheries and Wildlife, Perth.

Turnbull, J. W., 1980. Geographic variation in *Eucalyptus cloeziana*. Ph.D. Thesis, Australian National University, Canberra.

CHAPTER 13

North American Forests and Grasslands: Biotic Conservation

Robert F. Whitcomb[1]

Studies of eastern forest birds and grassland insects of North America are summarized with respect to their implications for conservation and reserve design. Large tracts (thousands of hectares) are required in both forest and grassland biomes for preservation of landscape heritage and conservation of mammal or bird species. Life history strategies of plant or insect species vary geographically, climatically, and along altitudinal, latitudinal, and longitudinal gradients. Development of insect conservation strategies therefore requires a thorough analysis of these habitat characteristics. Goals catering to insect or plant conservation may thus conflict with strategies designed solely for conservation of birds or mammals. In both grassland and forests, conservation strategies should emphasize (1) acquisition of representative large parcels in each major vegetational region, (2) identification and preservation of endangered species, habitats, or communities not protected in large reserves and (3) insistence that management strategies of large government-owned multiply used landholdings include conservation of biotic diversity as a major objective.

INTRODUCTION

IN the last two decades I have intensively studied two very contrasting habitat types: North America's eastern forest and its grasslands. In each of these types, I have studied biogeographic relationships, including the effects of habitat fragmentation. In this chapter I summarize features of forests and grasslands that determine distribution and abundance of bird and leafhopper species and discuss problems in biotic conservation of insects and birds from these perspectives.

EASTERN FOREST

As represented on simplified vegetation maps, the eastern forest of North America was homogeneous. Upland forest was dominated by deciduous trees such as oaks *Quercus* spp., hickories *Carya* spp., and American chestnut *Castanea dentata* and the lowlands by maples *Acer* spp. and elms *Ulmus* spp. As seen on larger maps (Fig. 1) that permit delineation of regional differences, however, the eastern forest biome was much less homogeneous (Küchler 1964; Brush *et al.* 1980). Longitude influenced the amount of precipitation received by the forest, and latitude influenced the retention of moisture once it was received. In particular, hemlock *Tsuga canadensis* occurred throughout much of the forest in moist areas. The composition of the eastern forest was especially affected by precipitation and climatic influences relating to the Appalachian Mountains. At the highest elevations, the climate approached that of more northern latitudes, and the vegetation of the forests, in its content of spruce *Picea* and birch *Betula* spp., resembled Canadian forests more than the deciduous forests to the east and west of the ridges. The northern and southern forests and those of the Appalachians had significant compositions of coniferous trees. In the coastal provinces, the forest changed through a transitional zone to Coastal Plain forests, which had an increasingly strong representation of pine *Pinus* spp. to the south. These southern forests were strongly influenced by periodic fires that initiated successional change. The net result of the latitudinal, longitudinal, and altitudinal heterogeneity of the forest was reflected in differentiation in forest types that was mapped by Küchler (1964) as eleven zones (Fig. 1).

Although the eastern forest was vegetationally rich, there must have been significant breaks in canopy continuity. At the smallest grain, forest continuity was punctuated by treefalls. In fact, the mature forests of precolumbian times could be viewed as mosaics of treefalls of different ages. At another grain, periodic disturbances such as storms,

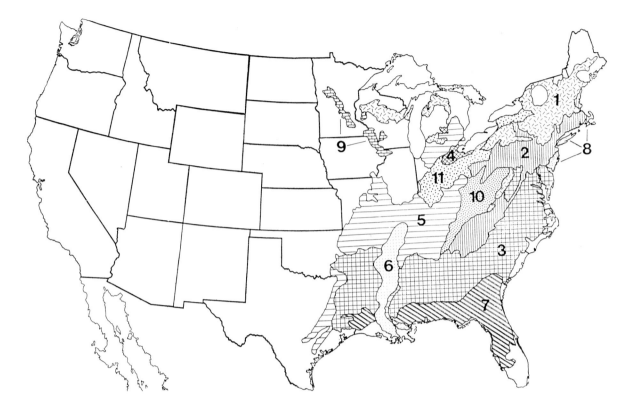

Fig 1. Eastern deciduous forest. 1. Northern hardwoods *(Acer-Betula-Abies-Tsuga)*; 2. Appalachian oak forest *(Quercus)*; 3. Oak-hickory-pine forest; 4. Elm-ash forest *(Ulmus-Fraxinus)*; 5. Oak-hickory forest *(Quercus-Carya)*; 6. Southern floodplain forest *(Quercus-Nyssa-Taxodium)*; 7. Southern mixed forest *(Fagus-Liquidambar-Magnolia-Pinus-Quercus)*; 8. Northeastern oak-pine forest *(Quercus-Pinus)*; 9. Maple-basswood forest *(Acer-Tilia)*; 10. Mixed mesophytic forest *(Acer-Aesculus-Fagus-Liriodendron-Quercus-Tilia)*; 11. Beech-maple forest *(Fagus-Acer)*.

cyclic insect infestations, or plant disease epidemics may have regularly opened the canopy. On the tops of the mountain ridges where the forest was exposed to weather extremes, or where slopes were periodically disturbed by erosion or rockslides, there was probably a continually changing mosaic of mature forest and various successional forest types. Flooding of rivers and streams was also capable of flattening forests, leading to successional change. It is likely that fires and other disturbances caused by precolumbian man produced and maintained forest openings. Furthermore, certain areas, such as shale or serpentine barrens, probably never supported mature forest in the conventional sense. Instead, the communities that occupied these barrens comprised unusual plant assemblages that consisted of the usual forest and/or savanna plant species growing in suboptimal habitat and a few endemic species that evolved with the barrens and grew optimally there. Finally, west of the mountains, over a region from Ohio (Transeau 1935) to the eastern sections of the plains states (Whitford 1958; Ranney and Johnson 1977), trees and grasses formed either a savanna or a mosaic of forest and prairie.

The concept of an equilibrial forest community is further eroded by considerations of recent geologic history (Wright and Frey 1965; Dort and Jones 1970). Although the forest is sometimes thought of as an equilibrial community, the boundaries of forest and prairie probably changed dynamically with climatic cycles. Recession of the ice sheets of the Pleistocene Wisconsinan maximum, after all, occurred only about 10,000 years ago. Pollen analyses in diverse regions of North America demonstrate that disturbance was by no means confined to areas actually covered by ice. Rather, massive climatic shifts occurred throughout temperate regions of the northern and southern hemispheres. With these climatic shifts came mass movements of entire vegetational regions. Iowa, for example, was covered by coniferous forests during the Wisconsinan (Wright and Ruhe 1965), a time when the prairie may have receded to a small area in south Texas and northern Chihuahua (Ross 1965, 1970). Pregill and Olson (1981) believed that effect of climatic changes extended as far south as the Caribbean islands.

In the east today there are many 'edge species,' apparently well adapted regionally, such as common yellowthroat *Geothlypis trichas*, rufous-sided towhee *Pipilo erythrophthalmus*, Eastern kingbird *Tyrannus tyrannus*, gray catbird *Dumetella carolinensis*, Northern oriole *Icterus galbula*, rose-breasted grosbeak *Pheucticus ludovicianus*, and chestnut-sided warbler *Dendroica pensylvanica*. All of these species are well adapted to successional or broken forests, suggesting that there must have been a substantial amount of successional habitat in pre-columbian North America. It is clear that some of these 'edge species' evolved with the savanna-forest-prairie mosaics of the midwest. Species that were probably endemic in this region include Bell's vireo *Vireo bellii*, yellow-breasted chat *Icteria virens*, blue-winged warbler *Vermivora pinus*, warbling vireo *Vireo gilvus*, red-headed woodpecker *Melanerpes erythrocephalus*, field sparrow *Spizella pusilla*, and indigo bunting *Passerina cyanea*. Another group of 'edge species' was associated with southeastern fire succession. These species included prairie warbler *Dendroica discolor*, Bachman's sparrow *Aimophila aestivalis*, and white-eyed vireo *Vireo griseus*.

Despite the considerable evidence for regular disturbance and patterns of irregularity in canopy cover, I believe that the eastern deciduous forest was essentially homogeneous from the point of view of forest interior bird species. Biogeographic criteria that I feel were important were the total fraction of the landscape that was covered by unbroken forests, the existence of copious corridors of forest among and between disturbed areas, the existence of huge archipelagos of forest island stepping stones, and dendritic corridors of forest along all stream valleys throughout the savanna-mosaic region. Structurally, at the level of habitat discrimination of bird species, different forest types were similar enough to permit adaptations or even moderate niche shifts in some cases. Disturbed or successional areas must have occurred as islands in a great sea of oak, hickory and chestnut trees of the eastern deciduous forest.

FRAGMENTATION

To make way for civilization, almost all of this forest was cut to make room for farms, houses, barns, bridges and other accoutrements of progress as European colonization pushed westward across the Appalachians toward the prairie. However, the rate of deforestation of virgin tracts permitted afforestation to follow in its wake. In a region with predictable and adequate rainfall and generally rich soils, new forests, if allowed to do so, rapidly replaced the mature forests. In the central Atlantic region, regrowth of oak forests with trees of 15 cm diameter at breast height occurred within 50-60 years. Even in coastal forests of the scrub pine-oak *Pinus virginiana-Quercus* spp. type, in which oak regrowth was slowed by a primary flush of broom-sedge *Andropogon virginicus* and scrub pine growth (Johnston and Odum 1956), oak forests emerged in less than 100 years. It has been estimated that afforestation of cut forest, or regrowth on abandoned agricultural land, prevented the total forest area from declining to more than 50% of the initial forested area at any one time.

The fragmentation of the forest did have the conspicuous effect of creating a new biome. A vast neosavanna appeared, in which forest fragments were interspersed with tall-grass prairie (pasture). Also, increased disturbance from all of man's activities caused gross increases in the amount of successional habitat. For example, in the central and southeastern Atlantic states, the amount of broomsedge-scrub pine habitat, normally restricted to post-fire succession, must have increased dramatically.

The existence of a vast new biome created opportunities for bird species that were adapted to the structurally similar habitat of the midwestern savanna-forest-mosaic region, or, in fact, of the prairie itself. Where trees were replaced by grass, warblers were replaced by upland sandpiper *Bartramia longicauda*, dickcissel *Spiza americana*, and grasshopper sparrow *Ammodramus savannarum*. And in savanna-like associations, field sparrow, indigo bunting, yellow-breasted chat, Bewick's wren *Thryomanes bewickii*, red-headed woodpecker, blue-winged warbler, and yellow warbler *Dendroica petechia* appeared. Successional habitats were colonized by rufous-sided towhee, prairie warbler, Northern yellowthroat, blue grosbeak *Guiraca caerulea*, or even Bachman's sparrow.

Before the reader jumps to the conclusion that fragmentation has desirably enhanced regional bird diversity, I hasten to point out that colonization of the eastern neosavanna and neoprairie now appears to have been a failed experiment for many of these species. Most of the species that rushed in to exploit the new habitats are now experiencing serious regional declines (Robbins *et al.* 1986). In West Virginia, for example, yellow-breasted chat is undergoing a pronounced population decline (Hall 1984) that seems inexplicable in terms of habitat change. Similarly, dickcissel and upland sandpiper are extremely rare in eastern pastures and may be headed for extirpation there; and Bewick's wren, once locally common in the Appalachians, is now rare there (indeed, it is gone from some regions). Only the unusual steps of recreating and perpetuating grassland, by active management, in a region where it is inappropriate as an equilibrial community, would save the eastern populations of these species.

EFFECTS OF FRAGMENTATION ON FOREST BIRDS

Whitcomb *et al.* reported in 1981 that birds of the eastern forest were drastically affected by habitat fragmentation. The initial observation of Shirley Briggs and Joan Criswell (reviewed by Briggs and Criswell 1979) that neotropical migrant species were suffering locally severe declines was confirmed and greatly extended. I would now like to recapitulate and update the main points of our 1981 account.

(a) We studied in detail the archipelagos of forest remnants in the Piedmont and Coastal Plains region in the vicinity of Washington, D.C., where deciduous woodland covered about 22% of the landscape. This woodland consisted of a complex archipelago of dendritic strips and fragments that varied in size, shape, successional maturity, vegetational composition, history of disturbance, and surrounding land use. Point surveys showed that small forest fragments were inhabited by a limited subset of common forest-dwelling bird species, together with a large number of species that preferred successional habitat; no bird species was restricted to small remnants.

(b) A large set of neotropical migrant bird species were forest interior specialists. These species were rare on small (1-5 ha) forest remnants and only somewhat less rare on fragments of intermediate (6-14 ha) size, but they were characteristic and abundant on larger (70+ ha) wooded tracts. Species capable of utilizing both forest-edge or successional habitats and forest interior were equally abundant in forest fragments of all sizes. Extensive forest systems, such as those of the U.S. Department of Agriculture, Beltsville Agricultural Research Centre and the U.S. Fish and Wildlife Service, Patuxent Wildlife Research Centre near Laurel, Maryland, supported typical densities of all small bird species that had declined in or disappeared from small tracts in the study region. Absence of sensitive bird species (e.g. worm-eating *Helmitheros vermivorus*, hooded *Wilsonia citrina*, and black-and-white *Mniotilta varia* warblers from entire 25-km^2 blocks surveyed for Breeding Bird Atlas studies indicated that these species were subject to regional as well as local extirpation. The State of Maryland is now in the fourth year of a revision of the Atlas. These recent studies of the Piedmont remnants suggest further local range retractions of neotropical migrant species (K. Van Ness, pers. comm.).

(c) Numbers of breeding forest interior bird species were significantly negatively correlated with fragment isolation. The one small remnant we studied that was situated in close proximity to a large mainland forest supported several forest interior species that were not found on any other fragment. In 1984 Wagner *et al.* recensused this remnant. Populations of sensitive species appeared stable, in keeping with the retention of the local biogeographic setting.

(d) Examination of published data showed that neotropical migrant bird species typically account for 80-90% of breeding individuals in extensive tracts of eastern deciduous forest. In contrast, neotropical migrants were consistently much less common on small tracts. Small forest remnants in Maryland, Illinois, Ohio, Delaware, Wisconsin, New Jersey, and Michigan all had similar avifaunas, conspicuously lacking in neotropical migratory species. Subsequent studies in Connecticut (Butcher *et al.* 1981) revealed similar patterns. In Figure 2 I have indicated the location of the various regions that share essentially the same pattern of avifaunal decay following fragmentation.

(e) In contrast to many neotropical migratory species, most residents were not restricted to forest interior. In the central Maryland study area only three resident species (pileated woodpecker *Dryocopus pileatus*, hairy woodpecker *Dendrocopus villosus*, and white-breasted nuthatch *Sitta carolinensis*) bred primarily in forest interior habitat. These species, all of which have large territories and require relatively mature forest, nevertheless remained common in the study area and appeared on suitable fragments that fulfilled their home range requirements. Forest maturity did not account for observed patterns of size sensitivity. In addition to the three resident species, only yellow-throated vireo *Vireo flavifrons* (of the neotropical migrants) was restricted to mature forest. In contrast, large tracts of early successional forest served as important population reservoirs for such forest interior species as acadian flycatcher *Empidonax virescens*, Northern parula *Parula americana*, ovenbird *Seiurus aurocapillus*, Louisiana waterthrush *Seiurus motacilla*, Kentucky warbler *Oporornis formosus*, hooded warbler, and American redstart *Setophaga ruticilla*.

(f) The impact of fragmentation was species specific. Forest fragments surrounded by fields or other nonforested habitat could be considered true habitat 'islands' for only about 19 of the 93 species that composed the regional pool of forest species. Remaining species either: (i) had very large home ranges and could combine several isolated tracts within a single territory (e.g., certain raptors and woodpeckers); (ii) inhabitated fields, hedgerows, residential areas, and other nonforested habitats and only occasionally 'spilled over' into forest interior, particularly in small woodlots (American robin *Turdus migratorius*, gray catbird, song sparrow *Melospiza melodia*); or (iii) were able to utilize a broad spectrum of habitats (small parks, suburbs, and other disturbed areas) in which some tree cover was maintained. Many of these habitat generalists were residents or short-distance migrants that did not depend on forest interior for their regional survival.

(g) Inherent attributes of the forest bird species accounted for their vulnerability to fragmentation, or lack of it. Statistical analyses (χ^2 and stepwise multiple regression) revealed a significant correlation between tolerance to fragmentation and migration strategy, degree of habitat specificity, nest type, and nest height. Species that were intolerant of fragmentation tended to be highly migratory, to be

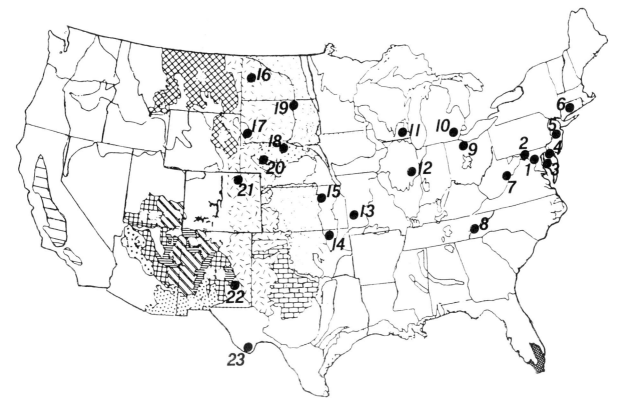

Fig. 2. Location of places discussed. 1. Maryland area near Washington, D.C. in which studies of forest remnants were conducted; 2. Hagerstown area, site of studies of Robbins et al. (1980); 3. Delmarva peninsula, site, in part, of studies of Lynch and Whigham (1984); 4, 5, 6, 9, 10, 11, 12. Sites of breeding bird censuses of remnants cited by Whitcomb et al. (1981); these remnants had similar avifaunal compositions; 7. Monongahela National Forest, West Virginia; 8. Great Smoky Mountain National Park, Tennessee and North Carolina; 13. Prairie State Park and Tsi Sho and Hunkah prairies, Missouri; 14. Proposed site of a tall-grass national park, Osage County, Oklahoma; 15. Konza Prairie, Kansas; 16. Theodore Roosevelt National Park, North Dakota; 17. Custer State Park, South Dakota; 18. Niobrara Prairie and Valentine National Wildlife Refuge, Nebraska; 19. Samuel Ordway Prairie, South Dakota; 20. Crescent Lake National Wildlife Refuge, Nebraska; 21. Pawnee National Grasslands, Colorado; 22. Bitter Lake National Wildlife Refuge, New Mexico; 23. Big Bend National Park, Texas.

specialized for forest interior habitat, to build open nests, and to nest on the ground. We were especially interested in examining differences in reproductive effort among species. Annual reproductive effort (estimated by multiplying mean clutch size by mean annual brood number) tended to be lower for forest interior species and for neotropical migrants generally. Neotropical migrant species that were also forest interior specialists had relatively low reproductive efforts, whereas residents and short-distance migrants that bred in open habitats had the highest reproductive efforts. On the other hand, neotropical migrants tended to be longer lived than other birds; their lifetime reproductive efforts were therefore similar to those of nonmigratory species of similar size.

(h) We concluded that the intolerance of neotropical migrants to fragmentation was multifactoral and probably resulted from unsuitability of several life history features to fragmented forest systems. Features of special importance, as determined by stepwise multiple regression analyses, included low annual reproductive effort, dispersal strategy, and nest type and location (and consequent vulnerability to nest predation and parasitism). Extirpations appeared to result from proximate stresses such as nest predation or parasitism or stochastic factors related to population size. Subsequent failure of populations to re-establish themselves reflected inherent tendencies toward faunal 'relaxation' (Diamond 1976; Wilcox 1978, 1980). Long-term census data, compiled by Criswell and her colleagues for Cabin John Island near Washington, D.C. (see review of Lynch and Whitcomb 1978 and Whitcomb et al. 1981), illustrated the relaxation of an island bird community following an increase in the degree of isolation. This isolation was produced by several circumstances. Criswell specifically commented on the effect of forest clearance for the George Washington Parkway but did not mention construction of the Cabin John Bridge of the Capital Beltway, 0.8 km west of the island. I believe that this event probably also had an impact on the local biogeographical setting of Cabin John Island. Six species of neotropical migrants responded similarly to the destruction of forest adjacent to the island. Densities of most of these species increased slightly in the year following habitat destruction, presumably as a result of an influx of birds displaced by deforestation. Subsequently, however, the species declined markedly in abundance, and some disappeared altogether from the census tract. Within a few years, new, lower equilibrium species richness and lower population levels of neotropical migratory species were approached.

A preliminary model was constructed explaining the dispersal dynamics involved in differential tolerance to fragmentation as a result of the combined effects of progeny 'leakage' following fledging and the difficulties of recolonizing forest tracts each year following a long-distance migratory flight. This model, based on actual data, continues to be a realistic representation of avifaunal decay after fragmentation. Banding data (Hann 1937; Robbins 1969) had shown that each spring recolonizing neotropical migrant fell into two categories. Territorial birds tended to be philopatric but suffered about 50% mortality during the nonbreeding season. On the other hand, banding data (Hann 1937; Nolan 1979) also showed that first-year birds rarely return to the same plot where they were raised. Thus, most birds that replace adults lost in the nonbreeding season must have fledged elsewhere. These 'new' territorial individuals comprise a second class of annual migratory colonizers of breeding habitat. From these data, we constructed a simple equation:

$$T_i = kT_{i-1} + C \quad (1)$$

where T_i, the number of territorial birds in year i, is the product of the number of territorial individuals in the previous year (T_{i-1}) times the annual survivorship rate (k), plus the number of 'new' colonists (C). At equilibrium, the number of territories will be approximately constant from year to year, and the number of 'new' birds will balance losses from winter mortality. Thus:

$$C = (1-k)T_{i-1} \quad (2)$$

or substituting, Eq. (1) becomes

$$T_i = \frac{kC}{1-k} + C = \frac{C}{1-k} \quad (3)$$

This relationship was presented formally by May (1981), with comments on the effect of fragmentation on the age structures of the bird populations.

GRASSLANDS

No one has claimed that North American grasslands were homogeneous. Even so, small maps tend to reduce descriptions of grassland to units such as 'tall-grass prairie', 'short-grass prairie', or 'desert plains'. Scrutiny of actual grassland communities tends to reveal inadequacies in such generalized descriptions. In our recent description (Whitcomb et al. 1986) of grassland-leafhopper associations, we mapped 30 regions of North American grasslands (Fig. 3). These regions were essentially those proposed by Küchler (1964). However, at a finer grain, close examination of grasslands of a complex region like the desert plains shows that even moderately detailed studies such as Küchler's are of limited use at a local level. Finer mappings are necessary to give meaning to the actual communities.

The nature of contemporary grasslands was determined by several important factors. First and foremost, they are predominantly a formation generated by the vast mountain chains, Tertiary in origin, that form a series of north-south-oriented backbones through the western part of North America. Moisture from the Pacific Ocean carried eastward by prevailing winds is trapped on the tops of these mountains, particularly on the west-facing slopes. In the immediate rainshadow of these mountains, the precolumbian vegetation consisted of semi-arid grasslands (about 320 mm annual precipitation) characterized by short grasses such as buffalograss *Buchloë dactyloides* and blue grama *Bouteloua gracilis*. In the eastern sections of the plains states (North and South Dakota, Nebraska, Kansas, Oklahoma, and Texas), tall-grass formations intermixed with a rich diversity of broad-leaved forbs predominated (Weaver 1954). The region between tall- and short-grass prairies was transitional, with different mixes of tall and short grasses that depended on interactions between soil depth and texture with climate and grazing (Risser et al. 1981). In the heart of this transitional zone lies a region whose vegetational nature is determined largely by the sandy substrate. This is the Nebraska Sand Hills region, a vast area formed at the southern interface of the glaciers during the Pleistocene.

The nature of the grasslands has been profoundly altered by European colonization (Whitcomb et al. 1986). Today the gently rolling Sand Hills (at least those that have not been decimated by centre-pivot irrigation cropping systems) are, for the most part, well-managed native range. The southwestern grasslands have been modified by overgrazing (Brown 1950; Buffington and Herbel 1965). Tall-grass prairie, except in some hilly regions, has been almost entirely converted to corn and soybean fields. Short-grass and transitional prairie regions have been converted to wheat fields or range, depending on the vagaries of local history. Where wheat agriculture was attempted in regions of especially low rainfall, or where overgrazing of the precariously balanced grasslands pushed the land too far, we now have National Grasslands, large tracts managed by the U.S. Department of Agriculture National Forest Service. In many of these grasslands, non-native grasses such as crested wheatgrass *Agropyron cristatum* have been planted to secure the blowing dust of the 'Great Drought' (1930-34), a natural disaster that is now seen as a foreseeable part of a climatic cycle.

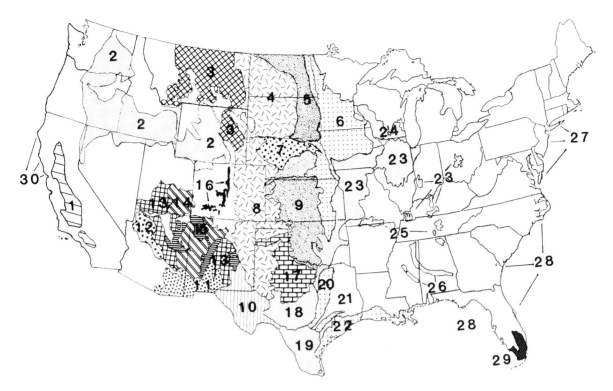

Fig. 3. Grassland formations. 1. California steppe: dense to medium-dense, low to medium tall grassland; 2. Sagebrush steppe; dense to open grassland with dense to open shrub synusia; 3. Grama-needlegrass-wheatgrass prairie; short, open to fairly dense grassland; 4. Wheatgrass-needlegrass prairie; moderately dense, short or medium tall grassland; 5. Wheatgrass-bluestem-needle-grass; dense, medium-tall to tall grassland; 6. Tall-grass prairie; dense vegetation of tall grasses and many forbs; 7. Nebraska Sandhills prairie; medium dense to open grassland, tall to medium tall, with inclusions of sand 'blowouts'; 8. Short-grass prairie; dense grassland of short grass with taller grasses in eastern sections; 9. Mixed prairie; dense medium-tall grassland with many forbs; 10. Trans-Pecos shrub savanna; dense to scattered shrubs with short grasses, with inclusions of grama-tobosa prairie and oak-juniper woodland; 11. Grama-tobosa shrubsteppe; short grasses with very open to dense shrub synusia; 12. Mosaic of grama-tobosa shrubsteppe (11) and pinyon-juniper woodland (scattered low evergreen trees with many grasses); 13. Mosaic of pinyon-juniper (see 11), coniferous forest and Great Basin shrub formation; 14. Mosaic of pinyon-juniper, coniferous woodland, blackbrush and galleta shrubsteppe; 15. Grama-galleta steppe; low to medium tall grassland with few woody plants; 16. Alpine meadows and barrens; medium tall moist grass meadows, and short tundra dominated by broad-leaved forbs; 17. Mesquite-buffalo grass; short grasses with scattered low trees and shrubs and low needleleaf evergreen shrubs; 18. Juniper-oak savanna; savanna with dense to very open synusia of broadleaf deciduous and evergreen low trees and shrubs; 19. Mesquite-acacia savanna; dense to open grassland with broadleaf shrubs scattered singly or in groves; 20. Blackland prairie; medium tall, moderately dense grassland; 21. Fayette prairie; medium tall, rather dense grassland with scattered open groves of broadleaf deciduous trees; 22. Coastal bluestem-sachuista prairie; medium tall to tall, dense to open grasslands; 23. Mosaic of tall-grass prairie (see 6) and oak-hickory forest; 24. Oak savanna; tall-grass prairie with broadleaf deciduous trees scattered singly or in groves; 25. Cedar glades; low to medium tall open grassland with scattered evergreen shrubs and groves of broadleaf deciduous trees; 26. Blackbelt: tall or medium tall broadleaf deciduous forest with concentrations of low needleleaf evergreen trees with patches of tall-grass prairie (see 6); 27. Northern cordgrass prairie; dense, medium tall grassland; 28. Live oak-sea oats; open grassland to dense shrubs and groves of low broadleaf evergreen trees; 29. Florida prairie; moist flats dominated by sawgrass with inclusions of hammocks and coniferous trees.

At their eastern periphery the grasslands merged with the eastern forest. This region, the midwestern savanna and prairie-forest mosaic, was the crossroads of the two biomes we discuss herein. The above description of grasslands is based on the extensive work of Weaver and his colleagues, whose extensive prairie research (Weaver 1954) was concentrated, as is much current research, in the rangeland of the Kansas Flint Hills. The Konza Prairie at Manhattan, Kansas (Fig. 2), represents one of the last large remnants of true tall-grass prairie. However, the concept of tall- and short-grass prairies most accurately describes grasslands of the central and perhaps north-central United States. An east-west transect across the northern grasslands of the United States would reveal a much different picture. This region, as far west as the western Dakotas, was covered by glaciers 12,000 years ago. At the time of European settlement (Küchler 1964), it was covered by grasslands that were dominated by 'cool-season' grasses of the subfamily Pooideae, such as Junegrass *Koeleria cristata*, needlegrasses *Stipa* spp., and wheatgrasses *Agropyron* spp. In the eastern portion, tall andropogonoid grasses (e.g. big bluestem *Andropogon gerardi* and indiangrass *Sorghastrum nutans*) were mixed with the cool-season grasses. To the west, blue grama was the major nonpooid grass of the northern grasslands. In Montana, however, the northern influence produced a vegetationally different region, the fescue grasslands, which occurred in the eastern foothills rainshadow of the Rocky Mountains of the northern

United States and southern Canada. In the north, the scarcity of buffalograss and its replacement by taller pooid grasses made the rainshadow prairie fundamentally different in structure from the homologous region in Colorado and New Mexico.

Whatever the vegetational divergences in the northern and central prairie, they are dwarfed by the great habitat heterogeneity of the southwestern 'desert plains' region. This region, covered at one time by mesic grasslands (Axelrod 1950; Kottlowski et al. 1965), is now undergoing a gradual process of desertification that has resulted in the creation of major arid and semi-arid regions such as the Mojave, Sonoran, and Chihuahuan Deserts and the Great Basin. Climatic variation in this region is generated by proximity not only to the Pacific Ocean but also to the Gulf of Mexico. Furthermore, there is sufficient altitudinal (from 1000 m to 4000 m) and latitudinal (from 32° to 37°) variation in the region to permit cool-season (pooid) grasses to be significant components at high elevations or in the north, but to assure that the main subfamily of warm-season grasses (Eragrostoideae) dominates the southern portion of the region. The interplay between the sources of precipitation results in very different rainfall patterns. Although the Chihuahuan region receives a small amount of winter precipitation, most of its 235 mm of annual precipitation falls during the 'rainy' season from mid-July to late September. In the Sonoran Desert there is a more equitable distribution of winter and summer rain (280 mm), with dry seasons between, whereas the Mojave Desert has its 'rainy' (150 mm) season largely in the winter. Locally, the timing of precipitation varies in the sense that higher elevations receive the first and most substantial precipitation. There are thus steep local elevational gradients in both amount of precipitation and the seasonal timing of receipt of the threshold amounts of precipitation required to terminate dormancy of the desert vegetation, whose growth is synchronized with precipitation cycles. For example, in the Chihuahuan region, the season 'moves' from lower elevations to higher elevations in the spring, because average temperatures are higher at lower elevations. However, the season then moves in reverse from higher elevations down as the low desert finally receives, in late summer, sufficient moisture to permit plant growth.

GRASSLAND LEAFHOPPERS

For two decades, we have studied the leafhopper fauna of North American grasslands. In the course of these studies, by standard collecting procedures (Blocker and Reed 1976), we have developed more than 2000 accessions, each an estimate of the assemblage [guild, in part (Denno et al. 1981)] of cicadellids appropriate to a dominant host plant of the grasslands. A recent monograph (Whitcomb et al. 1986) summarizes the results of some of these studies.

Host assemblages. More than 100 dominant grass species, and perhaps an equal number of forbs, are of considerable importance in explaining the diversity of leafhoppers in North American grasslands. Cicadellid assemblages were the predominant components of many guilds of homopterans exploiting plant phloem tissue. Perennial and dominant (but not annual or subdominant) grass and forb species tended to have appropriate, specific cicadellid assemblages in a given region, but the composition of assemblages varied geographically. Major biogeographic influences that affected guild composition on given grass or forb hosts included latitude, longitude, altitude, and/or amount and predictability of precipitation; geologic and evolutionary history; and the patch structure of host populations. The range of factors that influence cicadellid guild composition, vital in considerations of insect conservation (see final section of this chapter), are now reviewed.

Biogeographical patterns. Patch size and structure of host stands were of considerable significance, particularly at the periphery of the range of the species, where sizes of host stands decreased and isolation between patches increased. Disturbances that could cause major population reductions or extirpations of cicadellid species were common in all grasslands. Important disturbances included fire, flooding, predator or (especially) parasite overload, or unfavourable climatic patterns (especially drought). Therefore, extirpation rates were high, even in host patches of sufficient size to support reasonably large cicadellid populations.

Range-limiting isolation factors. The geographic ranges of cicadellid species were often limited by obvious biogeographic isolating mechanisms. The most important of these was availability of their host plant or plants. Most cicadellids that we studied were oligophagous. Phloem but not xylem feeders showed a high degree of host specificity that was expressed most often at the plant genus level, but not infrequently at the species level. Thus, one or a few dominant host species accounted for most occurrences of cicadellid species. For such specialists, host plants represent contemporary and evolutionary 'islands.' Plant species often had much wider ranges than the leafhopper specialists appropriate to them. Grasses that have achieved dominance or importance in several biomes (e.g. blue grama, little bluestem *Schizachyrium scoparium*, or buffalograss) were especially likely to have ranges much wider than their specialist leafhopper inhabitants. Climatic factors limited some species. Sharp isolating mechanisms were generated by freeze patterns in the southern United States. South of the freeze line, diapause might not be terminated; north of the line, freeze tolerance was required. A more gradual isolating mechanism was present in the north, where seasonal shortening led to an increase in univoltinism. Rainshadow areas associated with mountain ranges, in which sharp elevational increases were accompanied by sharp aridity gradients, were therefore doubly potent isolating mechanisms on a longitudinal basis. Local isolation factors were related to topography, especially in montane and semi-arid regions. As a consequence of such isolation factors, montane or desert regions had a speciose cicadellid fauna, featuring a high degree of endemism. Part of this endemism is derived from recent geologic history. Through successive ice ages, southwestern landscapes topographically similar to those of the present day have been covered by several different grassland types. The nature of the mountainous topography and epochs of desertification have assured insularization of many insect populations. Finally, the genera of chloridoid grasses, which dominate the region, are taxonomic islands in their own right. These are the circumstances that led, in my view, to accumulation of a rich leafhopper biota in the southwest.

Rarity. Conservation efforts often focus on rare species, so we were interested in ascertaining, when possible, the basis for cicadellid rarity. In many cases, this was explained simply by the regional rarity of the plant host. Actually, plants that were rare everywhere seldom or never accounted for rare insect species. Rather, rare cicadellids were often found to specialize on locally common plants occurring on uncommon substrates (e.g., gyp or sandy soils). In the savanna, we encountered another type of rarity. Insects that fed alternately on woody and forb hosts required juxtaposition of each of the hosts. Ovipositional preferences could account for many instances of species rarity of this type. Another explanation for insect species rarity is simply that the species had been isolated historically in a small geographic area. Examples of rare cicadellids and the isolating mechanisms that prevent their becoming common are presented in Table 1. Four [(*Flexamia arenicola* and *F. celata* (Lowry and Blocker 1987) and new species of *Athysanella* and *Laevicephalus*)]

Table 1. Rare grassland leafhoppers and the isolation mechanisms that prevent their becoming common.

Cicadellid species	Isolating circumstance
Limotettix n. sp.	Endemic in Maryland serpentine barrens
Flexamia arenicola	Endemic on *Muhlenbergia pungens* in sandhill grasslands
Flexamia celata	Endemic on *Redfieldia flexuosa* in sand blowouts of central Great Plains
Athysanella n. sp.	Restricted to *Bouteloua breviseta* on gyp soils in Chihuahuan grasslands
Laevicephalus n. sp.	Possibly endemic in Chisos Mountains of southwest Texas, on *Bouteloua ramosa*
Chlorotettix n. sp.	Endemic on *Leersia oryzoides* in fresh water marsh lacunae in the Maryland and Pennsylvania deciduous forest
Bandara aurata	Requires juxtaposition of *Lactuca canadensis* and *Pinus* spp.

appear to be specific on locally dominant grasses in sand, gyp or gravelly soils. Two (new species of *Limotettix* and *Chlorotettix*) are restricted to open islands (marsh or barrens) in the eastern forest. Finally, *Bandara aurata* occurs where its nymphal host, *Lactuca canadensis,* occurs in juxtaposition with *Pinus* spp.

Life history strategies. Changes in cicadellid assemblages in response to habitat turnover could be explained in terms of the life history strategies of grassland cicadellid species. In general, specialists that utilized perennial hosts showed evidence of K-selection and were sensitive to disturbance of the vegetational communities in which they occurred. Generalists that moved readily among host species showed evidence of *r*-selection; colonized crop, pasture, and lawn plantings readily; and were apt to be regarded as pests.

Regional differences in life history strategies. Life history strategies varied regionally. For example, tall-grass prairie, mixed prairie, sand hill prairie, short-grass prairie, and Chihuahuan grassland had different vegetational stratification, different types of dominance hierarchies, and greatly different amounts and predictability of precipitation. Accordingly, cicadellid assemblages in such contrasting biomes had different compositions of specialists and generalists (Whitcomb *et al.* 1986).

Evolutionary factors. Certain genera or subgenera of cicadellids were biome limited. Host specificity, habitat insularity, and contrasting precipitation patterns of various semi-arid biomes have promoted a high rate of speciation in certain cicadellid genera (e.g. *Athysanella* subgenus *Gladionura* and the *Laevicephalus parvulus* group). Other genera (e.g. *Flexamia*) were prairie endemics. In general, the genus *Athysanella* is characteristic and dominant in semi-arid grassland formations, whereas *Flexamia* is characteristic of mesic grasslands, including prairie. The region of overlap of the two genera is moderate. Pleistocene climatic cycles had a major influence on the cicadellid biota of North American grasslands (Ross 1970).

BIOTIC CONSERVATION

Conservation of bird species of the eastern forest is a matter that has now attracted considerable concern (Wilcove and Whitcomb 1984; Wilcove *et al.* 1986). Because documented extinctions that followed deforestation of only half of the forest and the documented population declines and extirpations that inevitably (Whitcomb *et al.* 1981) accompany forest fragmentation, conservation-minded scientists and organizations appear to be universally and justifiably concerned (see reviews of Harris 1984; McClellan *et al.* 1986; Wilcove *et al.* 1986). Since our earlier study, considerable work has been done on eastern forest birds (Robbins 1978, 1979,

1980; Butcher et al. 1981; Leck et al. 1981; Ambuel and Temple 1982, 1983; Howe 1984; Lynch and Whigham 1984). These recent data (see also chapter of Lynch, this volume) reinforce our conclusion that 'maintenance of forest bird communities requires wooded tracts of hundreds or even thousands of hectares' and that 'management of such preserves should be aimed at minimizing disturbance of the forest interior'. However, the studies of Lynch and Whigham (1984) emphasized vegetational features of forest that predict bird occurrence and the importance of regional events in determining the fate of bird populations. In the high-carrying-capacity mesic forests of eastern Maryland, where forest fragments were less isolated than in the Piedmont area we had studied, Lynch and Whigham found the effects of fragmentation to be less dramatic than those we had noted. In contrast, in the more fragmented and poorer archipelagos of Maryland's Ridge and Valley section, Robbins (1980) found the same patterns of avifaunal loss that we had noted in the Piedmont.

Particularly satisfying has been the demonstration of proximate factors that operate differentially in large and small forest systems. For example, Wilcove (1985a) using some of the same woodlots that we had censused in our study, showed that predation levels of open (but not hole) nests in wooded fragments greatly exceeded those in large forest systems. Also, the continued beneficial effect of brown-headed cowbird *Molothrus ater* control programmes on the precariously situated Kirtland's warbler *Dendroica kirtlandii* population in Michigan (Mayfield 1977; Brittingham and Temple 1983) demonstrates that nest parasitism is also a proximate factor contributing to extirpations of neotropical migrants.

During recent years, strategies for reserve design have been vigorously debated. Some theoretical biologists have attempted to reduce conservation decisions to simple models, based on assumptions that would be unrealistic or grossly oversimplified (Coleman et al. 1982) for either of the two biomes I discuss herein. A recent computer model (McClellan et al. 1986) addresses this problem. The model simulated the effects of habitat fragmentation on two species pools with different minimum area requirements and dispersal abilities. In this model, a realistic case was chosen: reduction of five very large tracts to an archipelago of more than 450 fragments totalling only 5% of the original area. Results of the model suggested that more species would persist at equilibrium if remaining habitat is concentrated into a single large patch rather than distributed over many small fragments. The model also suggested that insularization could cause extirpations independent of habitat reduction. Finally, even when most habitat had been fragmented into small islands, further fragmentation caused a rapid loss of the species that remained. The results of this model, unlike the hypothetical scenarios of less realistic models, are consistent with real processes that occur on the temperate forest islands I have studied.

Fortunately or unfortunately, there is no need, for either forest or grassland, to rely on models to predict the effects of habitat fragmentation. Strategies of reserves for these biomes can be designed by monitoring existing archipelagos of fragments, which are not in short supply. Some, like Illinois' 26 ha Trelease Woods (Kendeigh 1982), and in fact all small (<100 ha) woodlots, are obvious failures as avifaunal preserves. In contrast, Great Smoky Mountain National Park (8000 ha, plus many thousands of hectares of forest outside the park, Fig. 2) is a conspicuous success. In this park Wilcove (1985b) recently re-censused forest plots from transect lines along existing park trails that were first established and censused in 1948 (Kendeigh and Fawver 1981). Although Wilcove had expected to find declines in certain neotropical migrant species whose wintering grounds had been especially affected by tropical deforestation, his censuses showed remarkable stability in the avifaunal communities. The assemblages contained, in 1948 and today, a rich complement of neotropical migrant species. The success of large reserves like the Great Smoky Mountain National Park, Monangahela National Forest (West Virginia), and the forest systems of the Beltville Agricultural Research Centre and Patuxent Wildlife Research Centre (Fig. 2), standing as they do in stark contrast to archipelagos of disastrously small woodlots that have lost most of their forest interior bird species, projects a clear and consistent message.

In contrast to the clear messages concerning forest birds, strategies for leafhopper conservation would be difficult to devise, even in the light of my extensive studies. This is true even though grassland cicadellids provide an excellent 'indicator' taxon that could be used in part for formulation of grassland conservation strategy. My studies have suggested that many host-specific cicadellids (e.g., *Flexamia* or *Athysanella* spp.) are characteristic of undisturbed nonfragmented grasslands. In particular, the presence of these species is not only an indication of a functioning grassland ecosystem but is also an indicator of satisfactory use of management procedures such as fire (Cancelado and Yonke 1970) or mowing (Morris and Lakhani 1979) that, if utilized frequently or extensively, could produce extirpations. Before insects of any kind could be used in this way, however, support for the concept of insect conservation would have to be marshalled. Not only has the desirability of insect conservation not yet achieved general public acceptance (see Major, this volume), but the means for achieving such conservation would probably be controversial, even among conservation scientists. There would, at the very least, be arguments concerning the choice of insect taxa (Dagfinn 1980) to be monitored. Nevertheless, I hope that ecologists (if not legislators) might agree that because of (i) the species richness of insects, (ii) the diversity of their life history strategies, and (iii) the immense content of genetic information in the insect component of biotic communities, and (iv) because insects are vital elements at intermediate levels of trophic structure of almost all terrestrial communities, insect conservation is an important issue (New 1984). A major problem in formulating conservation strategies for invertebrates relates to the very diversity one might wish to protect. Conservation strategies that favour one kind of life history strategy are apt to disadvantage others. For this reason, use of insects as indicators would require careful selection of various taxa with a diversity of life history types (Arnold 1983). Life history strategies may vary along environmental gradients (Dearn 1977). My studies of grassland preserves in Missouri tall-grass prairie, Nebraska Sand Hill prairie, Kansas mixed prairie, and Chihuahuan Desert grasslands suggest that cicadellid preservation strategies in these regions would have to differ. Species in tall-grass prairie exhibit a strong tendency for seasonal shifts in food plant choice that probably involve only short-distance movement, whereas insect species in semi-arid biomes are either 'super-tramps' (*sensu* Diamond 1974) capable of seeking suitable resources by long-distance movement or specialists whose seasonal cycles are tied closely to their host's response to sparse and unpredictable precipitation (Whitcomb *et al*. 1986). Thus, as recognized by other students of insect conservation (Opler 1974; Jaenike 1978; Dingle 1981; Arnold 1983), the structural characteristics of the community, the patch structure of host stands, and the regional climate, would all be important factors in considering insect conservation.

Because some of our grassland types are disappearing rapidly, it is vital that strategies be developed for their conservation. Because economic incentives favour type conversion from native grassland to intensive agriculture, or, at the very least, to heavily grazed or overgrazed range, conservation requires intervention by government agencies or private conservation organizations. Grassland reserve design is made complex by the diversity of grassland formations. Diversity of plants in grassland is strongly influenced by soil and by elevational and climatic gradients. The existence of vastly different plant ecotypes in different regions (MacMillan 1959) or along altitudinal gradients (Clary 1975) underscores concerns related to preservation of genetic diversity of major plant species, rather than simple species conservation. Ecotypic differences may also have profound significance for their host-specific fauna. From the point of view of mammals and birds, large reserves are indicated (Terborgh 1974; Whitcomb *et al*. 1976).

Yet large but distantly spaced grassland reserves would fail to conserve many plant and insect species that did not happen to occur within their boundaries (see also Margules, this volume). Thus, conservation strategies that maximize conservation of one taxon or group of taxa may well be inadequate for conservation of other taxa.

With grasslands, as with forests, many remnants, with different successes, are available as reserve models. Through the auspices of the Katherine Ordway programme of The Nature Conservancy, some moderately sized grassland preserves exist. Together with certain National Parks and National Wildlife Refuges, their performance affords the best information on grassland preservation. The Konza Prairie (4000 ha) of Kansas State University, the Samuel Ordway Prairie (3000 ha) of The Nature Conservancy in South Dakota, Theodore Roosevelt National Park in North Dakota, the Crescent Lake National Wildlife Refuge (10,000 ha), and the Conservancy's Niobrara Prairie (8000 ha) in the Nebraska Sand Hills, Bitter Lake National Wildlife Refuge (New Mexico), and Custer State Park (South Dakota) (see Fig. 2) are examples of preserves that may have a chance to preserve all but the largest animals of the functioning ecosystems (Whitcomb *et al*. 1976) they represent. According to my indicator taxon, Cicadellidae, these reserves are conserving insects. Although I have collected little data, I also (inevitably) have impressions regarding avifaunal preservation in grasslands. It is my impression that sensitive species (e.g. Henslow's sparrow *Ammodramus henslowii*, Baird's sparrow *A. bairdii*, upland sandpiper, greater and lesser prairie chicken *Tympanuchus cupida* and *T. pallidicinctus*, McCown's longspur *Rhynchophanes mccownii*) are preserved only when natural vegetation is not pushed too far by overgrazing and in regions where native grasslands have not been badly fragmented by agriculture. It always seemed to be in the large and undisturbed areas where I heard the wild calls of the long-billed curlew *Numenius americana* or avocet *Recurvirostra americana* or the songs of Henslow's sparrows. This impression was strengthened, for example, in 1985 when I ran a Breeding Bird Survey route through Prairie State Park in Missouri, managed by annual burning for vegetational aspect; adjacent prairies managed astutely by the Missouri Department of Conservation for biotic conservation; and a long transect of private land adjoining secondary roads. Only the prairies managed for biotic conservation supported lesser prairie chicken; upland sandpiper and Henslow's sparrow were largely confined to the MDC praires. Remembering that the carrying capacity of grasslands, all of which are relatively dry, is much less than that of mesic forests, I feel that very large areas of the order of tens of thousands of hectares may be necessary to preserve complete, functioning grassland ecosystems. However, preserves of this size are apt to be viewed as under-utilized, even from a recreational point of view, unless they are multiply used. Even The Nature Conservancy, for example, recognizes the economic need to balance preservation with economic return in their Samuel Ordway Prairie in South Dakota. Similarly, tentative plans for a tall-grass prairie national park in Oklahoma (Fig. 2) provide for grazing. However, these large multiply used units, which will almost inevitably be grazed, should be supplemented by reserves, in each vegetational region, that are managed as nearly as possible to approximate presettlement vegetational conditions. For this purpose, grazing by cattle or (especially) sheep is inappropriate. These reserves should be selected for special features to supplement the total of preserved species or communities, and should minimize perimeter per unit area (Wilcove et al. 1986) to minimize disturbance from surrounding activity (Janzen 1983). Although smaller in size than the largest units, they must nevertheless be large enough to buffer against extirpations resulting from drought or fires. Wilcove et al. (1986) recently discussed conservation strategy in terms similar to those suggested here. In their view, acquisition of large reserves that preserve intact ecosystems and landscape heritage must be undertaken at the national level. Smaller organizations can most usefully focus on rare species or communities that can be preserved in smaller reserves.

Most of the above arguments refer to prairie grasslands. I would be baffled if I were asked to construct a preserve system for semi-arid or montane grasslands. There, climatic diversity and endemism derived by historical accident are so extensive that any scheme would sacrifice many significant biotic elements. Conservation of plant or insect biota in this region, which is a mosaic of hundreds of vegetational zones, cannot be accomplished by selection of only a few representatives. In such areas, the reserve concept itself may not be fully useful. Designation of reserves is predicated on a scenario in which reserves become increasingly insular, surrounded by alien, or at best neutral, seas of biotically depauperate land managed exclusively for human activities. Let us hope that in semi-arid or montane grasslands this nightmare never reaches fruition. The biota of the southwestern and montane regions is being preserved today by a large network of federal landholdings (e.g., Pawnee National Grasslands) administered by the U.S. Department of Agriculture National Forest Service and the U.S. Department of Interior Bureau of Land Management. The future of the biota of this region therefore depends to a large extent on appropriate management strategies on the part of the federal agencies that administer this land.

CONCLUSIONS

To conclude, I would like to contrast the two major vegetational formations and the biota that I have studied. Although both the grasslands and the eastern forest were vast formations, the deciduous forest was more uniform and can be thought of as a single biome. The precolumbian grasslands were so extensive that they cannot be thought of as a single biome. Rather, one might define at least eight major grassland biomes. The driving variables in the creation of these biomes include latitudinal, elevational, and climatic variation, particularly with respect to amount, predictability, and timing of precipitation. The eastern forest receives, over its full extent, sufficient rainfall to prevent the fires that maintained grasslands as treeless biomes. In the forest, latitudinal or elevational variation generates the chief differences that determine the various forest types. Rare taxa in the forest biome are apt to consist of (a) residents of essentially nonforest islands, such as serpentine or shale barrens, bogs, or swamps in which the canopy is not completely closed, or (b) plants, or invertebrate animals closely tied to plants, that are specialists in unusual soils. Conservation of forest biota thus logically involves (a) selection of major forests that preserve the landscape heritage of these immense formations and their largest mammals and avifauna and (b) identification of rare plants, insects, or other taxa that are not located in the large reserves, which were rare even without disturbance, and which only detailed autecology and monitoring can save. Conservation of grassland biota also logically involves these steps, with the *proviso* that we are now dealing with a more diverse assemblage of biomes and formations. Each of the 30 or more formations I have mapped (Fig. 3) deserves a major reserve, in which the goal of biotic conservation is a firm commitment. Again, plants or other taxa dependent on special conditions must also be identified, and smaller reserves established that meet their requirements. In the sequestering of reserves there is good news and bad news. The good news is that we have, in our National Park, National Forest, and National Grassland systems, a realistic base of federally owned land in which the option of management for biotic diversity still exists. The bad news is that proper management for conservation of species, genetic diversity, and communities is a relatively new concept. Much remains to be done in the realm of transferring information accumulated at such meetings as the Busselton conference to managers and planners of the agencies responsible for managing these lands.

ACKNOWLEDGEMENTS

I thank Paul Risser, Chandler Robbins, David Wilcove, and two anonymous reviewers for suggestions that greatly improved the manuscript.

REFERENCES

Ambuel, B. and Temple, S. A., 1982. Songbird populations in southern Wisconsin forests. *J. Field. Ornithol.* **53**: 149-58.

Ambuel, B. and Temple, S. A., 1983. Area-dependent changes in the bird communities and vegetation of southern Wisconsin forests. *Ecology* **64**: 1057-68.

Arnold, R. A., 1983. Ecological studies of six endangered butterflies (Lepidoptera, Lycaenidae): island biogeography, patch dynamics, and the design of habitat preserves. *Univ. Calif. Publ. Entomol.* 99, 161pp.

Axelrod, R. A., 1950. Evolution of desert vegetation in western North America *Contrib. Paleontol. Carnegie Inst. Publ.* **590**: 217-306.

Blocker, H. D. and Reed R., 1976. Leafhopper populations of a tallgrass prairie (Homoptera: Cicadellidae): collecting procedures and population estimates. *J. Kans. Entomol. Soc.* **49**: 145-54.

Briggs, S. A. and Criswell, J. H., 1979. Gradual silencing of spring in Washington. *Atlantic Nat.* **32**: 19-26.

Brittingham, M. C. and Temple, S. A., 1983. Have cowbirds caused forest songbirds to decline? *Bioscience* **33**: 31-5.

Brown, A. L., 1950. Shrub invasion of southern Arizona desert grassland. *J. Range Mgmt.* **3**: 172-7.

Brush, G. S., Lenk, C. and Smith, J., 1980. The natural forests of Maryland: an explanation of the vegetation map of Maryland. *Ecol. Monogr.* **50**: 77-92.

Buffington, L. C. and Herbel, C. H., 1965. Vegetational changes on a semi-desert grassland range from 1858 to 1963. *Ecol. Monogr.* **35**: 139-64.

Butcher, G. S., Niering, W. A., Barry, W. J. and Goodwin, R. H., 1981. Equilibrium biogeography and the size of nature preserves: an avian case study. *Oecologia* **49**: 29-37.

Cancelado, C. S. and Yonke, T. R., 1970. Effect of prairie burning on insect populations. *J. Kans. Entomol. Soc.* **43**: 274-81.

Clary, W. P., 1975. Ecotypic adaptation in *Sitanion hystrix. Ecology* **56**: 1407-15.

Coleman, B. D., Mares, M. A., Willig, M. R. and Hsieh, Y.-H., 1982. Randomness, area and species richness. *Ecology* **63**: 1121-33.

Dagfinn, R., 1980. Ecological analyses of carabid communities — potential use in biological classification for nature conservation. *Biol. Conserv.* **17**: 131-41.

Dearn, J. M., 1977. Variable life history characteristics along an altitudinal gradient in three species of Australian grasshoppers. *Oecologia* **28**: 67-85.

Denno, R. F., Raupp, M. J. and Tallamy, D. W., 1981. Organization of a guild of sap-feeding insects: equilibrium vs. nonequilibrium coexistence. Pp. 151-81 *in* Insect Life History Patterns ed by R. F. Denno and H. Dingle. Springer-Verlag, New York.

Diamond, J. M., 1974. Colonization of exploded volcanic islands by birds: the supertramp strategy. *Science* **184**: 803-6.

Diamond, J. M., 1976. Relaxation and differential extinction on land-bridge islands: applications to natural preserves. Pp. 616-28 *in* Proc. 16th. Int. Ornithol. Congr. (1974) ed by H. J. Frith and J. H. Calaby. Australian Academy of Science, Canberra.

Dingle, H., 1981. Geographic variation and behavioural flexibility in milkweed bug life histories. Pp. 183-94 *in* Insect Life History Patterns ed by R. F. Denno and H. Dingle. Springer-Verlag, New York.

Dort, S. W., Jr. and Jones, J. K. Jr. (eds.), 1970. Pleistocene and Recent Environments of the Central Great Plains. The University Press of Kansas, Lawrence/Manhattan/Wichita.

Hall, G. A., 1984. Population decline of neotropical migrants in an Appalachian forest. *Am. Birds* **38**: 14-8.

Hann, H. W., 1937. Life history of the ovenbird in southern Michigan. *Wilson Bull.* **49**: 145-240.

Harris, L. O., 1984. The Fragmented Forest: Island Biogeography Theory and the Preservation of Biotic Diversity. University of Chicago Press, Chicago.

Howe, R. W., 1984. Local dynamics of bird assemblages in small forest islands in Australia and North America. *Ecology* **65**: 1585-601.

Jaenike, J., 1978. Effect of island area on *Drosophila* population densities *Oecologia* **36**: 327-32.

Janzen, D. H., 1983. No park is an island: increase in interference from outside as park size decreases. *Oikos* **41**: 402-10.

Johnston, D. W. and Odum, E. P., 1956. Breeding bird populations in relation to plant succession on the piedmont of Georgia. *Ecology* **37**: 50-62.

Kendeigh, S. C., 1982. Bird Populations in East Central Illinois: Fluctuations, Variations, and Development over a Half-Century. Ill. Biol. Monogr. 52, Univ. of Illionis Press, Champaign. 136pp.

Kendeigh, S. C. and Fawver, B. J., 1981. Breeding bird populations in the Great Smoky Mountains, Tennessee and North Carolina. *Wilson Bull.* **93**: 218-42.

Kottlowski, F. E., Cooley, M. E. and Ruhe, R. V., 1965. Quaternary geology of the southwest. Pp. 287-88 *in* The Quaternary of the United States. Princeton University Press, Princeton, New Jersey.

Küchler, A. W., 1964. Potential Natural Vegetation of the Conterminous United States. Spec. Publ. 36. American Geographical Soc., Washington D.C. 116pp.

Leck, C. F., Murray, B. G., Jr. and Swinebroad, J., 1981. Changes in breeding bird populations at Hutcheson Memorial Forest since 1958. *William L. Hutcheson Memor. For. Bull.* **6**: 8-15.

Lowry, J. E. and Blocker, H. D., 1987. Two new species of *Flexamia* (Homoptera: Cicadellidae: Deltocephalinae) from the Nebraska Sand Hills. *Proc. Entomol. Soc. Wash.* **89** (in press).

Lynch, J. F. and Whigham, D. F., 1984. Effects of forest fragmentation on breeding bird communities in Maryland, USA. *Biol. Conserv.* **28**: 287-324.

Lynch, J. F. and Whitcomb, R. F., 1978. Effects of the insularization of the eastern deciduous forest on avifaunal diversity and turnover. Pp. 461-89 *in* Classification, inventory analysis of fish and wildlife habitat ed by A. Marmelstein. U.S. Fish and Wildlife Service, Washington, D. C.

MacMillan, C., 1959. The role of ecotypic variation in the distribution of the central grassland of North America. *Ecol. Monogr.* **29**: 285-308.

May, R. M., 1981. Modeling recolonization by neotropical migrants in habitats with changing patch structure, with notes on the age structure of populations. Pp. 207-13 *in* Forest Island Dynamics in Man-dominated Landscapes ed by R. L. Burgess and D. L. Sharpe. Springer-Verlag, New York.

Mayfield, H., 1977. Brown-headed cowbird: agent of extermination? *Am. Birds* **31**: 107-13.

McClellan, C. H., Dobson, A. P., Wilcove, D. S. and Lynch, J. M., 1986. Effects of forest fragmentation on New and Old World bird communities: empirical observations and theoretical implications. *In* Modelling Habitat Relationships of Terrestrial Vertebrates ed by J. Verner, M. Morrison and C. J. Ralph. University of Wisconsin Press, Madison (in press).

Morris, M. G. and Lakhani, K. H., 1979. Responses of grassland invertebrates to management by cutting. I. Species diversity of Hemiptera. *J. Appl. Ecol.* **16**: 77-98.

New, T. R., 1984. Insect Conservation: An Australian Perspective. Ser. Entomol. vol. 32, ed by K. A. Spencer. Dr W. Junk Publishers, Dordrecht/Boston/Lancaster. 184pp.

Nolan, V. Jr., 1979. The ecology and behaviour of the prairie warbler *Dendroica discolor*. Ornith. Monographs 26. American Ornithologists Union. Allen Press, Lawrence, Kansas. 595pp.

Opler, P. A., 1974. Oaks as evolutionary islands for leaf-mining insects. *Am. Sci.* 62: 67-73.

Pregill, G. K. and Olson, S. L., 1981. Zoogeography of West Indian vertebrates in relation to Pleistocene climatic cycles. *Ann. Rev. Ecol. Syst.* 12: 75-98.

Ranney, J. W. and Johnson, W. C., 1977. Propagule dispersal among forest islands in southeastern South Dakota. *Prairie Nat.* 9: 17-24.

Risser, P. G., Birney, E. C., Blocker, H. D., May, S. W., Parton, W. J. and Wiens, J. A., 1981. The true prairie ecosystem. US/IBP Synthesis Series 16. Hutchinson Ross, Stroudsburg, Pennsylvania. 557pp.

Robbins, C. S., 1969. Suggestions on gathering and summarizing return data. Migr. Bird Pop. Sta., USFWS, Laurel, Maryland (mimeo). 11pp.

Robbins, C. S., 1978. Determining habitat requirements of non-game species. *Amer. Wildl. Nat. Resourc. Conf. Trans.* 43: 57-68.

Robbins, C. S., 1979. Effect of forest fragmentation on bird populations. Pp. 198-212 *in* Management of North Central and Northeastern Forests for Nongame Birds. General Technical Report NC-51. USDA Forest Service, North Central Forest Experiment Station, St. Paul, Minnesota.

Robbins, C. S., 1980. Effect of forest fragmentation on breeding bird populations in the Piedmont of the mid-Atlantic region. *Atlantic Nat.* 33: 31-6.

Robbins, C. S., Bystrak, D. and Geissler, P. H., 1986. The breeding bird survey: its first fifteen years, 1965-1979. U.S. Dept. of Interior, Fish and Wildlife Service Resource Publication 157. 194pp.

Ross, H. H., 1965. Pleistocene events and insects. Pp. 583-96 *in* The Quaternary of the United States ed by H. E. Wright, Jr. and D. G. Frey. Princeton University Press, Princeton, New Jersey.

Ross, H. H., 1970. The ecological history of the great plains: evidence from grassland insects. Pp. 225-40 *in* Pleistocene and Recent Environments of the Central Great Plains ed by S. W. Dort Jr. and J. K. Jones, Jr. Dept. Geology, U. Kansas Spec. Publ. 3. The University Press of Kansas, Lawrence/Manhattan/Wichita.

Terborgh, J., 1974. Preservation of natural diversity: the problem of extinction prone species. *Bioscience* 24: 715-22.

Transeau, E. N., 1935. The prairie peninsula. *Ecology* 16: 423-37.

Wagner, J. Jr., Lowry, J. E., Webb, R. and Whitcomb, R. F., 1987. Avifaunal composition of a deciduous forest fragment and comparable mainland forest at the moving front of a gypsy moth infestation. *Maryland Birdlife* (in press).

Weaver, J. E., 1954. North American Prairie. Johnsen Publ. Co., Lincoln, Nebraska, 348pp.

Whitcomb, R. F., Kramer, J., Coan, M. E. and Hicks, A. L., 1986. Ecology and evolution of leafhopper-grass host relationships in North American prairie, savanna and ecotonal biomes. *Curr. Top. Vector Res.* 3: (in press).

Whitcomb, R. F., Robbins, C. S., Lynch, J. F., Whitcomb, B. L., Klimkiewicz, M. K. and Bystrak, D., 1981. Effects of forest fragmentation on avifauna of the eastern deciduous forest. Pp. 125-205 *in* Forest Island Dynamics in Man-dominated Landscapes ed by R. L. Burgess and D. L. Sharpe. Springer-Verlag, New York.

Whitcomb, R. F., Lynch, J. F., Opler, P. A. and Robbins, C. S., 1976. Island biogeography and conservation: strategy and limitations. *Science* 193: 1030-2.

Whitford, P. B., 1958. A study of prairie remnants in southeastern Wisconsin. *Ecology* 39: 727-33.

Wilcove, D. S., 1985a. Nest predation in forest tracts and the decline of migratory songbirds. *Ecology* 66: 1211-4.

Wilcove, D. S., 1985b. Forest Fragmentation and the Decline of Migratory Songbirds. Ph.D. Thesis, Princeton University, Princeton, New Jersey.

Wilcove, D. S., McClellan, C. H. and Dobson, A. P., 1986. *In* Habitat fragmentation in the temperate zone. Conservation Biology: Science of Diversity ed by M. Soulé. Sinauer Associates, Sunderland, Massachusetts (in press).

Wilcove, D. S. and Whitcomb, R. F., 1984. Gone with the trees. *Nat. Hist.* 92: 82-91.

Wilcox, B. A., 1978. Supersaturated island faunas for lizards on post-Pleistocene land-bridge islands in the Gulf of California. *Science* 199: 996-8.

Wilcox, B. A., 1980. Insular ecology and conservation. Pp 95-117 *in* Conservation Biology: an Evolutionary-Ecological Perspective ed by M. E. Soulé and B. A. Wilcox. Sinauer, Sunderland, Massachusetts.

Wright, H. E. Jr. and Frey, D. G., 1965. The Quaternary of the United States. Princeton University Press, Princeton, New Jersey. 922pp.

Wright, H. E. Jr. and Ruhe, R. V., 1965. Glaciation of Minnesota and Iowa. Quaternary geology of northern Great Plains. Pp. 29-41 *in* The Quaternary of the United States ed by H. E. Wright, Jr. and D. G. Frey. Princeton University Press, Princeton, New Jersey.

CHAPTER 14

Retaining Remnant Mature Forest for Nature Conservation at Eden, New South Wales: A Review of Theory and Practice

H. F. Recher[1], J. Shields[2], R. Kavanagh[2] and G. Webb[2]

At Eden, New South Wales, eucalypt forests are managed for the integrated production of pulpwood (woodchips) and sawlogs. During the first rotation alternate coupes are being logged on an approximately 40-50 year cutting cycle with corridors and patches of mature forest retained along creeks, in gullies and on steep or rocky slopes. Although these remnants of mature forest benefit wildlife, their shape and size limits their use in sustaining viable populations of some species dependent upon mature forest. The species most affected are those which the theory of central place foraging predicts would be affected by small patch size and linear environments. These include large animals, carnivores, animals which require multiple, non-interchangeable resources and social animals. Ways in which remnants can be improved for the conservation of such fauna are to buffer core areas by maximizing their size or width, by locating the boundaries of corridors along ecological boundaries, by linking remnants within a network of reserves and corridors, and to enhance the size of remnants by logging selected areas on a longer cutting cycle than the rest of the forest. The areas to be managed on a long rotation should be selected for their biological importance (e.g. high diversity of wildlife, presence of endangered species) and for their role as 'stepping-stones' along corridors between permanent reserves.

INTRODUCTION

THE Eden woodchip industry in southeastern New South Wales is part of an integrated logging operation yielding both sawlogs and pulpwood (woodchips) (Ovington and Thistlethwaite 1976; Bridges 1983). Logging commenced in 1968/69 and initially 800 ha blocks were cleared (Scott and Co. 1975; Bridges 1983). After a reduction in the size of the logging units (coupes) to approximately 200 ha, alternate small coupe logging was introduced in 1976 (Anon. 1982; Bridges 1983). During this phase of the industry the size and shape of coupes were highly variable ranging in size from 2 to 100 ha and averaging 15 ha (Bridges 1983). Since 1983, coupe size has remained variable, but has been increased to average 50-60 ha with a maximum size of 100 ha (Dobbyns *in litt.*). Logging is intensive and depending upon topography, forest type and tree size all non-reserved trees greater than 20 cm dbh (diameter at breast height, $\simeq$ 150 cm) are removed. The first cutting cycle should be completed by the year 2020. By then the forest managed for pulpwood will be a mosaic of coupes, variable in size and aged between 0 and 50 years post-logging. After the first cycle, the rotation period will be extended to 80-120 years depending on forest type and site conditions (Dobbyns *in litt.*). In this case the resulting mosaic will consist of variable sized coupes aged between 0 and 120 years with an increased amount of pulpwood production coming from thinnings at 40, 60, 80 and possibly 100 years post-clearing (Whitelaw *in litt.*). In practice the resulting mosaic is more complex with individual trees retained within the logged areas for seed production, for future sawlog production, and as habitat for wildlife (Anon. 1982; Bridges 1983). In addition patches of unlogged forest are retained within logging areas for a variety of reasons including low timber volumes, rocky terrain, steep slopes and fire damage. Mature forest is also retained or logged less intensively along creeks and in gullies to protect water quality and along roads, major streams or rivers, and on skylines to minimize the visual impact of logging, protect important recreation areas and provide wildlife habitat (Anon. 1982). Selective logging is permitted in the outer zone (>10 m from the creek bank) of

---

[1]The Australian Museum, P.O. Box A285, Sydney South, New South Wales, 2000.
[2]Forestry Commission of New South Wales, Wood Technology and Forest Research Division, P.O. Box 100, Beecroft, New South Wales 2119.
Pages 177-94 *in* NATURE CONSERVATION: THE ROLE OF REMNANTS OF NATIVE VEGETATION ed by Denis A. Saunders, Graham W. Arnold, Andrew A. Burbidge and Angas J. M. Hopkins. Surrey Beatty and Sons Pty Limited in association with CSIRO and CALM, 1987.

creek reserves. Exclusive of larger areas reserved for nature conservation as national parks, nature reserves and flora preserves, the extent of unlogged forest retained is variously estimated at 10 to 25% of the total forested area available for logging within the Eden region, but the *exact area* which will remain unlogged outside parks remains to be confirmed.

As logging proceeds these patches and strips of unlogged forest will represent a major resource for wildlife requiring mature forest (dependent fauna *sensu* Tyndale-Biscoe and Calaby 1975) and have already figured in the development of plans of management for nature conservation within the Eden Management Area (Recher *et al*. 1980; Dobbyns and Ryan 1983; Ryan 1984). Although initially reserved for erosion control, unlogged strips along creeks have been widened and extended over ridgelines linking catchments and larger reserves of unlogged forest in national parks. This has been done to improve their value for wildlife and to facilitate the movement of animals between habitats (Dobbyns and Ryan 1983; Ryan 1984). Nonetheless, mature forest within the Management Area will become highly fragmented, linked in a network of relatively narrow corridors along creeks and gullies, and surrounded by intensively logged and young forest. The large area of regrowth forest will provide some, but not all the needs of wildlife dependent on mature forest. The extent to which a forest mosaic of this sort can maintain viable populations of dependent fauna is an important question for wildlife biologists and forest managers to answer. In addition, foresters need to know the costs and benefits to wood production of reserving mature forest for wildlife.

Questions of coupe size, the linear nature of creek reserves and wildlife corridors, and the patchy distribution of wildlife make it difficult to predict the long-term effectiveness of managing forest wildlife in these ways. An alternative strategy may be to reserve more and/or larger areas specifically for wildlife (e.g. as national parks) and not make any particular effort to conserve wildlife within forest managed for pulpwood production. However, this is unlikely to sample the full range of genetic diversity of forest fauna within the region; it would isolate populations thereby affecting gene flow and hindering recolonization of logged or burnt forest; and may result in the populations of some fauna being too small for long-term survival. We have therefore rejected this approach as a viable option and, while accepting the need for large reserves, aim to conserve wildlife throughout the forest affected by integrated logging (e.g. Recher *et al*. 1980). Our emphasis has been on the conservation of dependent fauna and maximizing biotic diversity with the least cost to wood production. Thus we have based our management recommendations on those areas of mature forest already retained within the logging area as part of normal logging procedures and have sought to strengthen their values for dependent fauna. The forests retained along creeks and in gullies tend to have high wildlife value and at lower elevations (< 300 m) support the richest communities (Recher *et al*. 1980).

To evaluate the usefulness of small patches of mature forest and the linear network of creek reserves for wildlife conservation, we monitored the abundance of birds, mammals, reptiles and amphibians in small unlogged coupes (Kavanagh *et al*. 1985) and in creek reserves (Shields *et al*. 1986). The fauna in these retained areas were compared to populations in extensive areas of unlogged forest. The small coupes were located within forest managed as an integrated logging operation, while the creek reserves studied were in areas of forest cleared for the establishment of Monterey pine *Pinus radiata* plantations. We selected these strips for study to establish the minimum requirements for the survival of wildlife in a linear environment. The plantations provide an *in extremis* environment with most forest wildlife unable to use the pine plantations (e.g. Disney and Stokes 1976; Suckling *et al*. 1976; Friend 1979, 1980, 1982).

In this chapter we review the theoretical aspects of wildlife conservation in small patches and linear environments. We then compare the predictions derived from theory to the results of research into the effects on wildlife of logging and forest fragmentation. Taking a purely synthetic approach, the effectiveness of existing management guidelines for the woodchip area with respect to wildlife conservation are evaluated. Suggestions are then made as to ways that management procedures can be modified to improve the value of these intensively managed forests for wildlife. Our suggestions are made to stimulate thought and direct future research as much as they are intended to be a model for management.

MODELS OF RESERVE SIZE AND SHAPE

It is widely accepted among wildlife biologists that for most wildlife:

1. large reserves retain more species than small ones;

2. reserves with a small boundary in relation to area retain more species of the original natural habitat than those with long or irregular boundaries;

3. linking reserves with corridors is an effective way to enhance the size of reserves; and

4. multiple reserves provide a hedge against catastrophic loss (Diamond 1975, 1976; Simberloff and Abele 1976; Friend 1980; Kitchener *et al*. 1980; Suckling 1982). These concepts are derived

from the theory of island biogeography and are concerned primarily with communities and community diversity (MacArthur and Wilson 1967). What has been less thoroughly explained are the constraints imposed on individuals and populations. This is particularly so for wildlife constrained by the linear environments reserved along creeks or gullies within a logging area or plantation and by the area limits of small patches of isolated habitat. Clues as to the effects of linear environments and small patches on individuals or populations can be derived from 'central place theory' which is a part of the concept of 'optimal foraging theory' (Pyke et al. 1977).

Central place theory assumes that food is uniformly distributed in space and/or time and that animals forage in a single plane (Covich 1976a). These are reasonable starting points and allow us to postulate that the optimum home range or territory of an animal is a circle (or hemisphere) so that the time and energy expended in searching for food and returning to a central place (say a nest or den) is minimized (Covich 1976a; Dill 1978; Andersson 1981). In such a model not only is foraging efficiency optimized, but exposure to predation is minimized (Fig. 1). An animal with a circular territory or home range not only spends the least time searching for food, but on average retreats the shortest distance to shelter if threatened. There is empirical support for this model (Calhoun 1963; Grant 1968; Barlow 1974; Cody 1974; Covich 1976b). If resources are not uniformly distributed, then boundaries of foraging areas will describe the shape of an area in which foraging distances are on average minimized.

The cost of travel for foraging animals is shown by the ways they exploit food resources. Andersson (1981) showed that birds searched less thoroughly far from their nest than near it, took more of the available food near the nest than further away, and

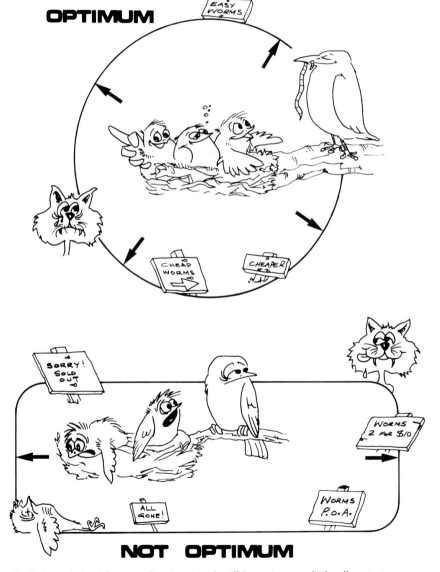

Fig. 1. *Central Place Theory* predicts that animals will forage in ways which will maximize energy gain, minimize energy and time expended, and reduce the risk of predation.

decreased the size of the area searched when food densities were increased. Animals also tend to take larger prey or 'better quality' prey when travel distances are great and to be less selective when distances are short (e.g. Andersson 1981; McGinley and Whitman 1986).

The model allows us to predict some consequences of constricting animals to linear reserves or small patches. The constriction of habitat to that available in creek reserves and corridors forces animals dependent upon mature forest to move in a linear direction (Fig. 1). If the retained area is too narrow or small, animals may not be able to search sufficiently far from a central place to obtain adequate food. In this event individuals may survive by occupying serial refuges along the strip, but ultimately the population will decline due to the inability of individuals to provide sufficient food for their young. This effect of increased foraging distances on the ability of animals to rear young has been demonstrated for black cockatoos in Western Australia where clearing for agriculture has restricted foraging areas to small, isolated patches of native vegetation and linear roadside reserves (Saunders 1980, 1982; Saunders et al. 1985). The rate of decline, as well as overall reproductive success, would also be affected by predators able to exploit edge environments or regrowth forest. Cockatoos feeding in native vegetation along roads suffer heavy mortality from automobiles (Saunders 1982), a situation not unlike predation. Even when animals normally occupy elongate territories there could be such an impact, if the surrounding habitat is changed and new predators introduced.

As illustrated by the cockatoos in Western Australia, similar problems are created when mature forest is retained as small, isolated patches forcing animals to move between them thus increasing travel distances and time. Although the aggregate area of such remnants may be enough to sustain viable populations, the problems of distance will reduce their value to dependent fauna.

The model also predicts adverse effects on social or gregarious animals or where animals require multiple resources. It follows that the greater the number of individuals occupying the same central place, the greater the area required for food for the group. Therefore social animals generally have to travel longer distances or search for longer periods than individuals occupying sole or pair territories (Covich 1976a).

If multiple, non-interchangeable resources are required, the average distances an animal must move increase (Covich 1976a). Conversely, if multiple resources are interchangeable, then average travel distances will be less. Examples of non-interchangeable resources are den or nest sites and food. Interchangeable resources might be different kinds of flowers for nectar, the foliage of different species of trees or multiple den and nest sites. In the linear and patchy habitats created by integrated logging, resources will be spaced farther apart and there will be fewer opportunities to interchange resources (e.g. within a given area there will be fewer kinds of food and fewer den or nest sites). Both situations will force foraging animals to move longer distances.

In summary, the ways animals use space, the kinds of resources they require and their social behaviour are relevant for predicting the impact of linear and/or patchy environments on fauna. According to the predictions of Dill (1978), animals which require larger territories or home ranges or have special requirements are most likely to be affected by a reduction in the available habitat or a change in its spatial configuration. These include:

1. large animals;
2. colonial or social animals;
3. animals that require more than one kind of resource;
4. animals whose resources are patchily distributed; and
5. animals whose resources are sparse.

The following review of our research at Eden is presented as a test of these predictions. Detailed analyses of this research has been presented in other papers.

STUDY SITES

All work was conducted in the South-east Forest Management Area of New South Wales (Bridges 1983) and adjacent parts of Victoria (Fig. 2). Research on the effects of patch size and studies of wildlife within eucalypt forests managed for pulpwood and sawlog production were carried out in the Nadgee, East Boyd, and Timbillica State Forests (50-300 m asl) near Eden, New South Wales (37° 34'S, 149° 44'E). Kavanagh et al. (1985) provide details of the study sites and associated vegetation. Studies of wildlife surviving in creek reserves of native forest were conducted in pine plantations established on the Bondi State Forest near Bombala, New South Wales (36° 54'S, 149° 14'E) (Fig. 3, Table 1). Data obtained from these reserves were compared to control plots of equal size located in unlogged forest in the Coolangubra State Forest and on Coast Range Road in Victoria. All plots were in similar types of forest and at similar elevations (450-850 m asl). Shields (1985) and Shields et al. (1986) provide details of these plots and their vegetation.

METHODS

Detailed accounts of survey and census procedures for birds, mammals, reptiles and frogs are given in Recher et al. (1980, 1983a, b, 1985a), Webb (1980),

Fig. 2. The South-east Forest Management Area of New South Wales showing the location of principal study areas.

Kavanagh and Rohan-Jones (1982), Kavanagh (1984), Kavanagh et al. (1985), Shields (1985) and Shields et al. (1986). The methods employed are briefly outlined below.

Creek Reserves

Birds. A transect procedure (Recher et al. 1983b) was used to census birds on six creek reserves each spring from 1976 to 1984 (Fig. 3, Table 1). Two control plots were located on Coast Range Road and censused at the same time as the reserves in 1976, 1980 and 1984. The reserves censused differed in width, topography, soil, forest type and year of establishment (Table 1). The widest reserve was affected by fire in 1980 and four others were burnt in 1983 (Table 1).

Arboreal Marsupials. Spotlighting procedures were used to census arboreal marsupials (possums and gliders) along 20 streamside reserves over 17 months from January 1982 to May 1983 (Shields 1985). Two control plots were established in the Coolangubra State Forest and surveyed at the same time as the reserve plots. As in the case of birds, reserves differed in width, distance from unlogged forest, soils and topography (Table 1). Some were affected by the 1980 and 1983 fires, others were not.

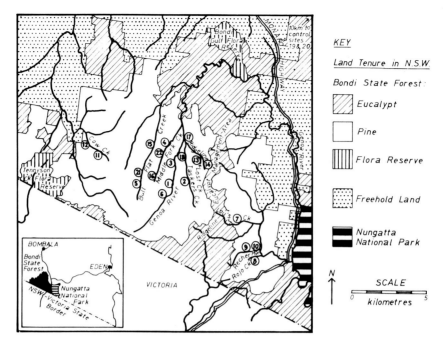

Fig. 3. Location of creek reserves surveyed for birds, mammals, reptiles and frogs in the Bondi and Coolangubra State Forests near Bombala, New South Wales. Numbers indicate study sites named in Tables 1 and 2. (From Shields 1985).

Table 1. Creek reserves surveyed for birds, mammals, reptiles and frogs in the Bondi and Coolangubra State Forests, New South Wales (from Shields 1985).

Study Site*	Mean Corridor Width (m)	Total Area (ha)**	No. Eucalyptus Species	Time Since Clearing (yr)	Distance to Unlogged Forest (km)	Crown Height (m)	No. Trees <20 cm D.B.H./ha	No. Trees <100 cm D.B.H./ha	Crown Vigour (1-4)***	Year Burnt
1. Genoa River	70	3.42	6	7.5	4.0	30.0	154	2.4	3.30	1983
2. Taskers Creek Upper	69	4.18	5	8.5	5.6	22.1	139	1.6	1.30	1983
3. Genoa River Lower	90	5.72	3	6.0	3.3	16.1	123	2.4	3.10	1983
4. Buffer Strip 4	95	4.79	3	6.5	3.2	26.3	157	3.2	3.15	1983
5. Bull Flat Creek Upper	32	1.68	4	9.0	3.8	26.1	63	2.4	1.23	1983
6. Genoa River Upper	70	3.41	3	8.5	4.5	24.4	83	2.4	2.35	1983
7. Lou's Creek	62	3.17	5	4.0	2.3	26.2	138	2.4	3.40	Unburnt
8. Rojo Creek	82	4.26	6	2.5	2.1	28.6	138	3.2	3.60	Unburnt
9. Recher River Upper	247	12.73	7	3.0	2.5	31.2	150	8.0	3.65	1980
10. Recher River Lower	121	5.97	7	3.5	1.8	26.2	145	4.8	3.65	Unburnt
11. Cow Creek Upper	56	2.54	2	10.0	10.0	12.0	559	2.4	1.20	1983
12. Cow Creek Lower	153	7.85	6	10.0	10.0	29.2	149	7.8	3.02	Unburnt
13. Lost Creek	76	3.76	2	7.0	3.8	15.5	178	2.4	2.75	1983
14. Ridgeline Creek	74	3.68	6	5.0	4.1	29.0	128	5.6	3.30	1983
15. WFAW Creek	48	2.31	2	7.5	3.7	23.8	90	1.6	1.30	1983
16. Middlefork Creek Upper	33	1.68	4	8.5	5.9	19.0	90	1.6	1.80	1983
17. Goldfields (Genoa River)	163	8.41	4	6.5	2.5	21.4	152	2.4	2.65	1983
18. Taskers Creek Lower	125	6.28	5	7.0	3.8	19.9	165	2.4	2.40	1983
19. Waratah Creek Upper	144	7.78	6	0	0	38.0	138	14.4	3.39	Unburnt
20. Waratah Creek Lower	130	6.76	7	0	0	31.1	134	13.6	3.57	Unburnt
21. Bull Flat Swamp	82	3.99	4	7.0	4.8	23.3	173	2.4	2.97	1983
22. Middlefork Creek Lower	58	3.07	3	6.5	3.7	21.4	64	2.4	1.16	1983

* All sites were surveyed for arboreal mammals. Plots 19 and 20 are unlogged control transects for arboreal marsupials. Plots 1-4, 8, 9 are the reserves on which birds were censused. Plots 1, 8 and 9 were also sampled for small, ground-dwelling mammals, reptiles and frogs. ** Area given is that of mature eucalypt forest surveyed for arboreal marsupials and birds. Bird transects had a constant area of 5 ha (120 m wide × 420 m long). *** Values given are the mean crown vigour rating of all trees sampled. Vigour values ranged from 1 (full crown, no sign of dieback) to 4 (complete crown dead).

Populations of arboreal marsupials were also studied in unlogged forest (Kavanagh 1984) and in wide corridors of mature forest retained within clear-felled areas (Kavanagh and Rohan-Jones 1982).

Small Ground-dwelling Vertebrates. Small mammals, reptiles and frogs were surveyed on three of the creek reserves (Table 1). Elliott live traps were used for mammals and pit traps (Webb 1980) for reptiles and frogs. Mammals were sampled during two summers and one winter in 1980 and 1981. Reptiles and frogs were surveyed from January 1980 through March 1981.

Patch Size: Small and Large Coupes

Birds. A point interval procedure was used to census birds on alternating small (< 20 ha), logged and unlogged coupes each spring from 1976 through 1980 (Kavanagh *et al.* 1985). Results from the small coupe plots are compared to censuses in 1977 and 1978 on large (> 100 ha) logged and unlogged coupes in the same forest.

Reptiles. The same pit trap procedures as used in the creek reserves were used to sample reptiles on logged and unlogged coupes used for the bird counts. Reptiles were sampled four years after logging in 1980/81.

RESULTS

Creek Reserves

Birds. Of the 113 species of terrestrial birds known from the Bombala district (Recher *et al.* 1980), 81 were recorded on the six census transects within the reserves. Forty-one species were found nesting and, of these, 30 species were sufficiently abundant for analysis. Of the forty species not recorded nesting, most were transient or irregular in occurrence and the others uncommon or rare. Initially birds were more abundant and there were more species on the widest reserves than on the narrowest (Fig. 4).

In the following analysis, we consider both the response of individual species and of the avian community as a whole. Species have been classified as either being dependent upon forest ('forest birds') for critical resources (e.g. food, nest sites) or as not being dependent on forest ('non-forest birds'). The latter include 'open-country' as well as 'generalist' species which frequent a wide variety of habitats including pine plantations, shrublands and edge habitats. We made this separation based on our extensive knowledge of the biology of each species in the Bombala district (e.g. Recher *et al.* 1980, 1983a, 1985b). We considered forest birds to be those species most likely to be adversely affected by the clearing and fragmentation of mature forest.

In the first year after clearing, there was no significant difference in total bird populations or populations of forest birds on the reserves when compared to the control sites ($p > 0.05$, Wilcoxon

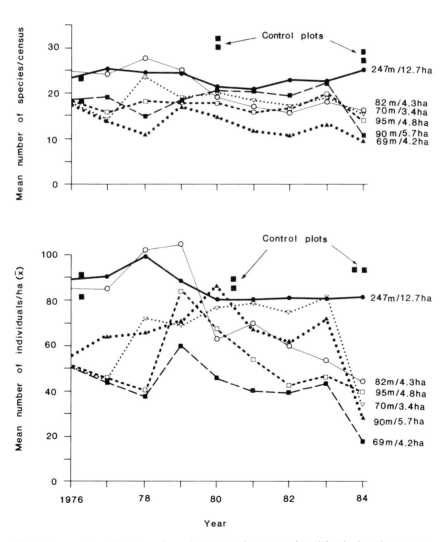

Fig. 4. The number of birds per ha and species richness are plotted for the breeding season censuses on the creek reserves in the Bondi State Forest pine plantations from 1976 to 1984. Two unlogged transects in mature forest on Coast Range Road, Victoria were used as controls. Numbers are the means of four counts of 2 h each with densities (birds/ha) adjusted for the differing widths of each reserve. Average width and the area of each census transect are shown alongside the 1984 census points.

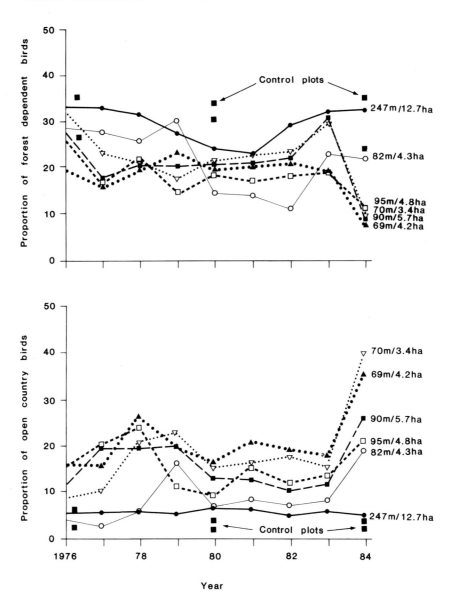

Fig. 5. The percent of forest dependent and non-forest or open-country bird species are plotted for each of the breeding season censuses from 1976 to 1984 for the creek reserves in the Bondi State Forest. Controls are the unlogged plots in mature forest on Coast Range Road, Victoria. Species which could not be assigned to either the 'forest bird' or 'open-country bird' categories ('generalist species') make up the remainder and are not plotted. The decrease in the species richness of forest birds and the increase of open-country species in 1984 follows the 1983 fire which burnt plots 1-4. Average width and the area of each census transect are shown alongside the 1984 census points.

paired sample test). In 1980, total bird populations remained similar on reserves and control plots. However, populations of forest birds on the five narrowest reserves were significantly smaller than on the controls or on the widest reserve ($p < 0.05$) (Fig. 5). Numbers of forest birds were similar on the widest reserve and the control plots. The similarity between total bird populations on the reserves and the control plots was due to the large number of open-country and generalist species (non-forest birds) recorded on the reserves.

In 1984, total bird populations and numbers of forest birds on the four reserves burnt in 1983 were significantly smaller than on the unburnt reserves or the controls ($p < 0.05$) (Figs 4 and 5). The unburnt narrow reserve maintained total bird numbers similar to the controls (Fig. 5), although forest bird populations were significantly smaller ($p < 0.05$) (Fig. 4). Despite being affected by fire in 1980, the widest reserve remained similar to the controls in both total numbers of birds and numbers of forest birds.

The precipitous decline in the number of birds between 1983 and 1984 on all but the widest reserve can only partially be attributed to the effects of fire. Bird populations over a wide area of eastern New

South Wales declined in 1983 and 1984 probably as a consequence of the severe drought which prevailed from 1978 to 1983 (Recher unpubl. data; Ford *et al.* 1985). It is not possible to fully separate the effects of drought from fire as the former often culminates in the latter. Regardless of the reasons for the decline in numbers on five of six reserves censused, it suggests that only very wide creek reserves are adequately buffered against climatic fluctuations. The issue merits detailed study.

Of the 30 species which were abundant on the reserves, about half showed little change while others fluctuated widely in abundance from year to year. Seven species increased in numbers from 1976 to 1983, and seven species decreased in abundance.

The species which increased were flame robin *Petroica phoenicea,* rufous whistler *Pachycephala rufiventris,* grey fantail *Rhipidura fuliginosa,* white-browed scrubwren *Sericornis frontalis,* superb blue wren *Malurus cyaneus,* yellow-faced and white-eared honeyeaters *Meliphaga chrysops* and *M. leucotis.* All are normally abundant in edge or open-forest environments and probably benefited from the edge environment created along the plantation/reserve boundary. Also all are small (< 25 g) and have small foraging areas.

The species which decreased in abundance were crimson rosella *Platycercus elegans,* gang-gang cockatoo *Callocephalon fimbriatum,* laughing kookaburra *Dacelo gigas,* crested shrike-tit *Falcunculus frontatus,* grey shrike-thrush *Colluricincla harmonica,* red wattlebird *Anthochaera carunculata,* and Australian raven *Corvus coronoides.* Apart from the shrike-tit (29 g), all are large forest birds (> 75 g). All have large foraging areas. In addition, all have specialized foraging or nesting requirements. The rosella and cockatoo feed mainly on ripening eucalypt capsules, the shrike-tit and to a lesser extent the shrike-thrush require loose strips of bark in which to hunt for insects. The wattlebird is a nectar-feeder. The two parrots, the kookaburra and shrike-thrush require tree hollows or cavities for nesting while the raven uses large trees.

The 1983 fire had a major impact on the composition of bird communities of the creek reserves. The number of individuals and species of forest birds decreased markedly, but the number of non-forest birds increased (Fig. 5). These are similar to changes reported for bird communities following extensive wildfires affecting unlogged forest (Recher and Christensen 1981; Recher *et al.* 1985a; Christensen *et al.* 1985), but the very large increase in open-country species is an artifact of the plantation environment.

Arboreal Marsupials. All species of arboreal marsupials expected in the area, except the pygmy possum *Cercartetus nanus* which is difficult to detect by spotlighting, were recorded in the creek reserves (Table 2). Greater gliders *Petauroides volans* were found on all strips, with densities from 0.08 to 1.36 individuals/ha ($\bar{x}$ = 0.67 individuals/ha). The density of greater gliders on the two control plots was 1.01 and 1.09 animals/ha (Table 2) (Shields 1985). During other surveys in similar types of unlogged forest in the Coolangubra State Forest, densities ranged from 0.5 to 1.3/ha with an average of 0.84 (Kavanagh 1984). The abundance of greater gliders in some reserves was therefore less than that in comparable unlogged forest.

The sugar glider *Petaurus breviceps* was found on 16 of the 20 reserves with a maximum density of 0.3 individuals/ha (Table 2). Brushtail possums *Trichosurus vulpecula* and mountain brushtail possums *T. caninus* were found on 10 and 9 of the reserves, respectively. Ringtail possums *Pseudocheirus peregrinus* were found on 8 of the 20 transects. Feathertail gliders *Acrobates pygmaeus*, another species which is difficult to survey, was found on only four of the reserves (Table 2).

The species least represented in creek reserves was the yellow-bellied glider *Petaurus australis* which was recorded on only the widest corridor in a section near unlogged forest (Table 2) (Shields 1985). This species is regular in forests of the region and has been studied in a wide (100-200 m) reserve of unlogged riparian forest in Nadgee State Forest near Eden (Kavanagh and Rohan-Jones 1982). This corridor was surrounded by forest regenerating following integrated logging 10 years previously. Yellow-bellied gliders were resident but it was necessary for them to obtain supplementary food outside this area during winter. The presence of retained mature trees in the adjacent logged areas allowed these gliders to forage outside of the reserve. The conclusion drawn is that yellow-bellied gliders depend upon a mosaic of forest resources, including mature forest. This fits the predictions of central place theory that large animals, social or gregarious animals, and those with specialized foraging habits such as the yellow-bellied glider will be affected by a reduction in the available habitat or a change in its spatial configuration.

Only the greater glider was abundant enough for analysis of the factors affecting its distribution within creek reserves. Linear regression of greater glider populations against site characteristics showed a significant association with the number of eucalypt species ($p < 0.05$), time since clearing ($p < 0.05$), mean tree height ($p < 0.025$), reserve

Table 2. Arboreal marsupial density (individuals/ha) on creek reserves and control plots in the Bondi and Coolangubra State Forests, New South Wales. See Table 1 for plot widths, numbers of trees and other site features. (from Shields 1985).

Study Site*	Ringtail possum	Greater glider	Yellow-bellied glider	Sugar glider	Brushtail possum	Mountain brushtail possum	Feathertail glider	Total arboreal marsupials
1. Genoa River	.22	1.25	–	–	–	.15	–	1.59
2. Taskers Creek Upper	–	.45	–	.18	–	.06	.06	.72
3. Genoa River Lower	.13	.52	–	.13	.09	.04	–	.91
4. Buffer Strip 4	.42	.43	–	–	–	.05	–	.89
5. Bull Flat Creek Upper	–	.15	–	–	.15	–	–	.30
6. Genoa River Upper	–	.66	–	.07	.07	–	–	.80
7. Lou's Creek	–	1.10	–	.08	–	–	–	1.18
8. Rojo Creek	–	1.11	–	.23	.06	–	–	1.40
9. Recher River Upper	–	.53	–	.13	.02	.05	–	.73
10. Recher River Lower	–	.83	.23	.15	.19	.09	–	1.45
11. Cow Creek Upper	–	.49	–	.20	–	–	–	.84
12. Cow Creek Lower	.19	.95	–	–	–	–	.05	1.19
13. Lost Creek	–	.66	–	.07	–	–	–	.73
14. Ridgeline Creek	–	1.36	–	.20	–	.07	–	1.63
15. WFAW Creek	–	.97	–	.21	–	–	–	1.18
16. Middlefork Creek Upper	1.49	.15	–	.30	.15	–	–	2.09
17. Goldfields (Genoa River)	–	.62	–	.09	–	.03	.03	.77
18. Taskers Creek Lower	.07	.83	–	.08	.08	–	–	1.06
19. Waratah Creek Upper	–	1.09	.58	.17	–	–	.04	1.88
20. Waratah Creek Lower	–	1.01	.30	.11	.05	–	.10	1.57
21. Bull Flat Swamp	.25	.19	–	.19	.19	.19	.06	1.07
22. Middlefork Creek Lower	.41	.08	–	.16	.24	–	–	.89

width ($p < 0.05$) and crown vigour ($p < 0.025$), but not with slope or distance from unlogged forest (Fig. 6) (Shields 1985). Multiple regression of habitat parameters showed that a combination of crown vigour, mean tree height, reserve width and trees per ha accounted for the greatest portion of variation in population density of greater gliders ($r^2 = 0.52$). Distance from unlogged forest and slope were not significant in accounting for the variability in the population density of greater gliders or total arboreal marsupials (Fig. 6), nor did they contribute to total variation in multiple regression analysis.

Total arboreal marsupial populations were significantly associated with number of eucalypt species, mean tree height, number of trees greater than 100 cm dbh (p's < 0.05) and marginally with crown vigour ($p < 0.1$) by linear regression (Fig. 6). Multiple regression of site characteristics showed that the number of eucalypt species, reserve width, number of trees greater than 100 cm dbh and crown vigour accounted for the greatest portion of variation in the population density of total arboreal marsupials ($r^2 = 0.49$).

Crown vigour and mean tree height are measures of stand condition, which is affected by soil type, topography, disturbance and moisture regime. The number of greater gliders and the total numbers of arboreal marsupials were greatest on reserves with productive soils, gentle slopes, abundant moisture and little disturbance.

The greater glider was the only arboreal marsupial recorded on all of the reserves (Table 2). This and its relative abundance suggests that the populations surviving on the reserves have, at least in the short-term, been little affected by the clearing of the adjacent forest. Greater gliders are solitary and occupy a small home range (Henry 1984). They feed primarily on the leaves of eucalypts, and can interchange resources by feeding on the foliage of different species (Kavanagh 1984). The greater glider is therefore a species which central place models predict should be less affected by forest fragmentation than species requiring larger areas or with special requirements.

In contrast, the yellow-belled glider, a species which appears to be strongly affected by forest fragmentation, occupies a large home range, is social, and has specialized foraging habits requiring a number of non-interchangeable food resources (e.g. invertebrates and honeydew found under loose and peeling bark, sap) (Henry and Craig 1984). The linear environment provided by creek reserves may be too restrictive because the home range of the yellow-bellied glider typically encompasses both ridge and gully habitats.

After the fire in 1983, the study sites were sampled for extant populations of arboreal marsupials. On the 15 sites which burnt in the fire (Table 1), only three greater gliders and one feathertail glider were observed. These animals survived in patches of forest which had escaped total crown scorch. Animal populations on the unburnt plots remained constant.

Small Ground-dwelling Mammals. The brown Antechinus *Antechinus stuartii*, the dusky Antechinus *A. swainsonii* and the bush rat *Rattus fuscipes* were the small mammals trapped on the three reserves sampled. Numbers were comparable

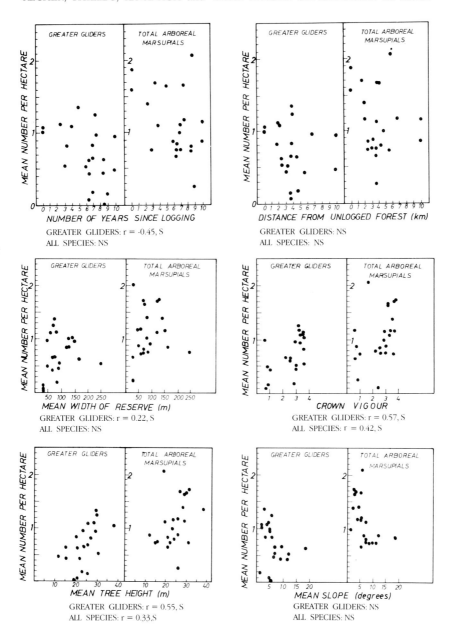

Fig. 6. Linear regression analysis of six plot variables against numbers of greater gliders and total arboreal marsupials. Analysis is based on total numbers of animals recorded over 12 months in 1982/83 with densities (individuals/ha) adjusted for variation in reserve width. Significant(s) regression values are shown. All others are not significant (NS).

to those expected in similar habitats in unlogged forest (see Recher et al. 1980), with nightly trapping success varying between 0.5% and 14% (mean 4.8%, total trap nights = 1599). The bush rat was trapped on all three reserves (27 captures), the brown Antechinus was trapped on two reserves (45 captures), and the dusky Antechinus was trapped on one reserve (the one without the brown Antechinus) (five captures).

Reptiles and Amphibians. Seven species of reptiles and 12 species of frogs were recorded on the reserves (Shields et al. 1986). These represent 30% of the reptiles and 75% of the frogs recorded for the region (Webb 1981), but include all the species expected for the habitats surveyed. All species except the lizard *Lampropholis guichenoti* are normally more abundant in riparian and adjacent forest than ridge forest. Apart from the loss of habitat, there was no evidence that any of these species were disadvantaged by the clearing of forest and the creation of creek reserves.

The abundance of small ground-dwelling mammals, lizards and frogs on the reserves fits the predictions of the model that small animals will be less affected by habitat fragmentation than larger species. However, populations of small ground-dwelling mammals were adversely affected by the fire that burnt two of the three reserves sampled

(Shields *et al.* 1986). Frogs were unaffected by the fire, increasing in numbers captured in the first summer after the burn. Only marginally fewer lizards were captured, but we consider that ground-dwelling animals will be sensitive to repetitive events that affect the forest floor. Re-colonization may be prevented or slowed significantly if the reserves are isolated by alien habitat or if sources of re-colonization are remote.

Patch Size: Small and Large Coupes

The integrated logging operation at Eden creates a mosaic of forest patches or coupes differing in size and time since logging. Although part of a continuous forest environment, logging coupes may be too small to provide all the requirements of some animals and the surrounding forest may be too young (or too old) to provide the missing resources. We investigated the effect of patch size on birds and reptiles by comparing populations on alternate logged and unlogged coupes (Fig. 7). A lack of resources prevented research on mammals.

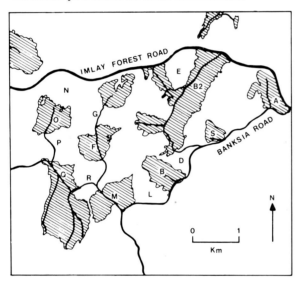

Fig. 7. At Eden integrated logging produces a mosaic of alternating logged and unlogged coupes (cutting compartments). Illustrated here are the small (< 20 ha) coupes censused for birds and reptiles in the East Boyd State Forest. Roads are on ridgelines. Logged coupes are cross-hatched. (From Kavanagh *et al.* 1985).

Birds. For most small coupes (< 20 ha) the immediate effect of logging was a significant reduction in the number of individuals and species of birds (Kavanagh *et al.* 1985; see also Loyn 1980; Recher *et al.* 1980 for details on logging effects). The impact was most severe during the first two years after logging when counts of individuals and number of species on logged small coupes averaged 50% lower than on unlogged coupes (Table 3). Coupe B2 (Table 3) was an exception to these trends and average numbers of species and individuals exceeded those recorded on unlogged plots. In contrast to other coupes censused, B2 incorporated a drainage line and was sited lower on the ridge. Possibly for these reasons and coupled with its length, it acted as a corridor for birds moving between areas of unlogged forest. Birds moving through this plot were recorded during censuses when they paused in small trees remaining within the census circle thereby inflating results.

As regeneration proceeded, some open-country and shrub-dwelling species increased in abundance. A number of forest species remained less abundant on the logged plots compared with unlogged small coupes for up to four years (the duration of the study). Four years after logging 78% of the species present in unlogged forest occurred on the logged plots. This rapid rate of recovery was attributed to the close proximity of unlogged forest (Kavanagh *et al.* 1985). Species found to be significantly more abundant on logged plots were superb blue wren, white-eared honeyeater, yellow-faced honeyeater, and pilot bird *Pycnoptilus floccosus* (Table 4). Species which were significantly more abundant in unlogged forest were grey fantail, rufous fantail *R. rufifrons*, striated thornbill *Acanthiza lineata*, white-throated treecreeper *Climacteris leucophaea*, yellow robin *Eopsaltria australis*, rufous whistler, spotted pardalote *Pardalotus punctatus*, fan-tailed cuckoo *Cuculus pyrrhophanus*, black-faced cuckoo-shrike *Coracina novaehollandiae*, scarlet robin *Petroica multicolor*, and grey shrike-thrush (Table 4). All these were either birds which forage in the canopy and subcanopy (grey fantail, rufous fantail, striated thornbill, white-throated treecreeper, rufous whistler, spotted pardalote, black-faced cuckoo-shrike, and grey shrike-thrush) or which foraged on litter invertebrates (yellow robin, scarlet robin, fan-tailed cuckoo) (Recher *et al.* 1985b).

The pattern of change in species abundances was similar when large coupes (> 100 ha) were compared. Logged coupes had fewer species and a smaller number of individuals than unlogged coupes. Again some species benefited from the logging, while others were affected adversely. Yellow robin, rufous whistler, striated thornbill, white-throated treecreeper, spotted pardalote, leaden flycatcher *Myiagra rubecula*, and Eastern spinebill honeyeater *Acanthorynchus tenuirostris* were less abundant on logged than unlogged coupes. Superb blue wren, white-browed scrub-wren, and brown thornbill *Acanthiza pusilla* were more abundant on logged than unlogged coupes.

In comparing logging areas of different sizes, small logged coupes were similar to large logged coupes in respect to numbers of species and individuals. However, individual species responded differently to the effect of logging. Grey fantails and brown thornbills were significantly more abundant on large coupes which had been logged than on small logged coupes. No species was more abundant

Table 3. Summary of Small Coupe Logging Bird Census Results 1976-1980. All logging took place in 1976. (from Kavanagh et al. 1985).

Logging Treatment	Adjacent matched small coupes	x̄ indivs/ha/hr					x̄ species/ha/hr					Total species/ha/4hrs				
		76	77	78	79	80	76	77	78	79	80	76	77	78	79	80
Years after Logging		0	1	2	3	4	0	1	2	3	4	0	1	2	3	4
Logged	B	1.0	2.0	4.0	9.5	14.0	1.0	1.8	2.5	5.5	4.8	2	7	6	14	12
Unlogged	D	–	14.5	20.0	19.5	20.0	–	6.0	9.5	9.0	9.0	–	10	16	17	14
Logged	B2	12.5	22.0	–	–	22.0	5.3	9.3	–	–	9.5	12	17	–	–	22
Unlogged	E	–	17.0	12.0	–	23.0	–	9.0	7.0	–	11.5	–	17	14	–	25
Logged	F	–	3.0	–	–	12.0	–	2.8	–	–	4.3	–	6	–	–	9
Unlogged	G	–	9.0	16.5	–	16.0	–	4.5	–	8.0	8.5	–	10	15	–	20
Logged	O	–	–	–	8.0	24.0	–	–	–	4.3	9.3	–	–	–	10	22
Unlogged	N	–	–	–	24.3	29.0	–	–	–	11.5	13.8	–	–	–	24	25
Logged	Q	–	–	–	28.5	26.5	–	–	–	10.2	11.0	–	–	–	22	24
Unlogged	P	–	–	–	29.5	27.0	–	–	–	14.8	11.5	–	–	–	29	20
Logged	M	–	–	–	30.3	30.0	–	–	–	11.0	12.0	–	–	–	20	22
Unlogged	R	–	–	–	32.5	30.0	–	–	–	13.3	12.5	–	–	–	27	25
Logged	S	–	–	–	16.5	20.0	–	–	–	6.3	5.5	–	–	–	12	13
Unlogged	L	–	–	–	23.3	24.5	–	–	–	11.5	9.0	–	–	–	25	19
Logged	A	17.0	*4.5	–	–	–	7.0	*3.3	–	–	–	14	*9	–	–	–
Logged Plots x̄		13.6	7.9	4.0	18.6	21.3	4.4	4.3	2.5	7.5	8.0	9.3	9.8	6.0	15.6	17.7
Unlogged Plots x̄		–	13.5	16.2	25.8	24.2	–	6.5	8.2	12.0	10.8	–	12.3	15.0	24.4	21.1

* Plot was burnt and later discontinued.

Table 4. Differences in abundance for selected species. Figures are rounded percentages concerning the number of 10 minute periods in which a species was observed. (1980, n = 168: 1979, n = 120). (From Kavanagh et al. 1985).

SPECIES ADVANTAGED BY LOGGING

	1980 (4th Year Post-logging)		1979 (3rd Year Post-logging)	
	Unlogged	Logged	Unlogged	Logged
Yellow-faced honeyeater	59	79	75	88
Pilot bird	5	20	3	13
Superb blue wren	2	11	6	17
White-eared honeyeater	1	7	3	18

SPECIES DISADVANTAGED BY LOGGING

	1980 (4th Year Post-logging)		1979 (3rd Year Post-logging)	
Striated thornbill	62	29	71	41
White-throated treecreeper	46	3	53	5
Rufous whistler	36	5	13	1
Spotted pardalote	20	11	22	6
Grey shrike-thrush	19	2	13	6
Yellow robin	29	6	23	6
Scarlet robin	7	1	7	0
Fan-tailed cuckoo	11	1	30	10
Grey fantail	57	33	42	35

on small logged coupes. As with logged areas, there were no differences between small and large unlogged coupes with respect to numbers of species and abundance of birds (density). However, rufous whistler, grey shrike-thrush, leaden flycatcher and white-naped honeyeater *Melithreptus lunatus* were significantly more abundant in large unlogged coupes than small unlogged coupes (Kavanagh et al. 1985 provide details). Eastern spinebill honeyeater, brush cuckoo *Cuculus variolosus*, orange-winged sittella *Daphoenositta chrysoptera*, and mistletoe-bird *Dicaeum hirundinaceum* were also disadvantaged by the size of small coupes relative to large unlogged coupes.

The response of birds to the effects of logging on different sized coupes shows that the major impact on the avifauna is the change in forest structure from a mature ecosystem with a diverse size and age range of trees to a young forest dominated by shrubs and saplings. Birds requiring large trees in which to nest and/or forage are absent or significantly less abundant on logged coupes. As predicted by our model, the size of the logging coupe (or the size of forest remnants) has an additional effect on the avifauna. No bird was more abundant on small coupes irrespective of logging history, but a number of species, including several which benefited from logging, were more abundant on large coupes. The species affected by coupe size were mainly medium sized birds requiring large home ranges (e.g. grey shrike-thrush), species with secialized foraging or food requirements (e.g. Eastern spinebill honeyeater, orange-winged sittella, mistletoebird), species with special breeding or nesting requirements (e.g. brush cuckoo), or were flocking birds and/or colonial nesters (e.g. orange-winged sittella, white-naped honeyeater). These fit the predictions of central place theory about the kinds of animals most likely to be affected by patch size and include many of the same birds affected by reserve width in the pine plantations.

We conclude that as logging proceeds, coupe size will be a significant factor influencing the abundance of these species and could be a major factor in developing plans of management for wildlife.

Reptiles. After four years, few differences were found between the reptile faunas of small (< 20 ha) and large (> 100 ha) logged coupes and adjacent paired unlogged coupes (Webb unpubl. data). Of the 13 reptiles recorded from these forests only one scincid lizard *Lampropholis guichenoti* was significantly reduced in numbers on small logged coupes (Chi Square = 10.45, p < 0.01), while no species significantly increased in numbers. No differences in species abundances were found between large logged and unlogged coupes.

In these forests, aspect appears to have a far greater influence on species abundances than disturbance by logging. Of the more abundant species, significantly fewer reptiles were recorded on warmer northern aspects than others, irrespective of logging (p < 0.01 for *L. guichenoti*, *Sphenomorphus tympanum* and total reptiles).

DISCUSSION

The immediate effect of integrated logging is a decrease in the abundance of birds and mammals dependent on mature forest (Loyn 1980; Recher *et al.* 1980; Smith 1985). Survival of these species therefore depends upon the retention of adequate areas of mature forest. At Eden the principal areas of mature forest retained within the logged forest are reserves along creeks and gullies and patches of unlogged forest on steep or rocky ground (Recher *et al.* 1980; Bridges 1983; Dobbyns and Ryan 1983; Ryan 1984). These clearly offer a refuge in the short-term for animals dependent upon mature forest, but because of their size and spatial configuration may not suffice for their long-term survival as an increasingly large proportion of the mature forest in the Eden district is logged.

The species most affected by the linear nature of creek reserves and patch size are those species predicted to be affected by central place theory. Thus, the birds and mammals which are absent from creek reserves and small coupes are relatively large, have specialized foraging or nesting habits, require multiple non-interchangeable resources and are social or gregarious. Management strategies designed to conserve these species and to enhance the value of creek reserves and patches of mature forest for their conservation should therefore attempt to minimize foraging distances, reduce edge effects and increase the kinds of available resources.

Natural processes occurring in large contiguous blocks of habitat may not take place in small reserves. Thus specific resources required by forest dependent species may not be available when ecological factors reach equilibrium in small reserves. Similarly, as evidenced by the effect on bird numbers in the creek reserves censused in the Bondi pine plantations, small reserves, and remnant populations of animals are more subject to change when climactic events occur (e.g. drought, fire).

Wildfires and the use of prescription burns in retained areas or as a pre- or post-logging practice to reduce fuel loads will have additional effects on wildlife which need to be considered when managing wildlife in these forests. Small populations and the restricted area of reserves will make fauna dependent on mature forest particularly sensitive to the effects of fire and changes in fire regimes.

Existing guidelines for the management of forest wildlife at Eden were based on studies carried out between 1975 and 1978 (e.g. Recher *et al.* 1980). These guidelines were useful in establishing wildlife management as an integral part of forest management in the Eden Management Area, but when proposed, continued review and adjustment of wildlife management procedures was envisioned (Recher *et al.* 1980; Recher 1984, 1985). Since 1978, a substantial body of research has been completed in the Eden district, part of which is reviewed here. Most of this work was specifically directed at improving our understanding of the effects of forest management (e.g. integrated logging, clearing for plantations, use of fire) on wildlife with the intent of refining management guidelines for the conservation of wildlife in intensively logged forests. It is appropriate in this chapter, where we make a first attempt to relate optimal foraging theory (i.e. central place models) to the impact of forest fragmentation, to propose additional guidelines for wildlife management at Eden. The following management procedures, which are either refinements of existing recommendations or are in addition to those proposed earlier, are suggested.

For wildlife conservation, creek reserves and movement corridors need to be as wide as possible. Based on observations in the pine plantations, the current prescription of 20 m to either side of a creek with logging (but not machinery) permitted in the outer 10 m (Anon. 1977) is too narrow for fauna conservation. Not only are such reserves too narrow to provide the resources required by wildlife, but the trees retained within them are subject to windthrow and dieback (loss of vigour). We therefore recommend that for wildlife conservation a minimum of 50 m be retained to either side of a creek or gully or drainage (irrespective of catchment area) and that no logging be allowed within this reserve. Preferably corridors and creek reserves adopted for wildlife management purposes should follow ecological boundaries, such as the edge of a flood plain or the natural interface between riparian and slope forests. Sufficient surrounding forest to buffer the reserve against windthrow and changes to microclimate (e.g. temperature, humidity, light) should be retained (see Harris 1984). Some of these procedures have already been adopted in the Eden Management Area and, at lower elevations (< 300 m), boundaries of creek reserves follow the distribution of monkey gum *Eucalyptus cypellocarpa* (Dobbyns and Ryan 1983; Ryan 1984). Monkey gum is an indicator species for high site quality and maximum faunal diversities at lower elevations in the forests south of Eden (Recher *et al.* 1980). Elsewhere in the Eden Management Area other indicator species are required. When coupled with the retention of habitat trees (trees providing hollows for dens and/or nests) within the logging coupes as a source of additional resources, such reserves and corridors will assist significantly in the conservation of sensitive species.

In addition to retaining mature forest along creeks and gullies and providing movement corridors between catchments and large reserves (e.g. national parks, flora reserves), as is currently practised at Eden (Dobbyns and Ryan 1983; Ryan 1984), large patches of mature forest need to be retained within the logging area. We reach this conclusion from the predictions of central place theory and from our observations of the response of fauna to integrated logging operations at Eden and the establishment of plantations near Bombala. Theory predicts that some fauna will be adversely affected by both narrow (linear) reserves regardless of total length and by the fragmentation of forest into a mosaic of even-aged and relatively small patches (coupes). Our observations confirm the predictions of the model.

To reduce these effects and enhance the value of creek reserves and movement corridors for sensitive fauna, we propose that some logging coupes be reserved from the first cycle of logging. As illustrated in Figure 8 these reserved coupes should adjoin the reserves along creeks and gullies and provide a large area of mature forest suitable for wildlife adversely affected by logging. The larger area of habitat provided reduces foraging distances and the risk of predation. By including slope and ridge habitats with the riparian vegetation reserved along creeks or in gullies, there would be an increase in the kinds of resources available to sensitive fauna as well as an increase in the amount of preferred habitat for dry forest fauna. Coupes reserved in this way would buffer permanent reserves ('core areas') against the environmental effects of clear-felling.

Reserved coupes would also act as 'stepping-stones' between larger reserves (e.g. national parks, flora reserves) and further enhance the value of creek reserves as movement corridors. In effect, we propose to string 'beads' on the existing reserve 'necklace' (Fig. 8). With the information currently available to us, we see this system as an adaptation of current logging procedures. Coupes reserved from logging during the first cutting cycle may be logged in the second or subsequent cycles after an adjoining logged coupe has matured sufficiently to provide the resources required by fauna dependent on mature forest. Logging selected coupes on a long rotation was first proposed by Recher *et al.* (1975). Harris (1984) has proposed a similar system of reserves and reserve selection for the conservation of old growth (mature) ecosystems in Douglas fir *Pseudotsuga menziesii* forests of North America.

Our studies of small coupe logging in these forests (Kavanagh *et al.* 1985) suggest that reserves of less than 50 to 100 ha are too small for some fauna. Unlogged reserves will be most effective if they are located in areas of preferred habitat for wildlife. Thus in addition to limits imposed by minimum area requirements, selection of coupes to be reserved from logging on the first or subsequent logging cycles should be based on such criteria as site quality, presence of rare or endangered species, high faunal or floral diversity, utility as stepping stones, and their contribution towards increasing the effective size of permanent reserves. Selection of such coupes needs to be preceded by surveys to determine which patches have the greatest wildlife and ecological values (e.g. Braithwaite 1983; Braithwaite *et al.* 1983, 1984; Loyn 1985). Such surveys should be conducted before logging, during the planning phase, to identify the coupes to be managed on a longer rotation for wildlife conservation.

The length of time required for a clear-felled forest to regenerate and mature so that it provides all the resources required by dependent fauna remains to be established. To a great extent the length of rotation for these reserved coupes will depend largely on the rate of production of hollows suitable for arboreal marsupials and large hole-nesting birds (e.g. cockatoos, owls). Suitable hollows may not develop in under 100-150 years (e.g. Disney and Stokes 1976; Saunders *et al.* 1982; Mackowski 1984) with considerable varation in hollow formation according to tree species, site conditions and the effects of fire and storm. There is an urgent need to obtain such information and to also establish the sizes and numbers of hollows required by forest wildlife at Eden to maintain particular population densities.

Although the Eden Management Area will remain predominately forested, the emphasis on pulpwood production means that the original forest ecosystem

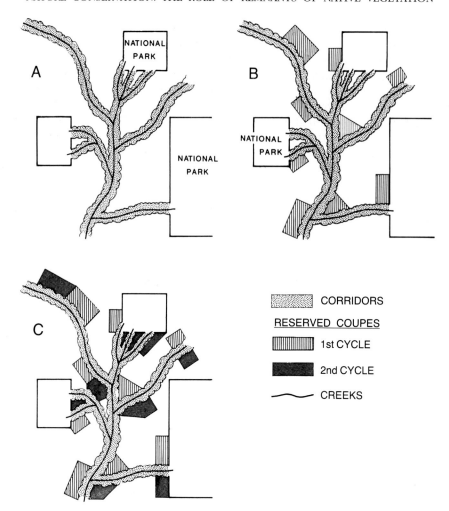

Fig. 8. A. The existing system of creek reserves and wildlife corridors at Eden is designed to protect riparian habitats particularly rich in species and to link large areas of mature forest permanently reserved in national parks, nature reserves and flora reserves. B. To improve the values of the system of creek reserves and corridors for wildlife dependent on mature forest and which may not be able to adapt to the constraints of a linear environment, we propose that carefully selected coupes be reserved from logging in the first cutting cycle. These coupes would be selected for their wildlife values and as 'stepping-stones' between the larger parks and reserves. C. As coupes logged in the first cycle mature and acquire the resources (e.g. hollows) required by dependent wildlife, the coupes originally reserved from logging may be logged and replaced by a second set contributing the same management values.

will be extensively, albeit internally, fragmented. Management of wildlife in these forests is essentially the management of remnants of the original forest ecosystem to provide essential resources for wildlife dependent on mature forest as well as providing some of the requirements of fauna which may be less sensitive to the effects of intensive logging. By combining theory with observation, it is possible to formulate plans of management which will minimize the impact of logging on wildlife and assist in their conservation. The effectiveness of these plans will require periodic review. Although there is an immediate cost to conserving the full complement of wildlife in these forests with an apparent reduction in the area that can be logged, the probable long-term benefits to forestry should not be ignored. Retention of the full complement of wildlife not only maximizes options for future forest management, but may provide immediate and tangible benefits in terms of pest control and retention of valuable genetic resources.

ACKNOWLEDGEMENTS

Greg Gowing, Wyn Rohan-Jones and Peter Smith among many others assisted with fieldwork and we would like to express our appreciation to all those who assisted for their support throughout this project. Ian Bevege, Ross Dobbyns, Dan Lunney and Frank Whitelaw provided useful comments on the manuscript, but should not be held responsible for its conclusions or recommendations. Illustrations were prepared by Perry de Rebeira. Participation by the Australian Museum in this work was made possible by a grant from Harris-Daishowa (Aust.) Pty Ltd.

REFERENCES

Andersson, M., 1981. Central place foraging in the Whinchat *Saxicola rubetra*. *Ecology* **62**: 538-44.

Anon, 1977. Standard erosion mitigation conditions for logging and clearing in New South Wales. Forestry Commission NSW, Sydney.

Anon, 1982. Eden Native Forest Management Plan. Forestry Commission NSW, Sydney.

Barlow, G. W., 1974. Hexagonal territories. *Anim. Behav.* **22**: 876-8.

Braithwaite, L. W., 1983. Studies of the arboreal fauna of eucalypt forests being harvested for woodpulp at Eden, New South Wales. I. The species and distribution of animals. *Aust. Wildl. Res.* **10**: 219-29.

Braithwaite, L. W., Dudzinski, M. L. and Turner, J., 1983. Studies of the arboreal marsupial fauna of eucalypt forests being harvested for woodpulp at Eden, New South Wales. II. Relationship between the fauna density, richness and diversity, and measured variables of habitat. *Aust. Wildl. Res.* **10**: 231-47.

Braithwaite, L. W., Turner, J. and Kelly, J., 1984. Studies of the arboreal marsupial fauna of eucalypt forests being harvested for woodpulp at Eden, New South Wales. III. Relationships between faunal densities, eucalypt occurrence and foliage nutrients, and soil parent materials. *Aust. Wildl. Res.* **11**: 41-8.

Bridges, R. G., 1983. Integrated logging and regeneration in the Silvertop Ash-Stringybark forests of the Eden region. Research Paper No. 2, Forestry Commission of NSW, 27pp.

Calhoun, J. B., 1963. The social use of space. Pp. 1-187 *in* Physiological mammalogy ed by W. V. Mayer and R. C. Van Gelder. Academic Press, New York.

Christensen, P., Wardell-Johnson, G. and Kimber, P., 1985. Birds and fire in southwestern forests. Pp. 291-9 *in* Birds of eucalypt forests and woodlands: ecology, conservation, management ed by A. Keast, H. F. Recher, H. Ford and D. Saunders. Surrey Beatty, Sydney.

Cody, M. L., 1974. Competition and the structure of bird communities. Princeton Univ. Press, Princeton.

Covich, A. P., 1976a. Analyzing shapes of foraging areas: some ecological and economic theories. *Ann. Rev. Ecol. Syst.* **7**: 235-57.

Covich, A. P., 1976b. Recent changes in molluscan species diversity of a large tropical lake (Lago de Petan, Guatemala). *Limnol. Oceanogr.* **21**: 51-9.

Diamond, J. M., 1975. The island dilemma: lessons of modern biogeographic studies for the design of nature reserves. *Biol. Conserv.* **7**: 129-46.

Diamond, J. M., 1976. Island biogeography and conservation: strategy and limitations. *Science* **193**: 1027-9.

Dill, L. M., 1978. An energy-based model of optimal feeding-territory size. *Ther. Popul. Biol.* **14**: 396-429.

Disney, H. J., de S. and A. Stokes., 1976. Birds in pine and native forests. *Emu* **76**: 133-8.

Dobbyns, R. and Ryan, D., 1983. Birds, glider possums and monkey gums: the wildlife reserve system in the Eden district. *Forest and Timber* **19**: 12-5.

Ford, H. A., Bridges, L. and Noske, S., 1985. Density of birds in eucalypt woodland near Armidale, northeastern New South Wales. *Corella* **9**: 97-107.

Friend, G. R., 1979. The response of small mammals to clearing and burning of eucalypt forest in southeastern Australia. *Aust. Wildl. Res.* **6**: 151-63.

Friend, G. R., 1980. Wildlife conservation and softwood forestry in Australia: some considerations. *Aust. For.* **43**: 217-24.

Friend, G. R., 1982. Mammal populations in exotic pine plantations and indigenous eucalypt forests in Gippsland, Victoria. *Aust. For.* **45**: 3-18.

Grant, P. R., 1968. Polyhedral territories of animals. *Amer. Nat.* **102**: 75-80.

Harris, L. D., 1984. The fragmented forest: island biogeography theory and the preservation of biotic diversity. Chicago Univ. Press, Chicago.

Henry, S. R., 1984. Social organization of the greater glider in Victoria. Pp. 221-8 *in* Possums and gliders ed by A. Smith and I. Hume. Aust. Mammal Soc., Sydney.

Henry, S. R. and Craig, S. A., 1984. Diet ranging behaviour and social organization of the yellow-bellied glider in Victoria. Pp. 331-41 *in* Possums and gliders ed by A. Smith and I. Hume. Aust. Mammal Soc., Sydney.

Kavanagh, R. P., 1984. Seasonal changes in the use of habitat by gliders and possums in southeastern New South Wales. Pp. 527-43 *in* Possums and gliders ed by A. Smith and I. Hume. Aust. Mammal Soc., Sydney.

Kavanagh, R. P. and Rohan-Jones, W. G., 1982. Calling behaviour of the yellow-bellied glider, *Petaurus australis* Shaw (Marsupialia: Petauridae). *Aust. Mammal.* **5**: 95-111.

Kavanagh, R. P., Shields, J. M., Recher, H. F. and Rohan-Jones, W. G., 1985. Bird populations of a logged and unlogged forest mosaic at Eden, New South Wales. Pp. 273-81 *in* Birds of eucalypt forests and woodlands: ecology, conservation, management ed by A. Keast, H. F. Recher, H. Ford and D. Saunders. Surrey Beatty, Sydney.

Kitchener, D. J., Chapman, A., Dell, J., Muir, B. G. and Palmer, M., 1980. Lizard assemblage and reserve size and structure in the Western Australian wheatbelt — some implications for conservation. *Biol. Conserv.* **22**: 127-63.

Loyn, R. H., 1980. Bird populations in a mixed eucalypt forest used for production of wood in Gippsland, Victoria. *Emu* **80**: 146-56.

Loyn, R. H., 1985. Strategies for conserving wildlife in commercially productive eucalypt forest. *Aust. For.* **48**: 95-101.

MacArthur, R. H. and Wilson, E. O., 1967. Island biogeography. Princeton Univ. Press, Princeton.

Mackowski, C. M., 1984. The ontogeny of hollows in blackbutt (*Eucalyptus pilularis*) and its relevance to the management of forests for possums, gliders and timber. Pp. 553-67 *in* Possums and gliders ed by A. Smith and I. Hume. Aust. Mammal Soc., Sydney.

McGinley, M. A. and Whitham, T. G., 1985. Central place foraging by beavers (*Castor canadensis*): a test of foraging predictions and the impact of selective feeding on the growth form of cottonwoods (*Populus fremontii*). *Oecologia* **66**: 558-62.

Ovington, J. D. and Thistlethwaite, R. J., 1976. The woodchip industry: environmental effects of cutting and regeneration practices. *Search* **7**: 383-92.

Pyke, G. H., Pulliam, H. R. and Charnov, E. L., 1977. Optimal foraging: a selective review of theory and tests. *Quart. Rev. Biol.* **52**: 137-54.

Recher, H. F., 1984. The 'conservation ethic' in practice. *Aust. Nat. Hist.* **21**: 152-4.

Recher, H. F., 1985. A diminishing resource: mature forest and its role in forest management. Pp. 28-33 *in* Wildlife management in the forests and forestry-controlled lands in the tropics and the southern hemisphere ed by J. Kikkawa. IUFRO, University of Queensland, Brisbane.

Recher, H. F. and Christensen, P. E., 1981. Fire and the evolution of the Australian biota. Pp. 135-62 *in* Ecological biogeography of Australia ed by A. Keast. Dr W. Junk, The Hague.

Recher, H. F., Rohan-Jones, W. and Smith, P., 1980. Effects of the Eden woodchip industry on terrestrial vertebrates with recommendations for management. Forestry Commission (NSW) Research Note No. 42.

Recher, H. F., Gowing, G., Kavanagh, R., Shields, J. and Rohan-Jones, W., 1983a. Birds, resources and time in a tablelands forest. *Proc. Ecol. Soc. Aust.* **12**: 101-23.

Recher, H. F., Milledge, D. and Clark, S. S., 1975. An assessment of potential impact of the woodchip industry on ecosystems and wildlife in southeastern Australia. Pp. 108-83 *in* A study of the environmental, economic and sociological consequences of the woodchip operations in Eden, New South Wales. W. D. Scott and Co., Sydney.

Recher, H. F., Milledge, D. R, Smith, P. and Rohan-Jones, W. G., 1983b. A transect method to count birds in eucalypt forest. *Corella* **7**: 49-54.

Recher, H. F., Allen, D. and Gowing, G., 1985a. The impact of wildfire on birds in an intensively logged forest. Pp. 283-90 *in* Birds of eucalypt forests and woodlands: ecology, conservation, management ed by A. Keast, H. F. Recher, H. Ford and D. Saunders. Surrey Beatty, Sydney.

Recher, H. F., Holmes, R. T., Schulz, M., Shields, J. and Kavanagh, R., 1985b. Foraging patterns of breeding birds in eucalypt forest and woodland of southeastern Australia. *Aust. J. Ecol.* **10**: 399-419.

Ryan, D., 1984. Keeping the options open. *Aust. Nat. Hist.* **21**: 155-9.

Saunders, D. A., 1980. Food and movements of the short-billed form of the White-tailed Black Cockatoo. *Aust. Wildl. Res.* **7**: 257-69.

Saunders, D. A., 1982. The breeding behaviour and biology of the short-billed form of the White-tailed Black Cockatoo *Calyptorhynchus funereus*. *Ibis* **124**: 422-55.

Saunders, D. A., Rowley, I. and Smith, G. T., 1985. The effect of clearing for agriculture on the distribution of cockatoos in the south-west of Western Australia. Pp. 309-21 *in* Birds of eucalypt forests and woodlands: ecology, conservation, management ed by A. Keast, H. F. Recher, H. Ford and D. Saunders. Surrey Beatty, Sydney.

Saunders, D. A., Smith, G. T. and Rowley, I., 1982. The availability and dimensions of tree hollows that provide nest sites for cockatoos (Psittaciformes) in Western Australia. *Aust. Wildl. Res.* **9**: 541-56.

Scott, W. D. and Co., 1975. A study of the envronmental, economic and sociological consequences of the woodchip operations in Eden, New South Wales. W. D. Scott and Co., Sydney.

Shields, J., 1985. Riparian forest corridors as refuge for arboreal marsupials in a pine plantation at Bombala, New South Wales. Dip. For. Thesis, Australian National University.

Shields, J., Kavanagh, R., Recher, H. F. and Webb, G., 1986. The effectiveness of buffer strips for wildlife management within pine plantations at Bondi S.F., New South Wales Forestry Commission New South Wales, Sydney.

Smith, P., 1985. Effects of intensive logging on birds in eucalypt forest near Bega, New South Wales. *Emu* **85**: 200-10.

Simberloff, D. S. and Abele, L. G., 1976. Island biogeography theory and conservation practice. *Science* **191**: 285-6.

Suckling, G. C., 1982. Value of reserved habitat for mammal conservation in plantations. *Aust. For.* **45**: 19-27.

Suckling, G. C., Backen, E., Heislers, A. and Newmann, F. F., 1976. The flora and fauna of radiata pine plantations in northeastern Victoria. Forests Commission of Victoria Bull. No. 24.

Tyndale-Biscoe, C. H. and Calaby, J. H., 1975. Eucalypt forests as a refuge for wildlife. *Aust. For.* **38**: 117-33.

Webb, G. A., 1980. A preliminary investigation of techniques for sampling herpetofaunal communities. Forestry Commission New South Wales unpubl. rep. no. 772, Pp. 1-14.

Webb, G. A., 1981. Geographical distribution of reptiles and amphibians in the southern Eden forestry region. Forestry Commission New South Wales unpubl. rep. no. 783, Pp. 1-115.

CHAPTER 15

Connectivity: An Australian Perspective

P. B. Bridgewater[1]

Connectivity is a developing area of study which attempts to identify the networks which form landscape structure. Most advances in connectivity have occurred in the northern hemisphere, where attention has been focussed on hedgerows as an important mechanism for wildlife to survive and travel in cultivated landscapes. Such structures are largely absent from Australia. Nonetheless a variety of landscape features act as links between vegetation remnants e.g. drainage lines, burn scars, fence-lines. Undisturbed landscapes also have elements of connectivity. At the local landscape level I term these natural corridors, as well as cultural artefacts (fences, hedgerows), 'ecolines'. Working at the continental scale connectivity is apparent between large landscape units. I use the term 'geoline' for these features. From the management view point, a clear understanding of the structure and function of ecolines is as important as understanding the nature of the remnants they connect.

INTRODUCTION

THAT remnants of native vegetation have a role to play in nature conservation is implicit in the title of this book. This chapter will examine the role of vegetation remnants in a landscape context, particularly in the light of recent work in the northern hemisphere on connectivity as an important landscape property.

To start, we should address the question, 'What is landscape, and how can it be analysed?' Risser *et al.* (1984) state, 'Understanding landscapes requires dealing with human impacts contributing to the landscape phenomenon, without attempting to draw the traditional distinction between basic and applied ecological science or ignoring the social sciences'.

In recent years the subject of landscape study has developed as landscape ecology, which involves an understanding of the effects of temporal change and spatial variety in maintaining and developing the landscape fabric — of which an important component is remnant patches from previous ecological eras. Flows of energy, material and biota are also key elements of landscape ecology. A knowledge of the ability of organisms to move between different elements of the landscape, particularly in highly patterned or textured landscapes, is necessary for effective conservation management.

Organism flux is an area of active research in North America and Europe, with particular emphasis on higher vertebrates and birds, and their use of hedgerows or fencerow corridors. e.g. Eldridge (1971), Fahrig *et al.* (1983), Baudry (1984) and McDonnell (1984).

Some populations of small vertebrates (e.g. *Peromyscus leucopus, Tamias* sp.) are able to survive only because they can move between woodlots on farmland through fencerow corridors which transect the intervening farmland (Fahrig *et al.* 1983). At the landscape scale such species survival depends on their ability to colonize habitat patches following frequent local extinctions in other patches. Although the intensive agricultural landscapes of eastern North America and western Europe offer the most dramatic examples of isolated patches linked by remnant lines or corridors, such networks apparently exist even in undisturbed vegetation. There is evidence that within apparently continuous forests some species require a mosaic of habitat patches. Cockburn (1978) documented this

[1]Bureau of Flora and Fauna, Canberra.

this for *Pseudomys shortridgei* in the Grampians, where species survival depends on the availability of a particular phase in the fire regeneration sequence of a plant community.

Historically, spatial and temporal variability in species attributes and population characteristics have been a problem for ecologists attempting to define communities regarded as homogeneous in space and time. Recognition that spatial and temporal dynamics are vital to the integrity and continuity of ecosystem processes may be essential to our understanding of landscape ecology, and the role of patches in landscape.

Many species of insects require resources in two or more landscape patches to complete their life cycles. For example, many herbivorous species feeding in agricultural systems move to semi-natural wooded areas for overwintering, and many predacious insects, such as vespid wasps, colonize hedgerows but forage in cultivated fields. Still others (e.g. Odonata) spend time in both aquatic and terrestrial communities. Thus, patchiness of natural and agricultural communities, phenology, as well as other environmental variation, have major influences on the flow and survival of insects across the landscape.

There is thus a common theme of spatial heterogeneity and energy/nutrient/population flux in landscape. Any landscape consists of patches, linked by corridors and a surrounding matrix. Forman and Godron (1984) note, 'Size, shape and the nature of the edge are particularly important patch characteristics. Corridor characteristics, such as width, connectivity, curvilinearity, narrows, breaks and nodes, control the important conduit and barrier functions of a corridor. Stream corridors play additional critical roles in controlling water and nutrient relations in a landscape. Networks are characterised by key differences in intersection types, reticulate structure and mesh size. The landscape matrix, as the most extensive and connected landscape element type, plays the predominant role in landscape dynamics'.

'Overall landscape structure ties some of these concept areas together, and incorporates the additional factors of scale and heterogeneity. The degree of contrast and the level of micro- and macro-heterogeneity, are key characteristics of landscape structure.'

'The origin and development of landscapes are controlled by both natural processes and human influences. Landscape modification by human influences leads to distinctive patterns of change in patches, corridors and matrix.'

These comments are primarily written from a northern hemisphere perspective, where landscapes have been increasingly modified by man, and consist of natural or semi-natural patches interspersed by agricultural land linked by semi-natural communities (hedgerows, fencerows). Although human impact on Australian landscapes has consisted of a recent severe alteration of land use, imposed on a landscape shaped largely by fire and Aboriginal inhabitants, the comments apply equally well.

METHOLOGY AND APPLICATION OF CONNECTIVITY

Connectivity, as a measurable attribute of landscape is defined by Merriam (1984) as 'a parameter of the interconnection of functionally related ecological elements of a landscape so that species can move among them'. It is essentially a measurement of the complexity of networks — and can be approached at qualitive and quantitive levels.

As an example of qualitative approach, Baudry (1984) noted a number of different possible connections between hedgerows (Fig. 1). The 'T' connection is most common in the agricultural landscape, 'O' is equivalent to destruction (or underconnected landscape). 'W' connections are those which facilitate most transfer. Using this approach, and building on work in both North America and Europe, his conclusions are that colonization by plants and animals in hedgerows is a step-by-step process, which proceeds most rapidly where the level of 'connectedness' is high (W>T>O).

Hedgerows, as barriers rather than corridors, create a grain-size in landscapes. Typically, reducing the length of hedgerows in a landscape (e.g. increasing the grain coarseness) initially results in an increase in biotic diversity of the grain patches, but ultimately creates a sharp decrease (Baudry 1984).

Quantitative approaches utilize graph theory to develop models of connectivity in the system (e.g. Rapoport 1982), and have been used by Rapoport and others in the development of ecogeographical theory. Such techniques have also been used to measure connectivity of networks at the landscape level. Merriam (1984) used the connectivity approach to simulate the population dynamics of *Peromyscus* populations. His analyses showed that increased connectivity;

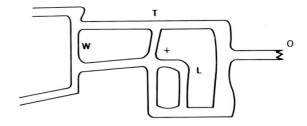

Fig. 1. Possible connection types between hedgerows (after Baudry 1984). For an explanation of the symbols please see text.

- lowered the frequency of local extinction,
- increased the *patch* population growth rate and population size, and
- decreased the coefficient of variation of patch population size.

Apart from that specific study, the analysis of network connectivity can help us understand the nature of landscape complexity and its likely impact on biota. Figure 2 shows two networks with nodes (where corridors begin, intersect or end) and corridors. Forman and Godron (1984) follow Taaffe and Gaulthier (1973) in defining a gamma index of connectivity — the ratio of links to the maximum number of links. The index ranges from 0 (no nodes linked) to 1 (every node linked to every other node).

Another parameter of value to undrstanding the ecological potential of networks is the measure of network circuitry (alpha index). With no circuitry (Fig. 2A) (alpha index = 0) biota have only 1 route through the network. With some circuitry (Fig. 2B) biota have a number of opportunities to move through the network, and thus avoid disturbances, predators, etc.

Forman and Godron (1984) citing several authors conclude that some species do move along corridors. They make the point, however, that to that date, no one had demonstrated the movement of species across a landscape using a network. They further caution that we need to know more of landscape ecological processes before uncritically applying connectivity indices.

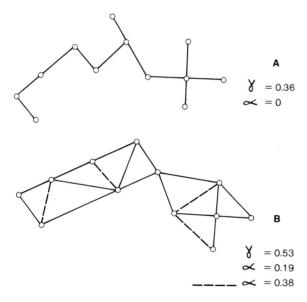

Fig. 2. Sample networks showing alpha and gamma indices of connectivity. A, shows a network with an alpha index of zero i.e. no circuitry. B, Shows a network with limited circuitry (solid lines), and one with greater circuitry (dashed and solid lines). For an explanation of the alpha and gamma indices please see text.

LANDSCAPE ECOLOGY IN AUSTRALIA

Although much of the work outlined above has developed in the northern hemisphere, there are many lessons from this work that can be applied to the Australian landscapes. The Russian School of landscape ecology takes modern landscapes as products of the co-creation of systems by man and nature, whilst combining the features of self-organization (nature) overlain by management (typical of human societies). Modern Australian landscapes are equally the product of the interactions between 'self-organizing' natural elements, 'management' prescriptions imposed by human societies, and more recently, the aggressive element of exotic species deliberately (and innocently) introduced by human agency.

That Aboriginal man imposed particular management regimes on Australian landscapes no longer needs questioning — only the length and frequency of intervention are open to dispute. Superimposed on the Aboriginal management/self-organization system, European colonization produced a violent shift to the technological management aspects of landscape, with many areas having most of their important self-organization aspects removed. Perhaps the most important example here is the Australian 'wheat-belt', development of which has had major, widespread impacts on the woodland/heathland mosaic landscapes from Queensland to Western Australia. Here is the ultimate in native/agriculturized landscape patchiness — with isolated 'native' vegetation remnants scattered throughout a highly mechanized agricultural landscape.

Unlike the northern hemisphere, there are no hedgerows hundreds of years old, connecting remnant patches of native vegetation. In parts of Tasmania native shrubs are able to grow in hedgerow situations with *Crataegus* and *Ulex* species — typical hedgerow species in northern Europe. Hedgerows of non-native shrubs species are known from southern Victoria and the longer settled tablelands of New South Wales. Vegetation remnants mostly exist as islands in a sea of agricultural land — or do they? In fact, roadsides, fencelines, drainage lines (linking both freshwater systems and isolated playas or salt lakes), non-utilizable land (sharp ridges, hollows, break-aways) and burn scars are common features of the landscape, and can act as continuous, or nearly continuous networks.

An important conclusion from work on hedgerows has been that it is not important that the hedgerows be the same vegetation type as the woodlands they connect, but that they are different from the surrounding croplands (Merriam 1984). Thus heterogeneous corridors, different from the agricultural matrix, can be very effective at supporting what Levins (1970) has called 'metapopulations' — a population of populations which go extinct at some points while colonizing at others.

Many management implications are raised by these concepts. In an Australian context it is perhaps useful to start at the continental level, and work to the local. At these two scales it is possible to distinguish lines of connectivity between patches. At the local level I propose the term 'ecoline' to cover any corridor, natural or otherwise. Such corridors are also recognizable, in a broad sense, at a continental level, and here I propose the term 'geoline'. Geolines are simply intended to indicate the trajectories along which the biota, energy, nutrients, etc. are able to flow, across and between major landscapes. Examples are the boundary line between *Atriplex-Maireana* shrubland and *Triodia* grassland in southern Australia, salt lake chains in central Australia, riverine edges along the major river systems, linking adjacent woodland systems. Along the eastern seaboard geolines run north-south, while throughout the rest of Australia the major trend is east-west.

There are few geolines within Australia, which emphasizes the overall homogeneity of landscape pattern at the continental scale.

At the local landscape scale, ecolines may be very abundant. Topographic and soil variation, with fire-related dynamic changes are the major determinants for patchiness in undisturbed Australian landscapes. In well-watered regions streamlines form classic corridors, although the high levels of moisture present in these environments may make them unsuitable as routeways for many species. In the semi-arid climatic regions typically dry watercourses tend towards high indices for circuitry — as well as providing numerous isolated 'islands' of dryland vegetation between their anastamosing branches.

Ecolines for areas of southeastern Victoria, and the Pilbara Region of Western Australia, are displayed in Figures 3-4. The lines represent corridors between landscape elements and offer a basis for understanding connectivity in relatively natural systems. The same exercise can be repeated easily in semi-natural and agricultural landscapes, as has been done for an area of coastal land, Westernport, Victoria. (Fig. 5). Ecolines are features of both disturbed and undisturbed vegetation, and should not be confused with ecotones or edges, which already have a defined meaning in ecological literature.

Features shown by the ecolines of south-east Gippsland and the Pilbara show similarities in having separated parallel ecolines with few cross connections and anastamosing networks, based on drainage lines. Both are strongly influenced by the basic land forms. The Pilbara ecolines, however, tend to have more nodes, with greater branching. In both examples, there are few cross-links between some ecolines, which would inhibit the movement of some biota freely across the matrix.

Cross-links between an agricultural matrix and 'natural' remnants from southern Victoria (Westernport) show ecolines with nodes more typical of these from the hedgerow/fencerow systems of the northern hemisphere. These ecolines are tracks and roadside verges with predominantly native vegetation, although a number of exotic species occur. A major conclusion is that in landscapes with little fragmentation ecolines have fewer branches and nodes than in fragmented landscapes (e.g. remnants in an agricultural matrix).

MANAGEMENT PRESCRIPTIONS

Implications for management are clear. Where 'natural' vegetation remnants exist without adequate linkages, such linkages or corridors need to be developed. Thus, Australia needs a homologue of the hedgerow system of Europe and North America, both to assist the movement of biota and create a high level of landscape heterogeneity. As well as management of agricultural land by re-introducing trees on farms through schemes such as the National Tree Programme (Anon. 1985), we need to develop and foster ecolines. Some can simply be developed by upgrading roadside reserves — others may need to be developed by reclaiming areas across agricultural land, recognizing the problem that such corridors may also offer shelter and transport lines for exotic species.

Taking the evidence offered earlier that corridors need to be different from the agricultural land, but not necessarily similar to the native remnants, many opportunities present themselves. In particular ruderal or disturbed systems with exotic species present or even dominant may nevertheless afford protection and movement opportunity for native biota. An example here is the endangered vertebrate *Perameles gunnii* which in Victoria uses the exotic shrub *Ulex europaeus* as shelter (P. Brown pers. comm.). The increasing tendency for suburban gardens to be planted with native Australian species has increased some avian populations, and is likely to favour populations of native invertebrates.

A referee questioned the viability of invertebrate populations in corridors with exotic species, as well as questioning the effect of exotics on the abiotic environment. At this time there are no studies which examine these problems in detail: more systematic research on such corridors is urgently needed to identify the problems and benefits of 'synthetic' ecolines.

Although not isolated remnants in the sense this volume is considering, rangeland vegetation, particularly that which has been overgrazed, needs careful management to ensure ecolines are free from intensive grazing — although total exclusion of stock may be neither practical nor desirable. Greater

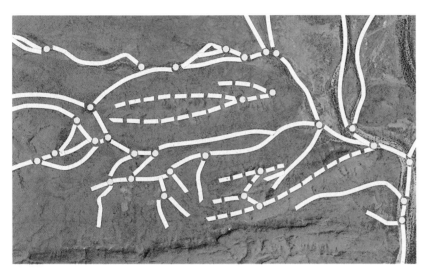

Fig. 3. Three ecolines in a region of coastal Gippsland, Victoria. Each symbol represents a different ecoline. Note the linearity of the ecolines with no obvious circuitry. Taken from Victorian State Aerial Survey photography: Mallacoota-Wingan Inlet Road Project, No. 35, 7/4/1969.

Fig. 4. Ecolines in the Pilbara region, Western Australia, near Turee Creek. Two different ecolines shown by solid and dashed lines, with nodes shown as open circles. Note limited circuitry in the ecolines represented as dashed line and potential transfer nodes between ecolines. (Taken from 1982 aerial photography obtained privately).

Fig. 5. Ecolines in the Westernport Bay region, Victoria. Two types of ecoline are represented, as dashed and solid lines, with nodes as dots. Ecolines represent linear residues of vegetation in the agricultural matrix, linking to natural ecolines in the remnants. Solid lines represent ecolines in dry systems, with dashed lines linking wetland remnants via linear residues of wetland vegetation. Note high level of circuitry in contrast to the ecolines of natural systems shown in Figures 3-4. Taken from aerial photography flown by Department of Crown Lands and Survey, Victoria; Westernport Project No. 1106, 29/7/73.

attention to these natural fluxlines will assist both native biota, and, by promoting better quality rangeland, the introduced stock.

A programme which is devoted to the identification and understanding of ecolines in natural and semi-natural systems is urgently needed — as is the identification of ecolines *within* remnant vegetation patches. External ecolines are likely to be most effective if connected to the internal network system of remnant vegetation stands.

Ecolines, although spatial in nature, also have dynamic aspects which need attention in any management prescription. Management of remnants may include firing at different times. A small number of remnants, linked by functional ecolines represents a lower risk management strategy than attempts to develop a fire regeneration mosaic in a larger remnant, unlinked to other vegetation patches.

Ultimately, the function and survival of a vegetation remnant and its contribution to nature conservation will depend on the linkages provided between it and other remnants, through the surrounding matrix. These linkages are as important for the promotion and maintenance of landscape diversity as they are for the movement of biota.

ACKNOWLEDGEMENTS

A number of colleagues contributed to discussions which resulted in the ideas expressed in this paper. The ideas in this paper are my own, and should not be taken as attributable to the Department of Arts, Heritage and Environment.

Mrs E. Oldfield deserves especial thanks for transforming my initial manuscript to the final product.

REFERENCES

Anon., 1985. Towards a greener Australia. Australian Government Publishing Service, Canberra.

Baudry, J., 1984. Effects of Landscape structure on biological communities: The case of hedgerow network landscapes. Pp. 55-66 *in* Methodology in Landscape Ecological research and planning. I ed by J. Brandt and P. Agger. Roskilde University Centre, Denmark.

Cockburn, A., 1978. The distribution of *Pseudomys shortridgei* (Muridae: Rodentia) and its relevance to that of other heathland *Pseudomys. Aust. Wildl. Res.* **5**: 213-9.

Eldridge, J., 1971. Some observations on the dispersal of small mammals in hedgerows. *J. Zool.* **165**: 530-4.

Fahrig, L., Lefkovitch L. and Merriam, G., 1983. Population stability in a patchy environment. Pp. 61-7 *in* Analysis of ecological systems: State-of-the-art in ecological modelling ed by W. K. Lauenroth, G. V. Skogerboe, and M. Flug. Elsevier, New York.

Forman, R. T. T. and Godron, M., 1984. Landscape ecology principles and landscape function. Pp. 4-15 *in* Methodology in Landscape Ecological Research and Planning. V ed by J. Brandt, and P. Agger. Roskilde University Centre, Denmark.

Levins, R., 1970. Extinction. Pp. 77-107 *in* Some mathematical questions in biology. Lectures on mathematics in the Life Sciences. 2. American Mathematical Society, Providence R.I.

McDonnell, M. J., 1984. Interactions between landscape elements: Dispersal of bird disseminated plants in post-agricultural landscapes. Pp. 47-58 *in* Methodology in Landscape Ecological Research and Planning. II ed by J. Brandt and P. Agger. Roskilde University Centre, Denmark.

Merriam, G., 1984. Connectivity: A fundamental ecological characteristic of landscape pattern. Pp. 5-16 *in* Methodology in Landscape Ecological Research and Planning. I ed by J. Brandt and P. Agger. Roskilde University Centre, Denmark.

Rapoport, E. H., 1982. Aerography. Pergamon Press, Sydney.

Risser, P. G., Karr, J. R. and Forman, R. T. T., 1984. Landscape Ecology: Directions and Approaches. Illinois Natural History Survey Special Publication, 2. Champaign, Illinois.

Taaffee, E. J. and Gaulthier, H. L., 1973. Geography of transportation. Prentice Hall, New Jersey.

CHAPTER 16

Monitoring Populations on Remnants of Native Vegetation

Paul R. Ehrlich and Dennis D. Murphy[1]

Most long-term studies of populations are now undertaken in remnants of what were once more extensive habitats. The unique characteristics of ecological systems on habitat islands must be considered in such studies. Those can include high edge to interior ratios, lacks of microhabitat heterogeneity, and other characteristics which can bias study results. Selective extinctions in habitat remnants may disrupt community interactions, effecting resource relationships, and the structures of surviving populations. Our long-term studies of the population biology of *Euphydryas editha* in remnant native grasslands in California are presented as examples of the importance and difficulty of studying populations in habitat remnants. We suggest that studies on habitat remnants should be designed to enhance understanding of population processes leading to extinction and to aid in the design and management of reserves.

INTRODUCTION

A⊤ present most temperate zone 'natural' habitats are actually relatively small remnants of what were once much larger ecosystems: bits of ravine in Spanish farmland too steep for ploughing where fragments of the original flora hang on: islands of second-growth woodland in northern Europe, Great Britain, or the eastern United States; tiny patches of disturbed prairie in vast crop lands in the central plains of North America; or bits of rainforest on extreme slopes in Queensland. Even where one might expect less habitat modification and fragmentation, such as in the montane regions of the western United States, the degree of disturbance and fragmentation can be, on close inspection, astounding. For example, even in the highest reaches of the isolated mountain ranges of the Great Basin region of the United States, the effects of overgrazing on the availability of foodplants critical to the survival of certain butterflies can be detected. The impact of livestock on riparian habitats in this region is often so severe that many of them could not even qualify as 'remnants' of natural vegetation. Clearing and overgrazing has had much the same effects in New Zealand and Australia, and in much of temperate South America. In short, scientists studying population dynamics in temperate zone terrestrial habitats are nearly *always* working with organisms that are persisting in fragments of modified habitat.

The situation is similar in many parts of the tropics, although large, unbroken stands of moist forest still remain in the Amazon Basin, the Congo Basin, and, to a lesser extent, in southeast Asia. But, if current trends continue, even these stands will be converted and fragmented in the next few decades. The decay of habitat quality in the temperate zone proceeds under impacts ranging from the conversion of native forests into plantations of exotic species (pines in Australia, Australian eucalypts in much of the rest of the world) through overgrazing to acid deposition. But the rate is comparatively slow and the countervailing forces of conservation are relatively well-organized, recognized and funded. There is at least reason to hope that substantial fragments of habitat can be kept in reserves, and that airborne threats to all habitats may be abated before their effects become irreversible. In the tropics there are no such hopeful signs. Population growth rates of three or four percent per year, extreme poverty combined with political repression and economic injustice, and exploitation of tropical resources by rich nations, all interact to make the destruction of tropical habitats appear a one-way freeway. Unless these trends can be reversed, the boundaries of most 'protected areas' in the tropics will become simply meaningless lines on maps within the next 50 to 100 years.

---

[1]Department of Biological Sciences, Stanford University, Stanford, CA 94305, USA.
Pages 201-10 *in* NATURE CONSERVATION: THE ROLE OF REMNANTS OF NATIVE VEGETATION ed by Denis A. Saunders, Graham W. Arnold, Andrew A. Burbidge and Angas J. M. Hopkins. Surrey Beatty and Sons Pty Limited in association with CSIRO and CALM, 1987.

In this light, why should we wish to monitor the populations found in remnants of native vegetation? There are compelling reasons beyond the obvious one that most accessible study systems are now restricted to remnants. For example, many species on remnants are rare, threatened or endangered, and an understanding of their dynamics is necessary in conservation efforts. Kirtland's warbler *Dendroica kirtlandii* in Michigan is a typical example. An invasion of the warblers' habitat by parasitic cowbirds was recognized as an important contributor to a decline in the warblers' population size. A programme of cowbird removal was instituted and the decline was halted (Walkinshaw 1983). Some species, although not threatened, may be keystones in remnant ecosystems or indicators of ecosystem condition. Still others may be of particular scientific or educational value, such as the long-studied populations of the moth *Panaxia dominula* in a 6 ha marsh at Cothill, Berkshire, England (Ford 1975) or of the checkerspot butterfly *Euphydryas editha* on Stanford University's Jasper Ridge Biological Reserve (Ehrlich *et al.* 1975; Ehrlich and Murphy 1981). And finally, some may be of recreational or aesthetic value. The beautiful golden gladiolus *Gladiolus aureus* from near Cape Town, South Africa, is being monitored as development threatens the remnants of habitat in which it survives. At last report, in 1980, only 45 remained: 22 full grown and the rest seedlings (Ehrlich and Ehrlich 1981).

Since studies these days nearly always will be undertaken in remnants, such studies should shed light on the process of extinction, and, perhaps, of ways to counteract it. Ecologists should select study organisms, sites, and methodologies to enhance long-term habitat protection. Remember virtually all threatened populations, species, and ecosystems require management; and while investigating questions of general interest to population biologists researchers should strive concurrently to seek information of use to managers.

To facilitate this, of course, selection of the organisms for study becomes a key consideration. Three groups of species seem especially desirable targets for study. The first are 'umbrella species', which are organisms whose protection confers protection on many other species by preserving valuable habitat. Umbrella species normally are large, and often have 'popular appeal' (in the latter respect, impalas *Aepyceros melampus* and wedge-tailed eagles *Aquila audax* make better umbrella species than hyenas *Hyaena* spp. or starlings *Sturnus vulgaris*). Note that the geographic scale of habitat remnants is a crucial factor here; small remnants often lack large species that can serve as umbrellas.

A second group of potential targets consists of 'ecologically significant species', such as the keystone mutualists or mobile links (Gilbert 1980), species whose roles in ecosystems are such that their loss would precipitate a cascade of extinctions or faunal collapse.

A third target group is 'indicator species' those which represent major elements in ecosystems, are closely tied ecologically to other organisms, are relatively well-known systematically and can be readily censused in the field (see e.g. Wilcox 1984). Butterflies are a particularly good example. they are herbivorous insects, the most speciose of all animals groups, and stenophagous, thus are tied closely to another group of organisms, their host plants. Butterflies are also the best-known taxonomically of all terrestrial invertebrate groups, are prominent, and are quite easily censused by simple transect techniques (e.g. Pollard 1977; Ehrlich and Wheye 1984). The diversity of butterflies in many parts of the world is often a good indicator of plant species diversity, while that of birds or mammals, more commonly the target of conservation efforts, are much less indicative of the nature of the plant communities in their ecosystems (Wilcox 1984).

TEMPERATE ZONES

Here we turn first to monitoring population dynamics in temperate zones where there appears to be some chance of doing long-term work to understand the causes of population size trends. We use as an example work done by our research group on an indicator species, a nymphaline butterfly that in the San Fransisco Bay area occurs on remnant patches of native grassland on serpentine soil; habitat fragments with a now geographically restricted flora and fauna. Since 1959 our research group has been monitoring three demographic units of the Bay checkerspot butterfly, *Euphydryas editha* (Nymphalidae: Nymphalinae) on the 500 ha Jasper Ridge Biological Preserve on the campus of Stanford University. These demographic units occur in three 'islands' of the serpentine-based grassland, covering a total of about 10 ha in an island of grassland surrounded by chaparral, but in any given year they probably occupy less than half of the total area of serpentine soil (Ehrlich 1965). The adult butterflies fly and lay eggs in March and April. Young larvae feed on two native California annual plants, *Plantago erecta* (Plantaginaceae) and *Orthocarpus densiflorus* (Scrophulariaceae) both of which bloom during the flight period and senesce soon afterwards. The larvae must reach a obligatory size to enter diapause and survive the California summer drought. The vast majority of the mortality in these populations occurs through larval starvation as a result of food plant senescence (Singer 1972; Singer and Ehrlich 1979).

The three demographic units of Jaspar Ridge have shown quite different dynamic histories through much of the study (Fig. 1). In the early years, the unit in area H experienced a population explosion, while the one in C, at the other end of the ridge, fluctuated. The population in area G, the smallest area, which is between the other two, declined to extinction. After being extinct for a year, the population in area G was re-established and persisted until the middle of the 1970s, at which point it again went extinct and the area has not been recolonized. The populations in area C and H have fluctuated more or less in unison subsequently, but have not fully recovered from the disastrous drought years of 1975-77.

The causes of the fluctuations in population size are a complex series of interactions between the availability of food plants, the timing of the life cycle of the butterflies, and macro- and micro-climates (Ehrlich *et al.* 1975; Singer and Ehrlich 1979). The key to population size seems to lie in the 'race' between pre-diapause larval development and the rapid senescence of their annual food plants. Survivorship can be high in places that are micro-climatically favourable or where *Orthocarpus* (which generally senesces later than the *Plantago*) is abundant. Also in years that are generally climatically favourable, survivorship can be high. But two or more unfavourable years may cause local extinction, especially in habitats in which the topography, and thus microclimates, are suboptimal for the macroclimatic conditions pertaining.

Our problems in sorting out the causes of observed trends in population size are increased by the realization that gradual habitat changes also play a considerable role. Until the early 1960s, the sites of the Jasper Ridge colony were grazed by cattle; then the cattle were removed. We suspect that the grazing helped to slow the infiltration of exotic weeds into and reduced the density of native bunchgrasses on serpentine soil. Thus in that way grazing was beneficial. But cattle also crush larvae and pupae and the nutrients in their droppings encourage the weeds. And at sites where cattle remained during the drought years, we believe that over-cropping of the annual food plants of *Euphydryas* helped push the butterfly populations to extinction.

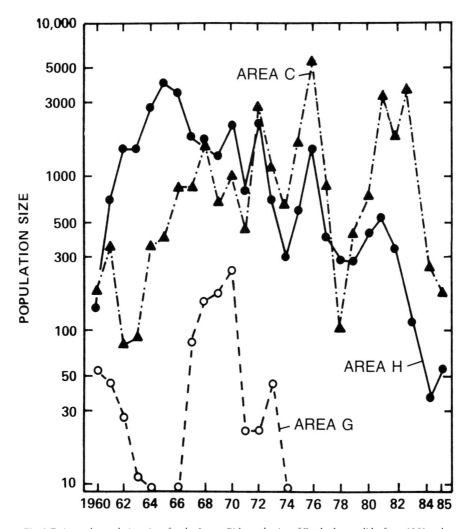

Fig. 1. Estimated population sizes for the Jasper Ridge colonies of *Euphydryas editha* from 1960 to the present.

During the drought it became apparent that the Jasper Ridge Biological Preserve is inadequate to protect *Euphydryas editha* populations indefinitely. Had there been a third consecutive year of drought, the populations in areas C and H almost certainly would have joined area G in extinction.

The history of the remnants of serpentine grassland that support *Euphydryas editha* in the San Fransisco Bay region is clouded. Due to its unique chemical constitution, serpentine soil has proven relatively resistant to invasion by the exotic herbaceous weeds (many introduced by the very first Spanish explorers) that have transformed the grassland flora of California. While both *Plantago* and *Orthocarpus* are most abundant on serpentine soils, neither is a serpentine endemic. It is probable that both of these plants were distributed extensively on non-serpentine soils three centuries ago, and that there were many more populations of *E. editha* at that time. The relatively low vagility of *E. editha* (Ehrlich 1961; Ehrlich *et al.* 1984; Murphy *et al.* 1986) suggests, however, that even when the native flora was in place, the species was more or less colonial. The hosts of the butterfly may only have been common in early successional stages of the primitive grassland and thus been patchily distributed before alien species became established in non-serpentine grassland. In any case, there is little doubt that for the recent past *E. editha* has been virtually restricted to serpentine-based grasslands in the San Fransisco area.

The dynamics of the entire suite of Bay Area populations for that period can be reconstructed as follows. Several large 'reservoir' populations have persisted in extensive areas of grassland on serpentine soil with varied topography. These populations have been buffered against extinction both by their large size and the varied topographies of their habitats. Topographic diversity provides diverse microclimatic situations some of which are favourable to the butterflies in every year, regardless of macroclimate. Two such large areas still exist (Fig. 2). One south of the Bay at Morgan Hill (MH Fig. 2), contains more than 1300 ha. A second, about 10 km north of Jasper Ridge (JR Fig. 2) at Edgewood Park (EW Fig. 2), is now about 25 ha, and is marginally capable of serving as a reservoir.

In addition to the large reservoir colonies, there are satellite colonies on relatively small patches of serpentine. The demographic units on Jasper Ridge are characteristic of these, and a handful of others still exist. The pattern for at least the last few centuries appears to have been continual population size fluctuations in these satellite colonies, with stochastic extinctions occurring perhaps as often as every fifty years or so depending, of course, on habitat size. On a similar time scale, extinct

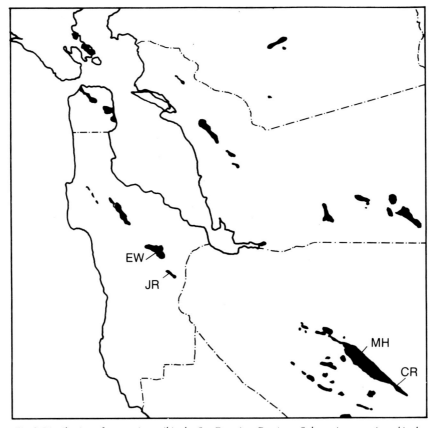

Fig. 2. Distribution of serpentine soil in the San Francisco Bay Area. Colony sites mentioned in the text are identified Edgewood Park (EW), Jasper Ridge (JR), Morgan Hill (MH), and Coyote Reservoir (CR).

population sites would be recolonized by migrants from nearby reservoir colonies. Thus the species has persisted in the Bay Area as a mosaic of colony sites, the populations in most of them periodically going extinct and then the sites being recolonized; only those populations in a very few sites have persisted as reservoirs. But over the past century or so, the trend has been one of disturbance, destruction and fragmentation of colony sites as development has gradually transformed the San Francisco Bay region.

The result of this loss of sepentine-based grassland habitat has almost certainly been an increase in the frequency of local population extinction and thus the average waiting time to recolonization following stochastic extinction events (Wilcox and Murphy 1985). Thus a current threat to destroy the Edgewood Park site (a remnant holding a major reservoir population that has already been fragmented) may portend the ultimate extermination of the Bay Checkerspot on the San Francisco Peninsula. The butterfly already appears to have been extirpated from all sites on the east side of San Francisco Bay, and the major reservoir colony south of the Bay (Morgan Hill) resides solely on private property. Therefore the prospects for the long-term survival of this organism are not bright. If the Jasper Ridge Preserve retains its integrity it is possible that the insect could survive there for perhaps hundreds of years, barring a very unfavourable weather sequence. But further infiltration of the serpentine by weeds (which appears to be occurring but has not yet been documented) could reduce that future to just a very few years.

TROPICAL ECOSYSTEMS

Our research group has had two experiences attempting to monitor butterfly populations in tropical situations. The first was in the mountains of northern Trinidad, near the Simla Research Station of the New York Zoological Society. We conducted mark-recapture experiments with the longwing butterfly, *Heliconius ethilla*, for several years and were able to demonstrate that the population size in these insects, which are very long-lived as adults, was much more constant than those of the temperate zone insects we had studied (Ehrlich and Gilbert 1973). We had started genetic investigations on those populations and were planning to monitor them over the long-term, when financial exigencies forced the new York Zoological Society to close the research station. Shortly thereafter landless peasants illegally burned part of the watershed forest in which the populations were located and the current status of *H. ethilla* there is unknown.

Subsequently our group began to monitor the dynamics of an entire butterfly mimicry complex in riverine forest in the Gombe Stream Reserve, Tanzania. That reserve was the site of research on chimpanzees conducted by Jane Goodall, who was at that time affiliated with the Human Biology Programme at Stanford. It was our hope that the presence of the chimpanzees would guarantee the integrity of the Reserve; unfortunately relatively few tropical forests have the large animal umbrella species that make efforts at habitat conservation so much more likely to succeed.

When Dr Goodall began her studies in the early 1960s, the forest stretched unbroken from the shores of Lake Tanganyika eastward for some 100 km. By the time we began our work in the early 1970s, 97 km had been cleared for cultivation and the forest stopped at the ridgeline boundary of the Reserve, 3 km inland from the lakeshore. Nonetheless, the boundaries seemed to be holding, and we thought the prospects were good for continuing our monitoring programme; we even constructed a laboratory building to use as a base of operations. But a couple of years after our operations commenced, rebels from Zaire attacked the Gombe Stream Station, attempted to kidnap Goodall, and actually kidnapped a small group of Stanford undergraduate students. After many months the students were recovered alive, but the incident effectively ended our work at Gombe Stream. Fortunately, the Reserve still exists and some chimpanzee work has continued by Tanzanian Nationals under Goodall's direction.

Of course not all third world situations are so grim. Costa Rica has placed some eight percent of its land surface in National Parks and plans to expand it to 10 percent (Myers 1980). Costa Rican reserves are under the very competent stewardship of Alvaro Ugalde, head of the Costa Rican National Park Service. But, that nation still faces difficult problems of park maintenance and boundary enforcement with inadequate financial resources. Brazil plans to place some 1.5 million km^2 in parks and reserves (about 20 percent of their territory), and has cooperatively run an elegant experiment on the biological costs of deforestation with the World Wildlife Fund (e.g. Lovejoy *et al.* 1983, 1984). And nations ranging from Kenya and Tanzania to India and the People's Republic of China have important, independently-run conservation programmes. But, in all these nations funds are short and human population pressures heavy.

Sadly, one of the places where one would not want to initiate a monitoring programme in lowland tropical rainforests would be in coastal Queensland. Most of those immensely rich, biologically unique forests are slated to be logged in the next few years. The tragedy is that this is not occurring in a desperately poor nation with little ecological expertise, but in a rich nation with what may be *per capita* the finest community of ecologists in the world.

All of this adds up to what might be called the first rule of monitoring *any* kind of population today. Do not count on being able to continue the work over the long-term, even though long-term studies are the most badly needed. One must start with the intention of carrying on for many generations of the organism under study, but be prepared for the worst.

What criteria should, then, be considered in the selection of a study site? One criterion, of course, is the size of remnant and other 'island biogeographic' considerations such as shape, degree of isolation, etc. Another is the habitat characteristics of the site. Is it, for example, uniform or heterogeneous and what is that likely to mean to the species or suite of species to be studied? Related to this is the successional stage of the site, since succession can rapidly and dramatically change habitat characteristics. Planning to study grassland birds over the long term in an old field would not be the best strategy. Often the decision will be dictated by the goals of the study. If one were interested in the effects of fire regimes on legume populations, then the remnants of jarrah *Eucalyptus marginata* forests would be ideal sites; however, they would be much less suitable for long-term investigations of mammal or bird populations, since the jarrah forests are being replaced by pines and other non-native plant species.

Evaluating the research potential of the site relative to its security is often a difficult challenge. A large site in the Namibian Desert might offer considerable safety from intrusion, yet not be the place to study highly coevolved mutualisms. In temperate zones, phenomena such as acid deposition and pesticide drift now threaten virtually all habitats whether legally protected or not. In the tropics, human population growth and forest clearing is going on so rapidly that sites that are both reasonably accessible for study and reasonably safe from intrusion are very difficult to locate. And, of course, the potential of stochastic natural extinction events eliminating significant portions of study populations and/or habitats is greatly enhanced in remnants, even without unwanted human interference. The best one usually can hope to do, however, is to locate a site that seems likely to remain relatively free of disturbance, and then trust to luck. But luck *will* be required.

WHY REMNANTS ARE DIFFERENT

Once a research site is located, the biologist must be cognizant of the characteristics of ecological systems on habitat remnants. Habitat remnants are insular ecosystems, more or less analogous to true islands. Since Darwin, biologists have used island ecosystems as unique windows to view ecological communities and evolutionary processes. Habitat remnants, however, offer few of the advantages of islands and a host of their disadvantages. Study systems on true islands are often comparatively simple and are often the result of long periods of coevolution isolated from more complex mainland community interactions. Island communities are thought by many to consist primarily of relatively unspecialized species and to be rather resistant to invasion (e.g. Lack 1976). Habitat remnants, in contrast, typically are by-products of relatively recent habitat fragmentation and disturbance, in which population and ecosytem processes may have been seriously disrupted. Hence, in habitat remnants, the integrity of the prospective study system often has been compromised to some degree and the entire system is frequently undergoing succession, simplification and invasion. Furthermore, populations on remnants are particularly sensitive to both extrinsic human-induced disturbance and to 'natural' stochastic events because they tend to exist as small demographic units.

Let us look in more detail at some of the characteristics of habitat remnants that can greatly influence the course of any dynamic investigations in them. Habitat remnants have proportionally higher edge to area ratios compared to intact habitats. The effects of edge on species diversity have been well-documented for a variety of taxa, particularly birds (e.g. Harris 1984). Recent studies monitoring tropical forest fragments show butterfly species numbers and diversity, and the proportional contributions of certain taxonomic groups (particularly grass-feeding species) to differ in patches with high versus low edge to interior ratios (Lovejoy *et al.* 1983, 1984).

Another way in which remnants differ from intact habitats is that decreases in area associated with habitat fragmentation often leave remnants lacking the microhabitat diversity necessary for population persistence or stability (Wilcox and Murphy 1985). The size of the fragments may not always have a simple relationship with the probability of extinction of populations being maintained within them. *Euphydryas editha* again provides an illustration. As noted above, the vagaries of climate in temperate grasslands make habitat containing a variety of slopes and exposures the most desirable. Generally, the largest habitat areas tend to be topographically the most diverse. But our studies of populations in areas ranging from less than one ha (Jasper Ridge area G) to more than a thousand ha (Morgan Hill) have indicated that extinction may be more closely related to lack of topographic heterogeneity than to small size. The colony site at Coyote Reservoir (CR Fig. 2), for instance, was large but consisted almost entirely of east-facing slope; and, while its population was among the largest known in 1971, it was extinct by 1976.

Populations residing in remnants lacking microhabitat diversity may differ from those in more diverse habitats by exhibiting 'atypical' phenologies. *Euphydryas editha* populations residing on habitat fragments of solely east-facing slopes, for instance, have delayed larval development times, emergence curves and peak-flight periods, shortened periods of larval host and adult nectar source availability. These conditions effect daily sex ratios (since females are larger and take longer to develop than males thus are differentially susceptible to the hazards of pre-adult existence) and residence times (adults tend to leave areas where nectar sources are no longer available) when compared to areas supporting a variety of microhabitats. Combinations of these factors cooperate to reduce the persistence of populations on habitat remnants.

There are great difficulties in drawing wide-ranging conclusions about the dynamics of populations or communities in habitat remnants. Extinctions of some species may lead to disrupted community interactions in habitat remnants. Where remnants are especially small and isolated, those species that do manage to persist may evolve atypical behaviours. Centuries of anthropogenic extinctions in central Europe have selected for populations that can survive on minute, disturbed habitat fragments on extreme slopes. Our investigations of *Euphydryas editha* on native grassland fragments suggest that studies on just one or a few such remnants would have greatly biased our view of its relationship with larval host plants and adult nectar sources. The major factors contributing to larval survival to diapause in *Euphydryas editha* populations on three habitat remnants within 10 km of one another, with both *Plantago erecta* and *Orthocarpus densiflora*, are distinct: at Jasper Ridge Area C, it appears to be from eggs laid on *Plantago* on gopher *Thomomys* spp.-tilled soil, at Jasper Ridge Area H it is from larvae transferring from senescing *Plantago* to *Orthocarpus*, at Edgewood B from eggs laid directly on *Orthocarpus* (the latter difference due to differing phases of the adult flight period with host plant availability [Singer 1972; Singer and Ehrlich 1979]). All three modes, on the other hand, contribute extensively to survival to diapause in the large, heterogenous Morgan Hill population.

Likewise, nectar source visitation of Jasper ridge differs between areas. The preferred nectar source in Area C, *Layia platyglossa*, is not present in Area H, which results in different visitation rates to nectar sources as well as locally differing effectiveness of the butterfly as a pollinator (Murphy 1984). The presence of preferred nectar sources may be a contributing factor to the more healthy recent status of the population in Area C and its quicker rebound from low population numbers following the drought. Certainly populations with the most complete array of preferred nectar sources have been the most robust in size during recent years.

Population structure, particularly intra- and inter-habitat dispersal, of organisms restricted to habitat remnants is greatly affected by the size of habitat remnants and their distribution. Our earliest studies on Jasper Ridge (Ehrlich 1961, 1965) showed *Euphydryas editha* to be an extremely sedentary butterfly; less than 2% of marked individuals were found subsequently in habitats less than half a km away. The Jasper Ridge colony site, however, was quite small and fragmented. With the discovery of the large habitat at Morgan Hill, we now have indications that *Euphydryas* butterflies move about freely in suitable habitat, and in large habitats this may mean home ranges of dozens of ha. Thus our recent studies force us to restate its behaviour as 'habitat specific' rather than sedentary *per se*.

These observations not only have compelled us to reconsider our 'remnant based' conclusions regarding movement in *Euphydryas editha*, but also to reconsider the value of mark-recapture studies themselves in providing information pertinent to conservation efforts (Murphy *et al.* 1986). Many 'endangered' or 'threatened' species are defined as such since they and their resources are restricted to tiny habitat remnants. The reasoning that 'sedentary' behaviour is associated with extinction proneness (e.g. Arnold 1983) thus is effectively circular. Indeed, 'habitat specific' behaviour may contribute to the survival of stenophagous species in highly fragmented habitat.

Studies on habitat remnants not only may provide a biased view of population structure and host use, but also of patterns of population regulation in the organisms under study. All present evidence points to survival of larvae to diapause in Jasper Ridge *Euphydryas editha* as the key factor determining population size. This is mediated by rainfall through its influence on plant senescence. Hence, population regulation is density-independent. Yet, as discussed above, we think that the isolation of *E. editha* and its host plants on porous, early drying serpentine soils is a rather recent condition. Prior to the introduction of Mediterranean grasses which have outcompeted most native Californian annuals (including *Plantago* and *Orthocarpus*) on non-serpentine soil, the hosts of *E. editha* were probably much more widespread on soil drying later in the season. Pre-diapause mortality may have been greatly reduced under such conditions. This suggests that population build-ups may have occurred, subjecting local colonies to severe competition among post-diapause larvae for host plants. Before populations became isolated on serpentine-based grassland remnants, generations may have been regulated intermittently in a density-dependent manner as has been observed in other 'ecotypes' of *E. editha* (Ehrlich *et al.* 1975; Murphy and White 1984), or parasitoids and disease may have played a larger role in regulation.

In remnants, especially those lacking full complements of microhabitats for a given species, populations may show an uncharacteristic lack of genetic diversity (as revealed by electrophoresis). Small populations on remnants are susceptible to a variety of deleterious effects due to inbreeding and/or genetic drift. Effective sizes (N_e) may be even smaller than simple census results suggest (e.g. Ehrlich *et al.* 1984). Furthermore, if Watt's (1977) results with variation in *PGI* along environmental gradients is representative of what happens in *E. editha*, the loss of habitat heterogeneity, particularly habitat 'extremes', could well translate into a loss of allozyme variation even if population sizes are not low. This would reduce the generality of conclusions about the population genetics of checkerspot butterflies based on studies of demographic units on remnants.

Finally, care must be taken to ensure that the population processes under investigation are not affected by the studies themselves. Small habitat remnants and the small populations they often contain may be particularly sensitive to the trampling, handling and other impacts concomitant to detailed studies, which may affect the results of such studies. For example, at the end of an extemely intensive mark-recapture study in Jasper Ridge Area H (Ehrlich *et al.* 1984; Murphy *et al.* 1986), there was highly visible trampling damage to grassland.

DISCUSSION

Beyond these considerations one must ask to what end monitoring is to be carried out. The most important goal, theoretically, is to help to achieve an understanding of population processes, both ecological and genetic. Many of the most important questions that motivated the work on *Euphydryas* more than a quarter-century ago remain unsolved. When it began, Ehrlich had hoped to be able to tie together the genetics and dynamics of the populations, gaining information on the magnitude and constancy of selection coefficients in nature, and seeing how the genetic attributes of the populations responded to changes in population size and *vice versa*.

Understanding the genetics of populations in remnant vegetation, potentially critical to designing management strategies, has proven especially difficult to achieve. In early years, the work was hampered by an inability to assay gene frequencies in the populations, thus causing our group to resort to studies of quantitative genetics (e.g. Ehrlich and Mason 1966). Then, subsequent to the work of Lewontin and Hubby (1968) and others in allozyme genetics, we thought the tools for answering the sorts of questions we wished to ask would be in hand. The neutrality controversy (e.g. Lewontin 1974; Kimura 1983) has largely dashed those hopes;

determining the interplay of ecological and evolutionary forces by evaluating allozyme frequencies would not be terribly productive if those frequencies largely reflect the operation of mutation, migration and drift, rather than selection. And determining which loci are under the influence of selection turns out to be exceedingly difficult. Recently, however, we have been able to document substantial differences between observed allele frequencies and computer-generated values that would be expected if drift alone were operating (e.g. Mueller *et al.* 1985).

An understanding of population processes is, however, not only an intellectually desirable goal, but one of enormous practical significance for *Homo sapiens*, a species that is attempting to suppress populations of its enemies and encourage populations of its resources. Therefore, careful observation over as long a term as possible should be carried out on a variety of organisms. Ideally *every* population biologist should be involved in the long-term monitoring of at least one population in a habitat remnant. Our experience shows both how much can be learned from long-term, intensive studies and also how difficult and time consuming it can be. The main constraints on developing an adequate, properly stratified sample of monitored populations is the number of trained population biologists able and willing to carry out such work, and the availability of funds to support it.

Many of the populations so monitored may go extinct before much of a long-term has past. But even such events, distressing as they may be to the researchers involved, can provide insight into the causes of population extinction. This is a process that is, in many ways, more important than species extinction. The loss of genetically distinct populations is occurring today at a rate far above that of the loss of entire species. Population extinctions reduce the availability of organisms to help with the delivery of ecosystem services (Ehrlich *et al.* 1977; Ehrlich and Mooney 1983). And, such extinctions reduce the chances that the species which are losing populations will have the genetic variability needed to evolve in the face of environmental change. Furthermore, population extinctions reduce the chances that geneticists will fully develop the economic potential of a species (Ehrlich and Ehrlich 1981).

Extinctions, however, can provide opportunities to use recolonization experiments as a tool for investigating population processes. Our group has attempted the recolonization of sites that suffered stochastic extinctions as one approach to resolving the neutrality controversy. Using material from surviving populations of the same ecotypes we have synthesized groups of colonists with gene frequencies at several loci far removed from what the frequencies were in the population that went extinct.

We have done this logistically difficult task at the sites of two population extinctions. One site was promptly destroyed by development. In the second, recolonization has apparently been successful, and we are in the process of analyzing the genetics of the first sample taken from the re-established colony. Should gene frequencies reconverge on the values that existed prior to extinction, we will have powerful evidence that, at least at those loci or loci closely linked to them, selection has been involved in maintaining those gene frequencies. If, on the other hand, the frequencies simply drift around the re-established values, it will indicate that selection was probably unimportant in maintaining the frequencies.

As indicated above, a major reason for monitoring populations on remnant vegetation should be to aid in the conservation of the organism being monitored. But exactly what kind of information is most valuable to those attempting to preserve a population? For example, there has been a long tradition of mark-recapture analyses of butterfly populations (e.g. Dowdeswell et al. 1940), and there has been a tendency for those interested in conserving butterfly populations to do detailed mark-recapture studies of the endangered insects. Such studies can provide information on population size and structure which may be very important in the conservation process, but our recent investigations suggest that they often do not get at the principle reasons for extinction proneness, and that perhaps other sorts of study should be given priority (Murphy et al. 1986). Few such studies are without impact on study organisms and their habitats, and care should be made to design projects to avoid 'studying populations to death'.

Finally, a word on monitoring the dynamics of multi-species populations (guilds or communities) on remnants. The difficulties here tend to be even more severe than for single species populations, but the rewards can be even greater. For example, long-term monitoring of *Anolis gingivinus* and *A. wattsi* populations on St Maartens in the Caribbean might permit a more definitive test of the theory of competitive coevolution (Roughgarden et al. 1983). But one wonders whether increasing overpopulation on already severely disturbed Caribbean islands would permit observation over a sufficiently long period, or satisfactory interpretation of data if they could be gathered. The same sorts of problems plague attempts to monitor community turnover rates in order to test the equilibrium theory of island biogeography.

It also seems unlikely that the equivalent of the carefully controlled long-term observations of community dynamics in the small Rothamsted grassland plots (Silvertown 1980) will be possible in the future. And the most thoroughly studied giant remnant, the Serengeti ecosystem (Sinclair and Norton-Griffiths 1979) seems likely to disappear before the effects of recent climatic changes and a rinderpest epidemic can be completely evaluated. Indeed, funding is so short in economically-depressed Tanzania that the required research probably will not be pursued even if burgeoning East African human populations do not totally invade and destroy the National Park in the next few decades.

In conclusion, let us say that all biologists concerned with the monitoring of populations must be alert that, in all likelihood, they are documenting the middle stages of the greatest biological calamity since the extinction episode at the Cretaceous-Tertiary boundary. And, in this case, the very species that is doing the monitoring is also responsible for the calamity. We suggest that every biologist should tithe to conservation, spending at least one-tenth of his or her time in working politically to avoid the loss of the remaining diversity on the planet. All biologists must work to educate those in power so that our few relatively intact ecosystems do not become habitat remnants.

ACKNOWLEDGEMENTS

We thank Anne H. Ehrlich and Bruce Wilcox for comments on this manuscript. Much of our group's work described here has been supported by grants from the National Science Foundation, the most recent of which have been DAR8022413 and DEB8206961, and by a grant from the Koret Foundation of San Francisco. This paper is a contribution from the Center for Conservation Biology.

REFERENCES

Arnold, R. A., 1983. Ecological studies of six endangered butterflies (Lepidoptera: Lycaenidae). *In* Island biogeography, patch dynamics and the design of preserves. University of California Publication in Entomology Vol. 99.

Dowdeswell, W. H., Fisher, R. A. and Ford, E. B., 1940. The quantitative study of populations in the Lepidoptera. *Ann. Eugen. Lond.* 10: 123-36.

Ehrlich, P. R., 1961. Intrinsic barriers to dispersal in a checkerspot butterfly. *Science* 134: 108-9.

Ehrlich, P. R., 1965. The population biology of the butterfly, *Euphydryas editha* II. The structure of the Jasper Ridge colony. *Evolution* 19: 327-36.

Ehrlich, P. R. and Ehrlich, A. H., 1981. Extinction: the causes and consequences of the disappearance of species. Random House, New York.

Ehrlich, P. R., Ehrlich, A. H. and Holdren, J. P., 1977. Ecoscience: Population, Resources, Environment. Freeman, San Francisco.

Ehrlich, P. R. and Gilbert, L. E., 1973. Population structure and dynamics of the tropical butterfly. *Heliconius ethilla. Biotropica* 5: 69-82.

Ehrlich, P. R., Launer, A. E. and Murphy, D. D., 1984. Can sex ratio be determined? The case of a population of checkerspot butterflies. *Am. Nat.* 124: 527-39.

Ehrlich, P. R. and Mason, L. G., 1966. The population biology of the butterfly *Euphydryas editha.* III. Selection and the phenetics of the Jasper Ridge Colony. *Evolution* 20: 165-73.

Ehrlich, P. R. and Mooney, H. A., 1983. Extinction substitution, and impairment of ecosystem services. *Bioscience* 33: 248-54.

Ehrlich, P. R. and Murphy, D. D., 1981. The population biology of checkerspot butterflies. *Euphydryas Biol. Zentralbl.* 100: 613-29.

Ehrlich, P. R. and Wheye, D., 1984. Some observations on spacial distribution in a montane population of *Euphydryas editha*. *J. Res. Lepid,* 23: 143-52.

Ehrlich, P. R., White, R. R., Singer, M. C., McKechnie, S. W. and Gilbert, L. E., 1975. Checkerspot butterflies: a historical perspective. *Science.* 188: 221-8.

Ford, E. B., 1975. Ecological Genetics. Chapman and Hall, London.

Gilbert, L. E., 1980. Food web organization and the conservation of neotropical diversity. Pp. 11-33 *in* Conservation Biology: An Evolutionary-ecological perspective. Sinauer Associates, Sunderland, Massachusetts.

Harris, L. D., 1984. The Fragmented Forest. University of Chicago Press, Chicago.

Kimura, M., 1983. The Neutral Theory of Molecular Evolution. Cambridge University Press, Cambridge.

Lack, D., 1976. Island Biology. University of California Press, Berkeley.

Lewontin, R. C., 1974. The Genetic Basis of Evolutionary Change. Columbia University Press, New York.

Lewontin, R. C. and Hubby, J. L., 1968. A molecular approach to the study of genic heterozygosity in natural populations. II. Amount of variation and degree of heterozygosity in natural populations of *Drosophila pseudoobscura*. *Genetics* 54: 595-609.

Lovejoy, T. E., Bierregard, R. O., Rankin, J. M. and Shubard, H. O. R., 1983. Ecological dynamics of forest fragments. Pp. 377-84 *in* Tropical Rain Forests. Blackwell, Oxford.

Lovejoy, T. E., Rankin, J. M., Bierregard, R. O., Brown, K. S., Emmons, L. H. and Van der Voort, M. H., 1984. Ecosystem decay of Amazon forest fragments. Pp. 296-325 *in* Extinctions. University of Chicago Press, Chicago.

Mueller, L. D., Wilcox, B. A., Ehrlich, P. R., Heckel, D. G. and Murphy, D. D., 1985. A direct assessment of the role of genetic drift in determining allele frequency variation in populations of *Euphydryas editha*. *Genetics* 110: 495-511.

Murphy, D. D., 1984. Butterflies and their nectar plants: the role of the checkerspot butterfly *Euphydryas editha* as a pollen vector. *Oikos* 43: 11-37.

Murphy, D. D., Menninger, M. S., Ehrlich, P. R. and Wilcox, B. A., 1986. Local population dynamics of adult butterflies and the conservation status of two closely related species. *Biol. Cons.* 37: 201-3.

Murphy, D. D. and White, R. R., 1984. Rainfall, resources and dispersal in southern populations of *Euphydryas editha*. (Lepidoptera: Nymphalidae). *Pan. Pac. Ent.* 60: 350-4.

Myers, N., 1980. Conservation of Moist Tropical Forests. National Academy of Sciences, Washington, D.C.

Pollard, E., 1977. A method for assessing change in the abundance of butterflies. *Biol. Cons.* 12: 115-32.

Roughgarden, J., Heckel, D. and Fuentes, E., 1983. Coevolutionary theory and the biogeography and community structure of *Anolis*. Pp. 371-410 *in* Lizard Ecology: Studies on a Model Organism ed by R. Huey, E. Pianka and T. Schoener. Harvard University Press, Cambridge, Massachusetts.

Sinclair, A. R. E. and Norton-Griffiths, N. (eds), 1979. Serengeti: Dynamics of an Ecosystem. University of Chicago Press, Chicago.

Singer, M. C., 1972. Complex components of habitat suitability within a butterfly colony. *Science* 176: 75-7.

Singer, M. C. and Ehrlich, P. R., 1979. Population dynamics of the checkerspot butterfly, *Euphydryas editha*. *Fortschr. Zool.* 25: 53-60.

Silvertown, J., 1980. The dynamics of a grassland ecosystem: botanical equilibrium in the park grass experiment. *J. Appl. Ecol.* 17: 491-504.

Walkinshaw, L. H., 1983. Kirtland's Warbler: The Natural History of an Endangered Species. Cranbrook Institute, Bloomfield Hills, Michigan.

Watt, W. B., 1977. Adaptation at specific loci. I. Natural selection on phosphoglucose isomerase of *Colias* butterflies: biochemical and population aspects. *Genetics* 87: 177-94.

Wilcox, B. A., 1984. In situ conservation of genetic resources: Determinants of minimum area requirements. Pp. 639-47 *in* National Parks, Conservation and Development ed by J. A. McNelley and K. R. Miller. Smithsonian Institution Press, Washington, D.C.

Wilcox, B. A. and Murphy, D. D., 1985. Conservation strategy: the effects of fragmentation on extinction. *Am. Nat.* 125: 879-87.

CHAPTER 17

The Response of a Small Insectivorous Bird to Fire in Heathlands

Ian Rowley[1] and Michael Brooker[1]

A series of minor fires appeared to reduce population density and productivity of the splendid fairy-wren *Malurus splendens*, especially in areas which were repeatedly burnt. Recovery was apparent three to four years after the last fire. A species of bird which shows high site tenacity and which depends mainly on the scrub layer can survive a hot fire and has shown no evidence of decreased survival during the nine months following this fire.

INTRODUCTION

SOME knowledge of the response of animal populations to different fire regimes is required, not only by conservation organizations who often find fire to be one of the most readily available management options, but also for various other authorities which use fuel reduction or prescribed burning to protect human life and property.

Although fire is an integral part of all heathlands (Specht 1981), the fire history of Australian heathlands is not well documented (Kikkawa *et al.* 1979). The role of fire and its effect on heath flora in the Jurien-Badgingarra region of south western Australia is reviewed by Bell *et al.* (1982) who found conflicts of interest in the preferred fire regimes required by the various land uses i.e. farming, apiculture and conservation.

Specht (1981) infers that the animals of heathlands are adapted to the consequences of frequent fire although Catling and Newsome (1981) could find few fire-specialists among the Australian vertebrate fauna. Studies on the effects of fire on the Australian avifauna are few, mostly opportunistic and short term according to Smith (1985, this volume), who believes that changes in the fire regime since the arrival of Europeans have contributed to the decline of at least three species of birds in southwestern Australia.

We here report on the effect of four fire regimes on a population of a 10 g sedentary passerine, the splendid fairy- wren *Malurus splendens*. This study has been opportunistic in as much as none of the fires were planned by us. However, as all the birds were colour-banded, the consequence of each fire can be examined at the population, territory and individual level.

METHODS

The Gooseberry Hill Study Area

The study area is steep, deeply dissected by gullies and is situated on the edge of the Darling Scarp (mean altitude 100 m a.s.l.) on the eastern outskirts of Perth, Western Australia. Perth has a typical Mediterranean climate with a hot dry summer and a wet winter during which, on average, 750 mm of rain falls between May and October. The vegetation is 1-2 m high heath dominated by species from the families Proteaceae and Myrtaceae with a patchy overstorey of trees especially marri *Eucalyptus calophylla* and wandoo *E. wandoo*.

The Bird

Malurus splendens belongs to the family Maluridae and has a wide distribution across southern Australia. A population on the Gooseberry Hill study area has been individually marked since 1973. Rowley (1981) found that the population occupied a saturated habitat and dispersed little. At the end of the 1984 breeding season, over 170 wrens occupied 32 territories totalling 120 ha in area (Fig. 1). Over 60% of the family groups contained one or more helpers which assisted with the feeding of nestlings and the care of fledglings. Helpers did not increase

---

[1] CSIRO, L.M.B. No. 4, Division of Wildlife and Rangelands Research, P.O., Midland, Western Australia 6056.
Pages 211-8 *in* NATURE CONSERVATION: THE ROLE OF REMNANTS OF NATIVE VEGETATION ed by Denis A. Saunders, Graham W. Arnold, Andrew A. Burbidge and Angas J. M. Hopkins.
Surrey Beatty and Sons Pty Limited in association with CSIRO and CALM, 1987.

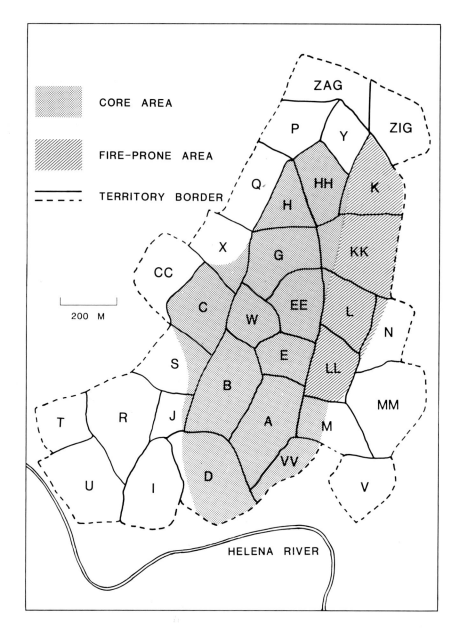

Fig. 1. Gooseberry Hill study area showing the 32 territories occupied in 1984, the Core Area and the Fire-prone Area.

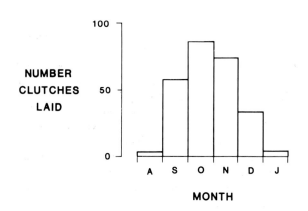

Fig. 2. Months in which *Malurus splendens* lay eggs on Gooseberry Hill (1980-1984 inclusive).

the productivity of a group but were in a position to ensure priority of opportunity to inherit their natal territory or to acquire a neighbouring territory. This species is a repeat-nester and its breeding season on Gooseberry Hill extends from August to January with a peak of egg-laying in October (Fig. 2).

The productivity of the wren population was measured at egg laying, hatching, fledging, independence (c.6 weeks old), the following July (c.8 months old) and the following September (c.10 months old). The survival of the young can be estimated from the number known to have fledged and the number seen in September as few disperse before then. Group size was the number of breeders plus adult helpers present in a territory during the nesting season (September to December). Adult

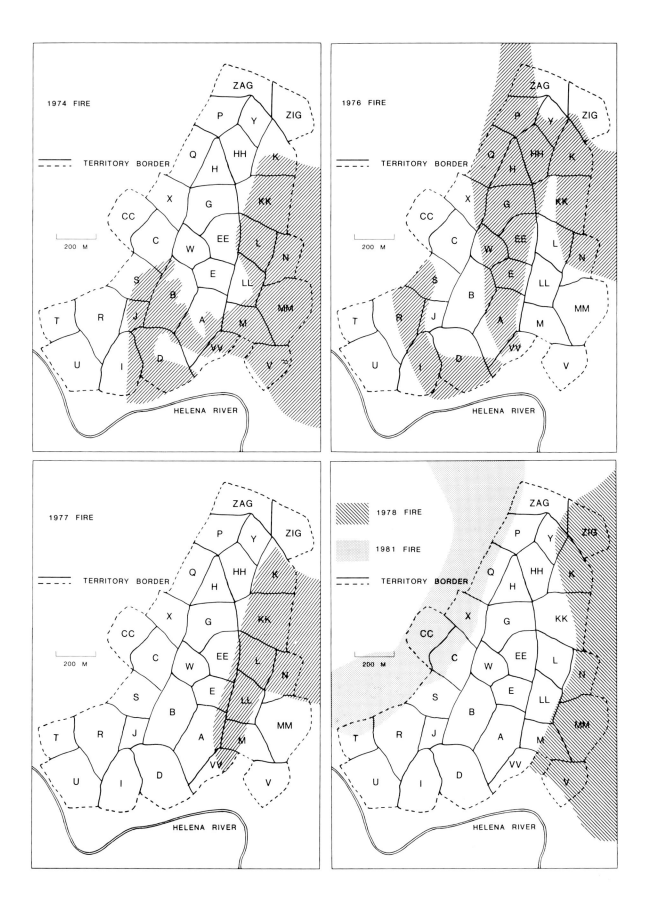

Fig. 3. The areas burnt by successive fires on Gooseberry Hill between 1973 and 1981.

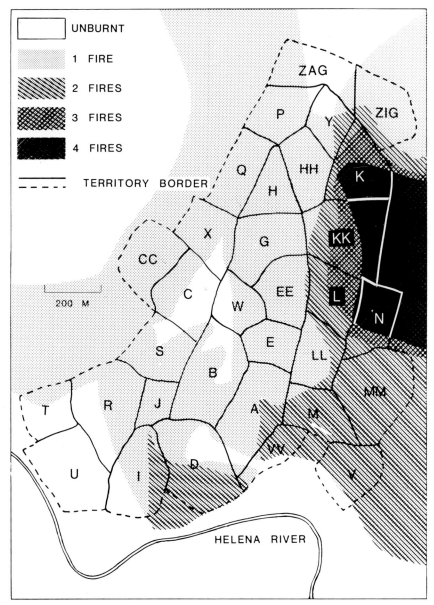

Fig. 4. The cumulative fire frequency on different parts of Gooseberry Hill over nine years (1973-1981).

survival was calculated from the number of adults alive in a nesting season which were present in the following September. This gives a minimum estimate of survival because some birds may disperse from the study area in the intervening period.

The Fires

Six fires have impinged to varying degrees on the wren study area since 1973. Several were lit on the northern or western margins and allowed to run their course on calm days. We subsequently mapped the extent of all fires and classified each as mild or hot depending on its effect on the vegetation.

The first five only burnt portions of the area (Fig. 3), but during this time (1974-81), most parts were burnt at least once and some were burnt four times (Fig. 4).

RESULTS

Minor Fires on Core Area

Wren data are available for the 45 ha Core Area (Fig. 1) over a 12-year period (1973-1984). During this time there were five minor fires (Fig. 3 and Table 1), two of which (1978 and 1981) are not considered further here because they only burnt a small proportion of the Core Area (Table 1). Two of the remaining fires (1974 and 1977) were mild and only that of 1976 was rated 'hot'. With all these fires, patches of vegetation were left unburnt in all territories except for territory F in 1974 and territory E in 1976.

Measures of wren density, productivity and survival are given in Table 1. Rainfall records are also included and show that two of the years (1976, 1977)

Table 1. Population data for *Malurus splendens* in the Core Area, Gooseberry Hill, 1973-85.

	1972	1973	1974	1975	1976	1977	1978	1979	1980	1981	1982	1983	1984
Date of Fire			16/12		14/12	12/12	27/12			11/11			30/1/85
Area Burnt (%)			40		80	30	10			5			100
Rainfall (mm)	695	1015	982	760	695	623	849	642	871	932	712	828	882
Number of Groups		8	8	7	7	6	6	7.5	7.5	8.5	12.5	12.5	11.5
No. of Adults per 100 ha		49	58$^+$	42$^+$	45$^+$	42	38	54	53	76	83	99	87
No. of Fledglings per 100 ha		73	40$^+$	38	48$^+$	51	71	24	58	90	73	69	60
Survival of Adults to next Breeding Season (%)		63.2$^\times$	43.5$^+$	61.5$^{\times\circ}$	72.2$^+$	80.0*	76.5	69.4	41.7	63.2	65.3	44.9	75.0
Survival of Fledgling to following September (%)		21.2	12.5$^+$	23.5	31.6$^+$	8.7	28.1	54.5	59.6	27.2	60.6	38.7	44.4
Groups parasitized by Cuckoos (%)		0	42.9$^+$	28.6	33.3$^+$	16.7	0	80.0	66.7	35.3	24.0	0	43.5

$^+$ No data Territory C. $^\times$ No data Territory H. $^\circ$ No data Territory D. * No data Territory B.

Table 2. Population data for *Malurus splendens* in the Fire-prone Area, Gooseberry Hill, 1975-85.

	1974	1975	1976	1977	1978	1979	1980	1981	1982	1983	1984
Date of Fire	16/12		14/12	12/12				11/11			30/1/85
Area Burnt (%)	75		70	100	40						100
Number of Groups	NR	2	2	2	1	2	1	3	5	4	4
Number of Adults per 100 ha	NR	24	29	29	12	41	24	47	106	100	100
Number of Fledglings per 100 ha	NR	NR	NR	0	47.1	52.9	41.2	41.2	94.1	105.9	88.2
Survival of Adults to next Breeding Season (%)	NR	NR	NR	NR	100	28.6	50.0	75.0	66.7	70.6	70.6
Survival of Fledglings to following September (%)	NR	NR	NR	0	50.0	33.3	42.9	87.5	56.2	44.4	26.7

in the fire-affected period were among the driest for the study.

In both territories (F in 1974: E in 1976) that were completely burnt most of the previous occupiers disappeared (four out of five) and since they were not found in neighbouring areas during intensive searching it must be presumed that those individuals perished either in the fire or shortly afterwards.

Fire would be expected to reduce both the availability and quality of suitable nest sites in the following year and this could make groups in burnt areas more vulnerable to parasitism by cuckoos. This was not the case in our population where the incidence of parasitism (Table 1) in the years following fire (1975, 1977, 1978) was low.

While a detailed examination of the effect of rainfall on the wren population is outside the scope of this chapter, the results shown in Table 1 do not suggest that rainfall rather than fire was responsible for the depression of density and productivity after minor fires. The survival of adults and juveniles was low in 1974, a year of above average rainfall but the same figures for 1976 were relatively high despite the low rainfall in that and the following year.

Minor Fires on Fire-prone Area

A 17 ha section on the western side of the study area (Fig. 1) experienced fires in four out of five years from 1974 to 1978 with some sections being burnt on four separate occasions (Fig. 3). Some data on wrens in this area are available from 1975 — the year after the first fire (Table 2). In the following six years it supported one or two groups of wrens. During the next four years (1981-1984), up to five groups have occupied this area.

Fire-free Period on Core and Fire-prone Areas

Apart from small portions of two territories (M and VV) in 1978 and one territory (C) in 1981 the Core Area was not burnt in the seven-year period 1978 to 1984 and there were no fires on the Fire-prone Area during a six-year period (1979-1984).

With the exception of adult survival (see Discussion) all measures of density, productivity and survival increased during the fire-free periods on both areas (Tables 1 and 2). The adult density doubled on the Core Area and quadrupled on the Fire-prone Area. In both cases, the major change occurred in the fourth year post-fire and appears to have been sustained thereafter. There is no evidence that rainfall variability (Table 1) contributed to this increase.

Major Fire

On 30 January 1985 the entire study area was burnt except for two territories (Zig and Zag, Fig. 1) on the southern edge. The cause of this fire is not known but it started on the river frontage north-east of the area at 1300 hours and was not controlled until it reached the southern side of the study area two hours later. The scorch height on marri reached 14 m and all leaves were burnt on some 10 m tall trees. The temperature of the fire was estimated from its effects on aluminium tags that had been used to label grid markers on the study area. Of forty-two tags examined, nine indicated a temperature of 600°C at 10 cm above ground level, and 10 a temperature in excess of 600°C. The rest were unaffected. A few small patches of heath located on the edge of roads, under electricity pylons and in rocky situations were not burnt.

It was not possible to study the behaviour of the birds during the fire which was endangering human life and property nearby, but on the day after the fire, few birds were sighted or heard. Those that we did identify were on or near their usual territory and one banded bird chased an unbanded intruder in a territory on the edge of the study area. The wrens were shy and secretive for about a week post-fire but after that their behaviour appeared normal. The only dead animals found after the fire were rabbits *Oryctolagus cuniculus* and bob-tail lizards *Trachydosaurus rugosus*.

We can but speculate on how the wrens found sufficient food immediately after the fire. Most of their foraging at this time was terrestrial and they adopted a strategy like that of the willy wagtail *Rhipidura leucophrys* in open habitats, namely active pursuit by running or flying after disturbed arthropods. There were the usual strong easterly winds during February and any insects blown onto the bare and blackened area would presumably have been more vulnerable than in unburnt heath.

Resprouting by many of the plants and epicormic growth on the trees followed quickly after the fire. Within a week there was fresh growth on *Xanthorrhoea preissii*, *Macrozamia reidlei* and *Mesomelaena tetragona*. After two weeks, grass such as *Themeda australis* was shooting. By four weeks, there was epicormic growth on marri and wandoo trees and *Grevillea bipinnatafida* was resprouting. Five weeks post-fire, marri, wandoo and a number of shrubs (*Allocasuarina humilis*, *Dryandra* spp., some *Hakea* spp., *Calothamnus quadrifidus*) had resprouted and there was epicormic response on Christmas trees *Nuytsia floribunda*. The recovery of the vegetation may have been assisted by an aseasonal 20 mm of rain on 10 February. However, the month following the fire was the hottest February recorded in Perth with an average maximum temperature of 35.1°C. (mean 29.9°C) (Bureau of Meteorology, Perth) so this rain probably had little beneficial effect.

The unburnt southern margin (Territories Zig and Zag) has been regularly searched and mist-netted since the major fire. We found no evidence of birds moving from burnt to unburnt areas.

Since the fire, the population has been censused in February, March, May, July and September. There was no evidence of mortality due to the actual fire (Table 3). The survival of adults and juveniles to September 1985 (Fig. 5) is not less than has been found in other years (Table 1). The survival figures of the 1984 young may have been enhanced by the fact that most were at least three months old at the time of the fire. This was because most of the nests in the latter half of that breeding season were parasitised by Horsfield's bronze-cuckoos *Chrysococcyx basalis*. Changes in family group compositions and territorial boundaries since the fire have been minor and similar to those that we would have expected without fire.

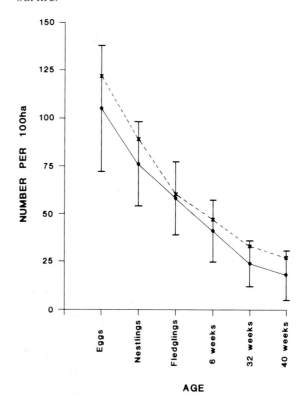

Fig. 5. The survival of the 1984 juvenile wrens of the Core Area throughout and following the severe fire of 30.1.85 (×---×) compared with the mean figures for the previous 11 years (●——●).

Table 3. Census of individual *Malurus splendens* on Gooseberry Hill study area before and after the intense fire* of 30.1.1985.

Age	Numbers of wrens present on Study Area						
	Jan	Fire	Feb	Mar	May	July	Sept
Adults	101	*	96	95	89	86	82
1984 Juveniles	60	*	60	51	48	42	32

Provenance of Colonizers

From 1975 to 1980 two territories on the Fire-prone Area were vacant for one breeding season (K in 1978; L in 1980). There are now (1985) four occupied territories in this area but only one of the progenitors of this sub-population was hatched here.

Twelve wrens recolonized territories which were vacated after being burnt. Six of these came from nearby territories. The other six were birds of unknown origin which colonized territories near the edge of the study area. This pattern of dispersal is in accordance with that of all dispersals (Rowley 1981).

DISCUSSION

Heathland vegetation is universally highly combustible and therefore very susceptible to wildfires. At the same time controlled burning can influence the density of the vegetation and probably the diversity of the constituent species.

Our study shows how a population of small insectivorous birds resident in heathland has responded to a succession of small fires, with a respite of six years followed by an intense widespread wildfire that consumed 95% of the vegetation on 30 January 1985.

The high adult survival of *M. splendens* even when burnt out (Tables 1 and 2) enables this species to maintain a population for some years despite low productivity. This occurred in the Core Area (Table 1). However, the frequent, overlapping fires on the Fire-prone Area reduced the population to such an extent that half this area was vacated in 1978 and 1980 (Table 2) and most subsequent recruitment came from outside.

The recovery period (four years) was the same for both areas. This period would probably be longer in heathlands which have poorer soils and/or lower rainfalls than our study area.

At this stage we can only examine the short term response of the wren population to the widespread intense wildfire of January 1985. The large population survived the fire remarkably well and the birds showed extreme site tenacity. We do not know how they survived the flames but they may have the same behaviour as the terrestrial sedentary marsupial *Bettongia penicillata* which Christensen (1977) found did not leave its home range during a hot forest fire but doubled back through the flames.

Birds behaved in a similar fashion during a burn in shrub woodland according to Recher and Christensen (1981). That wrens do not abandon a burnt area is a phenomenon that has been observed in other species of birds in Australia (e.g. Disney 1968; Cowley 1974; Wooller and Brooker 1980).

Survival rates for the Core Area of both adults (75.0%) and juveniles (44.4%) to the start of the next breeding season (September) were higher than the average for the preceding 11 years (61.9% and 33.3%) and for the previous three fire-free years (57.8% and 42.2%). Only two territories (J and Y) were not occupied in September 1985, and these were the smallest in the study area.

Why should four animals perish in smaller fires whilst very few died in the much more widespread burn? One answer is that where a few isolated individuals are rendered 'homeless' within a surrounding population that are less disadvantaged, they (the homeless) are stressed severely by being constantly rebuffed when they try and seek shelter in neighbouring territories. In contrast, where each bird is in the same situation, all having been burnt out, the universal drive to survive surmounts inter-group strife; territories, whilst still maintained, become more like home ranges.

Another explanation for the lower survival of adults and juveniles after frequent small fires when compared with the effect of the major 1985 fire could be the timing of the burns. Fires in mid-December (Table 1) during the nesting season (Fig. 2) would be more disruptive than a late-January fire because in December the adults would have to cope with the additional stresses of breeding and dependent juveniles. In particular, recent fledglings would be less mobile, inexperienced and very vulnerable both during and after the fire. There were few very young juveniles present at the time of the major fire in 1984 due to widespread parasitism by cuckoos that December (see Results). If the peak incidence of parasitism by cuckoos had been early in the nesting season, as it often was (Rowley 1981), rather than at the end as in 1984, the survival of the young might have been lower.

Our results demonstrate the deleterious effect of frequent burning for populations of small sedentary passerines (cf. Smith 1985). If parts of our study area (e.g. the Fire-prone Area) had been isolated remnants, the wren population on them could easily have become extinct after frequent fires and, without long distance dispersal which is rare in this species (Rowley 1981), the remnant would remain unoccupied.

The fire management required to maintain a *M. splendens* population at optimal density in this habitat must endeavour to protect the area from frequent burning, especially repeated firing of the same patch during the breeding season.

We have not been able (due to the 1985 fire) to estimate the minimum period between burns required to maintain an optimal density of wrens. This period appears to be greater than 12 years, as one territory (U) and parts of seven others (B, I, R, S, T and W) were not burnt for at least 12 years (Fig. 3) with no deleterious effect on the wrens. The preferred fire-frequency for conservation must also consider the needs of the other faunal elements and the flora. For example, Groves (1968) recommends a regular burning of heath every 10-15 years to maintain plant species diversity, while Bell et al. (1982) considered that a much longer time interval (25-50 years) may be required to ensure preservation of relic heath vegetation near Jurien, Western Australia.

Where native vegetation is surrounded by farms and suburbia (as ours is!), a compromise solution for managing the area is necessary. This could involve fuel reduction on the edges of the heath (e.g. effective fire-breaks), and the controlled burning of the heath at the longest possible time intervals (preferably not during the wren's nesting season, August-January).

ACKNOWLEDGEMENTS

The authors thank Craig Bradley, Graeme Chapman, Joe Leone, Bob and Laura Payne, and Eleanor Russell for field assistance, Graeme Smith and Gary Backhouse for commenting on the manuscript and Alcoa Australia for providing temperature information on the aluminium tags. Perry de Rebeira and Darren Baumgarten drew the illustrations and Claire Taplin typed the manuscript.

REFERENCES

Bell, D. T., Hopkins, A. J. M. and Pate, J. S., 1982. Fire in the kwongan. Pp. 178-204 *in* Kwongan: Plant Life of the Sandplain ed by J. S. Pate and J. S. Beard. University of Western Australia Press, Nedlands.

Catling, P. C. and Newsome, A. E., 1981. Responses of the Australian vertebrate fauna to fire: an evolutionary approach. Pp. 273-310 *in* Fire and the Australian Biota ed by A. M. Gill, R. H. Groves and I. R. Noble. Australian Academy of Science, Canberra.

Christensen, P., 1977. The biology of *Bettongia penicillata*, Gray 1837, and *Macropus eugenii* Desm. 1817, in relation to fire. Ph.D. Thesis. University of Western Australia, Nedlands.

Cowley, R. D., 1974. Effects of prescribed burning on birds of the mixed species forests of West Central Victoria. Pp. 58-65 *in* Third Fire Ecology Symposium. Forest Commission, Victoria.

Disney, H. J. de S., 1968. Bushfires and their effect on fauna and flora. *Aust. Nat. Hist.* 11: 87-9.

Groves, R. H., 1968. Nutrition of sclerophyll scrubs. *Proc. Ecol. Soc. Aust.* 3: 42-5.

Kikkawa, J., Ingram, G. J. and Dwyer, P. D., 1979. The vertebrate fauna of Australian heathland — an evolutionary perspective. Pp. 231-79 *in* Ecosystems of the World No. 9A Heathlands and Related Shrublands ed by R. L Specht. Elsevier, Amsterdam.

Recher, H. F. and Christensen, P. E., 1981. Fire and the evolution of the Australia biota. Pp. 137-62 *in* Ecological Biogeography of Australia ed by A. Keast. Junk, The Hague.

Rowley, I., 1981. The communal way of life in the Splendid Wren *Malurus splendens. Z. Tierpsychol.* 55: 228-67.

Smith, G. T., 1985. Fire, its effect on populations of the Noisy Scrub-bird *Atrichornis clamosus*, Western Bristle-bird *Dasyornis longirostris* and Western Whip-bird *Psophodes nigrogularis*. Pp. 95-102 *in* Fire ecology and Management in Western Australian Ecosystems ed by J. R. Ford. Environmental Studies Group, Western Australian Institute of Technology, Perth. Bull 13.

Specht, R. L., 1981. Responses to fires of heathlands and related scrublands. Pp. 395-415 *in* Fire and the Australian Biota ed by A. M. Gill, R. H. Groves and I. R. Noble. Australian Academy of Science, Canberra.

Wooller, R. D. and Brooker, K. S., 1980. The effects of controlled burning on some birds of the understorey in Karri forests. *Emu* 80: 165-6.

CHAPTER 18

Monitoring Population Densities of Western Grey Kangaroos in Remnants of Native Vegetation

G. W. Arnold[1] and R. A. Maller[2]

The densities of western grey kangaroos in remnants of native vegetation in the farming areas of the south-west of Western Australia were estimated by the Lincoln Index Method and from sightings along walking transects. There was a close correlation between these estimates and the weight of faecal dry matter deposited per unit area during the study.

Population estimates based on faecal accumulation were as accurate as the other two methods, and can be assessed about five times more quickly.

The faecal transect method was used to monitor fluctuations in kangaroo numbers in three remnants of wandoo woodland over four years. Faeces were collected from fixed transects of 100 m^2 at five-weekly intervals and data were pooled over seasons. With a density of one transect per 20 ha, differences of 20% in faecal accumulation per year were statistically significant. Differences of this order in faecal accumulation in different types of habitat can be detected where 5-6 transects are used in each habitat. The coefficient of variation between transects across different habitats in a site ranged between 200 and 600%, but within habitats between 22 and 75%.

Although weight per faecal pellet changes with season, faecal pellet numbers and faecal weight are correlated, and pellet numbers can be used to monitor population changes.

On the basis of our studies, we recommend that for long-term monitoring programmes, sampling of faecal output can be restricted to periods of the year when feed is of high quality and abundant.

INTRODUCTION

ESTIMATES of population numbers are basic to studies of population dynamics but prove difficult to obtain, particularly for large herbivores in woodland or heathland habitats. Aerial census is likely to miss a high proportion of the population and correction factors would be hard to obtain. Animals can be counted along transects on the ground, and a capture-recapture method, such as the Lincoln Index, can be used to estimate the population size if a sample of the population can be captured and marked. Both these direct methods are expensive in time. The Lincoln Index Method may be inaccurate since it assumes that there is no change in the composition of the population (through death, birth, dispersal or migration) during the period when the estimate is being made. The drive-census technique described by Raines (1982) can only be applied in habitats where the animals cannot move past the drivers without detection, i.e. on narrow islands of vegetation.

The most commonly used indirect method is the faecal accumulation method. This has been investigated and used extensively in studying the distribution of deer in different habitats (see review by Neff 1968), and has given reasonable estimates of populations (Dasmann and Taber 1955; White 1960; Bailey and Putman 1981). All deer studies have involved counting groups of pellets in fixed areas. Errors in population estimates result from variable defecation rates, counting faecal groups partially in the fixed area, lack of discrete groups, group overlap and variable pellet degradation. All these problems exist when applying this technique to kangaroos. However, the main one is that kangaroos rarely deposit faecal pellets in discrete groups.

[1] CSIRO, Division of Wildlife and Rangelands Research, L.M.B. No. 4, Midland, Western Australia 6056.
[2] CSIRO, Division of Mathematics and Statistics, Private Bag, Wembley, Western Australia 6014.

This chapter reports comparisons of two different estimates of populations of western grey kangaroos *Macropus fuliginosus* in wandoo *Eucalyptus wandoo* woodland in Western Australia (the walking transect method and capture-recapture estimation using the Lincoln Index) with faecal accumulation of dry weight per unit area. Data are presented showing the levels of precision with which comparisons can be made using this faecal accumulation method.

EXPERIMENTAL STUDIES

A. *Comparison of Methods*

Sites. Three adjacent remnants of wandoo woodland in the Baker's Hill area of Western Australia were used (Fig. 1). Site 1 was 305 ha, Site 2, 278 ha and Site 3, 314 ha. Each site was surrounded by farmland. A long term study of marked populations showed that very few individuals moved between the sites (Arnold *et al.* 1986).

Measurements

(a) *Capture-recapture method.* In Site 1 the density of kangaroos was estimated using the Lincoln Index Method in five successive years from 1981 to 1985. In each year animals were captured during summer and autumn and individually marked (See Arnold *et al.* 1986 for details). Spotlight counts and identification of animals were done at intervals during the capture period and soon after. The number of counts varied between two and five. Population densities were estimated and compared using an analysis of variance weighted for the variances of the numbers of animals observed:

$$W = C \frac{(m+1)^2(m+2)}{T^2(n+1)(n-m)}$$

where W is the weight and m is the number of marked kangaroos in a sample of size n from a population with a known number T of marked animals, and C is such that $\Sigma W^2 = 1$. The estimated density of kangaroos was calculated from

$$d = \frac{Tn}{m}$$

See Seber (1982, p59-61).

(b) *Walking transect method.* In all three sites, populations were estimated in 1984 using a walking transect method. Linear transects were marked out at each site using a compass and plastic tape. The total lengths of transects were 4.8 km (6 transects) in Site 1, 2.1 km (3 transects) in Site 2 and 3.3 km (3 transects) in Site 3. Two observers walked these transects at intervals over a four month period. All transects in a site were walked on every day that a set of observations were obtained. There were 25 sets of observations. The observers walked at about 2 km

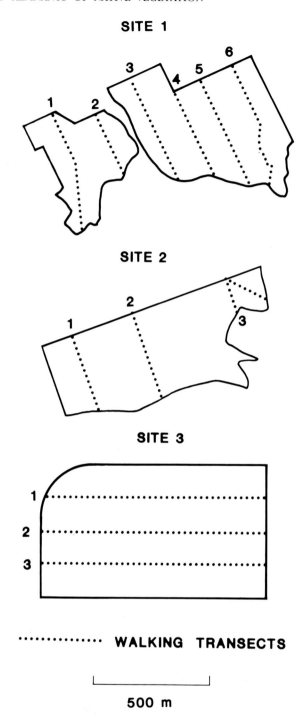

Fig. 1. Locations of study sites at Baker's Hill and of the walking transects.

per hour and recorded, each time kangaroos were seen, the numbers, their approximate distance from the observer and the angle made between the animals and the transect line.

Two density estimates were considered:

$$D_1 = \frac{(\Sigma n_i - 1).10^5}{2L\Sigma n_i r_i \sin\Theta_i} \quad \text{(Gates } et\ al.\ 1968\text{)}$$

$$D_2 = \frac{(\Sigma n_i/r_i).10^5}{2L} \quad \text{(Eberhardt 1978)}$$

where n = number of kangaroos at a distance r (in m) from the observer making an angle with the line of the transect, and the summations are over the number of groups of kangaroos observed on the transect at that date. When n = 0, D was taken as 0. Also L = length of transect (m), and the factor 10 converts the densities to kangaroos/ha. The data were split into values <1 and values >1 and these were analysed separately before combining (see Appendix). Method D_1 gave a higher estimate of population density than method D_2, and with a higher variability. A comparison for 1984 with the Lincoln Index gave a result closer to D_2 than to D_1 so we only discuss D_2 in the remainder of the paper.

(c) *Faecal transect method.* In all three sites, fixed belt transects were established for measuring faecal accumulation. Each transect was 100 m by 1 m. The faeces within the transects were collected every five weeks, oven-dried and weighed. Transects were allocated to vegetation communities in proportion to their total areas. There were 15 transects in each of Sites 1 and 2 and eight transects in Site 3. The transects covered 0.075% of each of Sites 1 and 2 and 0.025% of Site 3. Faecal accumulation was measured over the periods when the capture-recapture estimates and walking transects estimates were being made. The mean accumulation per 100 m² per five weeks was compared with the density estimates obtained by each of the two methods.

Results and Discussion

Time required to make estimates. The capture programme took about 110 person-hours a year to mark 60% of the population and each spotlight count took six person-hours. So 140 person-hours were required to make five estimates per year per site. Each year a new sample of the population had to be marked because we did not know whether all those marked in previous years were present.

The walking transects took four hours per site and 16 passes were needed to obtain an estimate with an error of ±15%, i.e. 64 hours per site were required.

The faecal transects took four hours for collection, drying and weighing. Allowing for three five-week collection periods to obtain an estimate, the time cost was 12 hours per estimate per site.

Comparison of estimates. Figure 2 shows the relationship for Site 1 between estimates by the Lincoln Index and faecal accumulation over the period March to May for each of five years. There is one highly discrepant point, which is probably aberrant. In that year, smaller numbers of animals, few of which were marked individuals, were seen than in all but the last year. It could be that the tagged animals were not evenly distributed over the site and so were under-represented in the spotlight counts. This would result in a high population estimate.

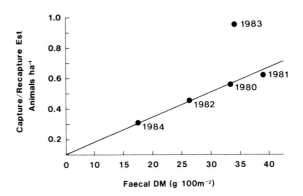

Fig. 2. Relationship between kangaroo densities estimated by the Lincoln Index Method and faecal accumulation over the capture-recapture periods (line hand fitted).

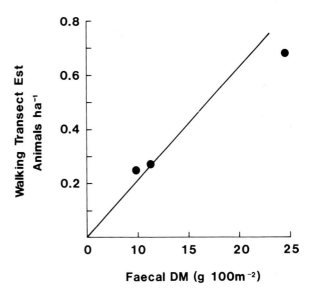

Fig. 3. Relationship between kangaroo densities estimated from walking transect analyses and faecal accumulation over the period when the transects were walked.

Figure 3 is based on limited data but suggests that estimates of population from walking transects on the three sites are related to faecal accumulation over the period August to October when the transects were being walked.

B. *Monitoring Changes in Population and in Differential Use of Vegetation Communities*

Methods

Study 1. Comparisons were made of faecal accumulation in different habitats. In the first, a comparison was made of accumulation in six locations (Table 1). All were woodland but their areas varied considerably. Transects were 1 m × 100 m and their number varied from 5-25 over the six locations. Faeces were collected every five weeks over periods from 11-30 months. These data were analysed to determine the size of the variation in faecal weights between transects, seasons and sites.

Table 1. Mean faecal accumulation rates in six locations.

Site	Number of transects	Mean faecal accumulation rate (g/100 m/ 5 weeks)
1. 135 ha Remnants of Wandoo	5	61.8[a]
2. 200 ha woodland at	15	32.1[b]
3. 86 ha Baker's Hill, W.A.	8	17.2[bc]
4. 2250 ha Tutanning Reserve near Pingelly, W.A.	25	13.7[c]
5. Wandoo woodland, Mt Observation, W.A.	15	28.4[b]
6. Jarrah forest, Forsythe's Mill, W.A.	10	11.9[c]

Values with superscript letter in common are not statistically different at the 5% level of probability.

Study 2. The aim of this study was to estimate the magnitude of differences in faecal accumulation rate which could be detected between sites, years and vegetation types when data were pooled over a season (10 or 15 weeks).

The faecal transects at Sites 1-3 were cleared every five weeks for three years. Two analyses were done. First, sites and seasons were compared by means of faecal accumulation over ten weeks. Secondly, a more detailed analysis was done on the data for Site 1, for which the 15 transects could be classified into the following vegetation categories: *Acacia-Casurina* low-open woodland; *Eucalyptus wandoo* open woodland; *E. accedens-Dryandra sessilis* open forest; and *E. wandoo-E. accedens* low open forest. For this analysis transect data for four years were pooled into seasons (December-February, March-May, June-August, September-November) by averaging data for either 10 or 15 weeks. Years, seasons and vegetation types were then compared by analysis of variance. Plots of residuals and variances showed that the assumptions of the analysis of variance were approximately satisfied.

Faecal transect data were also collected from a fourth site — the Tutanning Nature Reserve near Pingelly — over four years. There were 25 transects of 1 m × 100 m within the 2250 ha reserve. The transects were classified as being in one of five vegetation types: *Casuarina-Acacia,* woodland, shrub dominated formations (kwongan), *C. heugeliana-E. wandoo-E. accedens,* woodland, *E. wandoo-E. accedens* woodland with some *C. heugeliana,* and *E. wandoo-E. accedens* low open forest. An analysis of variance was done to compare years, seasons and vegetation types.

Results and Discussion

Study 1. Data from the five-weekly collection periods were highly skewed but a log-transform produced approximate normality. Analyses within the sites showed that faecal accumulation changed significantly with time in each site. But overall variability between transects was very high, much higher than the variability between times when compared with transect differences at each time. The coefficients of variation between transects ranged from 46 to 212% on the log-transformed data and from 199 to 582% on untransformed data. With this level of variability, differences between sites needed to be about 50% for statistical significance.

Study 2. In order to reduce the variability of the data, an analysis was done on 15 ten-week means. A transformation was not necessary for these. Faecal accumulation decreased significantly ($P<0.001$) from year 1 to years 2 and 3 (39.1 vs 22.5, 22.4 S.E. ± 2.5 g 100 m^2). Site 2 had significantly ($P<0.001$) more faecal dry matter (43.7 g/100 m^2) than Sites 1 and 3 (21.4, 22.3, S.E. ± 2.5). On these data a 25% difference in site and year means was significant at the 5% level of probability.

The analysis of vegetation type, year and season on Site 1 showed that a 19% difference in faecal accumulation within different vegetation types was significant. Accumulation was significantly higher ($P<0.001$) in *Casuarina-Acacia* woodland and *E. wandoo-D. sessilis* open forest than in the other two types of vegetation.

At the fourth site, faecal accumulation was lower than at Site 1 (17.9 g/100 m^2 vs 27.1 g/100 m^2). The transect density here was only 15% of that used at Site 1. A 27% difference in means over the four years was required for significance at the 5% level of probability. The variance in the two data sets was quite similar, which is surprising since it might be expected that quadrat density would need to increase with decrease in faecal density. We do not know whether the careful siting of transects to cover the variation in vegetation allowed precision to be maintained, or whether it is because the kangaroos are consistent in their distribution. (There were no season × vegetation or year × vegetation interactions at either site).

C. *Use of Weight or Number of Pellets*

Methods

For fourteen 5-week periods both the weight and numbers of pellets accumulated on Sites 1-3 were measured. After individually weighing the dried pellets from a quadrat, they were bulked and the total number of pellets counted. Analyses were done to see if weight per pellet varied between sites and if it was related to total faecal weight accumulation.

Results and Discussion

Data on pellet weight were pooled over 10-week periods (Fig. 4). There was a highly significant ($P<0.001$) change in average pellet weight with season, it being lower in winter and spring than in summer and autumn. Pellets from Site 2 were

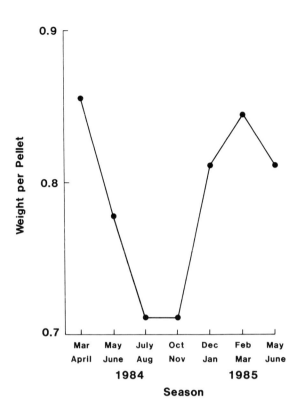

Fig. 4. Seasonal changes in mean weight per pellet at three sites.

heavier (P<0.01) than those from Sites 1 and 3. Differences between sites in mean pellet weight may reflect differences in population composition, i.e. Site 2 may have more larger animals. However, there are no data relating body size to pellet weight.

Within Sites 1 and 2 there was a weak, positive correlation between pellet weight and total faecal weights, but for Site 2 no correlation at all (Fig. 5). Assuming that within the period of study populations were not changing greatly, then differences in faecal accumulation reflect differences in faecal output per animal. Clearly, pellet weight does not increase as faecal output increases.

The overall relationship between pellet number and weight of pellets (Fig. 6) for the three sites has a correlation of 0.87 (P<0.001). There is a wider scatter amongst points from Site 2 than for Sites 1 and 2, probably reflecting the smaller number of transects used on this site. The tightness of the relationship suggest that it is possible to use number of pellets as an estimate of population size.

GENERAL DISCUSSION

Coulson and Raines (1985) used drive counts, line transect counts and faecal pellet counts to estimate the population of eastern grey kangaroos *Macropus giganteus* on a long narrow island. They found that the line transect counts, using Eberhardt's (1978) method, gave estimates similar to that of the drive count, whilst pellet counts (assuming 412 pellets were produced by each animal each day) gave similar estimates. However, they state that the errors were very large.

From our studies, faecal accumulation over 10-week periods is closely related to population size. The accumulation method can be used to monitor population size over long time periods; differences in populations of 20-25% can be detected with the type of sampling used.

Selection of the appropriate period of the year for measuring faecal accumulation is a problem. The faecal output per animal will depend on its food

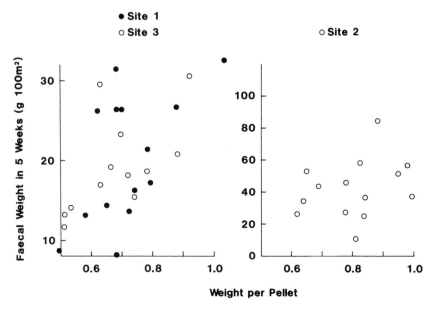

Fig. 5. Relationships between faecal weight per 100 m² collected over five weeks and weight per pellet for three locations.

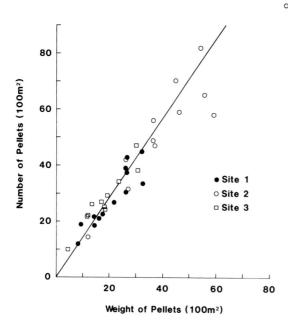

Fig. 6. Relationship between number of pellets collected per 100 m² over five weeks and the weight of those pellets.

intake and the digestibility of the food eaten. This is likely to differ from year to year in those seasons when food abundance and quality can vary widely, i.e. mid-summer to mid-winter in the woodland environments in Western Australia. Diets are most likely to be similar from year to year in late spring to early summer. Table 2 gives the relative faecal accumulation values for Sites 1-4 using these two periods. The two sets of figures tell different stories. For example, the relative values for Site 4 are higher in all years using September-February values, and the decline from year 1 to year 4 is greater in Site 2.

Table 2. Relative faecal accumulation at four sites at two periods of the year.

Site	Faeces collected March-August				Faeces collected September-February			
	1	2	3	4	1	2	3	4
Year								
1	56	100	–	29	68	100	–	40
2	48	86	–	22	62	82	62	38
3	39	77	41	23	27	65	43	31
4	20	73	23	18	45	63	30	42

Also, the relative values for years 3 and 4 in Sites 1 and 4 are quite different. Thus we favour the use of values taken when feed is most abundant to compare estimates from year to year and from site to site.

ACKNOWLEDGEMENTS

The diligent pellet picking of Olwen Brown, Dion Steven and Rex Elphick is gratefully acknowledged. John Weeldenburg and Dion Steven did the walking transect counts, and they, Rex Elphick and Gerald Clune did the spotlight counts. All these people and, over the years, many others, were involved in the animal capture programme. Without their assistance this work could not have been done.

REFERENCES

Arnold, G. W., Steven, D., Weeldenberg, J. and Brown, O. E., 1986. The use of alpha-chloralose for the repeated capture of western grey kangaroos *Macropus fuliginosus ocydromus*. *Aust. Wildl. Res.* 13: 527-33.

Bailey, R. E. and Putman, R. J., 1981. Estimation of fallow deer *(Dama dama)* populations from faecal accumulation. *J. Appl. Ecol.* 18: 697-702.

Coulson, G. M. and Raines, J. A., 1985. Methods for small-scale surveys of grey kangaroo populations. *Aust. Wildl. Res.* 12: 119-25.

Dasmann, R. F. and Taber, R. D., 1955. A comparison of four deer census methods. *Calif. Fish and Game.* 41: 225-8.

Eberhardt, L. L., 1978. Transect methods for population studies. *J. Wildl. Manage.* 42: 1-31.

Gates, C. E., Marshall, W. H. and Olson, D. P., 1968. Line transect methods of estimating grouse population densities. *Biometrics* 24: 135-45.

Neff, D. J., 1968. The pellet-group count technique for big game trend, census and distribution: a review. *J. Wildl. Manage.* 32: 597-614.

Raines, J. A., 1982. Density estimates of three herbivores on Rotomah Island, Gippsland. *Vic. Nat.* 99: 142-3.

Seber, G. A. F., 1982. The estimation of animal abundance. 2nd edition. Griffin and Co., London.

Smith, R. H., 1968. A comparison of several sizes of circular plots for estimating deer pellet-group density. *J. Wildl. Manage.* 32: 585-91.

White, K. L., 1960. Differential range use by mule deer in the spruce-fir zone. *Northwest. Sci.* 34: 118-26.

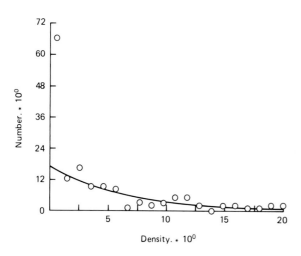

Fig. 7. Distribution of density of kangaroos estimated from walking transect data in 1984.

APPENDIX

Analysis of Walking Transect Data

The data consist of many zeroes (no animal sighted) along with occasional high values (groups of animals sighted). The distribution of density (kangaroos/ha) for all sites in 1984 is shown in Figure 7. (The distribution in 1979 was similar.) Ordinary analysis of variance to test for differences between Sites, Years and Dates of sampling is clearly not indicated. But Figure 7 shows that an exponential fit to values of density >1 looks reasonable. So the method adopted was to treat the data as a mixture of a gamma distribution with a point mass at 0.5, i.e., approximately at zero.

The gamma distribution is easily catered for using the statistical package GLIM. The estimated gamma parameter came out close to 1, showing that the initial guess of an exponential was reasonable. The 'zeroes' were analysed with a logistic analysis, i.e. essentially by contingency table methods. Having obtained estimates (with their standard errors) of mean densities and the proportion of zeroes, the overall densities with standard errors were calculated. This rather informal fitting of a mixture distribution could be replaced with a likelihood procedure, but this would be unlikely to increase the precision of the estimates significantly.

▷
PLATE 13
CHAPTER 17

Colour banded male *Malurus splendens*; he lived in Territory A for more than ten years. In this time he fathered 23 fledglings, five of which became breeders in the population. (G. S. Chapman).

◁
PLATE 14
CHAPTER 17

Gooseberry Hill study area 4 days after the fire of 30 January, 1985. The photograph was taken from Territory X, looking south towards the unburnt territories of Zig and Zag. (M. Brooker).

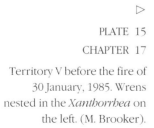
▷
PLATE 15
CHAPTER 17

Territory V before the fire of 30 January, 1985. Wrens nested in the *Xanthorrhea* on the left. (M. Brooker).

△ PLATE 16   CHAPTER 17
Territory V eleven days after the fire of 30 January, 1985, and the *Xanthorrhea* is already greening up. Compare this with Plate 15. (M. Brooker).

△ PLATE 17   CHAPTER 20
Experimental fire in a remnant area of *Banksia menzeissii/B. attenuata* woodland near Gingin, approximately 80 km north of Perth, Western Australia. Vegetation recovery following this fire in March 1985 is being compared with that in a similar area burned in September 1984. (R. Hobbs).

▽ PLATE 18   CHAPTER 20
Small-scale disturbance caused by gopher activity in an area of annual grassland on serpentine soil in N. California. Gopher mounds 30-50 cm in diameter have a pronounced effect on community patterns and dynamics. (R. Hobbs).

▽ PLATE 19   CHAPTER 22
Typical road reserve in the northern wheatbelt of Western Australia. Native vegetation is sometimes only one bush wide along these road verges. (D. Saunders).

CHAPTER 19

Three Decades of Habitat Change: Kooragang Island, New South Wales

R. T. Buckney[1]

Aerial photographs dating from 1954 show that mangroves have invaded saltmarsh areas of Kooragang Island in the Hunter estuary, New South Wales. In recent years the areas colonized by mangroves have shown a marked loss of vigour in virtually all cases and spectacular deterioration in some areas. Although artificial changes in the area may have contributed, the changes in the habitat characteristics of the island may also be interpreted in terms of published data which identify an increase in annual precipitation in the region from about 1946 and in terms of the recent drought in eastern Australia. The relevance of these changes to the fauna of the island are analysed. The importance of long-term climatic change in determining the variability of habitat characteristics in a proposed reserve centred on the island are considered. The general problem of the effect of climatic change on the viability of reserves of this type is discussed.

INTRODUCTION

KOORAGANG Island occupies 2560 ha in the mouth of the Hunter River near Newcastle, New South Wales. It formed from deposition of river-borne sediments and, together with Fullerton Cove on the northern side of the estuary, constitutes an extensive area of intertidal wetlands (mangrove and saltmarsh) noted for its bird fauna. Figure 1 shows the main details of the island and surrounding areas.

In 1960, Kooragang Island was zoned for heavy industrial use and reclamation and construction proceeded rapidly on its southern portion. By 1972 concern over pollution in the area resulted in an inquiry into the environmental problems; the report of the inquiry (Coffey 1973) recommended, among other things, that a part of Kooragang Island be preserved in its natural state.

In 1981 a study was commissioned to advise New South Wales Government departments on boundaries for a reserve (now known as the Kooragang Island Nature Reserve) centred on Kooragang Island and suitable for the protection of wetland birds, particularly migratory waders which are the subject of an agreement between Japan and Australia. The report of that study (Moss 1983) identified, but did not analyse, a long-term change in the remnants of natural vegetation of the area.

In this chapter a detailed analysis of the changes which have occurred on Kooragang Island since 1954 is attempted and their implications for the fauna are considered. The area considered in this analysis consists only of those parts of Kooragang Island not reclaimed or substantially modified artificially since 1954 (see Fig. 1).

Outhred and Buckney (1984) have described the vegetation of Kooragang Island; Clarke and van Gessel (1983) analysed habitat use by wetland birds of the area. Clarke and Miller (1983) recorded the insects and spiders found in brief collections made in a variety of habitats on the island.

MATERIALS AND METHODS

Vegetation maps for 1954, 1966, 1975 and 1979 were prepared from the available black and white aerial photographs:

Film	Photograph	Run No.	Date	Scale
NSW 252	5067, 8	3N	22.7.54	1:30000
NSW 1464	5189, 90	3N	14.8.66	1:40600
NSW 2314	91	6	27.5.75	1:42250
NSW 2830	100, 102	7	29.12.79	1:17000
NSW 2830	117	8	29.12.79	1:17000

(New South Wales Department of Lands, Sydney).

---

[1]School of Biological and Biomedical Sciences, New South Wales Institute of Technology, P.O. Box 123, Broadway, New South Wales 2007.
Pages 227-32 in NATURE CONSERVATION: THE ROLE OF REMNANTS OF NATIVE VEGETATION ed by Denis A. Saunders, Graham W. Arnold, Andrew A. Burbidge and Angas J. M. Hopkins. Surrey Beatty and Sons Pty Limited in association with CSIRO and CALM, 1987.

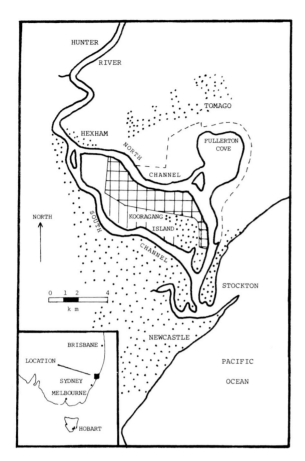

Fig. 1. Diagram showing the location of the study area (cross-hatched), the main areas of urban and industrial development (stipple) and areas of Kooragang Island to be filled and developed (vertical lines). The study area and that enclosed by the broken line represents the largest form of the reserve proposed for this area.

Maps were prepared on clear plastic overlays to the photographs, transferred to paper and photographically reduced or enlarged to the same scale (1:32000).

In preparing the vegetation maps, no attempt was made to subdivide the vegetation into floristic categories as done by Outhred and Buckney (1984). In this chapter only the following habitat categories have been recognized: vigorous mangroves (M1), mangroves lacking vigour (M2), saltmarsh (S), pasture (P), brackish swamps (SW), tidal flats (F) and bare ground (B). The map presented here for 1982 is the same as that published by Outhred and Buckney (1984), which had been based on the 1979 aerial photographs and modified on the basis of ground inspection, but without the floristic detail (as explained above). It should be noted that the data for 1982 presented below are thus based on slightly different source material from those of earlier years; it is felt, however, that the areas of habitats reported for 1982 would not be significantly different if based on aerial photographs.

Information from field studies in 1983 and 1984 have been used to assess the condition of vegetation on some areas of Kooragang Island since 1982. These studies utilized the same locations as used previously (see Moss 1983), but plant abundance was determined by a modified method, so only a qualitative comparison between the data sets is possible.

Areas of habitat types were determined from the maps using an overlay of random dots (density 10 per square centimetre); all areas reported are the mean of five measurements.

DESCRIPTION OF HABITAT TYPES

The floristic and structural details of the habitat types have been presented elsewhere (Moss 1983; Outhred and Buckney 1984); the following is a brief summary of their characteristics. Soil salinity data are taken from Moss (1983).

Vigorous Mangroves (Type M1)

Mangrove forests or woodlands with a dense canopy formed by *Avicennia marina* up to 12 m high. Along tidal channels a shrub layer of *Aegiceras corniculatum* may be present. Soil water salinity is near that of sea water (35 g l^{-1}).

Mangroves Lacking Vigour (Type M2)

These areas are dominated by the mangrove *Avicennia marina* up to 12 m high, but plants show stress symptoms such as defoliation, canopy dieback, stunting, peeling bark and epicormic growth. Numerous dead trees occur and seedlings are often absent. Occasional specimens of the saltmarsh species *Sarcocornia quinqueflora*, *Suaeda australis*, *Sporobolus virginicus* and *Triglochin striata* occur. Outhred and Buckney (1984) distinguish two sub-types: a well-drained type and one which appears to be permanently waterlogged. Soil water salinity is always high (in the range 50-70 g l^{-1}).

Saltmarsh (Type S)

A floristically varied group of low herblands, grasslands and rushlands (Outhred and Buckney [1984] recognised several distinct types). *Sarcocornia quinqueflora* occurs throughout most areas, often without other species but may be accompanied by *Sporobolus virginicus*, *Suaeda australis* and *Triglochin striata* (any of which may dominate the community locally), or replaced at slightly higher elevations by *Juncus kraussii* and *Sporobolus virginicus*. Stunted mangroves may be present sparsely in some areas near mangrove stands. Soil water salinity is as high as 70 g l^{-1} in areas of pure *Sarcocornia*, but is usually lower (20-35 g l^{-1}) in other areas.

Pastures (Type P)

These are predominantly grasslands containing many species of grasses and other herbs (many introduced). Saltmarsh species may be present where tidal water sometimes inundates these areas.

Swamp (Type SW)

Sedgelands with more or less permanent water of salinity about 17 g l^{-1}. Sometimes, the rush *Juncus kraussii* dominates the community. Aquatic herbs are very common.

Tidal Flats (Type F)

Areas of sand or mud subjected to diurnal tidal inundation and virtually free of tracheophytes. Mangroves fringe some such areas.

Bare Ground (Type B)

Roads, tracks, embankments or spoil dumps devoid of vegetation.

RESULTS AND DISCUSSION

Vegetation maps of the study area are shown in Figure 2.

Areas of the various habitat types for each map are summarised in Table 1.

The regression equation:

S = 714 − 0.956 M1 − 0.977 M2,

relating area of saltmarsh (S) to those of vigorous mangroves (M1) and mangroves lacking vigour (M2), accounts for 92.1% of the variability in the area of saltmarsh. This complies with the observation that saltmarsh was invaded by mangroves and that the mangroves in these areas later deteriorated.

Aerial photographs subsequent to 1979 are not available, but field work conducted between 1981 and 1984 has shown only relatively small changes in the area of the vegetation types since 1979, when compared to changes which occurred before that time. In 1984 it was apparent that many of the trees in the areas showing loss of vigour, including some

Table 1. Areas (ha) of the habitat types, 1954-82. See text for description of types. All areas are the mean of five determinations.

Year	Habitat Type						
	M1	M2	S	P	F	SW	B
1954	251.6	12.9	452.4	109.4	12.5	10.1	8.6
1966	382.1	3.5	370.4	123.1	41.8	19.5	7.4
1975	405.7	33.4	278.1	107.3	55.3	20.7	3.2
1979	298.8	106.1	309.7	109.2	46.3	23.1	21.3
1982	245.9	144.8	352.7	103.0	25.3	22.4	30.3

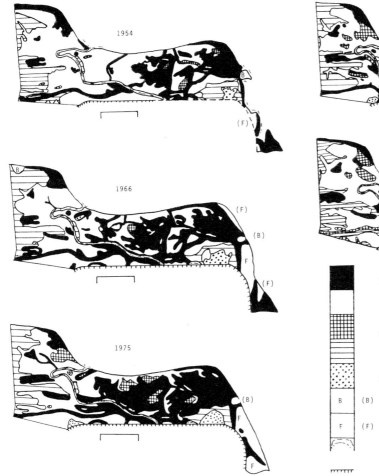
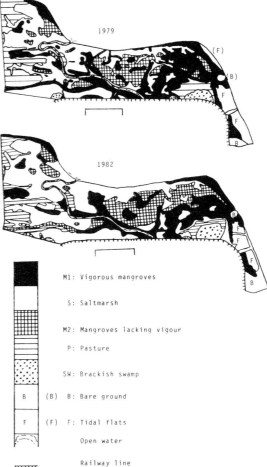

Fig. 2. Maps of the study area showing the change of distribution of the habitat types, 1954 to 1982. Scale lines represent one kilometre and are aligned approximately east-west.

identified in 1982 as 'stumps' by Outhred and Buckney (1984), were showing signs, such as the appearance of epicormic shoots, that the areas designated M2 were undergoing a revival.

The maps (Fig. 2), the area data (Table 1) and subsequent inspections show that the major changes in the vegetation of the study area have been:

1. a considerable increase in the area of mangroves since 1954 and a commensurate loss of saltmarsh;

2. a marked decrease in the area of vigorous mangroves since 1975 (probably commencing sometime between 1966 and 1975), with an equivalent increase in the area of dead or moribund mangroves in places previously occupied by saltmarsh; and

3. a return of some vigour to the dead areas of mangroves in 1984, the extent and permanence of which is yet to be determined.

Changes in the areas of other habitat components during the period studied are comparatively slight, but include variations in the areas identified as pasture (P), the appearance of some small areas of brackish swamp (SW) near the railway embankment and an increase in the area of bare ground (B).

The degree of floristic similarity between the habitat types is summarized in Table 2, which shows the percentage of species in common between them. These data show that the habitat types are very dissimilar; equivalent comparisons within the types yield values of approximately 60%. Floristically, the most similar habitat types are the mangrove associations which have only very few species present.

Causes for the Changes

Two hypotheses may be offered to explain the observed changes: that the vegetation has been extensively affected by man-made changes to the island and to the Hunter River catchment and estuary; or that the vegetation transformations reflect a long-term climatic change.

A number of artificial changes have been made to the study area and surroundings over many decades. Coffey (1973) provides a useful summary of these.

Some of those changes are apparent on the vegetation maps (Fig. 2) and include the construction of a railway line along the southern edge of the study area and the realignment of the eastern shoreline of the island to alleviate flooding problems. To the south of the study area, reclamation of mangrove, saltmarsh and channels has been extensive and continues at the time of writing. It is likely that these changes have had subtle effects on the hydrology and ecology of the study area in addition to those which are obvious in the immediate vicinity of the alterations. A detailed description of these changes is provided by Moss (1983). Close inspection of the aerial photographs failed, however, to identify extensive channel development or deterioration or marked vegetation changes which could be ascribed to these influences.

Other changes which could have affected the hydrology of the island include dredging of the South Channel of the Hunter River, land clearing and the construction of dams in the catchment. The last activity has significantly reduced the size of floods in the Hunter River and it could be inferred that the amount of (fresh) flood water available to the island has been decreased accordingly; this could explain the loss of vigour of the mangroves.

An alternative hypothesis, which explains the whole pattern of temporal changes in the vegetation more satisfactorily, is based on the long-term effect of climatic change recorded for this area since 1946.

Tucker (1975) and Cornish (1977) have reported on changes to the seasonal pattern and total annual rainfall in eastern Australia. The latter identified an increase in summer rainfall of about 60% in the Sydney area, 25% around Newcastle and an

Table 2. Percentages of species common to the habitat types taken in pairs. Data from Clarke and Miller (1983) and Clarke and van Gessel (1983); see text for details. – indicates no data.

Habitat	Type	M1	M2	S	P	F	SW	B	
PLANTS	M1		39	27	14	16	17	0	BIRDS
	M2	50		19	15	6	13	0	
	S	6	19		46	41	34	0	
	P	1	5	31		13	34	1	
	F	0	0	0	0		0	0	
	SW	0	9	28	16	0		0	
	B	0	0	0	0	0	0		
INSECTS	M1		20	17	20	–	11	–	SPIDERS
	M2	0		13	14	–	11	–	
	S	10	2		0	–	6	–	
	P	9	4	28		–	43	–	
	F	–	–	–	–		–	–	
	SW	9	0	29	46	–		–	
	B	–	–	–	–	–	–		

intermediate increase in the Hunter Valley; this was accompanied by a smaller increase in winter rainfall. These changes occurred about 1946. Bell and Erskine (1981) have examined some of the consequences of this increase in rainfall and identified an increase in flood frequency in the Hunter Valley since 1946. This climatic change would have made fresh water more plentiful in the study area, either as rain water or as flood water. One effect of this would be to leach salt from the saltmarsh soils and allow the growth of mangroves which are favoured by less saline conditions. An increase in the frequency of floods, particularly in the warmer months, would provide greater opportunities for the spread of mangrove propagules into areas away from the main channels on Kooragang Island.

The period 1978 to 1983 has been somewhat drier than usual. This is particularly illustrated by the prolonged sequence of relatively dry summers when evaporative losses of water would have been high. The five-year mean rainfall for December to February in this period was 217 mm compared to the average of 312 mm; the probability of obtaining a five-year mean of 217 mm or less is 0.05 (t-test). Large areas of eastern Australia were declared as drought affected in this time. The effect of this sequence of dry years would be to increase soil salinity, resulting in the stressing of mangroves and the reappearance of saltmarsh species in the areas colonized by mangroves (as was observed). The observation that mangroves in M2 areas were showing signs of rejuvenation in 1984 (after the drought had broken) provides support for this hypothesis.

Expansion and deterioration of mangrove communities in Fullerton Cove (see Fig. 1) have accompanied the changes apparent on Kooragang Island, but are not analysed here. The author is aware of, but unable to confirm, reports of similar changes on other parts of the New South Wales coast.

It is likely that both influences (artificial alterations and climatic change) combined to produce the pattern of vegetation change identified. Detailed studies on the vegetation over a long period and incorporating an investigation of the relationships between climatic events and the growth of mangroves and saltmarsh species would be needed to elucidate the mechanism by which a climatic effect operates.

Possible Impact of Changes on the Fauna

Data on insects and spiders in intertidal wetlands are scarce, and those used here (Clarke and Miller 1983) are based on very limited sampling which did not include mangrove canopy species, borers or fauna firmly attached to plants. Nevertheless, the low degrees of similarity shown by the habitat types for those groups recorded (Table 2) is probably typical of such areas (J. Clancy pers. comm.).

A total of 169 bird species use Kooragang Island (Clarke and van Gessel 1983). The data for birds show that open areas devoid of trees (such as S, P, F and SW) are utilized by a moderately high proportion of species in common (Table 2) and that the mangrove areas (M1 and M2) also have a similar proportion of species in common. Few species use both the wooded and open areas.

The information collected in 1982 provide a strong basis for concluding that the habitat types identified for this paper are biologically very distinct from each other. Consequently, the changes in the extent of the habitats can be expected to have had marked effects on the fauna of the study area.

The data on the avifauna allow more detailed analysis.

Twenty-two species of Palearctic waders utilize the area. All visit the saltmarsh areas, while only 15 species utilize mangroves. Clarke and van Gessel (1983) stress the importance of saltmarsh as a habitat component for these species and comment on the fact that the appearance of scattered mangroves in these areas reduces the value of the habitat by providing cover for land-based predators and by restricting take off and landing space. Despite these considerations, none of the Palearctic waders are likely to be seriously affected by the vegetation changes as long as other suitable areas are available.

None of the Australasian waders which are found in the area utilizes mangroves, but a high proportion of species feed, roost or breed in the saltmarsh. The suitability of the study area as a habitat for these species relies heavily on the presence of the saltmarsh areas.

Of 73 species of birds which breed on Kooragang Island, 14 species breed only in mangrove areas, so the breeding habitat for these has increased in the study area during the past thirty years. None of the breeding birds rely solely on saltmarsh for this purpose, though about 30% of species utilize saltmarsh for feeding (again, not exclusively). The impact of the vegetation changes on the species composition of breeding birds would probably have been restricted to an increase in the number of birds and/or species breeding in the mangrove areas.

Consequences of the Changes for Management of the Reserve

As has been mentioned above, the consequences of the vegetation changes for the birds of Kooragang Island probably have not been deleterious to total bird populations, since in most cases suitable habitat areas have been available elsewhere on the New South Wales coast. Changes to land use pattern along the New South Wales coast are occurring rapidly and in a piecemeal fashion, wetlands being severely altered (Adam 1984). It is worthwhile, then, to examine the likely effects of the changes on the fauna of the study area in order to assess the suitability of the reserve in the absence of alternative habitats.

The large decrease in the area of saltmarsh suitable for roosting by wading birds must be viewed as a deterioration in the quality of the habitat for these species. The appearance of mangroves lacking vigour has introduced a habitat component hardly present in 1954 and favourable for species which would otherwise find the island unsuitable because of a lack of dead trees in which to roost or nest. Thus, the appearance of the habitat type M2 probably has permitted the establishment of bird species which compete for food resources of the island with earlier-established species (see Clarke and van Gessel for details of bird utilization of habitats). An adequate survey of the abundance of food items would be needed to properly assess the importance of this factor and to allow rational management of such a reserve.

The appearance of mangroves in saltmarsh areas was rapid (an average rate of nine ha per year in the period 1954 to 1979, the saltmarsh/mangrove boundaries moving at an average rate of up to 300 m per year). A rapid reversion to conditions favouring saltmarsh (such as prolonged dry conditions) would not, however, see an equally quick restoration of saltmarsh suitable for waders because tree stumps and fallen logs would continue to provide cover for predators and to restrict flight paths for many years; under these circumstances, the rate of habitat change would be governed by the decomposition rate of the dead wood. Management of the reserve, to increase the area of suitable saltmarsh, might then require clearing of dead timber.

Wolanski *et al.* (1980) developed a model which related the geometry and hydrodynamics of mangrove channels to the vegetation density; there appears to be an equilibrium between mangrove density and channel size for a given set of hydrological conditions. If such a model adequately describes conditions on Kooragang Island, it is apparent from the present data that the establishment of any equilibrium may have taken about twenty years.

The changes may be presumed to have been accompanied by alterations in the invertebrate fauna of the area. Apart from the effects such alterations may have had on the avifauna, these organisms themselves may be worth special consideration. The data presently available are inadequate for that purpose, so a monitoring of selected species appears warranted.

The reserve is being progressively isolated from similar natural areas and, even as proposed in its largest form (see Fig. 1), has a very high edge to area ratio. It is, therefore, vulnerable to change from a number of causes. Given the variability displayed over three decades, it is apparent that a single survey of habitat conditions in the study area would be quite inadequate as a basis for the planning of long term management regimes. It is also apparent that a decision on the size of this reserve is fraught with difficulties. Ideally, the reserve should be sufficiently large to provide a variety of habitat types regardless of the variation that can be expected. If the observed changes have (as is suggested) a climatic component, future variations remain largely unpredictable. For this reason, the management objectives for reserves of this type ought not be very specific; it would be inappropriate to conserve such areas as refuges for particular species unless the areas are very large. Even when the area is as large as practicable, managers ought to be prepared to manipulate water flows and topography to regulate the changes in habitat characteristics from time to time.

ACKNOWLEDGEMENTS

The author is grateful for the co-operation of Dr Carina Clarke who commented on a draft of this paper, to Mr Peter Jones for photographic work and to Ms Sue Walsh for assistance with the figures. Rainfall data were kindly supplied by the Bureau of Meteorology of the Department of Science and Technology.

REFERENCES

Adam, P., 1984. Towards a wetland conservation strategy. *Wetlands (Australia)* 4: 33-48.

Bell, F. C. and Erskine, W. D., 1981. Effects of recent increases in rainfall on floods and runoff in the Upper Hunter Valley. *Search* 12: 82-3.

Clarke, C. J. and Miller, P. F., 1983. Habitat evaluation — invertebrates. Pp. 107-16 *in* Investigation of the Natural Areas of Kooragang Island ed by J. Moss. New South Wales Department of Environment and Planning, Sydney.

Clarke, C. J. and van Gessel, F. W. C., 1983. Habitat evaluation — birds. Pp. 117-44 *in* Investigation of the Natural Areas of Kooragang Island ed by J. Moss. New South Wales Department of Environment and Planning, Sydney.

Coffey, E. J., 1973. Inquiry into Pollution from Kooragang Island. State Pollution Control Commission, Sydney.

Cornish, P. M., 1977. Changes in seasonal and annual rainfall in New South Wales. *Search* 8: 38-40.

Moss, J., 1983. Investigation of the Natural Areas of Kooragang Island. New South Wales Department of Environment and Planning, Sydney.

Outhred, R. K. and Buckney, R. T., 1984. The vegetation of Kooragang Island, New South Wales. *Wetlands (Australia)* 3: 58-70.

Tucker, G. B., 1979. Climate: is Australia's changing? *Search* 6: 323.

Wolanski, E., Jones, M. and Bunt, J. S., 1980. Hydrodynamics of a tidal creek-mangrove swamp system. *Aust. J. Mar. Freshwat. Res.* 31: 431-51.

CHAPTER 20

Disturbance Regimes in Remnants of Natural Vegetation

R. J. Hobbs[1]

Natural communities are subject to disturbances of many types and at many spatial and temporal scales. Community dynamics and landscape patterns are often determined by the prevalent disturbance regime. Fragmentation of the natural vegetation leads to alterations to the disturbance regime which may alter the long-term functioning of communities. Alterations include changes in frequency, timing and intensity of fires or changes in grazing pressures. Additional disturbance may arise from new transfers of nutrients, water and biota (including non-native plant and animal species) across boundaries formed between natural vegetation and adjacent agricultural land. Conservation must involve the maintenance of natural community and ecosystem processes, as far as is possible in the fragmented landscape. Problems lie in determining what the natural disturbance regime was before fragmentation and how relevant that is to the present situation. The task for research and management is to find ways to perpetuate the essential elements of the natural disturbance regime, as far as is practical, while minimizing the effects of human-induced disturbance.

INTRODUCTION

THE importance of disturbance in shaping community and ecosystem dynamics has been discussed in a number of recent reviews (White 1979; Reiners 1983; Sousa 1984; Pickett and White 1985). Foster (1980) and Delcourt et al. (1983) have indicated that scales and frequencies of disturbances in natural communities can range over several orders of magnitude. For instance, disturbances may range from very local modifications of edaphic or microclimatic conditions to landscape-scale disturbances such as wildfires, floods and hurricanes. Pickett and White (1985) define disturbance in general terms as 'any relatively discrete event in time that disrupts ecosystem, community or population structure and changes resources, substrate availability, or the physical environment'.

In this chapter I examine the types of disturbance experienced by natural communities, firstly in the context of continuous vegetation cover, and secondly in cases where the natural communities have been fragmented by agriculture. In particular I examine the problems caused to management and conservation of remnants by various forms of disturbance.

As Janzen (1983) has pointed out, the purpose of nature reserves should not be the conservation of species *per se,* but the conservation of the interactions between species. Taking this further, managers should aim at the maintenance of natural processes within areas of native vegetation. If these natural processes break down, conservation of natural communities may be impossible. In this chapter I therefore draw a distinction between natural disturbances and those which are caused by human activities and interfere with the natural ecosystem and community processes.

DISTURBANCE REGIMES

The sum of disturbances operating in a given landscape is classed as the 'disturbance regime'. This regime is characterized by the areal extent, magnitude (or intensity), frequency, duration and predictability of the disturbances experienced (Pickett and White 1985). A natural disturbance regime is likely to consist of a variety of disturbance types occurring at different time and spatial scales. The disturbance characteristics are interlinked, and the most intense, large scale disturbances are liable to be infrequent, while less intense disturbances at a smaller scale may be more frequent. For example, rivers flood to a certain limited level every 10-20 years without causing widespread damage, but the rarer 100-200 year flood will reach much greater proportions and cause much more widespread damage. Such events are predictable within certain statistical limits, and thus differ from other types of disturbance which are generally unpredictable (e.g. volcanic eruptions). Hansen and Walker (1985) have

---
[1] CSIRO, Division of Wildlife and Rangelands Research, L.M.B 4, P.O. Midland, Western Australia, Australia 6056.
Pages 233-40 *in* NATURE CONSERVATION: THE ROLE OF REMNANTS OF NATIVE VEGETATION ed by Denis A. Saunders, Graham W. Arnold, Andrew A. Burbidge and Angas J. M. Hopkins. Surrey Beatty and Sons Pty Limited in association with CSIRO and CALM, 1987.

emphasized that a disturbance will be experienced primarily by those elements in the biota that are of a similar scale. Thus, for instance, small-scale soil disturbance in a forest will affect only the ground layer and not the forest as a whole.

Disturbance Types

Many types of natural event cause local or widespread destruction of the biota; e.g. fire, floods, severe storms, hurricanes, landslides, avalanches and, in some areas of the world, earthquake damage and volcanic eruptions. Also included would be the effects of extreme variations in climatic variables, such as droughts or sever winters. The effects of such events depend on their relation to the normal climatic regime: severe drought is potentially a much greater disturbance in areas not usually affected by water shortage than in those experiencing regular drought cycles. The question of whether a particular event represents a disturbance or is simply part of a long-term cycle is irrelevent if disruptive effects on the biota occur. Perhaps more important is the timing of disturbances with respect to each other, or with respect to climatic cycles. The incidence of fires may, for instance, be linked with specific climatic events (e.g. following high rainfall years in Central Australia; Griffin *et al*. 1983), and the vegetation response to fire will also depend on climatic conditions in the post-fire period.

In addition to relatively large scale disturbances, a given community may experience disturbances at a smaller scale. Individual treefalls often represent the main form of disturbance in forest types not subject to fire or wind damage. Animals cause various types of disturbance through grazing and trampling (e.g. Crawley 1983) or by digging or burrowing (e.g. Platt 1975; Hobbs and Mooney 1985). These types of disturbance are likely to occur more frequently than the larger scale disturbances discussed above. Populations of animals often fluctuate widely in any given area however, and hence the disturbance regime may change with time. This would then impose a secondary longer-term periodicity on the disturbance regime.

The important feature of all these types of disturbance is that they can all be considered part of the natural system, especially over a long time scale. While a fire or hurricane may temporarily destroy large areas of native vegetation, there is increasing evidence that the biota is well adapted to recover from these apparently major disruptions and that disturbances of various sorts are important in determining community dynamics. This is illustrated by studies of the effects of severe storms in various parts of the world (e.g. Doyle 1981; Sprugel and Borman 1981; Dittus 1985), and of many studies of fire (e.g. Heinselman and Wright 1973; Gill *et al*. 1981).

Although it is important to consider the present 'natural' disturbance regime of an area, it is equally important to remember that this regime may have been different in the past, and has probably varied with long-term climatic changes. In a recent review of the factors leading to the formation of the grassland biome in North America, Axelrod (1985) emphasizes the role of changing fire regimes with changing climate as a factor hastening the spread of grassland at the expense of forest.

HUMAN DISTURBANCE

Aboriginal Populations

There is much debate on the extent of environmental modification caused by Aboriginal peoples present before European settlement in the United States, Australia and elsewhere. There is, however, reasonable agreement that in many areas they used fire to clear land or for hunting purposes (e.g. Sauer 1950; Stewart 1956; Hallam 1975; Singh *et al*. 1981; Kimber 1983). This use of fire undoubtedly represented a departure from the disturbance regime prevalent before the arrival of Aboriginal people in these areas. By the time Europeans arrived, however, the Aboriginal influence was an important component of the regime. The question of what actually is the 'natural' disturbance regime then becomes a matter of timescale; should we aim at a pre-Aboriginal or pre-European 'naturalness'? Such questions have more than academic importance in many cases. For example, the incidence of Aboriginal burning and hunting in Central Australia may have led to the extinction of large marsupials (Merrilees 1968). On the other hand, there is growing support for the idea that many smaller marsupial species depended on the finegrain patchiness created by Aboriginal burning, and that the cessation of this type of burning led to their numbers being drastically reduced (e.g. Bolton and Latz 1978; Latz and Griffin 1978; Burbidge 1985).

Disturbance Following European Colonization

Whatever the effect of Aboriginals on disturbance regimes, there is little doubt that Europeans have done much to disrupt and alter these regimes wherever they have colonized. Adamson and Fox (1982) comment that 'The arrival of European man less than 200 years ago was an apocalyptic event for Australian ecosystems'. They go on to catalogue in detail the effects of European colonization, which included the displacement of Aborigines, grazing, cultivation, mining, urban development and the introduction of many non-native plants and animals.

Similar catalogues of destruction and modification can be produced for almost any part of the world. Godron and Forman (1983) recognized various levels and types of distrubance and landscape modification, ranging from limited land clearance to complete urbanization. Even in less altered areas disruption can occur due to human attempts to mitigate or remove threats of natural disturbance for economic or social reasons; e.g. flood control on rivers or control of wildfires to prevent loss of life and damage to property or forestry concerns. However, widespread attempts to suppress fires in natural areas in the United States during most of this century have led to many more problems than if a policy of controlled burning had been instigated earlier. Fire suppression has led to increased fuel loads and greater fire hazards (Dodge 1972; Kilgore and Taylor 1979; Minnich 1983), and in some cases has altered the vegetation structure and composition. In the Sierra Nevada mountains in California, where reduced fire frequencies have allowed coniferous species to encroach on open meadows and areas previously dominated by oaks (e.g. Whitney 1979). Further examples of changes in natural communities brought about by fire suppression are given by Tande (1979) and Henderson and Long (1984). The reinstatement of fire as a management tool has in some cases permitted a return to something approximating a 'natural' regime. A danger is, however, that communities in which fire has been suppressed for some time may have passed some sort of threshold and reintroduction of fire management will do nothing to alter community composition.

Even a more liberal approach to fire management, allowing controlled burning, may be a marked departure from the natural regime with profound effects on the biota. For instance, Malanson (1985) has suggested that frequent prescribed burning in coastal sage scrub in California may lead to the elimination of some dominant native shrub species. He suggests that maintenance of the natural vegetation and fire hazard reduction may be mutually exclusive management goals in this community. Similarly, in Australia the use of frequent fuel reduction fires and safer cool season fires may eventually lead to the disappearance of plant species dependent on longer fire free periods for reproduction (e.g. *Banksia ericifolia* in eastern Australia; Adamson and Fox 1982) or on strong fire cues for germination (e.g. leguminous species in *Eucalyptus marginata* forests in Western Australia; Shea *et al.* 1979). Griffin and Freidel (1985), on the other hand, discuss the effects of infrequent high intensity fires on central Australian plant communities and conclude that a return to more frequent low intensity burning may be necessary for long-term landscape stability. Clearly, for conservation purposes, the management regime must be set to match the requirements of the biota. On the other hand, not all fires are started by managing bodies, and the increased incidence of accidental fires caused by tourists, campers and surrounding farmers further complicates the current fire regime. Adequate control measures such as fire breaks and access roads are required to prevent such accidental fires causing major damage. This in turn leads to dissection of the natural vegetation with the potential for increased human disturbance.

EFFECTS OF FRAGMENTATION

One of the most obvious disturbances caused by man throughout the world is the widespread clearance of natural vegetation to make way for agriculture. This may be only a temporary clearance with secondary regrowth of natural vegetation after 2-3 year's cropping, as in various forms of shifting cultivation. However, in many parts of the world clearance is more or less permanent, resulting in a landscape dominated by agricultural land with remnants of native vegetation dispersed across it. These remnants may vary in size and degree of isolation, and these parameters are likely to change with time as more land is cleared. The important question to be addressed in such cases is to what extent the disturbance regimes experienced by fragmented natural systems differ from those in continuous vegetation. Whereas unaltered landscapes may be in dynamic equilibrium (White 1979), with local fluctuations cancelling out over the whole, fragmented landscapes are unlikely to be in such an equilibrium (Sousa 1984). Several authors have pointed out that remnants or 'habitat islands' differ from oceanic islands in their relation to the 'sea' by which they are surrounded (e.g. Burgess and Sharpe 1981; Harris 1984). Burgess and Sharpe (1981) conclude that the interplay between remnants and their matrix and between remnants themselves as mitigated by the intervening matrix adds new dimensions to their dynamics.

Effects on Community Composition

Curtis (1956) discusses the likely effects of fragmentation on natural woodland communities. Fragmentation leads to a breakdown in the normal dispersal and reproductive success of both plants and animals. This may lead to changes in species abundances, with some species being lost from the community and others assuming greater importance than in continuous vegetation. For instance, there is a selective effect on tree composition when edges are created, and gap-phase species increase in abundance (Ramney *et al.* 1981). This leads to the replacement of rarer 'interior' species by commoner edge species (Harris 1984). Plant species composition may also change because of changes in dispersal patterns due to fragmentation. For example, Auclair and Cottam (1971) found that oak remnants in Wisconsin were invaded by *Prunus serotina* rather than *Acer saccharum*, the species expected in uncleared areas. They concluded that this was due to the combination of isolation, disturbance and the higher dispersibility of bird-carried *Prunus* seeds.

Such changes represent a clear disturbance to the natural processes in remnants and render extremely difficult the conservation of 'natural' communities.

Effects on Animal Populations

Fragmentation affects the levels of native animal populations in a number of ways. Firstly, numbers are often reduced by the destruction of habitat or by interference by non-native species invading from surrounding areas. These may be species competing for the same resources, as for example in the interaction between native herbivores and introduced livestock. Introduced predators can also reduce populations of native species not normally subject to high levels of predation. Such problems also occur when species which are native but not normally found in a given community invade and interfere with the population processes of truly native species. Into this category fall many species associated with open or agricultural land which utilize remnant areas at certain times. An example of this is given by Saunders and Ingram (this volume), who discuss the interaction between the galah *Cacatua roseicapilla*, a cosmopolitan species, and Carnaby's cockatoo *Calyptorhyncus funereus latirostrus*, a species dependent on remnant areas. In the much-fragmented southern English heaths Webb and Hopkins (1984) found that small remnants were especially prone to invasion by invertebrate species from surrounding areas. These species actually increased the diversity of the invertebrate fauna in the smaller remnants, illustrating the danger of using simple total numbers of species as a guide to diversity in remnants. Janzen (1983) also contends that similar problems arise when pristine areas become surrounded with areas of secondary growth; here species of the secondary areas are likely to invade the undisturbed sections and disrupt the natural interactions within them.

Populations of native animals sometimes increase in response to fragmentation because of increased food or water resources provided by the surrounding agricultural land. The species which respond in such a way are likely to be the more opportunistic species which can adapt to the new habitat. Numbers of herbivores may also increase simply because of the compression of populations into smaller areas by increases in human occupation, as has happened with elephants in African savannas (Walker 1981).

Animal Populations and Community Dynamics

Such changes in animal population numbers must have profound effects on the remnant community as a whole. Where herbivore numbers change, vegetation structure and composition will be altered. Where grazing is particularly selective, the preferred species may be released and assume new importance in the community upon the removal of grazing or be lost entirely if grazing increases. Where predator numbers change, the effect on lower trophic levels will follow a similar path.

Further disturbances to normal community processes can occur through changes in animal abundances. For example, the disappearance of frugivorous species may prevent the regeneration of plants dependent on animal dispersal (Terborg and Winter 1980). Bond and Slingsby (1984) discuss a case where native ants have been replaced by invading non-native Argentine ants *Iridomyrmex humilis*. Whereas the native ants played a significant role in transporting and burying seeds of several plant species, thus protecting them from fire damage, the non-native ants did not. Rare species of Proteaceae may therefore be in danger of extinction. This example illustrates the importance of considering the interactions between various components of the disturbance regime. Fire on its own would not be damaging to the community, but when it is coupled with the invasion of non-native species it has a destabilizing effect.

The same may be true in many other cases. For instance, in Western Australia, fires sometimes allow the invasion of non-native annual species from the surrounding agricultural land (e.g. Bridgewater and Zammit 1979). However, this process is most marked where fires are very frequent (e.g. Wycherly 1984) and also possibly in areas where rabbit populations are high and their soil disturbances provide foci for weed invasion. The interaction between fire and grazing animals in general is important, and herbivores are known to graze preferentially on recently burned areas: e.g. native marsupials in south-east Australia (Leigh and Holdgate 1979), deer and gamebirds on heathland in Scotland (Hobbs and Gimingham 1986), moose in boreal forest (Peek 1974), large herbivores and insects in African savanna (Walker 1981). In southwestern Australia this effect may be increased if the fire area is small where kangaroos graze heavily on the regrowth in burned heath areas and significantly alter the vegetation composition (Hopkins 1985 and pers. comm.).

INTERACTIONS WITH SURROUNDING FARMLAND

Movement of Livestock

The interaction between remnant areas and surrounding farmland adds new types of disturbance to the existing regime. A major form of disturbance is the movement of livestock into remnant areas. Grazing by sheep or cattle drastically alters the structure of native vegetation and prevents regeneration of woody species. In oak woodland in Britain, Pigott (1983) found that oak *Quercus petraea* and *Betula* spp. could regenerate successfully only after the exclusion of sheep. A combination of cattle and rabbit grazing prevents the regeneration of the tree *Acacia papyrocarpa* and other woody species in South Australia (Crisp and Lange 1976; Lange and Purdie 1976). Similarly, reductions in establishment have been found for large cacti in desert communities subjected to cattle grazing in south-west United States (Steenberg and Lowe 1977).

Added to the direct grazing and trampling effect is the disruption to normal nutrient flows caused by defecation. Faeces of domestic animals, and of native animals that graze in surrounding farmland, are also sources of weed seeds which can subsequently establish in a ready-made nutrient rich microhabitat.

Transfers from Agricultural Land

There are many other possible transfers between agricultural land and areas of native vegetation. Of particular importance is the invasion of agricultural and weed species into the natural vegetation. Such species may disperse into the edges of remnants by wind or be carried further in by animals. Little is known about the factors affecting the 'invasibility' of natural communities, but it appears likely that some form of disturbance is required before weed species will establish. Once established, it is likely that the non-native species will have a detrimental effect on the community either through altered nutrient and water availabilities or through impedence of regeneration of the native species.

Transfers of nutrients, pesticides and organic matter (e.g. in the form of wind-blown stubble) also occur from field to remnant. Clearly, the method by which chemicals are applied to crops will affect the extent to which they spread. Aerial spraying, for instance, is liable to cause sizable drifts into adjacent vegetation. Little is known of the potential effects of such transfers. The spread of fire from fields to vegetation remnants is also possible where stubble burning takes place.

Water Regime

A further effect of land clearance is the alteration of the local water regime, possibly leading to changes in water availability and salt levels within remnants. An extreme example of this is found in the fens of eastern England, where drainage of peat areas has led to peat wastage and a drop in the land level of almost 4 m in 130 years (Hutchinson 1980). This created enormous problems in the conservation of the few remnants of natural fenland which remain (e.g. Wicken Fen; Rowell 1986).

While transfers from agricultural land to remnant areas are of prime importance to conservation, transfers in the other direction must also be considered. The possibility of the spread of fire or undesirable species from areas of native vegetation to surrounding farmland may be important in shaping the local community's attitudes to conservation. This is as important a problem for management as is the problem of agricultural impingement on natural areas.

Although I have emphasized the negative aspects of fragmentation on natural processes here, the manager can derive some benefit from dealing with remnants rather than continuous vegetation. Fire management is easier to cope with in discrete vegetation remnants and managing authorities may be more willing to apply more experimental types of fire regime to these. As Simberloff and Abele (1982) have discussed, there is less chance of the loss of species due to catastrophic disturbance if they are distributed over a number of remnants rather than in a continuous area, even though local extinctions will still occur.

FRAGMENTATION AND LANDSCAPE MOSAICS

Fragmentation of the natural habitat may lead to more subtle effects caused by the break-up of previously continuous landscape mosaics. For instance Hobbs and Gimingham (1986) argue that the landscape of the Scottish uplands was originally composed of a 'moving mosaic' of heath, birch woodland and pine forest. Birch and pine do not regenerate well under their own canopies and probably regenerated after fires, with heathland as a sucessional stage. In the original landscape various stages of development would be present at any one time and, although pines in one area may senesce and die or be destroyed by fire, younger stands would be present elsewhere. Today, however, the landscape is more fragmented and areas are set aside as reserves for individual vegetation types such as heath or pinewood or birchwood. There are very few places where large enough areas are available to encompass a whole landscape mosaic containing all community types. As a result, problems arise with lack of regeneration in old pinewoods. The few remaining old growth woods are too precious to be disturbed to allow them to regenerate, and remnants are now surrounded by altered vegetation where regeneration cannot occur. Here again, increased numbers of deer and sheep prevent pine regeneration in areas of heath that might otherwise develop to pine forest.

This example highlights one of the major problems confronting conservation through reserves. Heinselman and Wright (1973) and White (1979) have discussed the concept of biotic stability at the landscape level despite great fluctuations caused by disturbances at a local level. A 'shifting locus' of disturbance may result in a relatively stable distribution of patches among various age classes within the landscape mosaic of community states. This concept depends, however, on the existence of an entire unfragmented landscape. Where only remnants of the natural vegetation remain, it may be impossible to maintain overall landscape stability. Pickett and Thompson (1978) have suggested that there is a minimum area within a landscape that has a natural disturbance regime which maintains internal re-colonization sources, and hence prevents extinctions from occurring. Henderson et al. (1985) have applied this concept of landscape mosaics and minimum areas to the problem of animal population survival in a fragmented landscape, and suggest that such mosaics should be the elementary units in much conservation planning and management.

After fragmentation, remnant areas are likely to contain only a subset of the overall community states. Loss through disturbance of a particular community will no longer necessarily be matched by its reappearance somewhere else in the landscape. This problem is increased where only a subset of the original community states are represented in the system of remnants. This is likely since clearing was probably highly selective and only the agriculturally poorer areas were left uncleared. This is certainly true in the wheatbelt of Western Australia, where *Eucalyptus loxophleba* and *E. salmonophloia* woodlands, which grow on richer soils, were cleared preferentially and are thus severely under-represented in the remaining collection of remnants. The resultant simplification of landscape mosaics will have important consequences for the maintenance of ecosystem processes, especially where key species such as canopy trees are lost.

While it may be impossible to maintain intact landscape processes, there is a danger that natural communities may be lost regardless by lack of suitable management. Failure to recognize the importance of the natural disturbance regime will maintain communities in a reasonable state in the short term but lead to the absence of important community states in the future. Pyle (1980) emphasizes the need to maintain all successional stages within reserves. For instance, lack of fires in ageing vegetation may produce a shortage of younger stages important for short-lived plant species and as habitat for animals. The time scales involved must also be recognized and changes in vegetation after disturbance seen as successional rather irreversible. Simply because it takes a woodland 200-300 years after fire to return to mature woodland through other non-woodland stages is not a reason for preserving all of it as is. The intermediate stages may be important in themselves and without disturbance the mature wood might last for another 50 years but there would then be nothing to replace it. An example of the potential loss of tree species through lack of fire is given by Withers and Ashton (1977).

CONCLUSION

The major task facing the manager trying to conserve native communities is to define the natural disturbance regime and separate it from the altered regime caused by clearance and its associated features. It is clear that the fragmented system cannot retain all the attributes of the original system and the manager must work with the present situation where remnant and surrounding farmland have to get on together. In many cases it may be impossible or undesirable to re-establish a natural disturbance regime even if information on the natural regime is available. Management constraints and overall objectives will determine the type of management intervention possible, but it is nevertheless important to consider the role of disturbances in the maintenance of natural communities. Clearly, considerably more information is required on the features of the overall disturbance regime in each area and on the effects of human-induced disturbances. However, decisions have to be made now without adequate information before species or entire communities are lost.

In this chapter I have tried to illustrate that disturbance of some sort is a vital component of most natural systems, and that many of the processes operating in natural communities are based on disturbance. In that respect, some form of management retaining the major aspects of the natural disturbance regime must be aimed at. At the same time, however, other forms of disturbance through imbalances caused by man's activities have to be minimized. Where the two types of disturbance act synergistically (e.g. where fire leads to weed invasion) problems arise. However, these should be regarded as problems to be solved rather than excuses for inactivity. Failure to view natural communities in the light of their overall dynamics will result only in the failure of the long-term goals of conservation.

ACKNOWLEDGEMENTS

I thank A. J. M. Hopkins, S. D. Hopper, K. Tinley and B. H. Walker for their constructive comments on the draft manuscript.

REFERENCES

Adamson, D. A. and Fox, M. D., 1982. Change in Australasian vegetation since European settlement Pp. 109-46 *in* A History of Australasian Vegetation ed by J. M. B. Smith. McGraw-Hill, Sydney.

Auclair, A. N. and Cottam, G., 1971. Dynamics of black cherry (*Prunus serotina* Ehrh.) in southern Wisconsin oak forests. *Ecol. Monogr.* **41**: 153-77.

Axelrod, D. I., 1985. Rise of the grassland biome, Central North America. *Bot. Rev.* **51**: 163-201.

Bolton, B. L. and Latz, P. K., 1978. The western hare-wallaby, *Lagorchestes hirsutus* (Gould) (Macropodidae), in the Tanami Desert. *Aust. Wildl. Res.* **5**: 285-93.

Bond, W. and Slingsby, P., 1984. Collapse of an ant-plant mutualism: the argentine ant (*Iridomyrmex humilis*) and myrmecochorous proteaceae. *Ecology* **65**: 1031-7.

Bridgewater, P. B. and Zammit, C. A., 1979. Phytosociology of south-west Australian limestone heaths. *Phytocoenologia* 6: 327-43.

Burbidge, A. A., 1985. Fire and mammals in hummock grasslands of the arid zone. Pp. 91-4 *in* Fire Ecology and Management in Western Australian Ecosystems ed by J. Ford. Western Australian Institute of Technology, Bentley.

Burgess, R. L. and Sharpe, D. M., 1981. Introduction. Pp. 1-5 *in* Forest Island Dynamics in Man-Dominated Landscapes ed by R. L. Burgess and D. M. Sharpe. Springer, New York.

Crawley, M. J., 1983. Herbivory: The Dynamics of Animal-Plant Interactions. University of California Press, Berkeley.

Crisp, M. D. and Lange, R. T., 1976. Age structue, distribution and survival under grazing of the arid-zone shrub *Acacia burkittii*. *Oikos* 27:86-92.

Curtis, J. T., 1956. The modification of mid-latitude grasslands and forests by man. Pp. 721-36 *in* Man's Role in Changing the Face of the Earth ed by E. L. Thomas. University of Chicago Press, Chicago.

Delcourt, H. R., Delcourt, P. A. and Webb, T., 1983. Dynamic plant ecology: the spectrum of vegetational change in space and time. *Quat. Sci. Rev.* 1: 265-71.

Dittus, W. P. J., 1985. The influence of cyclones on the dry evergreen forest of Sri Lanka. *Biotropica* 17: 1-14.

Dodge, M., 1972. Forest fuel accumulation — a growing problem. *Science* 177: 139-42.

Doyle, T. W., 1981. The role of disturbance in the gap dynamics of a montane rain forest: an application of a tropical forest succession model. Pp. 56-73 *in* Forest Succession: Concepts and Application ed by D. C. West, H. H. Shugart and D. B. Botkin. Springer, New York.

Foster, R. B., 1980. Heterogeneity and disturbance in tropical vegetation. Pp. 75-92 *in* Conservation Biology: An Evolutionary-ecological Perspective ed by M. E. Soulé and B. A. Wilcox. Sinauer, Sunderland, Massachusetts.

Gill, A. M., Groves, R. A. and Noble, I. R., 1981. Fire and the Australian Biota. Australian Academy of Science, Canberra.

Godron, M. and Forman, R. T. T., 1983. Landscape modification and changing ecological characteristics. Pp. 11-28 *in* Disturbance and Ecosystems: Components of Response ed by H. A. Mooney and M. Godron. Springer, Berlin.

Griffin, G. F. and Freidel, M. H., 1985. Discontinuous change in central Australia: some implications of major ecological events for land management. *J. Arid. Env.* 9: 63-80.

Griffin, G. F., Price, N. F. and Portlock, H. F., 1983. Wildfires in central Australia, 1970-1980. *J. Env. Manage.* 17: 311-23.

Hallam, S. J., 1975. Fire and Hearth: a Study of Aboriginal Usage and Usurpation in South-western Australia. Australian Institute of Aboriginal Studies, Canberra.

Hansen, A. J. and Walker, B. H., 1985. The dynamic landscape: perturbation, biotic response, biotic patterns. *Bull. S.A. Inst. Ecol.* 4: 5-14.

Harris, L. D., 1984. The Fragmented Forest: Island Biogeographic Theory and the Preservation of Biotic Diversity. University of Chicago Press, Chicago.

Heinselman, M. L. and Wright, H. E., 1973. The ecological role of fire in natural conifer forests of western and northern North America. *Quat. Res.* 3: 317-8.

Henderson, M. T., Merriam, G. and Wegner, J., 1985. Patchy environments and species survival: chipmunks in an agricultural mosaic. *Biol. Conserv.* 31: 95-105.

Henderson, N. R. and Long, J. N., 1984. A comparison of stand structure and fire history in two black oak woodlands in northwestern Indiana. *Bot. Gaz.* 145: 222-8.

Hobbs, R. J. and Gimingham, C. H., 1986. Vegetation, fire and herbivore interactions in heathland. *Adv. Ecol. Res.* 16: (in press).

Hobbs, R. J. and Mooney, H. A., 1985. Community and population dynamics of serpentine grassland annuals in relation to gopher disturbance. *Oecologia (Berlin)* 67: 342-51.

Hopkins, A. J. M., 1985. Fire in woodlands and associated formations of the semi-arid region of south-western Australia. Pp. 83-90 *in* Fire Ecology and Management in Western Australian Ecosystems ed by J. Ford. Western Australian Institute of Technology, Bentley.

Hutchinson, J. N., 1980. The record of peat wastage in the East Anglian fenlands at Holme Post, 1848-1978 A.D. *J. Ecol.* 68: 229-49.

Janzen, D. H., 1983. No park is an island: increase in interference from outside as park size decreases. *Oikos* 41: 402-10.

Kilgore, B. M. and Taylor, D., 1979. Fire history of a sequoia-mixed conifer forest. *Ecology* 60: 129-42.

Kimber, R. G., 1983. Black lightning: Aborigines and fire in central Australia and the western desert. *Archaeology in Oceanea* 18: 38-45.

Lange, R. and Purdie, R., 1976. Western myall (*Acacia sowdenii*), its survival prospects and management needs. *Aust. Rangel. J.* 1: 64-9.

Latz, P. K. and Griffin, G. F., 1978. Changes in Aboriginal land management in relation to fire and to food plants in central Australia. Pp. 77-85 *in* The Nutrition of Aborigines in Relation to the Ecosystem of Central Australia ed by B. S. Hetzel and H. J. Frith. CSIRO, Melbourne.

Leigh, J. H. and Holdgate, M. D., 1979. The responses of the understorey of forests and woodlands of the Southern Tablelands to grazing and burning. *Aust. J. Ecol.* 4: 25-45.

Malanson, G. P., 1985. Fire management in coastal sage-scrub, Southern California, USA. *Environ. Conserv.* 12: 141-6.

Merrilees, D., 1968. Man the destroyer; late Quaternary changes in the Australian marsupial fauna. *Proc. R. Soc. W.A.* 51: 1-24.

Minnich, R. A., 1983. Fire mosaics in Southern California and Northern Baja California. *Science* 219: 1287-94.

Peek, J. M., 1974. Initial response of moose to a forest fire in northern Minnesota. *Am. Midl. Nat.* 91: 435-8.

Pickett, S. T. A. and Thompson, J. N., 1978. Patch dynamics and the design of nature reserves. *Biol. Conserv.* 13: 27-37.

Pickett, S. T. A. and White, P. S., 1985. The Ecology of Natural Disturbance and Patch Dynamics. Academic Press, New York.

Pigott, C. D., 1983. Regeneration of oak-birch woodland following exclusion of sheep. *J. Ecol.* 71: 629-46.

Platt, W. J., 1975. The colonization and formation of equilibrium plant species associations of badger disturbances in a tallgrass prairie. *Ecol. Monogr.* 45: 285-305.

Pyle, R. M, 1980. Management of Nature Reserves. Pp. 319-27 *in* Conservation Biology: An Evolutionary-ecological Perspective ed by M. E. Soulé and B. A. Wilcox. Sinauer, Sunderland, Massachusetts.

Ramney, J. W., Bruner, M. C. and Levenson, J. B., 1981. The importance of edge in the structure and dynamics of forest islands. Pp. 67-95 *in* Forest Island Dynamics in Man-Dominated Landscapes ed by R. L. Burgess and D. M. Sharpe. Springer, New York.

Reiners, W. A., 1983. Disturbance and the basic properties of ecosystem energetics. Pp. 83-98 *in* Disturbance and Ecosystems: Components of Response ed by H. A. Mooney and M. Godron. Springer, Berlin.

Rowell, T. A., 1986. The history of drainage at Wicken Fen, Cambridgeshire, England, and its relevance to conservation. *Biol. Conserv.* 35: 111-42.

Sauer, C. O., 1950. Grassland, climax, fire and man. *J. Range Manage.* 3: 16-22.

Shea, S. R., McCormick, J. and Portlock, C. C., 1979. The effects of fires on regeneration of leguminous species in the northern jarrah (*Eucalyptus marginata* Sm) forest of Western Australia. *Aust. J. Ecol.* 4: 195-205.

Simberloff, D. and Abele, L. G., 1982. Refuge design and island biogeographic theory: effects of fragmentation. *Am. Nat.* 120: 41-50.

Singh, G., Kershaw, A. P. and Clark, R., 1981. Quaternary vegetation and fire history in Australia. Pp. 23-54 *in* Fire and the Australian Biota ed A. M. Gill, R. A. Groves and I. R. Noble. Australian Academy of Science, Canberra.

Sousa, W. P., 1984. The role of disturbance in natural communities. *Ann. Rev. Ecol. Syst.* 15: 353-91.

Sprugel, D. G. and Bormann, F. H., 1981. Natural disturbance and the steady state in high-altitude balsam fir forests. *Science* 211: 390-3.

Steenberg, W. F. and Lowe, C. H., 1977. Ecology of the Saguaro II. Reproduction, Germination, Establishment, Growth and Survival of the Young Plant. National Park Service Monogr. 8. U.S. Govt. Printing Office, Washington, D.C.

Stewart, O. C., 1956. Fires as the first great force employed by man. Pp. 115-33 *in* Man's Role in Changing the Face of the Earth ed by W. L. Thomas. University of Chicago Press, Chicago.

Tande, G. F., 1979. Fire history and vegetation pattern of coniferous forests in Jasper National Park, Alberta. *Can. J. Bot.* 57: 1912-31.

Terborg, J. and Winter, B., 1980. Some causes of extinction. Pp. 119-33 *in* Conservation Biology: An Evolutionary-ecological Perspective. ed by M. E. Soulé and B. A. Wilcox. Sinauer, Sunderland, Massachusetts.

Walker, B. H., 1981. Is succession a viable concept in African savanna ecosystems? Pp. 431-47 *in* Forest Succession: Concepts and Application ed by D. C. West, H. H. Shugart and D. B. Botkin. Springer, New York.

Webb, N. R. and Hopkins, P. J., 1984. Invertebrate diversity on fragmented *Calluna* heathland. *J. Appl. Ecol.* 21: 921-33.

White, P. S., 1979. Pattern, process, and natural disturbance in vegetation. *Bot. Rev.* 45: 229-99.

Whitney, S., 1979. The Sierra Nevada. Sierra Club Books, San Francisco.

Withers, J. and Ashton, D. H., 1977. Studies on the status of unburnt *Eucalyptus* woodland at Ocean Grove, Victoria. 1 The Structure and regeneration. *Aust. J. Bot.* 25: 623-37.

Wycherly, P., 1984. People, fire and weeds: can the vicious spiral be broken? Pp. 11-7 *in* The Management of Small Bush Areas in the Perth Metropolitan Region ed by S. A. Moore. Dept. Fisheries and Wildlife, Perth.

CHAPTER 21

Characteristics of Problem Weeds in New Zealand's Protected Natural Areas

Susan M. Timmins[1] and P. A. Williams[2]

Problem weeds of remnant protected natural areas in New Zealand are characterized. Features such as taxonomy, life form and height, life span and growth rate, dispersal mechanism, seed longevity and the communities that weeds invade are discussed. Many problem weeds are short trees, with a rapid growth rate, a specialized dispersal mechanism, a relatively short life span and an ability to invade a range of communities. This characterization could well be different if New Zealand had a broader spectrum of reserves. Further, some of the worst weeds do not fit this characterization.

INTRODUCTION

THE concept of a weed in the context of the New Zealand protected natural area system has been discussed by considering aspects of weediness (Williams 1984). Their introduction and ability to spread, the influence of habitat and other environmental factors on their invasiveness and control, and the vulnerability of native vegetation to invasion by weeds were considered.

This chapter looks only at *problem* weeds within reserves. Problem weeds are defined as those naturalized plants which are a major management problem. They are invariably difficult to eradicate and substantially change the indigenous vegetation of the sites in which they occur.

Our investigation began with the question 'Are the problem weeds of New Zealand's protected natural areas a random selection of the naturalized flora of New Zealand, or is there a bias towards certain types of plants or certain plant characteristics?' Our aim was to establish a set of plant features that described problem weeds in reserves and that could be used to predict the potential weediness of new plant species.

From discussions with reserve managers (Timmins 1984) a list of 73 species of problem weeds was derived. Our data for this chapter are the characteristics of those 73 species under the following headings: taxonomy, life form and height, life span and growth rate, dispersal mechanism, seed longevity and the communities they invade. The following section uses the same headings.

RESULTS AND DISCUSSION

Taxonomy

The 73 problem weeds are distributed among 32 plant families but about half of the species belong to just five families (Table 1). The percentage of problem weeds which are legumes (14%) is high compared to the representation of legumes in the naturalized flora of New Zealand as a whole (8%). This could imply that features such as hard seediness and ability to fix nitrogen enable a plant to be a successful weed. Legume weeds are found across the range of vegetation types, from disturbed forest to ruderal situations. By contrast, the Compositae and Gramineae are under-represented compared to their contribution to the naturalized flora. Many plants in these families are abundant in open habitats but seldom threaten the native biota.

Table 1. Distribution of families among problem weeds in protected natural areas.

Family	% of problem weeds
Leguminosae	14
Compositae	8
Rosaceae	8
Pinaceae	8
Gramineae	8
Other 27 families	1-3% each

[1]Department of Lands and Survey, Head Office, Private Bag, Wellington, New Zealand.
[2]Botany Division, DSIR, Private Bag, Nelson, New Zealand.

Life Form

The complete range of life forms is represented among problem weeds examined. There are tall, evergreen conifers such as lodgepole pine *Pinus contorta*, deciduous broadleaved trees such as sycamore *Acer pseudoplatanus*, various shrubs both leafy-armed such as barberry *Berberis glaucocarpa* and semi-leafless such as broom *Cytisus scoparius*, tussock grasses — nassella tussock *Stipa trichotoma*, non-woody herbs — periwinkle *Vinca major*, and soft woody climbers — banana passionfruit *Passiflora mollissima*.

Even though the range of life forms is represented, half the problem weeds are trees or tall shrubs (Fig. 1). Plants such as green wattle *Acacia decurrens*, brush wattle *Albizia lophantha*, tree lucerne *Chamaecytisus palmensis*, Douglas fir *Pseudostuga menziesii*, *Pinus* species, holly *Ilex aquifolium* and tree privet *Ligustrum lucideum* fall into this tree/tall shrub category. Many of these tree or tall shrub species are members of Leguminosae or Pinaceae.

Low shrubs are unimportant as a group but there are several large, perennial herbs. Non-woody dicotyledonous herbs are unimportant as problem weeds in reserves, despite their abundance in the New Zealand countryside. Tall grasses are generally less of a problem than they are in tropical climates, but the two pampas grasses *Cortaderia jubata* and *C. selloana* are spreading rapidly in northern New Zealand.

The low percentage of climbers and lianes (8%) disguises their significance as weeds in New Zealand reserves. Among this group are species which are having a large impact on the native vegetation in reserves, e.g. old man's beard *Clematis vitalba* is smothering tall forest trees in some of New Zealand's forest remnants.

The advantage a particular life form confers on a weed often depends on the vegetation type it is invading. For example, lodgepole pine invading a shrubland or disturbed forest will eventually be outcompeted by native trees in most climatic zones. However, when it invades native tussock grassland or subalpine herbfield, its tree form generally facilitates its ultimate dominance. Similarly, the tree buddleja *Buddleja davidii* causes little concern

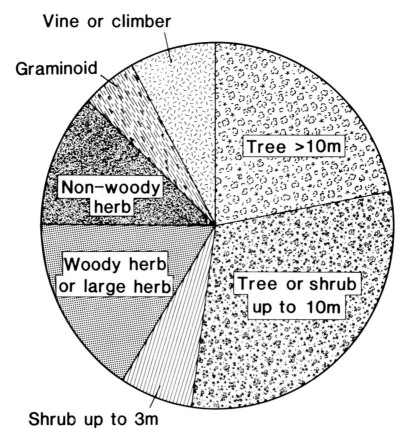

Fig. 1. Life form and height of problem weeds in protected natural areas.

when growing in native shrubland but it can shade out low growing native species in stream beds (Williams 1984).

In direct contrast to the above examples, in some disturbed forest remnants, low creeping herbs which reproduce mainly by vegetative reproduction may prevent establishment of native seedlings. One such plant is wandering Jew *Tradescantia fluminensis*. Only forest plants taller than 10 cm at the time when wandering Jew invades can survive and no seedlings can grow through the dense carpet it forms (Esler 1962). Periwinkle *Vinca major* affects native vegetation in a similar way.

Life Span and Growth Rate

Life span was categorized into four broad groups: greater than 80 years, between 30 and 80 years, between two and 30 years and less than two years i.e. the annuals and biennials. Growth rate was also subjectively assessed as; fast, medium or slow, based on the impact of each species on the environment it invades. Table 2 shows the percent of problem weeds in different life span and growth form categories.

Table 2. Life span (and growth form) of problem weeds in protected natural areas.

Growth form and life span	% of problem weeds
Tree >80 years	12
Tree or vine 30-80 years	44
Tree or vine <30 years	6
Shrubs, perennials and monocot herbs	9
Annuals and biennials	14
Vegetative reproduction	15

About half of the weeds of reserves are trees, shrubs or vines which live between 30 and 80 years. Weedy trees and shrubs tend to be early to mid-successional species with short life spans. By contrast, the native trees with which they predominantly compete have longer life spans and thus ultimately dominate the site (Druce 1957; Williams 1983). However, when short-lived weed species establish in sparsely vegetated, harsh sites, e.g. broom on river beds, they can usurp native vegetation.

There are similar, small numbers of annual or biennial herbs and long-lived trees in the problem weed list. Long-lived trees, for example sycamore, are likely to have a severe impact on forest remnants in which they become established. Once established, not only the original individuals persist but they continue to contribute seed.

A feature associated with life span is the ability of some weeds to respond to damage; many vine, shrub and even tree weeds can resprout after damage, e.g. heather *Calluna vulgaris* will resprout after fire, old man's beard and sycamore will resprout after cutting.

Of greater importance perhaps than life span of an individual plant is the longevity of the population at a particular site. Some plants have only a moderate life span yet they persist at a site for more than one generation either because of a large seed bank or for lack of competition, e.g. tree lupin *Lupinus arboreus* on sandy soils or broom on a rocky face.

Unfortunately, it is difficult to generalize about the persistence of weed species in reserves because persistence is as much a function of site characteristics as species characteristics. By definition, persistence tends to be a feature of problem weeds.

Contrary to expectation, problem weeds are not characterized by rapid growth rate. About half the weedy trees, shrubs and vines with a life span of between 30 and 80 years have a relatively fast growth rate e.g. woolly nightshade *Solanum mauritianum*, tree lucerne *Chamaecytisus palmensis* and banana passionfruit *Passiflora mollissima*. However, many problem weeds have only a medium or even slow growth rate. Heather, for example, has a slow growth rate but it is taking hold of areas of native red tussockland and herb-shrublands in Tongariro National Park. It does not rely on a rapid growth rate for its success but rather changes its micro-environment to favour establishment of its own seedlings (Chapman 1984).

Dispersal Mechanism

Most problem weeds have specialized dispersal mechanisms, be it for bird, wind or explosive dispersal (Fig. 2). One of the most important dispersal mechanisms is succulent, bird-dispersed fruit. Of the species dispersed in this way, 91% have either blue-black or red-orange fruit. These two colours have been shown to be particularly attractive to frugivorous birds (Gautier-Hion *et al.* 1985).

A large group of weed species are wind dispersed e.g. species in the families Compositae and Gramineae and a smaller group have explosive seed pods. Gorse *Ulex europaeus* is an example of the latter, with seeds being propelled up to 5 m from the parent plant (Moss 1960).

Even those species which do not have obviously *specialized* mechanisms are still well adapted for dispersal e.g. Spanish heath *Erica luscitanica* has light seed which is readily wind dispersed. Wandering Jew, which reproduces vegetatively, is easily dismembered and grows from small segments which may be dispersed by water or animals (Kelly and Skipworth 1984).

Although 13% of the weed species reproduce predominantly by vegetative reproduction very few of these species reproduce by vegetative means alone. Most species spread by seeds as well e.g. *Hieracium* spp. A notable exception seems to be wild ginger *Hedychium gardnerianum*, a major weed problem in small reserves of North Auckland. Those that

Dispersal mechanism of problem weeds

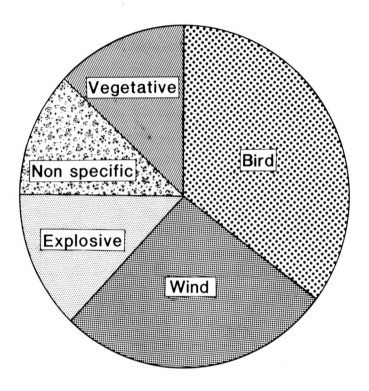

Fig. 2. Major dispersal mechanism of problem weeds in protected natural areas.

spread only by vegetative reproduction are all lowland species with a disproportionately high number occurring on the floors of forest remnants e.g. wandering Jew and periwinkle.

Dispersal mechanisms are not randomly distributed by vegetation types. Succulent bird dispersed fruits are most abundant in woody vegetation in areas of equable climate. Wind dispersal is more important in herbaceous vegetation in more severe climates. Vegetative reproduction is widespread from forest floor habitats to mountain tops, but on a proportionate basis, it is particularly important in very specialized habitats like sand dunes and salt marshes.

Humans have been instrumental in planting and spreading some species which are now problem weeds in particular protected natural areas in New Zealand. One group of such plants is those with colourful flowers e.g. buddleja, heather and Russell lupin *Lupinus polyphyllus*. The present distribution of these three weeds is as much a function of human activity in response to their 'pretty' flowers, as to the efficiency of the species' own dispersal mechanisms. Similarly, wild ginger, which is grown by some home gardeners for its scent, has been spread by these same gardeners dumping the plant as rubbish when it starts taking over their garden.

Seed Longevity

Many problem weed species produce vast numbers of seeds (e.g. broom and buddleja) but few data are available to compare production with similar species which are not problem weeds. Seed longevity is probably a more critical characteristic of weeds but again data on entry of seeds into the seed bank and seed viability are poor as many of the species have not been studied in temperate habitats. While we did not have accurate data on a third of our list of species, we distinguished three groups with respect to seed longevity and inclusion in the seed bank.

In the first group, the seeds have a limited life and seeds tend to germinate immediately or within a year or two after dispersal. This group includes species of *Pinus* (Schopmeyer 1974), certain fleshy fruited species, for example, elderberry *Sambucus nigra* (Donelan and Thompson 1980) and some herbs (Donelan and Thompson 1980).

Gautier-Hion, A., Duplantier, J.-M., Quris, R., Feer, F., Sourd, C., Decoux, J-P., Dubost, G., Emmons, L., Erard, C., Hecketsweiler, P., Moungazi, A., Roussilhon, C. and Thiollay, J. M., 1985. Fruit characters as a basis of fruit choice and seed dispersal in a tropical forest vertebrate community. *Oecologia* **65**: 324-37.

Healy, A. J., 1945. Nassella tussock (*Nassella trichotoma* (Nees) Hack). Field studies and their agricultural significance. *NZ DSIR Bull.* **91**: 90pp.

Healy, A. J., 1958. Contributions to a knowledge of the naturalized flora of New Zealand. 4. *Trans. Roy. Soc. NZ* **82**: 263-9.

Healy, A. J., 1959. Weeds of parks and reserves. *Proc. Ann. Conf. Inst. Park Admin.* **1959**: 18-22.

Healy, A. J., 1961. The interaction of native and adventive plant species in New Zealand. *Proc. NZ Ecol. Soc.* **8**: 39-43.

Healy, A. J., 1973. Introduced vegetation. Pp. 170-89 *in* The Natural History of New Zealand ed by G. R. Williams. Reed, Wellington.

Kelly, G. C., 1972. Scenic reserves of Canterbury. *Biological Survey of Reserves 2*. Dept. Lands and Survey, Wellington.

Kelly, D. and Skipworth, J. P., 1984. *Tradescantia fluminensis* in a Manawatu (New Zealand) forest: 1. Growth and effects on regeneration. *NZ J. Bot.* **22**: 393-7.

Martin, K. W., 1976. The concise British flora in colour. Edbury Press and Michael Joseph, London.

Moss, G. R., 1959. The gorse seed problem. *Proc. 12th NZ Weed and Pest Control Conf.* 59-64.

Moss, G. R., 1960. Gorse, a weed problem on thousands of acres of farm land. *NZ J. Agriculture* **100**: 561-7.

Schopmeyer, C. S., 1974. Seeds of wood plants in the United States. *Agricultural Handbook* 450. United States Department of Agriculture, Forest Service, Washington.

Taylor, R. L., 1981. Weeds of roadsides and waste ground in New Zealand. R. L. Taylor, Nelson.

Thompson, K. and Grime, J. P., 1979. Seasonal variation in the seed banks of herbaceous species in ten contrasting habitats. *J. Ecol.* **67**: 893-921.

Timmins, S. M., 1984. Weeds in National Parks and Reserves. Summary of responses to a questionnaire. Report presented to National Parks and Reserves Authority, 30 November 1984.

Williams, P. A., 1983. Secondary vegetation succession on the Port Hills Banks Peninsula, Canterbury, New Zealand. *NZ J. Bot.* **21**: 237-47.

Williams, P. A., 1984. Woody weeds and native vegetation — a conservation problem. Pp. 61-6 *in* Protection and Parks compiled by P. R. Dingwall. *Proc. Section A4e, 15th Pacific Science Congress,* Dunedin, February 1983.

CHAPTER 22

Factors Affecting Survival of Breeding Populations of Carnaby's cockatoo *Calyptorhynchus funereus latirostris* in Remnants of Native Vegetation

D. A. Saunders[1] and J. A. Ingram[1]

The breeding biology of the short-billed white-tailed black cockatoo or Carnaby's cockatoo was studied intensively at Coomallo Creek and Manmanning between 1970 and 1976. This species was also studied less intensively at Nereeno Hill, Tarwonga and Moornaming. By 1977 it had become extinct at Manmanning as a result of food shortage caused by habitat fragmentation. The population at Coomallo Creek was monitored twice each breeding season from 1977 to 1984 (with the exceptions of 1979 and 1980) and the area was subject to clearing for agriculture in the mid-1970s. The breeding population fell by half but breeding success and nestling growth rates remained at levels existing before the population decline. Comparison of breeding results from all five populations studied and the amount of native vegetation remaining in each area showed that Carnaby's cockatoo can breed successfully in areas which have been extensively cleared providing there are corridors of native vegetation connecting patches of remnant vegetation. A mosaic where the remnants are not visually isolated from neighbours will enhance the prospects for Carnaby's cockatoo. In areas where extensive development has been undertaken, it will be necessary to protect remaining nesting habitat (woodland) and feeding habitat (heath and shrubland) to ensure the continued survival of this cockatoo.

INTRODUCTION

MacARTHUR and Wilson's (1963, 1967) theory of island biogeography proposed that the number of species existing on an island is in dynamic equilibrium between immigrations and extinctions and that this process results in the changing or turnover of the species existing on that island. A number of workers have expanded these ideas and applied them to habitat islands created by a change in surrounding vegetation, usually as a result of the activities of humans (see Diamond 1975 for support of this application) while others have cautioned against the indiscriminate acceptance of these ideas and their application to the design of nature reserves (see Margules *et al*. 1982). The equilibrium theory of island biogeography has been used to derive linear relationships between number of species and area of habitat and species collapse models (see Boecklen and Gotelli 1984 for a critical examination of this approach) on the assumption that once an area of habitat is isolated (for whatever reason) it has more species than it is capable of supporting and, as a result, species will be lost from that area (species collapse or relaxation). As pointed out by Boecklen and Gotelli (1984) these models ignore many factors including species identity, habitat heterogeneity and population size. The analogy is sometimes made that changes in surrounding vegetation create 'habitat islands' or remnants surrounded by a 'sea of change'. The models also ignore the impact of the disturbance resulting from this 'sea of change'.

Some of the discussion on various aspects of the equilibrium theory has centred on turnover rates in avifauna on islands and the methods of evaluating these rates (Diamond 1969, 1971; Power 1972; Terborgh and Faaborg 1973; Hunt and Hunt 1974; Lynch and Johnson 1974; Jones and Diamond 1976; Diamond and May 1977; McCoy 1982; Saunders and

[1]CSIRO Division of Wildlife and Rangelands Research, L.M.B. No. 4, Midland, Western Australia 6056.

de Rebeira 1985). Only land birds are considered in discussions about turnover of island avifauna, yet marine species do have an impact on islands in various ways (nutrient enrichment via guano deposits, trampling, burrowing, predation, etc.). When the island analogy is applied to reserves isolated by land clearing, little account is taken of the species which move into the area as a result of the creation of suitable habitat. Some of these species, called 'marine' species in this chapter, do have a significant impact on some of the species which existed in the area before disturbance created a habitat isolate.

The short-billed white-tailed black cockatoo *Calyptorhynchus funereus latirostris* Carnaby (now called Carnaby's cockatoo) is a species which has been extensively studied and has been adversely affected by the fragmentation of its habitat by land clearing and by interaction with some of the 'marine' species which fragmentation has brought into contact with it. This chapter discusses the biology of this cockatoo, investigates the effect of habitat fragmentation on the species and presents a method for monitoring populations on remnants of native vegetation.

Carnaby's cockatoo is endemic to the south-west corner of Western Australia with the bulk of its breeding population occurring in the area receiving between 300 and 750 mm of annual average rainfall (Fig. 1). This region is now the area of intensive agriculture and extensive tracts of native vegetation have been cleared for cereal cropping and sheep grazing. As a result, there are thousands of remnants of native vegetation scattered throughout this area, isolated to varying degrees and surrounded by a sea of exotic species in the form of cereal crops (mainly wheat) and pasture grasses as well as agricultural weeds like *Erodium* spp. and doublegee *Emex australis* (Fig. 2).

Before Europeans changed this region (now called 'the wheatbelt'), this cockatoo was distributed throughout the area with the exception of the

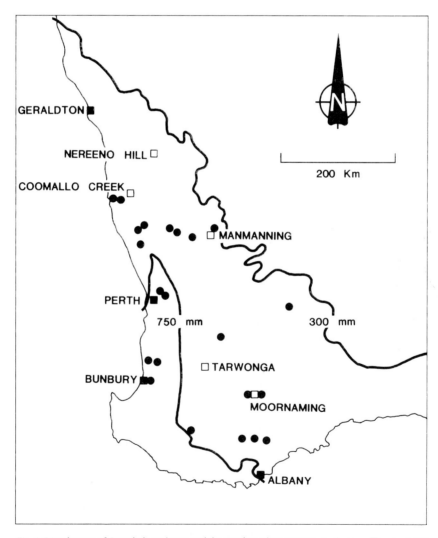

Fig. 1. Distribution of Carnaby's cockatoo with known breeding areas ●, study areas □ and rainfall isohyets indicated.

▷

PLATE 20

CHAPTER 22

Carnaby's cockatoos feeding in roadside native vegetation. The narrow linear distribution of native vegetation (see Plate 19) means that animals feeding in these situations are exposed to 'predation' by automobiles. (L. Moore).

◁

PLATE 21

CHAPTER 22

Wandoo woodland at Coomallo Creek. Recent clearing has removed the native heath surrounding the woodland prior to burning to prepare the area for agriculture. (D. Saunders).

▷

PLATE 22

CHAPTER 22

Carnaby's cockatoo. (G. Chapman).

△ PLATE 23   CHAPTER 25
Euros on Barrow Island using the shade provided by part of the oilfield installations. (H. Butler).

△ PLATE 24   CHAPTER 27
Brigalow remnant surrounded by farmland in northwest New South Wales. Note clearing on the periphery and directly through the remnant. (R. Dick).

▽ PLATE 25   CHAPTER 4
The Long-nosed Potoroo, *Potorous tridactylus* can persist in a fragmented and disturbed forest provided there are patches of suitable habitat linked by forest to allow dispersal between patches. (G. Coulson).

▽ PLATE 26   CHAPTER 2
Female Noisy Scrub-bird, *Atrichornis clamosus,* leaving the nest carrying a faecal sac. (G. Chapman).

Fig. 2. Aerial photograph of Manmanning in the north-central wheatbelt of Western Australia. Note the various habitat isolates created as a result of land clearing for agriculture. Photograph supplied by the Department of Lands and Surveys.

extreme south-west corner of the state between Bunbury and Albany (Fig. 1) which was occupied by the long-billed white-tailed black cockatoo *C. baudinii* Lear (now called Baudin's cockatoo) (Saunders 1979a). Carnaby's cockatoo bred in hollows in *Eucalyptus* spp. (Saunders 1979b) and fed predominantly on the seeds of Proteaceous species *(Banksia, Dryandra, Grevillea* and *Hakea)* (Saunders 1980) which were widely distributed over the region. Since European settlement, the widespread clearing of native vegetation has seriously depleted the food supply of this bird as it does not feed on human introduced cereal crops. Seed of *Erodium* sp. is the only agricultural weed which provides a significant food source for this bird. Unfortunately this seed is only available for a limited period during the breeding season and is not a reliable food supply in the breeding areas (Saunders 1980).

METHODS

The breeding biology of Carnaby's cockatoo was studied in detail at Coomallo Creek and Manmanning between 1969 and 1976 (Fig. 1). These study areas have been described in detail and the results of this part of the investigation have been published elsewhere (Saunders 1979b, 1980, 1982). Both study areas were established in 1969 when several visits were made late in the breeding season. In 1970 the study area at Manmanning was searched completely and all suitable hollow trees examined at least twice per month for evidence of nesting by Carnaby's cockatoo. During each breeding season from 1971 until 1976 the area was visited at approximately weekly intervals and the area was searched three times in 1977 but the species had become extinct in the district and has not bred there since.

Coomallo Creek had a greater number of suitable nesting hollows than Manmanning and was visited on the same basis as Manmanning between 1970 and 1977 but the number of trees being examined was increased from 1969, with all available hollows being surveyed from 1972. From 1978 Coomallo Creek was visited twice during each breeding season, with the exceptions of 1979 and 1980 when the area was not visited.

In addition to these two intensive study areas, populations of Carnaby's cockatoo at three other areas have been examined less intensively. During the breeding seasons of 1969 and 1970 a total of 27 nests at Tarwonga and 25 at Moornaming were examined as were a total of 16 nests at Nereeno Hill during 1974 and 1976 and again in 1979 (Fig. 1). The

Tarwonga and Moornaming sites were described by Saunders (1979b) and the Nereeno Hill site by Saunders et al. (1982).

In the five study areas the contents of nest hollows were noted each visit, nestlings were weighed and the length of the folded left wing measured. Nestlings were aged by comparing length of folded left wing with growth curves for nestlings of known age (Saunders 1982). Laying dates were extrapolated from estimated hatching dates.

RESULTS

The results of the intensive study of the breeding biology of Carnaby's cockatoo at Coomallo Creek and Manmanning between 1969 and 1976 are summarized in Table 1. The birds at Coomallo Creek were significantly more successful ($0.01 > P > 0.001$) and produced almost twice as many offspring per breeding unit as the birds at Manmanning (i.e. two young produced every three years compared with only one). At no stage did the range of breeding success overlap between the two areas (Table 1). There was no difference between areas in the length of time between hatching and the nestlings leaving the nests (fledging) but nestlings at Coomallo Creek were larger and heavier at fledging than those at Manmanning. During the first 45 days after hatching (when rate of increase in weight or body mass was linear) nestlings at Coomallo Creek averaged daily weight gains which were nearly 33% greater than those at Manmanning (Table 1).

Table 1. Breeding data for populations of Carnaby's cockatoo breeding at Coomallo Creek and Manmanning between 1970 and 1976. Data from Saunders (1982).

Breeding success*	Coomallo Creek	Manmanning
7 year average	0.66	0.35
Range	0.56-0.86	0.07-0.50
Average fledgling weight of nestlings (g)	584	501
Average fledgling weight as % of mean adult weight	89	76
Average daily weight gain for nestlings up to 45 days from hatching (g). #	13.0	9.8
Average length of folded left wing at fledging (mm)	324	300
Average length of folded left wing at fledging as % of mean adult wing length	89	82

*Breeding success is $\frac{\text{Number of nests producing free-flying young}}{\text{Total number of nests in area}}$

\# Over the first 45 days from hatching rate of increase in weight was linear.

Females in both areas performed all the incubation. At Coomallo Creek, during this stage of the nesting cycle and the first 14 days of the nestlings' life, the females spent little time outside the nest hollows, being dependent on the males for food. At Manmanning females did leave the nest hollows during the incubation and early nestling stages to forage for themselves. In addition, males at Coomallo Creek returned to the nest site at mid-morning and dusk to feed the females. During the remainder of the nestling stage, both adults foraged and returned to feed the nestling at mid-morning and dusk. At Manmanning, males often did not return at mid-morning, feeding the females only at dusk. Later in the nestling stage when both adults were foraging together they often did not return at mid-morning, feeding the nestling only once per day.

The modifications of behaviour and the differences in growth rates and breeding success were directly attributable to limitations in food supply at Manmanning compared with Coomallo Creek and this aspect of the work is discussed in detail in Saunders (1980, 1982).

At Manmanning between 1969 and 1976 the breeding season (based on the date the first egg was laid in each nest hollow) was from the last week in July until the second week in October (76 nests). At Coomallo Creek between 1969 and 1984 it was from the first week in July to the third week in November (532 nests), nearly six weeks longer than at Manmanning (Fig. 3). In fact the breeding season was long enough for one nest hollow to be used successfully twice (by two different pairs) in the same breeding season. During the breeding season of 1974 one nest hollow was in continuous use from the third week in July until the third week in March the following year (eight months). The start of the breeding season varied with the season: from 29 July (1969) to 2 September (1972) at Manmanning and 9 July (1984) to 26 August (1978) at Coomallo Creek.

Since 1977 Coomallo Creek has been visited at least twice during each breeding season (with the exceptions of 1979 and 1980); once during mid-September and again during mid-November. These two survey periods are shown in Figure 3 and by the first survey period about 67% of all 532 nests had been started. This obviously varied between seasons, but all nests had started by the latter survey period. These survey periods were chosen to provide the maximum information from two visits per breeding season. The first period sampled the oldest nestlings and they could be measured. By the second period these nestlings had fledged (or died) but the youngest ones could be measured.

The same two survey periods could have been used at Manmanning to monitor the population as 58% of all nests had started by mid-September and the earliest nests had young nestlings. All laying had ceased by the second period with the oldest nestlings about to fledge or dead (Fig. 3).

Laying dates for the 27 nests at Tarwonga, 25 at Moornaming and 16 at Nereeno Hill (Saunders 1986) show that the breeding seasons at each of

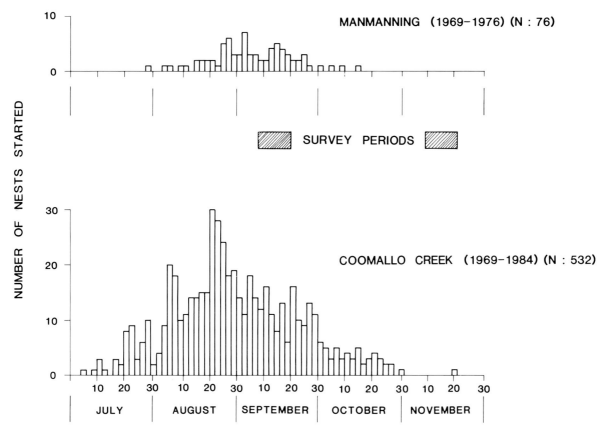

Fig. 3. Breeding season of Carnaby's cockatoo at Manmanning and Coomallo Creek based on the dates of the first egg laid in each nest hollow for which laying date could be determined. The timing of the two survey periods (cross-hatched rectangles) is shown in relation to breeding season.

these areas fell within those of Coomallo Creek and Manmanning, indicating that these two survey periods could be used to monitor breeding populations over the entire range of Carnaby's cockatoo (Fig. 1).

The nesting data for both Coomallo Creek and Manmanning for the intensive period of the study (1970-1976) have been re-examined as if the areas were only monitored twice per breeding season and the results of this analysis compared with results obtained from visits made at weekly intervals throughout each breeding season (Fig. 4). The estimate of total nests in an area based on two visits per breeding season averaged 93% of the total found by visiting an area weekly at Manmanning (seven year average, range 81-100%) and 92% at Coomallo Creek (seven year average, range 88-97%) (Fig. 4) and demonstrated the same trends between years at both areas. At Manmanning there was a decline in the number of nests in the area from 1971 and the birds became extinct as a breeding species in the district in 1977. At Coomallo Creek the apparent increase in number of nests in the area between 1970 and 1972 was due to the increased amount of the study area being searched but there was a real fall in the number of nests in the area between 1975 and 1977 and since then the population has remained at around 44 pairs, just under half of the number using the area in 1975.

The number of successful nests at Manmanning and Coomallo Creek between 1970 and 1976 based on weekly visits and two visits per breeding season is shown in Figure 4. Indices based on two visits per breeding season overestimated the number of successful nests because nests which appeared to contain healthy nestlings may have failed after the second visit, but the indices did show the same trends between years as demonstrated by number of successful nests based on weekly visits. These figures have been converted to show breeding success (number of successful nests divided by the total number of nests, expressed as a percentage) for each area between 1970 and 1976 for weekly surveys and two visits per breeding season (Fig. 5). At Manmanning breeding success fluctuated widely, averaging 35% (range 7-50%) based on weekly visits and 55% (17-83%) based on two visits per season. At Coomallo Creek breeding success averaged 66% (56-86%) based on weekly visits, 80% (68-92%) based on two visits per season and fluctuated around the average. In both areas the trends demonstrated by the different sampling techniques were the same (Fig. 5).

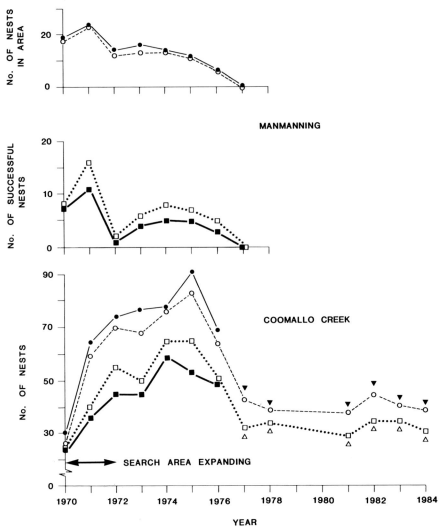

Fig. 4. The total number of nests based on weekly visits ● and on two visits per breeding season ○ are shown for Manmanning (1970-1976) and Coomallo Creek (1970-1984); predicted total number of nests at Coomallo Creek (1977-1984) ▼. The number of successful nests based on weekly visits ■ and on two visits per breeding season □ are shown for Manmanning (1970-1976) and Coomallo Creek (1970-1984); predicted number of successful nests at Coomallo Creek (1977-1984) △.

The data obtained from two visits per breeding season to Coomallo Creek from 1977 have been used to calculate corrected total numbers of nests and numbers of successful nests for each breeding season since 1977 (excluding 1979 and 1980 for which there were no data). The figures for the corrected total number of nests are calculated by dividing the total number of nests for the area (based on two visits) by 1.11 (which is the average value of the overestimate given by two visits per breeding season). The corrected number of successful nests is calculated by dividing the number of successful nests (estimated by two visits) by 0.92 (which is the average underestimate given by two visits) (Figs 4 and 5). Between 1970 and 1984 there was no change in breeding success for the population at Coomallo Creek, despite the fact that the total breeding population dropped to around 44 pairs, just less than half of the peak population of 1975. Most of that decrease took place over 1976 and 1977.

EXTENT OF CLEARING FOR AGRICULTURE IN EACH STUDY AREA

Each of the five areas in which populations of Carnaby's cockatoo were studied had different histories of farming settlement. Manmanning had been settled in the late 1920s and by 1969 had been extensively cleared for agriculture (Fig. 2). Examination of aerial photography taken in August 1962 revealed that 91% of the study area had been cleared of native vegetation and of the remainder, 7% was heath and 2% woodland. By January 1982 farmland and development occupied 92% with heath reduced to 6%. Remnants of native vegetation were isolated from each other and road verges were narrow (10-15 m) with little or no native vegetation along them.

Coomallo Creek presented a completely different pattern of development. Because of poor quality soils (low in nutrients and deficient in trace elements) this area was not developed until

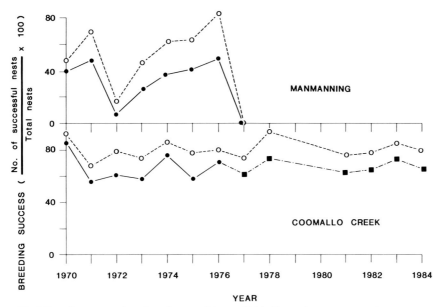

Fig. 5. Breeding success (number of successful nests divided by total nests, expressed as a percentage) based on weekly visits ● and two visits per breeding season ○ is shown for Manmanning (1970-1976) and Coomallo Creek (1970-1984). Breeding success calculated from Coomallo Creek data (1977-1984) based on two visits per breeding season projects the curve for weekly visits ■

comparatively recently. In June 1959 about 10% of the area was cleared of native vegetation and of the remainder, 8% was woodland and 82% heath. By December 1969 (when this study started) this had changed to 33% developed, 5% woodland and 62% heath. During the mid-1970s there was accelerated development in the area and by June 1982, 66% of the area was cleared, 4% woodland and 30% heath. Despite this extensive clearing, road verges were wide (200 m) with uncleared native vegetation linking areas of uncleared vegetation around the district. The other three areas show patterns intermediate between those of Coomallo Creek and Manmanning in two cases, or identical to Manmanning in the third (Nereeno Hill). Nereeno Hill was settled in the mid-1920s and extensively cleared. By December 1969 it consisted of 93% cleared land, 5% heath and 2% woodland with narrow verges with little or no vegetation on them. Tarwonga consisted of farmland bordering State Forest and it contained 73% cleared land, 19% woodland and 8% heath by October 1970. Road verges were narrow but had native vegetation lining them and patches of native vegetation were not visually isolated one from another. The Moornaming district had been settled early this century and by January 1971 contained 83% developed land, 8% woodland and 9% heath. Road verges were much wider (40-50 m) than those at Manmanning, Tarwonga and Nereeno Hill (10-15 m) with some native vegetation along them and patches of remnant vegetation were not visually isolated.

The breeding success for the populations of Carnaby's cockatoo at Manmanning, Coomallo Creek (1970-1976), Coomallo Creek (1977-1984), Nereeno Hill, Tarwonga and Moornaming was graphed against the percentage of uncleared native heath or shrubland remaining (Fig. 6). It is obvious that those populations with the lowest breeding success (Manmanning and Nereeno Hill) were in areas with the least and the most isolated uncleared native heath or shrubland. Both of the populations at Tarwonga and Moornaming enjoyed breeding successes of the order enjoyed by the population at Coomallo Creek but they occurred in areas with less native vegetation. In both areas there were roadside verges connecting larger remnant areas and most of the patches were not visually isolated.

DISCUSSION

Like most animals, Carnaby's cockatoo requires two important resources; breeding sites and food. Unfortunately the habitat that provides one does not usually provide the other. The birds nest in any species of *Eucalyptus* which provides a hollow of suitable dimensions in their breeding area (Saunders 1979b). *Eucalyptus calophylla* was the only species which also provided food but it only occurred at Tarwonga. In the other study areas the seeds of the nesting trees were not eaten. In fact, the woodland had an impoverished understorey and was not a suitable feeding area. Before the native vegetation was cleared, the patchy distribution of woodland interspersed with areas of heath and scrub favoured Carnaby's cockatoo and it was the dominant cockatoo throughout this region (Saunders *et al.* 1985). At present there are ample areas of suitable nesting habitat but as pointed out by Saunders *et al.* (1982) this is deteriorating and the

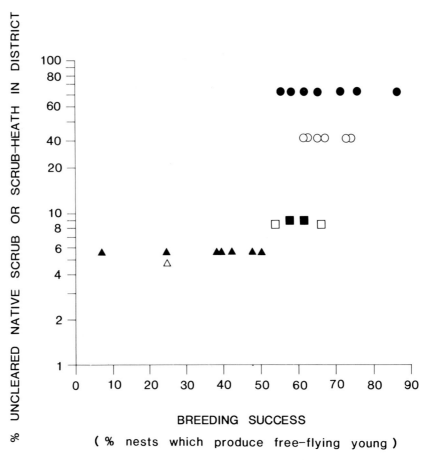

Fig. 6. Breeding success for populations at Manmanning (1970-1976) ▲, Coomallo Creek (1970-1976) ●, Coomallo Creek (1977-1984) ○, Nereeno Hill (4 year average) △, Tarwonga (1969, 1970) □ and Moornaming (1969, 1970) ■ is shown against percentage of uncleared native heath and scrub remaining in the study area.

lack of regeneration means that in future the supply of nest sites will not be sustained. There is an urgent need to investigate this potential loss of woodland and look at ways of preventing further degradation and to re-establish regeneration.

The presence of woodland was often taken as an indication of good agricultural soil and these areas were among the first taken up as settlement spread. Heath and scrub were easier to clear, leading to the establishment of many patches of woodland, isolated to varying degrees. The original patchy distribution of woodland forced hole nesting species to concentrate on these areas during the breeding season (Saunders 1980). Clearing for agriculture has compounded this situation by decreasing the areas of available woodland and allowing other species ('marine' species) to move into areas which were formerly unsuitable for them. Coomallo Creek, for example, has seen a decrease in area of woodland from 8% of the total area in 1959 to 4% in 1982. As pointed out by Saunders (1979b) prospecting females of Carnaby's cockatoo tend to interact with each other forcing a dispersal over the available nesting habitat. The breeding season is limited and the birds must breed at that time, but as clearing destroys suitable nesting habitat the population which may be supported, falls. This was demonstrated at Coomallo Creek where the population dropped significantly following a period of habitat destruction.

Before settlement by Europeans, the galah *Cacatua roseicapilla* was a bird of the water courses of arid and semi-arid Australia (Saunders *et al.* 1985). With the extensive clearing for agriculture, the planting of cereal crops and pasture and the provision of watering points for stock, large expanses of 'riverine plain' have been established. As a result, 'islands' of native vegetation have been created, surrounded by extensive areas which are suitable for colonization by 'marine' species: species such as the galah, which can successively exploit this 'sea of change'. These changes have created the situation where species which are entirely dependent on native vegetation for survival have to contend with changes in distribution of habitat and compete with some of the species attracted by these changes. Some of these 'marine' species, like the galah, obtain their food from the agricultural land but nest in the remnants and so they exert some influence on the remnants or their inhabitants. As more agricultural land is developed and sown to cereal crops so the population of the galah increases as has been happening over the last 15 years at Coomallo Creek (Raffan pers. comm.). Although the galah and Carnaby's cockatoo do prefer hollows of slightly different size and shape (Saunders et al. 1982) there is an overlap and as the population of the galah increases in an area, so the potential for conflict over nesting hollows increases. This was minimal at Coomallo Creek during the period 1970-1976 but at Manmanning there were conflicts between the two species and several nesting attempts failed as a result. When a population is already experiencing difficulties as a result of food shortages, conflict over nesting resources may hasten its decline.

Despite the halving of the population at Coomallo Creek there had been no change in breeding success (Fig. 5) because there were still large tracts of heath over which the birds foraged. Saunders (1986) demonstrated that the nestlings being produced over this period were a similar size and weight to those produced before the population decline, which indicated that food was not a limiting factor. Saunders (1986) also demonstrated a method for assessing the health of nestlings. This was done by measuring and weighing each nestling when the contents of the nest hollow were examined. Nestlings were aged from a curve for growth of folded left wing and the weights of the nestlings plotted on a growth curve for weight constructed from data from 246 nestlings measured at Coomallo Creek between 1970 and 1976. If the weight of the nestlings being examined fell within that of the Coomallo Creek nestlings of the same age, the population was producing healthy nestlings and was not facing shortages of food. This method was used to assess the population at Coomallo Creek breeding between 1977 and 1984 and the nestlings were all within the growth curve for the population breeding there between 1970 and 1976. Despite the halving of the population, breeding success and nestling growth remained the same and this population is probably secure in the long-term, provided its nest sites are protected. The populations at Tarwonga and Moornaming were also producing healthy nestlings but when the method was applied to nestlings produced at Manmanning and Nereeno Hill, both were shown to be producing nestlings which were far lighter than the population norm as a result of shortage of food. Since then the population at Manmanning has become extinct and that at Nereeno Hill is on the way to extinction.

This study has demonstrated that habitat fragmentation has posed problems for Carnaby's cockatoo but that despite a decrease in the size of populations after development for agriculture this need not have resulted in the local extinction of the species. Even in more recently developed areas the future is still not secure, as nest sites and food sources need to be protected, particularly where they are on private land. Corridors linking remnant patches and retention of a patchwork which is not visually isolated will enhance the chances of the long-term survival of this cockatoo.

The degradation of native vegetation on remnants and the lack of regeneration, particularly of woodland in the wheatbelt of Western Australia (Saunders et al. 1982) means that the future of many 'island' species like Carnaby's cockatoo is insecure. There is an urgent need for research on revegetating with native species, areas of land needed for various purposes: to reduce erosion by wind or water; to reduce the salinity of ground water; and to enhance the conservation potential of the system of remnants already in existence by linking them with corridors or providing 'stepping stones' within sight of each other so the more mobile species may move between remnants. This research should concentrate on the re-establishment of ecosystems rather than the planting of one or two species which have been selected for their known ability to thrive. It is only by working on this problem now that there is any chance of being able to reverse this degradation of our natural resources.

ACKNOWLEDGEMENTS

C. P. de Rebeira drew the figures and C. Taplin typed the manuscript; both are members of CSIRO Division of Wildlife and Rangelands Research. R. Raffan, P. Paish, B. Smith, E. Avery, D. and J. Wilson allowed access to their properties throughout the study. A. H. Burbidge and O. Nichols provided constructive criticism of an earlier draft of the manuscript. To these and many more, we are extremely grateful.

REFERENCES

Boecklen, W. J. and Gotelli, N. J., 1984. Island biogeographic theory and conservation practice: species — area or specious — area relationships? *Biol. Conserv.* 29: 63-80.

Diamond, J. M., 1969. Avifaunal equilibria and species turnover rates on the Channel Islands of California. *Proc. Natl. Acad. Sci. USA* 64: 57-63.

Diamond, J. M., 1971. Comparison of faunal equilibrium turnover rates on a tropical and a temperate island. *Proc. Natl. Acad. Sci. USA* 68: 2742-5.

Diamond, J. M., 1975. The island dilemma: lessons of modern biogeographic studies for the design of natural reserves. *Biol. Conserv.* 7: 129-46.

Diamond, J. M. and May, R. M., 1977. Species turnover rates on islands: dependence on census interval. *Science* 197: 266-70.

Hunt, G. L. and Hunt, M. W., 1974. Trophic levels and turnover rates: the avifauna of Santa Barbara Island, California. *Condor* 76: 363-9.

Jones, H. L. and Diamond, J. M., 1976. Short-time-base studies of turnover in breeding bird populations on the California Channel islands. *Condor* 78: 526-49.

Lynch, J. F. and Johnson, N. K., 1974. Turnover and equilibrium in insular avifaunas, with special reference to the California Channel Islands. *Condor* 76: 370-84.

MacArthur, R. H. and Wilson, E. O., 1963. An equilibrium theory of insular zoogeography. *Evolution* 17: 373-87.

MacArthur, R. H. and Wilson, E. O., 1967. The Theory of Island Biogeography. Princeton Univ. Press, Princeton.

Margules, C., Higgs, A. J. and Rafe, R. W., 1982. Modern biogeographic theory: are there any lessons for nature reserve design? *Biol. Conserv.* 24: 115-28.

McCoy, E. D., 1982. The application of island-biogeographic theory to forest tracts: problems in the determination of turnover rates. *Biol. Conserv.* 22: 217-27.

Power, D. M., 1972. Numbers of bird species on the California Islands. *Evolution* 26: 451-63.

Saunders, D. A., 1979a. Distribution and taxonomy of the White-tailed and Yellow-tailed Black Cockatoos *Calyptorhynchus* spp. *Emu* 79: 215-27.

Saunders, D. A., 1979b. The availability of tree hollows for use as nest sites by white-tailed black cockatoos. *Aust. Wildl. Res.* 6: 205-16.

Saunders, D. A., 1980. Food and movements of the short-billed form of the white-tailed black cockatoo. *Aust. Wildl. Res.* 7: 257-69.

Saunders, D. A., 1982. The breeding behaviour and biology of the short-billed form of the white-tailed black cockatoo *Calyptorhynchus funereus*. *Ibis* 124: 422-55.

Saunders, D. A., 1986. Breeding season, nesting success and nestling growth in Carnaby's cockatoo *Calyptorhynchus funereus latirostris* over 16 years at Coomallo Creek and a method for assessing the viability of populations in other areas. *Aust. Wildl. Res.* 13: 261-73.

Saunders, D. A. and de Rebeira, C. P., 1985. Turnover in breeding bird populations on Rottnest Island, Western Australia. *Aust. Wildl. Res.* 12: 467-77.

Saunders, D. A., Rowley, I. and Smith, G. T., 1985. The effects of clearing for agriculture on the distribution of cockatoos in the south-west of Western Australia. Pp. 309-21 *in* Birds of Eucalypt Forests and Woodlands: Ecology, Conservation, Management ed by A. Keast, H. F. Recher, H. Ford and D. Saunders. Surrey Beatty & Sons, Sydney.

Saunders, D. A., Smith, G. T. and Rowley, I., 1982. The availability and dimensions of tree hollows that provide nest sites for cockatoos (Psittaciformes) in Western Australia. *Aust. Wildl. Res.* 9: 541-56.

Terborgh, J. and Faaborg, J., 1973. Turnover and ecological release in the avifauna of Mona Island, Puerto Rico. *Auk* 90: 759-79.

CHAPTER 23

Management of Remnant Bushland for Nature Conservation in Agricultural Areas of Southwestern Australia — Operational and Planning Perspectives

K. J. Wallace[1] and S. A. Moore[2]

Within inland agricultural areas of southwestern Australia, land set aside for nature conservation is largely contained in a system of nature reserves. Over much of this region little further land can be acquired for conservation, and management of the present reserve system has become an important consideration. Currently, management can be typified as 'crisis' management. However, to achieve nature conservation objectives, management must become pre-emptive in character. If this change in management style is to be achieved, then the following are required:

1. a research data base;
2. a technical data base;
3. an informed and sympathetic public;
4. staff and financial resources; and
5. an accepted philosophy/methodology for drawing together 1-4 and implementing management.

Management problems in relation to resources and the general public are highlighted. Written management plans provide one means of solving some of these problems and this point is illustrated by describing two case studies of the development of management plans.

INTRODUCTION

WITHIN Western Australia, broadscale clearing of natural vegetation has almost entirely been restricted to the south-west where farming is the predominant land use (Fig. 1). Between 1949 and 1969 the area of cleared land on farms increased from 6.48 million ha to 13.77 million ha (Burvill 1979a) and land clearing for agriculture has continued until the present. Therefore depletion of natural vegetation in the south-west has been a comparatively recent event which has resulted in many remnants of natural vegetation scattered within the cleared, agricultural areas.

The high density of rare and geographically restricted plant species (Hopper and Muir 1984) and the presence of fauna which are rare or at risk [such as the woylie *Bettongia penicillata ogilbyi*, numbat *Myrmecobius fasciatus* and red-tailed wambenger *Phascogale calura* (Strahan 1983)] within natural vegetation in agricultural areas emphasizes the importance of these remnants for nature conservation. While a few fauna species have readily adapted to or been favoured by the new, largely agricultural habitat, most depend on the remaining pockets of natural vegetation for their survival. Already 13 species of mammals have disappeared from agricultural districts and less than half of those which originally occurred are regarded as common (Kitchener *et al*. 1980; Saunders 1985). Also, in a recent study of 277 plant species recorded from these areas, Patrick (1985) concluded that 58 are possibly extinct, 56 extremely rare, and 154 species poorly known or from very small populations.

---

[1]Department of Conservation and Land Management, P.O. Box 811, Katanning, Western Australia 6317.
[2]Department of Conservation and Land Management, Murdoch House, 5 The Esplanade, Mount Pleasant, Western Australia 6153.
Pages 259-68 *in* NATURE CONSERVATION: THE ROLE OF REMNANTS OF NATIVE VEGETATION ed by Denis A. Saunders, Graham W. Arnold, Andrew A. Burbidge and Angas J. M. Hopkins. Surrey Beatty and Sons Pty Limited in association with CSIRO and CALM, 1987.

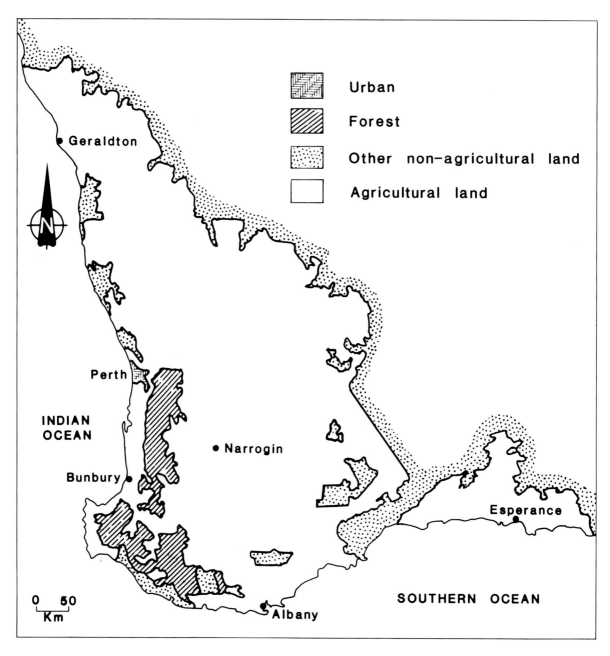

Fig. 1. Land use in southwestern Australia. Adapted from Murray (1979, Fig. 11.2).

Within most agricultural areas away from the coast and State forest (Fig. 1), land set aside for nature conservation is restricted to a system of nature reserves which was largely established during the 1960s and 1970s. The majority of these reserves have the purpose of 'Conservation of Flora and Fauna' and are vested in the National Parks and Nature Conservation Authority. However, others have various purposes which include nature conservation together with some other purpose, such as water catchment, and control may not be vested in the National Parks and Nature Conservation Authority. While a management research officer was appointed by the State Government in 1968, operational management and management planning with respect to nature reserves did not gain impetus until the mid to late 1970s (A. A. Burbidge pers. comm.).

That land acquisition for nature conservation and wildlife refuges has preceded management capability is not unique to Western Australia. The Nature Conservancy in Britain (Stamp 1974, page 51) and State wildlife agencies in the United States (Kruckenberg 1985) have had similar histories.

This chapter examines both operational and planning approaches to management of remnant vegetation in southwestern Australia. The first section, dealing with operations, examines the requirements for successful management and briefly details management objectives and the management process. Planning solutions to some of the management problems raised are provided in the second section of the chapter which describes management plan preparation and development through an examination of two case studies.

OPERATIONAL PERSPECTIVE

The agricultural areas of southwestern Australia can be broadly divided into two types — areas of high rainfall near the south and west coasts, and drier inland areas (Fig. 2). To provide a statistical boundary, and to define a relatively homogenous area with respect to management operations, discussion in this section will be restricted to the Wheatbelt Region (Fig. 2) as defined by the Western Australian Department of Conservation and Land Management. This Region covers some 14.4 million ha and includes a large section of the agricultural land in Western Australia.

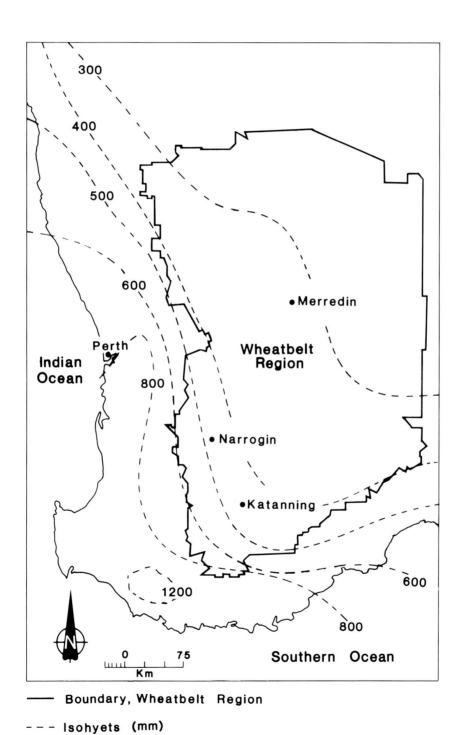

Fig. 2. Wheatbelt Region of the Department of Conservation and Land Management. Isohyets for mean annual rainfall adapted from Burvill (1979b, Fig. 6.1).

Parts of seven botanical districts (Beard 1980) occur in the Region which is situated within a broad transition zone between the inland arid zone and the wetter south-west (Hopper 1979). Major vegetation formations include woodlands, mallee and kwongan scrub. The most obvious physical features in the gently undulating topography are granite outcrops, laterite breakaways and salt lake systems.

Land set aside for nature conservation within the Region is contained within 639 nature reserves (numbers are based on data as at September 1985). There are also two areas of State forest which are in part managed for the conservation of flora and fauna. While other reserves and privately-owned bushland are important for nature conservation, the long term tenure and usage of such remnants is less certain.

The 639 nature reserves within the Region range in size from 0.4 to 309,000 ha, and have a median size of 114 ha. The latter figure emphasizes that most reserves are small. Together the 639 reserves total about one million ha and account for some 6.7% of the Region's surface area, however, this is reduced to 2.4% if the three largest reserves are excluded. Within the western part of the Region little further vegetation remains which could be set aside for nature conservation. With few exceptions, nature reserves fall clearly into the category of remnant vegetation.

Management Objectives and Management Process

The management objectives for nature reserves within the Wheatbelt Region are:

1. to maintain representative samples of the natural landscape and biota occurring within the Region. Within this objective it is recognized that —

   (a) replicate samples of each habitat type are required, and
   (b) where natural land important for flora and/or fauna exists outside reserves, then inclusion of these areas into the reserve system should be sought provided certain criteria are met;

2. to conserve rare species;

3. to maintain ecosystem processes; and

4. to develop in local communities an appreciation of nature conservation values and, in particular, these values as expressed in the system of nature reserves.

The management process operating to achieve these objectives is summarized in Figure 3. In this process management plans, tactics and objectives are derived on the basis of existing data subject to broad scale (for example State or National) and local social, economic and political constraints. Proposed management is also influenced by previous management action, and the current philosophy and methodology of management.

It is important to stress that all elements listed in the process are inter-related and that the management process is continuous. Furthermore broad scale social, economic and political constraints may differ greatly from those operating at a local level. Failure to recognize this may lead to inappropriate management decisions, especially when dealing with remnant vegetation scattered through a large region. Finally, process diagrams such as that in Figure 3 do not indicate where emphasis is placed; that is, whether or not emphasis is on planning; or research; or implementation; or social, economic and political constraints. The current emphasis within the Wheatbelt Region is on management action in relation to local social, economic and

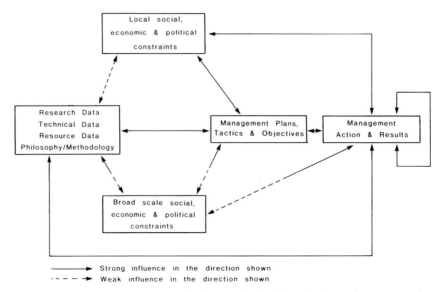

Fig. 3. Management Process. Note that the link between local and broad scale social, economic and political constraints is not shown. The interaction between the two varies considerably.

political constraints. While these local factors will always have an important influence on management, the current situation also reflects the fact that research and technical data bases are at an early stage of development.

Requirements For Successful Management

To successfully implement the management process outlined and achieve management objectives, the manager requires:

1. a research data base;
2. a technical data base;
3. an informed and sympathetic public;
4. resources (staff and financial); and
5. an accepted philosophy/methodology for drawing together 1-4 and implementing management.

Research Data Base. That successful management requires a broad research data base, particularly from the ecological fields, is apparent from the variety and content of chapters presented in this volume. Furthermore the important role for managers in research has been discussed in this volume by Hopkins and Saunders. With one exception, sociological research, the main research themes relevant to management of remnant areas are discussed in this book. Sociological research has the potentially important function of suggesting methods for managing human usage of reserves and informing the public of conservation issues. Remnant areas, with their high perimeter/area ratio, have a large interface with human activities in comparison with extensive tracts of vegetation. Therefore, sociological research may be particularly relevant to managers of these areas.

Two further points should be raised. Firstly, it is generally recognized that close co-operation between management and research staff is essential to ensure both the relevance of applied research and optimum involvement of managers. However, it must be emphasized that this relationship between research workers and management staff should occur as early as possible in the development of research projects. For example, two equally attractive sites for research may have quite different implications for management. Given that sites are often selected during the early stages of project development, liaison is essential to ensure that both management and research gain the maximum benefits from a particular study.

Secondly, while most researchers are constrained to short term projects, an understanding of long term changes is basic to successful management. Therefore short term studies involving survey should be conducted within the context of a long term data acquisition strategy, for example works should be site specific. This is of added importance in the case of remnant vegetation which may be drastically affected by a single event, such as a wildfire. Often nearby vegetation containing similar habitat types does not exist, thus long term data collection provides the only means of assessing the impact of these events.

Technical Data Base. The function of the technical data base is to provide the most effective and economic means of implementing management strategies. Information includes data on the most suitable machinery, fire protection techniques, methods of controlling pest species, and so on. While the distinction between research and technical data bases is blurred, managers will most often be responsible for generating technical data on the basis of experience in the field. Two important elements of an adequate technical data base are a core of experienced managers, and a written record of management experience. These requirements are only partially fulfilled in the Wheatbelt Region, and their development is a high priority.

An Informed and Sympathetic Public. The fact that managers of Nature Reserves, at least in the Wheatbelt Region, allocate a significant proportion of their time to working with people would surprise most of the general public, but few managers. There are several reasons for this heavy expenditure of time.

Firstly, the many small reserves in the Region result in a large number of landholders with land adjoining departmental land. This, combined with 42 different Local Government Authorities, entails an important and demanding liaison function.

Secondly, islands of vegetation attract both desirable and undesirable public attention. The negative aspects are demonstrated by a recent survey which recorded illegal usage of nature reserves in part of the Region for gravel and sand extraction, rubbish dumping and timber cutting. While some of the usage recorded is historical and pre-dates the current conservation purpose of the land, it was found that 86% of the 90 reserves sampled have been used by people to the detriment of the reserve. This indicates the potential for reserves to deteriorate through time as a result of neglect. Of even greater concern is that 48% of the reserves had received usage detrimental to their conservation values within the 12 months preceding the study. The need for management is apparent, and the study reinforces the fact that if no-one cares for a particular area of vegetation and its management, then neglect and abuse will follow. It must be stressed that, on an area basis, the proportion of a small reserve which may be affected by human usage is likely to be high in comparison with large parks or reserves.

Thirdly, recreation on reserves is placing an ever-increasing workload on management personnel. As with other forms of human usage, the impact of recreation on small remnant areas is likely to be greater than on large areas of vegetation.

Working with people is in fact an important part of wildlife management because:

1. increased political support for nature conservation is needed;

2. local communities must be better informed and encouraged to be sympathetic to nature conservation. If rural communities are opposed to conservation, then it is doubtful whether reserves can be successfully managed (Burbidge and Evans 1976); and

3. managers must be attuned to community thinking and aspirations.

All the above are more difficult to achieve for remnants than for large areas such as national parks. For example, it is easier to win support for areas with well known scenic attractions or recreation values.

Resources. In the Wheatbelt Region there is one management person, including clerical staff, to about 64,000 ha of nature reserves. Given the level of biological management required for remnant bushland, and the human usage problems discussed above, there are not enough staff available to achieve the management objectives listed earlier. This point is emphasized by comparison with the International Union for Conservation of Nature and Natural Resources guideline of one management person for every 10,000 ha of land managed where the population density is less than 50 inhabitants per square kilometre (cited in Burbidge and Evans 1976), as is the case in the Wheatbelt Region.

Inadequate staff and funds for management can largely be attributed to two factors: the low political priority accorded to nature conservation; and given the large area and sparse population of Western Australia, the small resource base for the State as a whole.

Also relevant are four facts listed by Wright (1982) as contributing to resource management problems for small (less than 400 ha) historic sites in the United States National Park System. These are that:

1. they are often overshadowed in budget priority by larger, more renowned parks;

2. their visitor use is often low, meaning less political leverage;

3. their resource management problems tend to be small and are viewed, often wrongly, with less concern than those encountered in larger parks; and

4. they often do not have the staff and expertise to deal with the problems they face.

Each of these factors contributes to problems experienced by managers of remnant areas in Western Australia.

Finally, managing remnant vegetation scattered over a large area is inherently inefficient in that the amount of time spent travelling is proportionately much greater than is the case for those managing large, consolidated areas of vegetation from nearby or on-site. The effects of this are often underestimated or not recognized by those unfamiliar with land management in regions such as the Wheatbelt.

The importance of increasing public and political support for nature conservation has been stated above. However, even with increased public support, staff and funds available for management of remnant areas are likely to be inadequate to achieve all management objectives, at least within the short to medium term.

When resources are inadequate, three broad strategies for changes are available. These are to: change management objectives; seek alternative resources; and increase management efficiency. Management objectives are currently under review within the Region, however it is unlikely that they will be changed to any significant degree. The second strategy, that of seeking alternative resources, is currently being pursued. One means of gaining resources is to involve other organizations and groups and this is discussed below under planning. Finally, while managers must always strive to use resources with maximum efficiency, this is imperative when resources are scarce. In this context it is essential that management priorities are established and adhered to, and in particular, *it is important to do a little well, rather than a lot badly.*

Management Philosophy/Methodology. A set of management philosophies and methods, agreed to by researchers, planners and managers, is essential for effective management. Without such a set, those ingredients for successful management previously discussed will be inefficiently used, and the public face of management will be less effective. Unfortunately management philosophies may be assumed rather than written, and are sometimes esoteric from the viewpoint of field staff. Clearly senior staff, including field managers, have an important function in producing realistic management philosophies which field workers can accept.

The current review of management objectives within the Region should result in an improved management philosophy. However, to improve the overall approach to management will require a change from crisis management to pre-emptive management. The latter is typified by:

1. clear management objectives and methods;
2. management philosophies which are largely agreed to by all involved in management;
3. written plans of management;
4. management action which prevents problems arising rather than reacting to them; and
5. a written rather than oral history of management practices, techniques, and results.

Bringing about the transition to pre-emptive management is an important challenge facing managers in the Wheatbelt Region.

PLANNING PERSPECTIVE

Management plans are an effective means of resolving some management problems of remnant vegetation. This point is illustrated below by describing two case studies of planning for remnant vegetation in the south-west of Western Australia. The description also provides an overview of current planning procedures for areas of remnant vegetation.

Management plans are defined in this chapter as documents containing resource information, a statement of management objectives, strategies to achieve these objectives, and operational prescriptions. The plans described in the following case studies were prepared using the management planning process developed in 1980-81 by the Department of Fisheries and Wildlife, Western Australia (the Wildlife Section of this Department is now part of the newly formed Department of Conservation and Land Management). This process has the following features:

1. provision for a number of reserves, either of a similar habitat type or in the same general area (such as within a single Local Government Authority) to be covered by one management plan;
2. provision for extensive consultation with the relevant Local Government Authorities;
3. provision for full consultation with the public and for promotion of draft plans to encourage public comment;
4. special provision for consultation with reserve neighbours and members of the local community;
5. procedures for the assessment of plans, accompanied by analysis of public submissions, by the vested authority (i.e. the authority responsible for the management of land to which the plan refers); and

6. procedures to fulfil statutory requirements.

This process emphasizes the importance of involving the public, especially the local community, in the formulation and development of management plans.

Each case study described below refers to a particular management plan and both are Local Authority based, covering groups of nature reserves. The case studies provide examples of how the development of management plans has:

1. enabled a lack of management resources to be overcome (Case Study No.1); and
2. informed the public about management issues and encouraged a more sympathetic attitude to conservation (Case Study No. 2).

Case Study No. 1: Management With Inadequate Resources

One factor limiting management is a lack of staff and funds, a problem which has already been discussed in this chapter. One way such resources can be increased is by working with local communities and Local Government Authorities to obtain additional man-power, or funding for management works, or both.

One objective of the management plan for nature reserves in the Shire of Wyalkatchem (Moore and Williams 1986) was to address this problem of inadequate resources. The Shire of Wyalkatchem, 170 km north-east of Perth, lies in the heart of the Western Australian wheatbelt (Fig. 4). At the time of plan compilation (October 1984) nature reserves occupied 1.0% of the Shire area, however, 9.5% of the Shire was occupied by Crown reserves (including water, timber, railway, road and public utility reserves). The management plan covers seven nature reserves, ranging in size from 14 ha to 259 ha with a median size of 130 ha.

These nature reserves include samples of most of the vegetation formations that occur in the Shire, ranging from salt-tolerant communities growing along drainage lines, through eucalypt woodlands, to a mixed kwongan scrub of *Acacia*, *Allocasuarina* and *Melaleuca*. Occasional granite outcrops and their associated vegetation are also a significant part of the landscape.

Development of the plan was instigated by a decision to vest most of the nature reserves under consideration in both State and Local Government (in the National Parks and Nature Conservation Authority and the Shire of Wyalkatchem respectively). Thus two vested agencies are jointly responsible for management of these nature reserves. In general, most nature reserves are vested only in State Government.

Formulation of the plan began with the definition of management objectives. This was followed by brief resource surveys, with only 1-2 days in the field

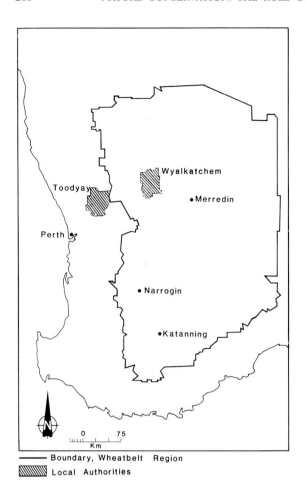

Fig. 4. Location of Planning Case Study Areas.

on each nature reserve. The surveys involved mapping the vegetation into broad categories, general descriptions of the soils, listings of birds sighted, and general notes on past uses and management practices. The development of strategies to achieve the management objectives, plus formulation of operation prescriptions, followed. Very often, staff working on the reserves were approached by adjacent landholders and this provided an ideal opportunity for information exchange.

One of the main aims of plan development was to work with the Shire Council towards a joint management commitment. The initial step involved discussing with a group of councillors the conservation values of the reserves on the one hand, and mis-use on the other. This provided an opportunity for representatives of the Shire Council and the Department to discuss problems of nature reserve management on-site. Therefore, both adjacent landholders and the Shire Council provided important input to the draft management plan.

Following publication of the draft, a display was erected and manned during business hours in the main street of Wyalkatchem as part of the effort to canvass public comment. This provided a further opportunity for information exchange between the Department, the Shire, and members of the local community. The exercise proved highly successful as it provided an excellent opportunity for the public to discuss the management and conservation values of the nature reserves. In one evening this technique enabled the Department to discuss management with approximately 90% of the adjacent landholders. It also allowed management issues that had remained unresolved for years, such as horse-riding and kangaroo-shooting, to be rapidly clarified.

At this time a firm financial commitment was made by the Shire Council to begin rehabilitation works in conjunction with the Department. These works were detailed in the draft plan, and included removal of rubbish and rehabilitation of gravel pits.

Furthermore, discussions between the Shire Council and landholders with land adjacent to the reserves led to a rationalization of fire protection. Adjacent landholders suggested that the boundary firebreaks on the nature reserves be left to regenerate on the *proviso* that they adequately maintain their adjoining boundary breaks to a standard considered satisfactory by the Shire Council. Once regeneration has occurred, this will add to the vegetated area of the reserves, as well as reducing the costs of fire protection.

Thus, development of a management plan has provided solutions to a lack of management resources in two ways:

1. by sharing management responsibilities and management costs with the local Shire Council; and
2. by close liaison and interaction with the local community, for example the decision by adjacent landholders regarding firebreaks on reserves; a decision which will result in financial savings.

Case Study No. 2: Planning As Education

This case study demonstrates how a management plan can both inform the public and generate a more sympathetic attitude to nature conservation. The example used is the management plan for nature reserves in the Shire of Toodyay (Moore *et al.* 1985). The Shire of Toodyay lies approximately 90 km north-east of Perth (Fig. 4) and is the northernmost of four Shires which encompass the fertile Avon Valley. At the time of plan compilation (October 1984) nature reserves occupied only 1.8% of the Shire area, however 35.1% of the Shire was occupied by Crown reserves (including State forest, army training land, national parks and timber, water, railway, road and public utility reserves). The management plan covers nine nature reserves, ranging in size from 39 ha to 1991 ha with a median size of 87 ha. These reserves encompass samples of most of the vegetation types that occur in the Shire. The vegetation ranges from eucalypt forests in wetter areas to various eucalypt woodland types and kwongan scrub.

The Toodyay area is an old, well-established agricultural district settled early in the State's European history. Most of the land suitable for agriculture was taken up at least 100 years ago, although some of the rugged upland country is still being cleared. Most of the remnants in the existing reserve system have been isolated for many years, and therefore form a prominent part of the landscape. In addition, these remnants are the only record of once extensive vegetation associations and as such are invaluable for nature conservation. It was against this background, and with strong interest from the local community, that the management plan was developed.

As with the Wyalkatchem management plan, plan preparation was based on a definition of management objectives, followed by brief resource surveys and the development of management strategies and prescriptions. The collection of resource information was greatly enhanced by the work of amateur naturalists based locally and in Perth. The interest shown by the local community in 'their' nature reserves made it desirable to encourage public involvement in all stages of plan preparation.

After informal discussions with the local community, a draft plan was distributed to reserve neighbours, the Toodyay Naturalists' Club and Shire Council. This was accompanied by invitations to a public release of the plan. This release was based on a slide show, followed by discussion during which people were free to ask questions of the Minister, Departmental head, senior research staff and the senior author of the management plan. The presence of the Minister and senior staff enabled all questions to be answered immediately, rather than having to refer the issue elsewhere. The ensuing response was excellent, with many submissions received. Before such an approach is taken it is essential to ensure that there is agreement within the Department on the objectives of the plan, plan implementation and the dedication of resources to implementation.

Following publication of the draft, the Department led an informal tour for members of conservation groups around the Toodyay nature reserves. Again, the response was excellent. The draft plan was amended in the light of comments received, both from the tour and in formal submissions.

One of the most important ways in which the development of management plans has aided management has been through informing and educating the public. One example of this, taken from the Toodyay planning process, illustrates the value of this educative role. Prior to completion of the final plan, a section of the local community demanded frequent fuel reduction burning of two small nature reserves (each with areas less than 50 ha) in the Shire. These people had been stimulated to air their views by the production of the draft plan and its circulation throughout the local community. Prior to this they had either spasmodically burnt sections of the reserves, or canvassed their local member to stir the relevant government department into action to look more closely at the issue. An on-site meeting was arranged. Once the reasons for fire exclusion were explained, most of the people accepted the management prescriptions given in the plan. At a later date the same prescriptions were defended by one of the former protagonists!

Public participation in planning is a natural extension of informing the public and increasing their awareness of nature conservation. There are three reasons for encouraging public participation.

1. By an open approach to land management, members of the community will accept provisions and restrictions with which they may not initially agree, or which may affect them personally, providing the provisions are applied fairly and the reasons for them are logically argued and explained. Also, people respond to being asked, rather than told. This approach has been employed by ensuring that management plans contain full explanations and rationalizations of the provisions they contain, and by publishing them for free distribution to all interested parties. It is a logical extension of this approach to invite the community to contribute to the development of management provisions.

2. As mentioned above in operational considerations, successful management is dependent on the active support of the local community. Nature reserves should be considered as much a part of the local community as they are of the landscape, and it is well accepted that the successful management of natural areas for conservation requires, at the very least, the understanding and sympathy of neighbouring landowners and local authorities.

3. Decisions resulting from consideration of a wide range of viewpoints increases the probability that conficts of interest will be resolved.

The final role of the Toodyay management plan was to ensure the dedication of resources to management. The detailing of management works in a management plan convinces community members and Departmental staff alike that a firm commitment has been made to the implementation of management prescriptions. However, it must be emphasized that management plan development must proceed within the constraints of operational resources, and be guided by realistic objectives.

CONCLUSIONS

This chapter has given an overview of operational and planning aspects of management for remnant vegetation in southwestern Australia. The importance of fostering a positive attitude to conservation in both local communities and the public at large is emphasized.

Kruckenberg (1985) has written that, in the case of State wildlife agencies in the United States, the 'acquire now manage later' approach to land acquisition will no longer succeed. He also infers that improved planning, communication and management are essential to assure the future of wildlife resources. We believe that Kruckenberg's comments are generally applicable to southwestern Australia, and that unless we inform and educate local communities and the public at large, and actively involve them in planning and management, then remnant areas will become increasingly degraded. As shown in the case studies above, management plans are one means of achieving this objective; provided of course that operational resources are sufficient for implementation.

That nature conservation reserves in southwestern Australia require management has often been stated (e.g. Burbidge and Evans 1976; Main 1979; Hopper and Muir 1984). Given the modest and recent allocation of staff and finances to management of remnant areas, progress towards achieving requirements for successful management and pre-emptive management practice has been satisfactory.

If a moderate increase in staff and finance for management can be obtained, then the transition to pre-emptive management methods is attainable in the short term. Granted also that, at the same time, further research can be successfully translated into effective management, then the short term persistence of remnant vegetation in the Wheatbelt Region seems assured with few, if any, losses in the biota. The longer term view is less certain, and all those involved in management of remant areas have a responsibility to clearly document their work to provide a basis for improving management and planning methods in the future.

ACKNOWLEDGEMENTS

We wish to thank Mr H. Butler, Dr A. A. Farrar and Mr A. J. M. Hopkins for their valuable comments on an early draft of the paper, and Drs G. W. Arnold and A. A. Burbidge for their helpful editorial comments.

The work of Mrs S. S. Howe in collating reserve statistics, Mrs M. Walker and Mrs J. Newbey in typing, and Messrs C. P. de Rebeira and M. Graham in preparation of figures is gratefully acknowledged.

REFERENCES

Beard, J. S., 1980. A new phytogeographic map of Western Australia. *West. Aust. Herb. Res. Notes.* No. 3: 37-58.

Burbidge, A. A. and Evans, T., 1976. The management of Nature Reserves in Western Australia. Report No. 23. Department of Fisheries and Wildlife, Western Australia.

Burvill, G. H., 1979a. The last fifty years, 1929-1979. Pp. 47-86 *in* Agriculture in Western Australia 1829-1979 ed by G. H. Burvill. University of Western Australia Press, Nedlands.

Burvill, G. H., 1979b. The natural environment. Pp. 91-105 *in* Agriculture in Western Australia 1829-1979 ed by G. H. Burvill. University of Western Australia Press, Nedlands.

Hopper, S. D., 1979. Biogeographical aspects of speciation in the south-west Australian flora. *Ann. Rev. Ecol. Syst.* 10: 399-422.

Hopper, S. D. and Muir, B. G., 1984. Conservation of the Kwongan. Pp. 253-66 *in* Kwongan: Plant Life of the Sandplain ed by J. S. Pate and J. S. Beard. Univeristy of Western Australia Press, Nedlands.

Kitchener, D. J., Chapman, A., Muir, B. G. and Palmer, M., 1980. The conservation value for mammals of reserves in the Western Australian Wheatbelt. *Biol. Conserv.* 18: 179-207.

Kruckenberg, L. L., 1985. Public perceptions of land acquisition. *Renewable Resources Journal* 3: 10-3.

Main, A. R., 1979. The fauna. Pp. 77-99 *in* Environment and Science ed by B. J. O'Brien. Univeristy of Western Australia Press, Nedlands.

Moore, S. A., Williams, A. A. E., Crook, I. G. and Chatfield, G. R., 1985. Nature Reserves of the Shire of Toodyay. West. Aust. Nature Reserve Manage. Plan No. 6. Department of Fisheries and Wildlife, Perth.

Moore, S. A. and Williams, A. A. E., 1986. Nature Reserves of the Shire of Wyalkatchem. Management Plan No. 2. Department of Conservation and Land Management, Como.

Murray, D., 1979. Land-use and farming regions. Pp. 254-90 *in* Western Landscapes ed by J. Gentilli. Univeristy of Western Australia Press, Nedlands.

Patrick, S. J., 1985. Possibly extinct and extremely rare plants of the wheatbelt of Western Australia. Rare and geographically restricted plants of Western Australia, Report No. 28. Confidential unpublished report. Department of Conservation and Land Management, Como.

Saunders, D. A., 1985. Human impact: The response of forest and woodland bird communities. Whither the future? A Synthesis. Pp. 355-7 *in* Birds of Eucalypt Forests and Woodlands: Ecology, Conservation, Management ed by A. Keast, H. F. Recher, H. Ford and D. Saunders. Surrey Beatty and Sons, Sydney.

Stamp, D., 1974. Nature Conservation in Britain. Second edition, Collins, London.

Strahan, R. (Ed), 1983. The Complete Book of Australian Mammals. The Australian Museum and Angus and Robertson, Sydney.

Wright, R. G., 1982. Problems of natural resource management at small historic sites. *Parks* 7: 10-2.

CHAPTER 24

The Changing Environment for Birds in the South-West of Western Australia; Some Management Implications

G. T. Smith[1]

The major change to the environment in the south-west of Western Australia has been the large scale destruction of native vegetation since 1890. All together, 65% of the South-West Botanical Province has been cleared, half of this area being cleared since 1945. The percentage cleared in the Botanical Districts ranges from 44 in Roe to 93 in Avon. The worst effected vegetation associations have been the woodlands which have had 89% of their area cleared. Least affected have been the forests with 44% of their area cleared. The effect of these changes on the avifauna, especially the 151 species of land birds has been small for most species. While populations have been reduced and changes in distribution have taken place, only one species *(Amytoris textilis)* is no longer resident in the region. However, 83% of the land birds are dependent on native vegetation for all or part of their annual requirements and it is clear that the slow, ongoing changes taking place in the remnants of native vegetation will be detrimental to many species. Management will be necessary to counter the effect of these changes and is best considered in terms of regional communities.

Some of the problems of the management of habitat are discussed in relation to the Two Peoples Bay Nature Reserve. The need to monitor populations of birds in order to understand the effect of different fire frequencies on these populations as well as understanding the vegetation dynamics in relation to their habitat requirements is stressed in this chapter.

INTRODUCTION

DURING the course of their evolution, birds have come to occupy all major habitats during all or part of their annual cycle. While man has always had some impact on bird populations, the rapid growth of the human population and the range and magnitude of his activities in the last two centuries have resulted in him affecting most, if not all, species of birds. For some species, such as the galah *Cacatua roseicapilla*, man's activities have resulted in an increase in abundance and distribution (Saunders *et al*. 1985) while for others it has resulted in seriously reduced populations (noisy scrub-bird *Atrichornis clamosus*, Smith 1977) or extinction (moas, Dinornithidae, Anderson 1984).

King (1980), reviewing avian extinctions since 1600, concluded that the main causes were predation, habitat destruction and hunting. Further, of those species now considered to be in danger of becoming extinct, habitat destruction is the main cause in 65% of cases (King 1977). In the future, habitat destruction will continue to be a major cause of extinction or endangerment of birds in many parts of the world. However, in areas such as the south-west of Western Australia, the major phase of agricultural development has finished and outright destruction of habitat is no longer a problem. Instead, the problems of maintaining the remaining native vegetation, centre around the effects of fragmentation, small reserve size, chronic external disturbances (grazing, salting, nutrient input, etc.) and the effect of major disturbances such as fire.

This chapter focuses on the non-forested areas of the south-west of Western Australia which are equated with the South-West Botanical Province of Beard (1981) and particularly the Irwin, Avon, Roe and Eyre Botanical Districts (Fig. 1). This is broadly

---

[1]CSIRO, Division of Wildlife and Rangelands Research, L.M.B. 4, Midland, Western Australia 6056.
Pages 269-77 *in* NATURE CONSERVATION: THE ROLE OF REMNANTS OF NATIVE VEGETATION ed by Denis A. Saunders, Graham W. Arnold, Andrew A. Burbidge and Angas J. M. Hopkins. Surrey Beatty and Sons Pty Limited in association with CSIRO and CALM, 1987.

Fig. 1. South-west of Western Australia showing the botanical districts in the South-West Botanical Province, the botanical sub-districts in the Darling district, and the area called the South-West Region by Blakers et al. (1984) ------

comparable with the south-west region used in the Atlas of Australian Birds (Blakers et al. 1984) (Fig. 1). In this chapter I broadly examine the timing, extent and distribution of agricultural clearing as a basis for determining those areas most in need of management, given the habitat requirements and distribution of the avifauna. Finally I present a more detailed study of the Two Peoples Bay Nature Reserve to highlight some of the factors that need to be considered before attempting to manage a reserve where the management priorities specify the conservation of territorial passerines.

HABITAT CHANGES IN THE SOUTH-WEST OF WESTERN AUSTRALIA

The initial development of the agricultural and pastoral industries and the growth of urban centres in the south-west were slow. By 1889 the population was 53,000 and 530 km^2 of the 310,000 km^2 in the region had been cleared (Bolton and Hutchison 1973; Anon 1979). Pastoral activities were extensive, with most of the non-agricultural land being leased for grazing. The effect of these developments on the avifauna was probably small and local in extent. However, indirectly, considerable changes were made to the environment. The arrival of Europeans resulted in a rapid breakdown in Aboriginal society and by the 1880s traditional life had virtually disappeared (Berndt 1979). The cessation of traditional burning practices resulted in an increase in the fuel load which, together with the Europeans' use of fire for clearing, led to an epidemic of intense fires in the early years of settlement (Cameron 1979). In coastal areas the practice of burning areas of heath and thicket every two or three years to provide new growth for stock became widespread. The few observations on Aboriginal use of fire suggests that they burnt areas in rotation and that the fires were not intensive. Europeans, on the other hand, either burnt areas more frequently or did not burn the native vegetation deliberately. In the latter case, this resulted in infrequent but intense fires.

After 1890 there was increasing agricultural development. This is illustrated by the increase in the area of land alienated (basically farming land); 26,790 km^2 in 1900, 130,100 km^2 in 1940 and 190,910 km^2 (61.6%) in 1982 (Bartlett 1984). Except for State Forests, coastal areas and a few National Parks the south-west is now intensely farmed.

Beard and Sprenger (1984) have calculated that 65% of the South-West Botanical Province has been cleared. The extent of clearing between botanical districts varies from 44% in the Roe district to 93% in Avon. Within districts the amount of clearing is also variable; for example in the Darling district, 31% of the land has been cleared in the Warren sub-district, but 78% in the Drummond sub-district (Table 1). It should be noted that Beard and Sprenger equate alienated land with cleared land; while not all alienated land is cleared, it is a reasonable assumption for this analysis. Also, the extent of clearing of the vegetation units previously mapped by Beard (1981) have been calculated (Table 2) and 44% of karri *Eucalyptus diversicolor* and jarrah *E. marginata* forests have been cleared, 89% of woodland, 64% of low forest and woodland, 63% shrubland and 82% of the remaining vegetation units (Beard and Sprenger 1984). Within these broad structural units, particular vegetation associations have been cleared more than others. Marri/wandoo (*E. calophylla/E. wandoo*) and york gum/wandoo (*E. loxophleba/E. wandoo*) have had 94% and 97% respectively of their original area cleared. Shrubland units called 'Scrub with scattered eucalyptus or cypress pines' and 'thicket, *Acacia-Casuarina-Melaleuca* alliance' have had 80% and 83% respectively of their area cleared.

Table 1. Area and the percentage of land cleared for the botanical districts and the subdistricts of the Darling district in the South-West Botanical Province (Beard and Sprenger 1984).

Botanical Unit	Area (km^2)	% Cleared
Avon	93,520	93
Darling	69,005	60
Drummond	14,637	78
Menzies	26,572	61
Dale	19,473	57
Warren	8323	31
Irwin	39,656	59
Eyre	28,702	52
Roe	78,957	44

Table 2. Area and the percentage cleared for various vegetation formations in the South-West Botanical Province (Beard and Sprenger 1984).

Vegetation	Area (km^2)	% Cleared
Forest	35,806	41
Karri	4004	14
Jarrah	31,802	44
Woodland	80,146	89
Marri/wandoo	16,697	94
York gum/wandoo/ salmon gum	41,126	97
Salmon gum/gimlet	17,580	78
Low forest and woodland	10,499	64
Shrublands	150,063	63

In general there has been a massive destruction and fragmentation of the native vegetation throughout the south-west. The worst affected area is the wheatbelt (mainly the Avon district, plus the western half of Roe and the southern portion of Irwin) which covers an area of some 140,000 km^2. Of this, 6.7% is reserved in some 625 reserves, with a median area of 117 ha (Wallace and Moore, this volume). In addition there are numerous small patches (<20 ha) on private land, but these are often badly degraded, as in fact, are many of the reserves.

BIRDS OF THE SOUTH-WEST

The Atlas of Australian Birds (Blakers *et al.* 1984) lists 315 birds, including 151 land birds, as having been recorded in the south-west. Of these 151 species, the south-west is the major area of distribution in Western Australia for 81 species. Ten species are endemic and there is one endemic genus (*Purpureicephalus*).

The distribution of land birds in relation to the bioclimatic zones in the south-west (Nix 1982), show that one species (*Atrichornis clamosus*) is confined to the Bassian zone (approximately the area of forest), three species (*Pachycephala inornata, Aphelocephala leucopsis, Corvus orru*) are restricted to the Eyrean zone and no species are confined to the wide interzone. Forty-eight species are evenly distributed throughout the region; for 53 species the distribution declines from Bassian to Eyrean, in 37 the decline in distribution is from Eyrean to Bassian, while in eight species the widest distribution is found in the interzone. Clearly, the broad intermingling of Bassian and Eyrean elements suggest that the present bioclimatic zones have had only minor influence on the distribution of the avifauna.

The broad habitat requirements of the land birds can be summarized as follows: twenty-four species are territorial, 61 require native vegetation for all phases of their life cycle, 41 for breeding and only 25 live in man-altered habitat. Thus the majority (83%) of species are dependent on native vegetation for all or part of their annual requirements.

The effect on the avifauna of the environmental changes in the south-west during the last 156 years has been documented poorly (Serventy and Whittell 1976), however, it is obvious from the extent of land clearing that most species have suffered severe reductions in population. The changes in burning practices following European colonization are thought to be the main reason for the rapid reduction in the distribution of the noisy scrubbird, Western bristlebird *Dasyornis longirostris* and Western whipbird *Psophodes nigrogularis* (Smith 1977, 1985a, b) and possibly the ground parrot *Pezoporus wallicus*. Also, fire may have been responsible for the possible extinction of the rufous bristlebird *D. broadbenti* in Western Australia (Smith 1977, 1985a, b; Meredith *et al.* 1984; Muir 1985). Land clearing is the most likely cause of the disappearance from the south-west of the thick-billed grasswren *Amytornis textilis* and the considerable changes in the distribution of the cockatoos. The distributions of Carnaby's cockatoo *Calyptorhynchus funereus latirostris* and Major Mitchell's cockatoo *Cacatua leadbeateri* have declined while those of the galah *C. roseicapilla* and inland red-tailed black cockatoo *Calyptorhynchus magnificus samueli* have increased. The long-billed corella *Cacatua pastinator pastinator* showed an initial decline in distribution which may have resulted from poisoning by farmers. A later increase in the two surviving populations probably was a response to the increased food provided by cereal farming (Saunders *et al*. 1985). Kitchener *et al*. (1982) list a number of other species whose distributions are thought to have been changed in the wheatbelt. Further, they note that many species, especially insectivorous passerines, have suffered population reductions.

While no species have become extinct in the south-west, the majority of clearing has taken place during the last 20-60 years and it may be that the loss of species in the region will be a slow process. This may be a consequence of the low rates of vegetation growth in drier areas, where regeneration after fire may take 50 to 100 years (Hopkins and Robinson 1981; van der Moezel and Bell 1984). There are other slow, ongoing processes that will affect many birds. The widespread destruction of woodland noted above has severely reduced the number of nest sites available for hole nesting species. The remaining woodland is in a poor condition and there is little regeneration. Saunders *et al*. (1982) found that in a 15 ha woodland (mainly *E. salmonophloia*) patch near Three Springs no regeneration had taken place since the farm was developed in 1929. In 1978 23% of the trees were rated healthy while 19% were dead. By 1981 only 5% were rated healthy and 40% were dead. While this rate of change may be atypical, it indicates the rapidity with which changes may take place. Certainly, the condition of the woodland in 1978 was typical of many areas in the wheatbelt and the future of hole nesting species in the wheatbelt is in jeopardy. Similar problems may arise in the forests with the continued loss of old and dead trees from forestry practices and control burning (Saunders *et al*. 1985).

The long term changes taking place in other vegetation types are poorly documented, especially in small remnants which are particularly vulnerable to periodic but devastating fires, droughts or the chronic disturbances from nutrient input, salting, grazing, etc. The visible but undocumented degradation of roadside corridors will have an effect on populations of nomadic species and possibly disrupt the dispersal phase of more sedentary species.

It is clear that there are slow continuing changes in remnants of native vegetation that will have deleterious consequences for many species of birds.

The broad distribution of most species of land birds indicate that, with the exception of some rare species, intensive habitat management is not required. Rather, a regional approach with less intensive management of a number of reserves may be a more efficient policy to ensure the survival of the species within their present distribution. Such an approach will need detailed knowledge of the birds' habitat requirements, vegetation dynamics, especially the time scale of changes in different habitats and, in particular, the effect of fire on birds and their habitats.

TWO PEOPLES BAY NATURE RESERVE

In the above section I have presented a broad survey of the land bird community and alterations of their habitat to illustrate that habitat management is a necessity for some species now and will be needed for other species in the future. The Two Peoples Bay Nature Reserve is a useful example to examine in detail some of the problems of managing habitat for birds.

The reserve (4637 ha) is 40 km east of Albany on the south coast of Western Australia. It was established in 1967 for the purpose of 'Conservation of Fauna', in particular the noisy scrub-bird, which was rediscovered there in 1961 after not having been recorded for 72 years.

The reserve is dominated by a hilly granite headland around Mt Gardner (408 m a.s.l.) with deeply incised gullies descending to the sea. The headland is separated from a low lateritic plateau on the northern boundary by a series of dunes, the younger dunes creating the lakes and swamps in the northwest area of the reserve. The variation in topography and soils has allowed the development of a rich flora (617 vascular species) which has been described in terms of 34 plant associations (A. J. M. Hopkins, pers. comm.). These may be broadly grouped as *Agonis* forest around the margins of the lakes and swamps, *Eucalyptus-Melaleuca-Agonis* forests in the wetter gullies and *Eucalyptus-Casuarina* forest on the laterite plateau. Thicket associations (shrubs >2 m high) occur below outcrops, in dune swales and on swamp margins. Heath occupies most of the remaining area and is the most extensive formation. The plant associations make up a complex mosaic which provide a wide range of habitats for birds.

In the period 1968 to 1985, 144 species of birds were recorded on or around the reserve including 45 species of resident land birds. The diversity of birds compares favourably with other south coastal reserves and is typical of the wet south coastal region. The reserve is situated towards the eastern boundary of the region and as such, forms an important link in the chain of south coastal reserves.

Habitat Management

Ideally, management should consider the needs of all species occurring on a reserve, including regular visitors and those that may use the area as a staging post. Also, human values must be considered. At Two Peoples Bay, the rights of the public were recognized when the reserve was created (Anon 1971). Provision of recreational facilities and educational material have created considerable public goodwill which will be necessary for the future maintenance of the reserve.

In this chapter, however, I will consider only the requirements of the avifauna. The management objective for the reserve is to conserve the three rare species present and maintain bird species diversity. The stated management priorities are listed below.

Noisy scrub-bird. This is the only area with a natural population. Even with the development of the translocation programme (Burbidge *et al.* 1984) the reserve will continue to have the largest and best protected population.

Western bristlebird. The reserve is the only area where the species is protected from fire (McNee 1986; Smith 1987). A large fire around Mt Manypeaks in the summer of 1978/79 severely reduced the population in this area and a fire in February 1985 in the Fitzgerald River National Park wiped out the originally discovered population.

Western whipbird. A rare species but more widely distributed. Other locations are not managed and may be subject to fire and land clearing (McNee *loc. cit*).

Species diversity. Although the maintenance of species diversity is of less importance on this reserve, it is important in the regional context and a public asset for the reserve. Further, nectivorous birds may play an important role in pollination and thus help in maintaining plant species diversity.

In order to manage the reserve for the conservation of these species, their habitat requirements must be known. The preferred habitat of the noisy scrub-bird is *Eucalyptus* or *Agonis* forest with a dense understorey of shrubs and rushes. Thicket and heath associations provide suboptimal or survival habitat because breeding is rare in thicket and probably never occurs in heath (Smith 1985 a and c). The western whipbird occupies thicket but uses heath for nesting; the western bristlebird occupies closed heath, open heath may be used if the component clumps are large and dense (Fig. 2). These three rare birds occupy the major vegetation formations as do 41 of the 45 resident species of land birds. The other four species are associated with disturbed habitat, farms or picnic areas on or near the reserve. Thus, in broad terms, management of the habitat for the rare species is not incompatible with maintaining species diversity. Obviously, the specific habitat requirements of the other species will be different, but it is a reasonable assumption that managing the habitats for rare species will also maintain the habitat of most, if not all, of the other species.

Another important management consideration is the history of the major disturbances and their effect on bird populations and habitats. At Two Peoples Bay the only recorded major disturbance has been fire. The reserve would have been subject to a fire regime similar to that at Albany prior to European settlement (Smith 1977, 1985b). Later, fires caused by Europeans would have affected the reserve, but they were not documented and all that can be said is that they did not cause the extinction of any species. The 1946 aerial photographs are the earliest record of fires on the reserve. They show that there had been numerous small fires around the Mt Gardner headland with more extensive fires on the rest of the reserve. Later photographs and eye witness accounts from 1960 onwards indicate fewer and larger fires mainly away from the Mt Gardner headland. Since 1970, the management policy for the reserve has been one of fire prevention and there has been only one small fire in the north-west corner of the reserve which did not burn any of the habitat used by rare birds (Smith 1985b).

For at least the last 50 years the reserve has had numerous fires and all areas of the reserve have been burnt at one time or another but it is unlikely that all the reserve, and in particular the Mt Gardner headland, was ever burnt at the same time. This area, with its highly dissected topography and extensive bare granite ridges provides a natural network of firebreaks which prevents the whole area from being burnt at one time. Thus, there have always been patches of suitable habitat for the majority of the resident species and 34 of the 45 resident land birds are found in this area. For five species there is no suitable habitat and for a further six species, the areas of suitable habitat are too small and too isolated. In general, the latter species require the early stages of growth after fire. Clearly, the headland acted as a refuge for the rare species.

The effect of fire on the rare species can be illustrated by considering the changes in their populations in relation to the changing fire pattern (Fig. 3). Population data are from censuses and photo-interpretation of the availability of habitat. In 1946 the populations of the rare species were of the order of 20-30 pairs; by 1961 they had increased but were confined to the Mt Gardner headland. Fires in 1962 and 1964 were followed by a fall in the populations. The absence of fires since 1970 has allowed the growth of all populations. The decline in the population of the noisy scrub-bird from 1968 to 1971 and from 1983 to 1984 coincided with years having a very

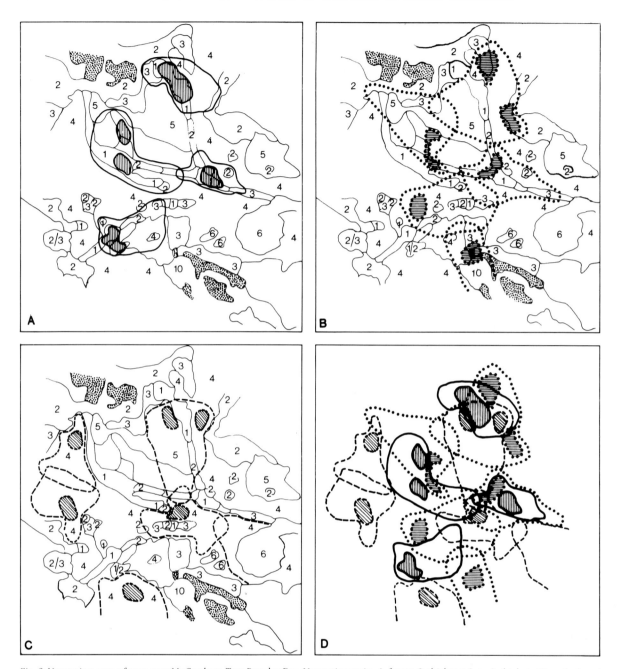

Fig. 2. Vegetation map of area near Mt Gardner, Two Peoples Bay. Vegetation units: 1, forest; 2, thicket >2 m; 3, thicket <2 m; 4, closed heath; 5, open heath; 6, Casuarina woodland. Maps A to C with outline of territory and core area of noisy scrub-bird, western whip-bird and western bristlebird respectively. Map D shows territory data for all species combined.

low rainfall (bottom decile), however, no decline was recorded after 1972, the driest year on record (Fig. 3). This suggests that timing and frequency of dry years rather than their absolute magnitude may be the critical factor.

There is no population data for other species, but broad abundance rating in burnt areas indicates marked changes in populations and suggests that the rate of the recovery phase differs among the species. An important aspect of the fire regime is the fire frequency. The periods taken by rare species to re-occupy areas indicates that the vegetation growth is sufficiently well advanced for all species somewhere between four and ten years after fire. The exact period at a particular site depends on the edaphic and moisture conditions, the proximity of sources of potential colonists and the presence of adequate refuge areas. In the case of the western bristlebird, re-occupation took place in two 20 ha patches, four to six years after they were burnt. In these areas heath attains maximum above ground biomass six years after fire (A. J. M. Hopkins pers. comm.).

The fire frequency most appropriate for the management of the habitat of each species can be determined by calculating how long after a fire the habitat remains suitable for each species. There are

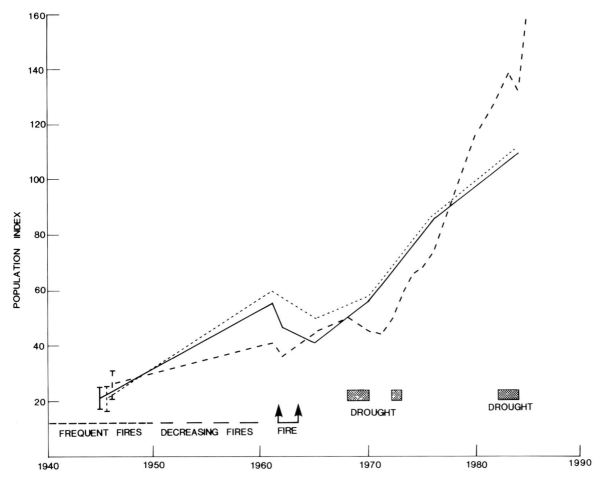

Fig. 3. Population changes in the noisy scrub-bird (number of singing males) – – – –, western bristlebird (number of singing pairs) - - - -, and western whipbird, (number of singing pairs) ——— from 1946 to 1984.

nine noisy scrub-bird territories that have not been burnt for at least 45 years. Six of these have been occupied continuously since 1970, while the other three have been used intermittently. In all of these areas there have been obvious structural and floristic changes in the vegetation since 1970, particularly in the lower and middle stories. These areas have remained suitable for the noisy scrub-bird because of the small scale heterogeneity, especially in relation to nesting areas. Small areas (about 100 m^2) within the territory appear to be going through successional changes in different phases, so that when one area becomes unsuitable, the birds can move to another. There is no indication of how long this will last, but to date no areas with apparently optimal habitat have become vacant (Smith 1985b). This situation shows that it is possible to manipulate the vegetation on a small scale to provide habitat suitable for the noisy scrub-bird. There are also areas of western bristlebird habitat that have not been burnt for at least 45 years. The density of birds in these areas is less than in areas that were burnt 20 years ago. It may be that while the heath remains structurally suitable for the birds, its productivity in terms of seed production and insect abundance (the main items in the diet) may be decreasing. A similar situation may apply to the western whipbird.

These observations suggest that fires occurring more frequently than 10 years would lead to a reduction and possible extinction of the populations. On the other hand, fire frequencies of the order of 50 or more years would not be deleterious and would have the advantage of allowing the growth of larger populations. The best management practice for maintaining habitat for these species is probably one of excluding fire by maintaining fire breaks that will prevent any fire from burning the whole area at one time.

Long-term exclusion of fire may have other consequences. Birds such as the pipit *Anthus novaeselandiae* and calamanthus *Sericornis fuliginosus* which are adapted to the early regrowth stages may be excluded. Further, there may be significant changes in the area of the different structural formations that may lead to serious declines in the populations of some species. For example, heath in some areas grows into thicket. This would be suitable for western whipbirds but not western bristlebirds. It is conceivable that such changes could lead to a situation where many patches of heath are too small for a pair of bristlebirds and result in a serious decline in the population. Examination of map C on Figure 2 indicates how areas of heath could be divided into areas too small for the bristlebirds. Since this area was mapped in 1976, the areas of heath have decreased and the area of open heath is now closed heath. It is therefore important to be aware of the succession of bird habitats and their spatial relations when determining fire frequencies.

CONCLUSIONS

The effect on the land birds af the environmental changes caused by Europeans in the south-west of Western Australia has been minimal. Possibly, only the rufous bristlebird has become extinct and significant reductions in distribution have been documented for only eight species. The effect on the other species, while not as dramatic, has resulted in reductions in the populations and local changes in distribution. Possibly only 17% of the land birds have benefited from the changes. The reason that the massive destruction of the native vegetation since 1829 has had such a minimal effect can be attributed to the low level of habitat specialization and the broad distribution of most species within the region.

The widespread fragmentation of the native vegetation in much of the non-forested areas of the south-west has left the remaining habitat more vulnerable to catastrophic disturbances such as fire, and exposed it to chronic, long-term disturbances from nutrient input, grazing, salting, erosion, etc. At present, only a few species, such as the noisy scrub-bird need specific management of their habitat. For the other 78% of species of land birds which rely on native vegetation for all or part of their annual requirements, the need for future management of their habitat is clear. In broad terms, management policy should be aimed at reducing the impact of disturbances on the local and regional avifauna. Further, the fragmentation of the vegetation will require a system of corridors be maintained or created to facilitate recolonization of areas affected by disturbances and thus avoid costly translocation programmes.

Data from Two Peoples Bay highlight three sets of data that are needed for the long-term management of habitat in a reserve. Firstly, detailed information is required on the habitat requirements of all species. This includes not only data on the preferred habitat (optimal) but on habitat in which the species can survive, with or without some level of breeding, while waiting for habitat to be replaced. The longevity of most Australian birds means that some individuals who survive a fire, for example, will live long enough to recolonize the area burnt, and breed.

Secondly, it is important to have data on the type and rate of change in the vegetation. With a knowledge of the habitat requirements, these data will provide the temporal framework for management actions.

The third factor is fire, which is the most likely major disturbance to affect a reserve and also the most useful management tool. Data are required on both the minimum and maximum intervals between fires that will allow the survival of all species, the changes in density with time since a fire, and the differences in the effect of different fire regimes on the birds and their habitat. If fire is to be used as a management tool then the optimum interval between fires will be a compromise between the various needs of the species present, the need to maintain the largest populations, and maintain an acceptable fire risk. For example, at Two Peoples Bay the optimal interval is of the order of 30 to 50 years, a period that is longer than would be acceptable if reducing the fire hazard was the only priority. However, the needs of the rare species on a reserve require a higher level of risk to be taken.

While there is no apparent need for extensive management of avian habitats at the present, there will be a need in the future. Given the paucity of data, such as that listed above, for most species, now is the time to acquire the basic data needed to meet future requirements.

ACKNOWLEDGEMENTS

I would like to thank Mike Ellis and Les Moore for their help with the field work at Two Peoples Bay. Also, Mike Brooker, Ken and Brenda Newbey, Ian Rowley and Denis Saunders for comments on earlier drafts of the chapter.

REFERENCES

Anderson, A., 1984. The extinction of Moa in Southern New Zealand. Pp. 728-40 *in* Quaternary Extinctions, a Prehistory Revolution ed by P. S. Martin and R. G. Klein. The University of Arizona Press, Tucson.

Anon., 1971. Two Peoples Bay management plan. *Swans* 2: 52-3.

Anon., 1979. Western Australia. An Atlas of Human Endeavour 1829-1979. Western Australia Government Printer, Perth.

Bartlett, W. M., 1984. Western Australian Year Book No. 22. Australian Bureau of Statistics, Perth.

Beard, J. S., 1981. The vegetation of Western Australia at the 1:3000000 scale. Explanatory notes. Forest Department of Western Australia.

Beard, J. S. and Sprenger, B. S., 1984. Geographical data from the vegetation survey of Western Australia. Vegetation Survey of Western Australia, Occasional Paper No. 2. Vegmap Publications, Perth.

Berndt, R. M., 1979. Aborigines of the south-west. Pp. 81-9 *in* Aborigines of the West, their Past and their Present ed by R. M. Berndt, and C. H. Berndt. University of Western Australia Press, Perth.

Blakers, M., Davies, S. J. J. F. and Reilly, P. N., 1984. The Atlas of Australian Birds. Melbourne University Press, Melbourne.

Bolton, G. C. and Hutchison, D., 1973. European man in south-western Australia. *J. R. Soc. West. Aust.* **56**: 56-64.

Burbidge, A. A., Folley, G. L. and Smith, G. T., 1984. The Noisy Scrub-bird. Wildlife Management Programme No. 2 (Draft) Department of Fisheries and Wildlife, Western Australia.

Cameron, J. M. R., 1979. Ambitions Fire. The Agricultural Colonization of Pre-convict Western Australia. University of Western Australia Press, Perth.

Hopkins, A. J. M. and Robinson, C. J., 1981. Fire induced structural change in a Western Australian woodland. *Aust. J. Ecol.* **6**: 177-88.

King, W. B., 1977. Endangered birds of the world and current efforts towards managing them. *In* Endangered Birds. Management Techniques for Preserving Threatened Species ed by S. A. Temple. University of Wisconsin Press, Madison.

King, W. B., 1980. Ecological basis for extinction in birds. Pp. 905-11 *in* 17th International Ornithological Congress, Berlin.

Kitchener, D. J., Dell, J., Muir, B. G. and Palmer, M., 1982. Birds in Western Australian wheatbelt reserves — implications for conservation. *Biol. Conserv.* **22**: 127-63.

Meredith, C. W., Gilmore, A. M. and Isles, A. C., 1984. The Ground Parrot (*Pezoporus wallicus* Kerr) in southeastern Australia: a fire adapted species? *Aust. J. Ecol.* **9**: 367-80.

Muir, B. G., 1985. The dibbler (*Parantechinus apicalis*: Dasyuridae) found in Fitzgerald River National Park, Western Australia. *West. Aust. Nat.* **16**: 48-51.

van der Moezel, P. G. and Bell, D. T., 1984. Fire in Western Australian Mallee. Pp. 151 *in* Proceedings of the 4th International Conference on Mediterranean Ecosystems ed by B. Dell. Perth, W.A.

McNee, S., 1986. Surveys of the Western Whipbird and Western Bristlebird in Western Australia, 1985. Royal Australasian Ornithologists Union. Report No. 18. Moonee Ponds.

Nix, H., 1982. Environmental determinants of biogeography and evolution in Terra Australia. Pp. 47-66 *in* Evolution of the Flora and Fauna of Arid Australia ed by W. R. Barker and P. J. M. Greenslade. Peacock Publications, Adelaide.

Saunders, D. A., Smith, G. T. and Rowley, I., 1982. The availability and dimensions of tree hollows that provide nest sites for cockatoos (Psittaciformes) in Western Australia. *Aust. Wildl. Res.* **9**: 541-56.

Saunders, D. A., Rowley, I. C. and Smith, G. T., 1985. The effects of clearing for agriculture on the distribution of cockatoos in the south-west of Western Australia. Pp. 309-21 *in* Birds of the Eucalypt Forests and Woodlands: Ecology, Conservation, Management ed by A. Keast, H. F. Recher, H. Ford, and D. Saunders. Surrey Beatty & Sons, Sydney.

Serventy, D. L. and Whittell, H. M., 1976. Birds of Western Australia. 5th edition. University of Western Australia Press, Perth.

Smith, G. T., 1977. The effect of environmental change on six rare birds. *Emu* **77**: 173-9.

Smith, G. T., 1985a. The Noisy Scrub-bird (*Atrichornis clamosus*), does its past suggest a future? Pp. 301-8 *in* Birds of the Eucalypt Forests and Woodlands: Ecology, Conservation, Management ed by A. Keast, H. F. Recher, H. Ford and D. Saunders. Surrey Beatty & Sons, Sydney.

Smith, G. T., 1985b. Fire effects on populations of the Noisy Scrub-bird (*Atrichornis clamosus*), Western Bristlebird (*Dasyornis longirostris*) and Western Whipbird (*Psophodes nigrogularis*). Pp. 95-102 *in* Fire Ecology and Management of Ecosystems in Western Australia ed by J. R. Ford. Environmental Studies Group. Western Australian Institute of Technology, Perth. Report No. 14.

Smith, G. T., 1985c. Population and habitat selection of the Noisy Scrub-bird (*Atrichornis clamosus*), 1962 to 1983. *Aust. Wildl. Res.* **12**: 479-85.

Smith, G. T., 1987. Observations on the biology of the Western Bristlebird *Dasyornis longirostris. Emu* (in press).

CHAPTER 25

Management of Disturbance in an Arid Remnant: The Barrow Island Experience

W. H. Butler[1]

Barrow Island, a 259 km² island off the coast of the arid zone of Australia, contains one of the richest assemblages of native fauna on an offshore Australian island. The island is an 'A-class' conservation reserve and since 1966 it has also been a major commercial oilfield. Conflict between conservation values and demands for oil production has been avoided by careful considered management of development and conservation. Four principles have emerged from the management of Barrow Island. These are:

1. education and involvement of the community;
2. clearly defined goals for management;
3. active management is essential; and
4. the control of exotic species is necessary.

The lessons learnt from the 20 years of management of Barrow Island may be applied to the management of disturbance on other remnants of native vegetation after the goals for the management of such areas have been defined.

INTRODUCTION

BARROW Island is a 25,900 ha island conservation reserve off the mid-west coast of the arid zone of Western Australia. It contains one of the richest assemblages of wildlife found on any Australian offshore island. Despite this fact, little biological work had been carried out until a visit by CSIRO in 1958. The collective scientific results of that and previous visits were published in 1964 (Serventy 1964), the same year that I first visited Barrow Island. During that time I was collecting specimens of behalf on the West Australian Museum and the American Museum of Natural History, and the number of vertebrate specimens then known to reside on Barrow Island was doubled.

Following the declaration of the commercial oilfield on Barrow Island in June 1966 development of this resource was made possible by a special Act of Parliament (Act No. 85 of 1966). The conservation value of Barrow Island as a home for several endangered species of marsupial and the need for Australia to exploit its own oil resources posed an apparent conflict of interests. I was commissioned by West Australian Petroleum Pty Ltd, (WAPET), the company which discovered the oil field, with carrying out studies to resolve the dilemma.

NATURAL HISTORY BACKGROUND

The surface of the island is predominantly limestone and the major vegetation association is *Triodia*, dotted with occasional clumps of *Ficus*, *Acacia* and other woody shrubs. More than three hundred species of plant have been identified. They are grouped into nine main communities which consist of:

1. *Triodia wiseana*, hummock grassland on limestone upland;
2. *Triodia angusta*, hummock grassland on water courses and lowland loams;
3. *Triodia pungens*, hummock grassland on red or white sand;
4. coastal dune complexes, primarily *Spinifex longifolius* on white calcareous foredunes;
5. a short forb community on floodout flats;
6. saltflats;
7. mangroves;
8. coastal rock assemblages; and
9. disturbed areas.

---

[1]Dinara Pty Ltd, Conservation Consultants, G.P.O. Box C1580, Perth, Western Australia 6001.

The island supports 13 species of native mammal, including eight marsupials; 105 species of birds, (including one endemic); 49 species of reptile (at least one endemic) and one species of frog (Butler 1970; Butler 1975; Smith 1976; Sedgwick 1978).

Barrow Island's importance as a remnant is that representatives of all these species were originally found on the adjacent mainland, but today, many populations are depleted, some becoming extinct.

Barrow Island was never settled prior to the development of the oilfield and so there were no introduced species of plants or animals on the island and none of the damage created by the practices of historic (Aboriginal) and recent times (European).

PLAN OF MANAGEMENT

Faced with the apparently conflicting requirements of oilfield development and wildlife conservation I was retained to establish a plan for the management of the island. The first stage in this process was to obtain accurate baseline data on various aspects of the ecology of the island. To that end I and a number of others visited the island to carry out scientific surveys of its natural resources. The flora was not considered to be of special rarity or in danger of extinction, and so most of the emphasis was placed on the preservation of the fauna on the island. Some species were endemic with others rare and endangered, particularly on the adjacent mainland.

Since the fauna was dependent upon the existing plant communities a programme of vegetation retention was quickly established. This focused attention on the dilemma facing the West Australian Government: should wildlife be protected or oil be developed? In a report to the Explorers' Club of New York (Butler 1970) I suggested that development could take place and that although changes would inevitably take place, with appropriate management to maintain conservation values, no serious or lasting damage would result.

The plan for the management of the island was based on the following factors:

1. to recognize and preserve the irreplaceable segments of the natural environment which are essential to the maintenance of the island's wildlife;
2. to preserve representative samples of all recognized habitats and plant associations in their natural state;
3. to rehabilitate all developed areas once they were no longer required;
4. to avoid unnecessary effects of the oilfield development and its workforce on the natural environment;
5. to minimize those effects which could not be avoided; and
6. to establish strict environmental rules and ensure that the workforce and visitors accept and adhere to those rules.

The initial identification of problems was difficult because many were compound or intrinsically related to necessary development. As a result, little time was spent endeavouring to identify individual problems, but it was accepted by all that a wide range of problems existed which could be reduced by proper planning. Most of these problems related to disturbance of the natural environment.

PROBLEMS AND SOLUTIONS

Some of the range of problems and their solutions are listed below.

1. The disturbance of fauna by humans and vehicles is unavoidable. The imposition of speed limits, the education of the workforce in not molesting or disturbing animals, and the reduction of all unnecessary vehicle and personnel movements are partial answers. Disturbance of fauna is viewed so seriously that anyone found to have broken the rules risks being ordered off the island. As a result of these controls little fear of man or machine is shown by island fauna.

2. Soil disturbance through active development is unavoidable. By June 1985 about 700 wells had been drilled for the purposes of producing oil or gas, water injection, water disposal or for obtaining water. This programme is supported by a workforce of up to 300 men occupying three camps and using over a thousand km of roads. Other developments include oilfield stations, flow lines, product lines, airports, landing beaches and the rest of the infrastructure that accompanies a major oilfield.

Gravel is one of the most important natural resources used on the island. A programme of employing a single extraction area at one time was established with restoration as part of that programme. Restoration includes topsoil stripping and replacement in the proper horizon, water control, erosion control and compaction relief.

3. The loss of vegetation through disturbance of the soil is a major problem of the field. This is countered by strict programmes of restoration which ensured rootstock is left in *situ* wherever possible. Stripped topsoil containing seed load is deployed on restoration areas before loss of viability takes place.

Some of the wildlife need shelter during the day and so artificial covers were built for their use. Once regrowth is successfully established artificial covers are no longer necessary and they are removed. In the field it is common to see fauna resting in the shade of oilfield fixtures despite the proximity of men and machinery.

4. The introduction of plants and animals is one of the greatest problems of remnant areas. The isolated and insular nature of Barrow Island allowed the establishment of a rigorous quarantine system barring domestic animals, pets and introduced plants. Incoming cargoes on aircraft and barges are checked for plant material and animal invaders, and machinery is washed down if it is thought to contain weed seed. No jetty was built and all material comes ashore on landing barges or aircraft.

   Historically *Rattus rattus* became established on some neighbouring islands (one of which connects to Barrow Island at low tide) and so eradication programmes in conjunction with the then Department of Fisheries and Wildlife were undertaken resulting in the destruction of these populations. Constant checks of warehouses and stores for the presence of introduced species, such as mice *Mus musculus* and rats, are mandatory, as are ground checks of camp areas and unloading bays for weed growth. Routine checks revealed that mice were imported to the island in one cargo and a successful eradication programme immediately followed. Capeweed *Arctotheca calendula* and black berry nightshade *Solanum nigrum* were introduced in the footwear of temporary workforce but were eradicated by hand weeding.

5. The possibility of road casualties is high considering the extensive road system. Speed limits were introduced and effectively maintained while off-duty use of vehicles is limited and restricted to particular roads. Approximately half of the island is totally closed to vehicles apart from planned service access roads. The closed areas are essential in maintaining a natural re-population source as they contain adequate representation of all the vegetation assemblages on the island.

6. Lights and flares and their effect on phototrophic species is of concern in changing food chains as well as the direct destruction of individuals attracted to the light. The reduction of flare sizes, coupled with filters on major lights, has reduced this problem. The effect of lights on breeding turtles, both adults and hatchlings is a potential problem and no lights are placed on beaches where turtles breed.

7. Many types of chemicals are used during the production process which, in addition to those produced as a result of extraction (a combination of saltwater, oil and gas) are regarded as a potential problem for the environment should spills occur.

   WAPET has established the best handling procedures presently applying in international oilfields. There are contingency plans which can be immediately implemented should the need arise. Should flowlines leak or accidental chemical spills occur, procedures for clean-up and restoration include ripping of all surfaces to allow volatile substances to escape followed by sheeting with suitable material to encourage regeneration. Residual spilled crude oil is organic material and readily breaks down into nutrients which are used by island plants.

8. The effect of dust from unsealed works or roads is a major problem in arid Australia for adjacent vegetation. My experience on Barrow Island has shown that dust coated vegetation is less palatable to the native grazing species. Dust also reduces water loss in roadside vegetation by coating the leaves. It was noted that during periodic thunderstorms dust was washed off the vegetation into the natural mulch material at the base of each plant which led to an increase in the mulch zone and a higher available nutrient supply. As a result, dust is not an environmental problem on Barrow Island with its limited traffic.

9. The construction of roads, pipelines and other facilities has changed surface runoff patterns. To minimize these effects culverts and diffusion drains are established to ensure continuation of original runoff patterns, or roadways are designed to allow runoff across the road into natural drainage areas. Of particular importance is the deliberate interference with natural drainage patterns through the creation of small check banks across the island's ephemeral water courses. The purpose of these is to recharge freshwater aquifers, rather than have the runoff, increased by the impervious surfaces of roads and developments, scour out the natural valleys and run to the sea.

10. Man is basically a hunter and gatherer and the workforce on Barrow Island is no exception. This can lead to conflict on a wildlife reserve. On the island there is a total ban on the taking of terrestrial fauna and flora. Fishing, although permitted, is strictly controlled by setting bag limits, controlling the size of species allowed to be taken and banning the harvest of certain species.

11. As restoration areas developed it was evident that differential water and nitrogen present in the regrowth caused preferential grazing by island fauna. This occasionally resulted in the total destruction of regrowth. This effect is minimized by establishing a pattern of restoration which ensures that regrowth is similar to that which periodically follows a natural

disaster, such as a fire or cyclone. Care is taken to confine permanent facilities to small areas while ensuring that restoration areas are as large as possible at any one time.

The key to the systematic restoration of areas of the island disturbed by the oilfield is the understanding that the dynamic cycles of plant/animal communities are a result of climatic extremes. These range from hot and dry to cold and wet with episodic natural catastrophes which include cyclones, floods, droughts and wildfires. Those plant and animal species that could not cope with these cyclic patterns and erratic climatic events became extinct on Barrow Island long before the oilfield commenced operations. The management plan took cognizance of the climatic variability and specified that the impact of the oilfield's operations on the island should not exceed the effect of natural disturbances and thus not place any species under further stress.

12. The disposal of waste is a problem in most human ventures particularly where waste may be eaten by fauna. On Barrow Island all inert wastes are buried in pits and these areas are subsequently landscaped to resemble the original profile, covered with topsoil and allowed to regenerate. Putrescible wastes are incinerated so that no extraneous food is available to the native animals and all sewerage undergoes secondary treatment before being discharged out to sea.

13. As mentioned earlier, gravel is used extensively on the oilfield, and creek beds are very productive areas for the extraction of this resource. As a result, rigid control over the development of these areas is exercised. No new development of a creek bed gravel pit is permitted until rehabilitation is completed on an area of similar vegetation composition which has been used for the supply of gravel. The effect of this activity has been reduced by manufacturing gravel by crushing rock. This has meant that less stress is placed on the limited amount of the *Triodia angusta* complex which occurs in the creek beds.

14. The barrier effects created by roads, pipelines and other surfaces are unavoidable, but as far as possible, over- and under-passes are developed through single line barriers while large temporary development areas are rehabilitated as soon as possible. The main barrier is created by the extensive network of flowlines and pipes lying on the surface. It was decided to leave these on the surface, recognizing that the barrier effect would exist, but offsetting this by providing adequate under-passes. There is little doubt that the effect of burying pipelines would result in more damage to the vegetation, more changes in drainage patterns and damage to the structure of the soil than pipeline laid on the surface. In order to minimize damage to the vegetation all such pipelines are welded in long lengths (1 km) at a line assembly yard, then dragged along roads behind a soft tyred vehicle to a point near where they are to be placed. No vegetation is cleared for the pipeline, but the lengths of pipe are dragged into position over the low vegetation and welded in *situ*. When no longer required, the pipes are broken up and dragged out in long sections. The effect on the vegetation is the same as that seen when a hose is left lying in a garden for an extended period.

15. Fire and its control are an essential part of any oilfield management programme with total fire prevention being the aim. A training programme of fire prevention is an essential part of the induction course which every new worker must attend.

The fire history of Barrow Island is one of episodic, extensive catastrophic wildfires rather than ones which were regular or patchy creating a mosaic pattern. Conventional management wisdom is that mosaic burning of remnant areas is desirable, but since the wildlife of the island is adapted to a catastrophic fire regime, any change from that regime could place an additional burden on the survival of some species.

Despite the rigorous fire prevention policy, accidental fires do occur. When this happens, the fires are contained by using existing roadways as fire breaks and burning back from these roads. No fire breaks are cut unless a fire breaks out in one of the large areas which is not buffered by a system of roads. In such cases a break may be cleared to allow a back burn and reduce the size of the fire. Under these circumstances fire breaks are made by grading off the vegetation with minimal disturbance to the soil.

RESULTS

Given that the extraction of oil is being undertaken in an important conservation area (which is a remnant of flora and fauna, some of which is extinct on the adjacent mainland) it was essential that a management plan be established. This plan sets out the problems associated with such developments and provides solutions to minimize the risks to the conservation values of the island. As mentioned earlier, these include regeneration of disturbed areas and, if necessary, re-population of these areas with individuals taken from areas not directly affected by the oilfield or its associated infrastructure.

The management plan and the programme controlled by it was started in 1964 and it is still in force today. This plan must recognize the differing needs of conservation and development and take into account the changing state of technology in both oilfield development and conservation. To achieve this, flexibility is inherent in it. To ensure that the management programme for Barrow Island reacts to new information as it becomes available and protects the conservation values it was evolved to retain, a system of monitoring the flora and fauna is carried out. This is based on transects, trapping, spotlight counts and field grid searches to determine if any changes are taking place. If any change is identified, the management programme can be modified either to change the trend if it is undesirable, or take note of the change if it is part of the natural balance of the ecosystems of the island.

EXTERNAL ASSESSMENT

I have found that there is a general difficulty of many outside the oil industry to accept that a company and private consultants should be so concerned in the preservation of conservation values. This is compounded in a situation such as Barrow Island where access by people other than those concerned with the oilfield is strictly limited. To counteract this, government departments concerned with conservation conduct biennial inspections of the island. These were begun in 1969 under the auspices of the West Australian Wildlife Authority (now the National Parks and Nature Conservation Authority of the Department of Conservation and Land Management). These inspections critically examine every aspect of the management programme on Barrow Island and make suggestions where problems are seen. Most criticism relates to litter and garbage disposal, both the result of human frailty. Today WAPET proudly claims that the roads of Barrow Island are the cleanest in Australia, and as mentioned above, the problem of garbage disposal is adequately resolved.

DISCUSSION

As the Environmental Consultant to WAPET, I advise on and am responsible for the management programme for Barrow Island. This includes plans for the cessation and removal of the company's activities on the island. Every new member of the workforce and visitor to the island must go through an induction course on arrival. During this course the strict environmental controls in force on the island, the reason for them and the commitment of the company to the preservation of the island's flora and fauna are all stressed.

The company also provides a system of Research Grants, and studies have been carried out to ensure that the widest range of information is available on the island's flora and fauna. So far, baseline studies of the major fauna and vegetation plus specialized studies on some groups of fauna have been carried out. Currently individuals and organizations are being invited to express interest in research work on Barrow Island from which future programmes of value will be determined. A list of references relating to the natural environment of Barrow Island is given in the Bibliography at the end of this chapter. Education has been a major component of the environmental management programme since 1964. More than 10,000 people who have worked on the island during those years have learned that development can work in harmony with conservation if it is planned and controlled. They have learned that there are alternatives to the notion that development is totally destructive and each has left the island and taken that lesson to other areas of Australia and the world.

The development of the oilfield began in 1964 and, apart from the obvious visual intrusion of the oilfield fixtures, there have been no significant or discernible changes in plant associations, ecosystems or faunal population structures on Barrow Island outside the fluctuations that follow seasonal and climatic variations.

The basis of the management programme on Barrow Island is the preservation of habitat through the protection of air, water and soil quality and the protection and where necessary, the regeneration of native vegetation. The emphasis has been placed on ecosystem protection rather than the protection of individual species. With the bones and sinews of the environment of Barrow Island present, the survivors of thousands of generations of natural selection can adjust to the short term intrusion of man and his works. I must stress that such intrusion must be actively managed because the extinction of some species could result from comparatively minor man-made perturbations in the fragile ecosystems of this remnant of the arid zone.

APPLICATION TO OTHER REMNANTS OF NATIVE VEGETATION

The concepts of the management programme for Barrow Island have applicability in other island or remnant vegetation areas. They have been successfully used in other Australian hydrocarbon and mining projects, but each case was tailored to recognize the specific environmental problems of the area. While philosophies and concepts may be constant, methods must vary with place and time. *Essentially there are no absolute answers for environmental management and preservation.*

The prime lessons from Barrow Island that can be applied elsewhere are listed below.

Education. This is essential so that the community involved with an area is aware of its conservation values, why it is being reserved and of the management programme. The local community should always be actively involved in both the formulation and implementation of the management programme.

Goals. These must be clearly defined. Far too often reserves are set aside under some blanket statement such as 'conservation of flora and fauna'. Without explicit goals 'benign neglect' may result in the destruction of such areas.

Active management is essential. It must be recognized that government has neither the prerogative nor the resources to manage all remnants vested in it. The community must be involved in management to the fullest extent.

Exotic species control is essential. In many cases eradication programmes are required to attempt to restore the equilibrium existing before their introduction.

In addition to these points a host of technical lessons are available which can be directly applied to the management of disturbance in remnant vegetation, e.g. topsoil preservation and restoration, surface runoff control and, by inference, the control of edge effects. Taken in isolation they are of little value in remnant management until the baseline philosophy and concept for each area is established. With such establishment the Barrow Island techniques can be significant contributions to the future of remnant conservation elsewhere in Australia and the world.

ACKNOWLEDGEMENTS

I wish to acknowledge: the interest and support of WAPET management and workforce, without which this programme would not have been possible; Drs G. Arnold, D. Saunders and two unknown referees who provided critical appraisal and constructive criticism of the manuscript; and Ms G. Banks and C. Taplin who painstakingly typed the various drafts of the manuscript.

BIBLIOGRAPHY

Archer, M., 1976. Revision of the Marsupial genus *Planigale* Troughton (Dasyuridae). *Memoir Queensland Museum* **7**: 341-65.

Bannister, J. L., 1966. A list of the species of mammals collected by W. H. Butler for the Archbold collections of the American Museum of Natural History and for the West Australian Museum 1963-1966. *The West Australian Museum 1966-67 Annual Report.*

Bakker, H. R. and Main, A. R., 1983. Renal function in the Spectacled Hare-Wallaby, *Lagorchestes conspicillatus:* effects of dehydration and protein deficiency. *Aust. J. Zool.* **31**: 101-8.

Beard, J. S., 1975. Pilbara: the vegetation of the Pilbara area. 1:1,000,000 vegetation survey of Western Australia Memoir 5. Perth, University of Western Australia Press.

Buckley, R. C., 1983. The flora and vegetation of Barrow Island, Western Australia. *J. Roy. Soc. West. Aust.* **66**: 91-105.

Burbidge, A. A., 1971. Report on a visit of inspections to Barrow Island November, 1969, by A. A. Burbidge and A. R. Main, Perth. Department of Fisheries and Fauna Report No. 8.

Butler, W. H., 1964/65. Field notes on Barrow Island 1964/65. Unpublished report.

Butler, W. H., 1970. A formal report on conservation and industry based entirely on Barrow Island oilfield. Unpublished report.

Butler, W. H., 1970. A summary of the vertebrate fauna of Barrow Island, Western Australia. *West. Aust. Nat.* **11**: 149-60.

Butler, W. H., 1975. Additions to the fauna of Barrow Island, Western Australia. *West. Aust. Nat.* **13**: 78-80.

Butler, W. H., 1982. Barrow Island. Perth, Western Australia, West Australian Petroleum Pty Ltd.

Gould, J., 1841. *Osphranter isabellinus. Proc. Zool. Soc.* (Lond.): 81.

Gould, J., 1841. *Lagorchestes conspicillatus. Proc. Zool. Soc.* (Lond.): 82-3.

Hartert, E., 1905. List of birds collected in North-Western Australia and Arnhem Land by J. T. Tunney. *Novitates Zoologicae.* **12**: 194-242.

Hartert, E., 1906. Additional notes on birds from North-West Australia. *Novitates Zoologicae.* **13**: 754-5.

Heatwole, H. and Butler, W. H., 1981. Structure of an assemblage of lizards on Barrow Island, Western Australia. *Aust. J. Herpetol.* **1**: 37-44.

McNamara, K. J., 1982. A new species of the Echinoid *Rhynobrissus* (Spatangoida: Brissidae) from the North-West of Australia. *Rec. West. Aust. Mus.* **9**: 349-60.

McNamara, K. J. and Kendrick, G. W., 1983. Middle Miocene Echinoids and the Molluscs of Barrow Island Western Australia. A report to West Australian Petroleum Pty Ltd and the Western Australian Wildlife Authority. West Australian Museum.

Main, A. R., 1967. Islands as natural laboratories. *Australian Natural History.* **15**: 388-91.

Main, A. R. and Yadov, M., 1971. Conservation of *Macropods* in reserves in Western Australia. *Biol. Conserv.* **3**: 123-33.

Main, A. R. and Bakker, H. R., 1981. Adaptation of macropod marsupials to aridity, *in* Ecological biography of Australia ed by A. Keast. Vol. 3. Junk, Amsterdam.

Perry, D. H., 1972. Some notes on the termites (Isoptera) of Barrow Island and a check list of species. *West. Aust. Nat.* **12**: 52-5.

Sedgwick, E. H., 1978. A population study of the Barrow Island avifauna. *West. Aust. Nat.* **14**: 85-108.

Serventy, D. L., 1964. A natural history reconnaisance of Barrow and Montebello Islands, 1958, by D. L. Serventy and A. L. Marshall. C.S.I.R.O. Divison of Wildlife Research, Technical Paper No. 6. Melbourne, C.S.I.R.O.

Smith, L. A., 1976. The reptiles of Barrow Island. *West. Aust. Nat.* **13**: 125-36.

Smithers, C. N., 1982. The Neuroptera of Barrow and nearby islands off the west coast of Western Australia. *Aust. Entomol. Mag.* **11**: 61-8.

Smithers, C. N., 1982. The Psocoptera of Barrow Island and Boodie Islands, Western Australia. *Entomol. Scand.* **15**: 215-26.

Smithers, C. N., 1985. New records of *Pogonella bispinus* (Stal) (Homoptera: Membracidea) from Eastern Australia and Barrow Island, Western Australia. *Aust. Entomol. Mag.* **12**: 35-6.

Smithers, C. N. and Butler, W. H., 1983. The butterflies (Lepidoptera: Hesperioidea and Papilionoidea) of Barrow and nearby islands, Western Australia. *West. Aust. Nat.* **15**: 141-5.

Smithers, C. N. and Butler, W. H., 1985. Dragonflies and Damselflies (Odonata) from Barrow and nearby islands off the coast of Western Australia. *Aust. Entomol. Mag.* **12**: 9-12.

Thomas, O., 1901. On some kangaroos and bandicoots from Barrow Island, North-West Australia and the adjoining mainland. *Novitates Zoologicae* **8**: 394-6.

Thomas, O., 1902. Two new Australian small mammals. *Annals and Magazine of Natural History* **10**: 491-2.

Waite, E. R., 1901. A description of *Macropus isabellinus*, Gould. *Records of the Australian Museum* **4**: 131-4.

Whitlock, F. L., 1918. Notes on North-Western birds. *Emu* **17**: 166-79.

Wooller, R. D. and Bradley, J. D., 1981. Consistent individuality in the calls of spinifex birds *Eremiornis carteri* on Barrow Island, Western Australia. *Emu* **81**: 40.

Wooller, R. D. and Calver, M. C., 1981. The diet of three insectivorous birds on Barrow Island, Western Australia. *Emu* **81**: 48-50.

CHAPTER 26

Management of Remnant Habitat for Conservation of the Helmeted Honeyeater *Lichenostomus melanops cassidix*

G. N. Backhouse[1]

In response to the concern over declining numbers of the helmeted honeyeater *Lichenostomus melanops cassidix*, the Yellingbo State Nature Reserve was created in 1965 to protect one of the few remaining populations of the bird, and remnants of habitat where it now survives. Prolonged and wide-spread vegetation clearing throughout the bird's restricted range has substantially reduced the numbers of the helmeted honeyeater, which now survives only near the township of Yellingbo, in the Dandenong Ranges, east of Melbourne. The narrow, extended reserve covers remnant riparian vegetation (the bird's preferred habitat) along three creeks in the Yarra River catchment. Management activities aimed at conserving the helmeted honeyeater have included land purchase, tree planting, pest control and fencing in an attempt to consolidate and enhance the bird's remaining habitat. The helmeted honeyeater is under constant pressure from internal and external ecological processes and disturbances because of the remnant, isolated nature of the habitat. Environmental disasters such as bushfires, possible winter food shortages and competition from another aggressive honeyeater, the bell miner *Manorina melanophrys* have contributed to recent declines in the number of helmeted honeyeaters, which casts doubt on the future survival of the bird.

INTRODUCTION

THE Yellingbo State Nature Reserve is an area of 500 ha set aside for the conservation of the surviving population of the helmeted honeyeater, *Lichenomus melanops cassidix*, Victoria's bird emblem. The reserve lies in the Dandenong Ranges, 50 km east of Melbourne, and contains remnants of the bird's formerly more widespread habitat.

This chapter examines the reasons for the decline of the helmeted honeyeater and describes management practices designed to halt the population decline and consolidate the bird's remaining habitat. The chapter also describes the problems encountered, formulation of solutions, and the implementation of a recovery strategy for managing endangered bird populations which exist in remnant patches of habitat.

BIOLOGY, STATUS AND DECLINE

The preferred habitat of the helmeted honeyeater is tall, closed riparian forest, dominated by manna gum *Eucalyptus viminalis* or swamp gum *Eucalyptus ovata* over a dense understorey (Wilson 1933; Cooper 1974; Woinarski and Wykes 1982). The helmeted honeyeater inhabits vegetation along stream margins and surrounding swampland, and is rarely found far from water.

Helmeted honeyeaters feed primarily by gleaning insects from leaves, bark and branches of trees, especially manna gums and swamp gums. Nectar from flowering plants is also obtained, an activity which appears to be more important in winter than in summer, and flying insects are hawked occasionally. Feeding activities are largely confined to the upper strata of very tall trees, with understorey vegetation utilized less frequently (Crome 1969; Woinarski 1981; Wykes 1982). The helmeted honeyeater appears to be essentially sedentary, establishing territories and maintaining a high breeding site fidelity, but may undertake some localized movement following flowering plants, especially in autumn and winter (Wilson 1933; Wykes 1982).

---

[1]Department of Conservation, Forests and Lands, Dandenong Regional Office, P.O. Box 21, Upper Ferntree Gully, Victoria 3156.

Since Gould described and named the helmeted honeyeater as *Ptilotus cassidix* in 1867, the exact taxonomic status of the bird has been the subject of debate. Crome (1973) reviewed the available data and his view that the helmeted honeyeater is merely the largest and most colourful sub-species of the yellow-tufted honeyeater *Lichenostomus melanops* complex is now generally accepted. The helmeted honeyeater is listed as a sub-species in many recent publications (e.g. Schodde 1975; Reader's Digest 1976; Ahern 1982; Blakers *et al*. 1984; Simpson and Day 1984), although Slater (1978) and Pizzey (1980) assign specific status to the helmeted honeyeater.

The past distribution of the helmeted honeyeater appears to have been the region covering the tributaries of the Upper Yarra River and Westernport Bay drainages (Fig. 1). Unconfirmed sightings were reported from the Strzelecki Ranges through to Yarram (southeastern Victoria), but due to almost total habitat clearing neither helmeted honeyeaters nor yellow-tufted honeyeaters exist in this area today. Sightings from elsewhere in Gippsland were regarded as the Gippsland form of the yellow-tufted honeyeater, *L. m. gippslandica* (Wakefield 1958; Crome 1973).

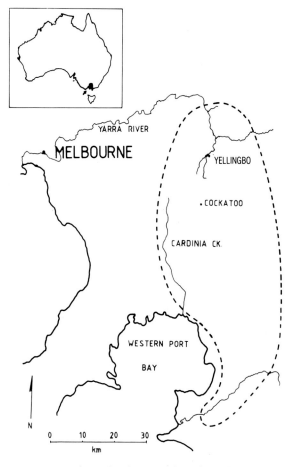

Fig. 1. Presumed past distribution of the helmeted honeyeater (------).

The distribution now appears to be almost entirely restricted to four discrete areas within the Yellingbo State Nature Reserve (I. Smales, pers. comm.; Fisheries and Wildlife Service Records) with one small group possibly still persisting on Macclesfield Creek, about 1 km west of the reserve (R. Loyn, pers. comm.).

Concern was expressed in the early 1900s at the bird's decline in the wake of widespread habitat destruction throughout its range, particularly the Westernport region. The helmeted honeyeater apparently disappeared completely from the Westernport region well before 1947, possibly before 1900 (Lee and Bryant 1948).

Wilson and Chandler (1910) commented on the rareness of the bird in the Dandenongs (specifically at Cardinia Creek), although Barrett (1933) described a flourishing population in the region. Wilson (1933) counted 29 birds along Cardinia Creek, and claimed the helmeted honeyeater was rare. The Royal Australasian Ornithologists Union (R.A.O.U.) organized a survey in 1947, which recorded 50 birds and estimated the population at 100 (Lee and Bryant 1948). Wakefield (1958) noted a decline in population numbers and commented that the bird might not survive in the area.

Prompted by the concern over the decline and threatened extinction of the helmeted honeyeater, a group called 'Survey Cassidix' was formed in 1952, which carried out searches for the bird over much of eastern Victoria, and monitored populations in the Dandenong Ranges until the early 1970s.

The 'Survey Cassidix' group recorded only seven birds in the area in 1951 and 1961 (Wheeler 1967). Surveys in 1964 produced an estimate of 300 birds in the two groups at Yellingbo and Cardinia Creek, and in 1967, 166 helmeted honeyeaters were counted at Yellingbo (Hyett 1976). Cooper (1967a) reported a decline in numbers of the Yellingbo population in the period 1963-1967.

Woinarski and Wykes (1983) document the further decline of the Cardinia Creek population, recording a maximum of 10 birds in early 1979, two birds in early 1980, and none in late 1981. They surmised that this population had become extinct. Surveys by Victorian Fisheries and Wildlife Service staff in early 1982 found a small group of 6-8 birds still present along Cardinia Creek. Another small group of helmeted honeyeaters (about 20 birds) near the township of Cockatoo (in the Dandenong Ranges) was brought to our attenion at about this time. The population had been known to exist for some years (I. Smales, pers. comm.), although there is apparently no published reference to its existence.

The dramatic decline in range and abundance of the helmeted honeyeater is undoubtedly due to extensive habitat destruction (Mack 1933; Lee and Bryant 1948; Cooper 1967a,b, 1974). Nest parasitism by cuckoos (Wilson 1933) and egg collecting (Barrett 1933) have also been reported as possibly being reasons for declining numbers, although this appears unlikely.

The severe bushfires of Feruary 1983 ('Ash Wednesday') burnt through the Cockatoo and Cardinia Greek areas destroying much of the habitat and both populations of helmeted honeyeaters have subsequently disappeared.

Within the Yellingbo population, the largest group (the subject of intensive study by Wykes 1981, 1982) has all but disappeared because a relatively large area (about 15 ha) of privately owned forest adjacent to the reserve, and used by the helmeted honeyeater, was illegally cleared (I. Smales, pers. comm.). The remaining groups of helmeted honeyeaters, now entirely restricted to isolated, remnant habitat near Yellingbo, may be facing pressures from winter food shortages (Wykes 1982) and competition from bell miners *Manorina melanophrys* (Woinarski and Wykes 1983).

MANAGEMENT OF THE RESERVE

Public concern at the plight of the helmeted honeyeater led to the State Government being urged strongly to take action to protect the bird and its habitat. The government reacted in May 1965 by designating a reserve of 170 ha of crown land containing populations of helmeted honeyeaters, along three creeks near the township of Yellingbo as a 'Site for Public Purposes (Conservation of Wildlife)'. Management of the reserve became the responsibility of the Victorian Fisheries and Wildlife Department, and regulations were proclaimed in 1967 to give legal backing to efforts to protect the reserve.

Efforts to conserve the helmeted honeyeater would only be successful if the remaining habitat could be effectively managed. However, the reserve, in its then current condition was largely unmanageable because of its size and shape. It stretched for more than 10 km along the three creeks, but was only 10 m wide in places. Remaining habitat consisted of narrow bands of streamside vegetation, largely surrounded by cleared land.

Management access in places was very difficult as the reserve was almost totally surrounded by private land. Most boundaries were unfenced, and remaining habitat was under pressure from stock grazing and illegal clearing. In addition, some of the helmeted honeyeater communities occurred on private land adjoining the reserve.

A recovery plan, centering on management of the remaining habitat, was developed. The plan consisted of the following major components:

1. publicity;
2. land Purchase;
3. staffing;
4. access;
5. habitat Management;
6. fire Protection;
7. external Influences; and
8. research.

Publicity

The demise of the helmeted honeyeater had already attracted considerable public attention, especially from ornithological groups. Ongoing publicity was important not only to provide the general public with information on efforts to conserve the helmeted honeyeater, but also to increase awareness of conservation issues in general. An important step in this process occurred in 1971 when the helmeted honeyeater was declared Victoria's state avifaunal emblem. Meetings were held with conservation groups to explain the proposed management of the reserve, and, importantly, with local shires and community groups. Adjoining landowners were visited individually to explain management aims and the proposed land purchase programme, and generally to seek their co-operation on matters such as firewood collection and stock grazing within the reserve.

Land Purchase

Land purchase was identified as an urgent priority. It was needed to consolidate and increase habitat area, provide a buffer zone around the critical streamside habitat, and facilitate management access to, and within, the reserve. Plans were prepared to rationalize the reserve boundary, and 350 ha of private land was identified as being required for purchase and incorporation into the reserve. Approaches were then made to the adjoining property owners, outlining the management plans for the reserve and explaining the need for land purchase. Sale of land was to be on a voluntary, negotiated basis, and most of the owners agreed to sell.

The rural-urban nature of the land surrounding the reserve dictated an expensive purchase price, so a staged purchase programme, on the basis of management priority, was devised. The Federal Government was approached for funds to assist the land purchase, and in February 1975 announced funding on a co-operative basis, and purchase commenced. Land purchase and further gazettal of Crown Land have increased the reserve size from 170 ha to approximately 500 ha and the reserve boundary now stretches for 50 km (Fig. 2).

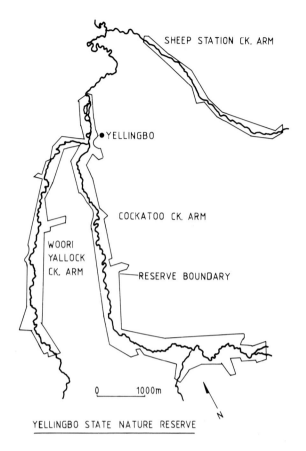

Fig. 2. Yellingbo State Nature Reserve.

Management Staff

Active management of the reserve was not feasible without full-time staff. A ranger/naturalist and two support staff were appointed to implement the works programme, supervise land purchase and provide security. Because of the keen interest in the helmeted honeyeater by amateur ornithologists, much expertise in locating the bird and identifying suitable habitat was external to Government Departments. Continued involvement of 'outside' groups in the management of Yellingbo State Nature Reserve was actively encouraged, and in 1977 the Yellingbo State Nature Reserve Advisory Committee was formed with representatives from Fisheries and Wildlife, the Bird Observer's Club, and the Society for Growing Australian Plants (Sprinvale and District Group). The working relationship with this Society was particularly important as its members collect seed and propagate plants for revegetation works within the reserve. This not only provides a supply of low-cost seedlings (all of local seed provenance) for the replanting programme but also encourages wider community involvement and assistance with a project such as the helmeted honeyeater conservation programme. The Bird Observer's Club have recently formed a technical advisory group which assists with survey and census work on helmeted honeyeaters within the reserve.

Access

The narrow, linear nature of the reserve renders it extremely susceptible to disturbance. What might otherwise be considered a 'minor' disturbance, such as anglers digging for worms along the stream banks, assumes a major significance in Yellingbo State Nature Reserve when such activities may cause bank erosion and the disturbed soil makes an ideal site for invading weeds to establish. Similarly, walkers along the stream edge may cause extensive vegetation disturbance in narrow reserves. For these reasons, public access to most areas of the reserve is deliberately discouraged except in one small part of the reserve, adjacent to the township of Yellingbo, where a small community of helmeted honeyeaters occurs. Management access throughout most of the reserve was facilitated by constructing tracks and installing several creek crossings.

Habitat Management

Once land purchase had commenced, habitat protection and enhancement was identified as the next major task. Boundaries were located and surveyed, and fencing of the entire reserve commenced. Tree planting was planned following an analysis of vegetation types within the reserve, especially of those species utilized by helmeted honeyeaters. Riparian habitat was re-established by critical plantings of *Eucalyptus viminalis* and *E. ovata*, followed by successful plantings of understorey trees and shrubs, including *Melaleuca, Prostanthera, Leptospermum, Acacia* and *Casuarina*. Further plantings will include *Helichrysum, Lomatia, Banksia* and *Coprosma*. Local seed provenances have been used for all plantings within the reserve.

An important consideration in the revegetation programme was an analysis of vegetation types adjoining the preferred riparian habitat. Much of these slope and hillside communities have been cleared and exist only in isolated patches where they adjoin the nature reserve. Wykes (1981) reported occasions when helmeted honeyeaters have been reported visiting areas in search of flowering plants. Although comparatively sedentary for a honeyeater, *L. m. cassidix* probably searched out trees and shrubs flowering locally, especially during the non-breeding season (autumn and winter). Indeed Crome (1969) observed a higher dependence on flowering plants during winter by helmeted honeyeaters, than did Woinarski (1981) and Woinarski and Wykes (1983) during summer, although the populations were at different locations.

Much of the foothills and slopes adjoining the streamside vegetation communities near Yellingbo were forested in messmate stringybark *E. obliqua*, silverleaf stringybark *E. cephalocarpa*, peppermints *E. dives* and *E. radiata* with some box *E. goniocalyx*.

With a different diverse shrub/understorey layer, the total number of flowering trees and shrubs would have been great, with various species in flower throughout the year. It is conceivable that with the loss of much of the adjoining vegetation communities the food resource available to helmeted honeyeaters has been much reduced. Crome (1969) and Wykes (1981) considered that food shortages, especially during winter may be limiting the numbers of some populations. With this in mind, areas of purchased land above the floodplain have been planted with those species described. Given the birds' preference for tall trees, and the small size of recent plantings, use of these areas by honeyeaters has not yet been observed. However, these areas perform an important function in diversifying the available habitat within the reserve, and providing a buffer zone from the surrounding cleared land.

Approximately 80,000 trees and shrubs have been planted since 1978, and further plantings are planned on recently purchased, cleared land, and of understorey species in vegetation degraded by previous stock grazing.

A major consequence of the reserve having a large edge: area ratio is that weed species can penetrate and establish rapidly within the reserve. Consequently, a great deal of time and effort (not to mention cost) is expended in controlling weed invasion and establishment. The worst repetitive invaders include blackberries *Rubis* spp. willows *Salix, Pittosporum*, St Johns wort *Hypericum* and ragwort *Senecio*. These are mainly controlled by extremely judicious use of contact systemic and knockdown herbicides, and manual control such as slashing and hand pulling. Many species that are considered weeds within the reserve are well established in the surrounding rural-urban landscape, and constantly invade the reserve. Complete control and eradication will, realistically, not be possible, so a considerable, ongoing, commitment has to be made to weed control

Feral vertebrates within the reserve include the ubiquitous dogs and cats, goats, blackbirds *Turdus merula* and starlings *Sturnus vulgaris*. These are all opportunistically trapped or shot wherever possible. Again, because of continued invasion from the surrounding land, complete control can never be possible, short of an unrealistic commitment of time and resources.

Fire Protection

The issue of fire protection of a remnant habitat area such as the reserve is a vexing one. The loss of two separate populations of helmeted honeyeaters in the 'Ash Wednesday' wildfires of February 1983 demonstrated the vulnerability of isolated endangered wildlife populations in remnant habitat. It is highly unlikely that any management measures could have prevented the loss of these areas under the extreme environmental conditions of that particular day. However, it remains necessary to undertake fire prevention measures for protection of the bird's habitat. Because of the narrow extended nature of the reserve, extensive fire prevention works are likely to cause major disturbances, creating more problems than they solve. The solution is necessarily one of compromise. A ploughed, mineral-earth break two metres wide is maintained around much of the reserve, and strategic, transverse fire breaks 10-15 m wide are maintained over water pipeline easements. Slashing is carried out on those grassed areas not yet planted.

External Influences

The reserve boundary of 50 km for an area of 500 ha creates considerable edge effects which are difficult to overcome. In an effort to contain the adverse influences of the surrounding land uses on the reserve, restrictions controlling land use have been implemented under Town and Country planning schemes. These include controls on clearing vegetation, land use activities, intensity of subdivision, and minimum subdivision size. However, these controls have little effect in controlling the external influences already operating on the reserve; they are primarily designed to prevent them from getting worse.

Research

As the early development work to consolidate the reserve proceeded, information for the reserve management plan was compiled. Much necessary baseline data has been gathered since active management at Yellingbo commenced. As is common with much of our native wildlife (and its habitat), little was known of the helmeted honeyeater's life history. In an effort to gain this much needed information, which is crucial to informed and successful reserve management, the Fisheries and Wildlife Division financially supported several research programmes designed to obtain information on aspects of life history including habitat requirements, feeding strategies, interaction with competitors and existing or potential threats to the birds' conservation. The results of this research are being used to devise management strategies for incorporation into the reserve management plan. An ongoing department research programme involves monitoring the birds' activities within the reserve, especially reactions to management activities, so that the management programme can be modified and improved where necessary. The aims of this programme include determining the number and total range of helmeted honeyeaters within the reserve; establishing if the total population is divided into sub-populations, or communities, and, if so, how discrete these are; determining the numbers of breeding birds within the population; and investigating some aspects of the population dynamics of helmeted honeyeaters including population structure and numbers, breeding periodicity, fledging and recruitment success. In addition, the opportunity is also being taken to collect morphometric data for further taxonomic studies.

DISCUSSION

The helmeted honeyeater population at Yellingbo appears to have stabilized at 100-150 birds (Woinarski and Wykes 1983) after declining over the last few decades. It is tempting to conclude that this might be due to the area's management as a nature reserve, but this conclusion is perhaps premature. Further data are required before such a conclusion is substantiated.

In spite of this apparent population stabilization, and the consolidation and improvement of habitat within the reserve, the recent losses of several small groups of helmeted honeyeaters are alarming, and must place the future of the bird in jeopardy. The losses caused by habitat clearing and wildfire are at least understood, but the reasons behind the decline and extinction of the Cardinia Creek population, as described by Woinarski and Wykes (1983) give much greater cause for concern.

Vegetation remnants along Cardinia Creek were similar to those at Yellingbo; narrow bands of streamside vegetation largely surrounded by cleared agricultural land. However, the observed decline in helmeted honeyeaters along Cardinia Creek could not be correlated to any further habitat disturbance or destruction. Although there was an apparent slow and steady decline in numbers, the helmeted honeyeater habitat in that area had been little disturbed in recent years. The evidence suggested the decline and apparent extinction of this population was largely due to a dense and expanding population of bell miners, whose aggressive, pugnacious habits, in common with other *Manorina* miners, often successfully completely exclude other birds, especially honeyeaters, from areas occupied by bell miners (McCulloch and Noelker 1974; Dow 1977; Smith and Robertson 1978; Wykes 1982; Loyn *et al.* 1983; Loyn in press). Bell miners have apparently increased in numbers throughout the Dandenong Ranges since the 1940s (McCulloch and Noelker 1974). Woinarski and Wykes (1983) documented the decline of helmeted honeyeaters as the bell miner population expanded and occupied former helmeted honeyeater territories at Cardinia Creek. They also recorded the observation that at many of the former sites where helmeted honeyeaters have been found, bell miners are now present in large numbers. It appears that extensive clearing and fragmentation of habitat has favoured the expansion of bell miner populations in the Dandenong Ranges. Clearing of surrounding slope and hillside vegetation appears to induce stress on the remaining streamside vegetation by changing runoff patterns, resulting in widely fluctuating runoff peaks and troughs. Removing surrounding vegetation may also subject the remaining vegetation to weather induced stress. One of the consequences of increased stress levels in eucalypts is their increased susceptibility to infestations of eucalypt defoliating psyllid insects (White 1969). These outbreaks result in extensive areas of diseased, dead and dying trees in the Dandenong Ranges (Ward 1981). These areas apparently are highly suitable for, and favoured by, bell miners, but are much less suitable for helmeted honeyeaters (Woinarski and Wykes 1983). Loyn *et al.* (1983) and Loyn (in press) describe how, by excluding virtually all other birds from their territories, bell miners maintain higher levels of psyllids than in surrounding areas, resulting in severe defoliation of trees in bell miner's territories. Psyllid outbreaks reduce the level of flowering in eucalypts (Moore 1961), possibly exacerbating winter food shortages faced by helmeted honeyeaters (Woinarski and Wykes 1983).

Bell miners and helmeted honeyeaters occupy complimentary territories within the reserve, and aggressive encounters are frequently observed along mutual boundaries. Competitive interactions with bell miners possibly results in reduced breeding success of helmeted honeyeaters because of increased territory defence. Similarly, juvenile helmeted honeyeaters would face extreme difficulties dispersing from the adult territories and establishing territories of their own. Tree defoliation caused by bell miners 'farming' high numbers of psyllids (Loyn in press), and bell miners preferentially occupying areas of swamp gum which is an important winter food source for helmeted honeyeaters (Wykes 1982) are also likely to have detrimental effects on the bird.

Several authors have raised doubts on the future survival of the helmeted honeyeater. Wakefield (1958) suggested that the species was 'senile' and decaying accordingly, and Crome (1969) suggested that *L. m. cassidix* arose as an outlying 'founder population' of *L. m. gippslandica*, with never more than a few hundred birds in existence, the fate of which always has been extinction. More recently, Brown (1983) stated that 'the merit of *cassidix* lies in the fact that it is a good example of a natural extinction process'. It appears more likely that, in the short term at least, the factors outlined earlier were responsible for the extensive decline in range and numbers of helmeted honeyeaters, rather than the genetic considerations raised by these authors. Other factors, including environmental disasters, environmentally stressed habitat and competition are acting against the helmeted honeyeater.

Other researchers have documented declines and extinctions of bird species in remnant or isolated habitat. Willis (1974) recorded cases of local declines and extinctions amongst groups of ecologically similar birds for no apparent external reason, other than competitive interactions. The American heath hen *Tympanuchus cupido* became extinct in 1932, following a large population decline, range restriction to one locality, some improvement in population numbers because of management action, an environmental disaster (a fire) again reducing numbers, then a steady decline with reduced fertility in the breeding population until extinction (Simon and Geroudet 1970). Petterson (1985) documented the decline to extinction of an isolated population of the middle spotted woodpecker *Dendrocopos medius* in Sweden. Again, this was a small population restricted to isolated, remnant habitat. Although there was no appreciable deterioration of the remaining habitat in recent years, the isolated population apparently suffered declines due to a series of severe winters, then inbreeding depression resulting in reduced breeding success. Extinction of this population followed rapidly.

The example of the heath hen demonstrated that, with intensive management, severely depleted populations may undergo temporary recovery in numbers. However, with both the heath hen in America and middle spotted woodpecker in Sweden environmental disasters severely affected the remaining small populations, reducing them to a size where other factors (possibly genetic) caused inbreeding depression, leading to extinction.

The future of the helmeted honeyeater remains very much in doubt. The reserve's size and shape impose considerable constraints on efficient and effective management. Pressures acting on the birds' remnant habitat (stress, defoliation, weed competition) and the bird itself (competition, food shortages) are extremely difficult to overcome. The reserve's increase in size and management programmes have been considerable and yet are, at best, only short-term efforts involving a small population in a remnant, heavily altered, semi-controlled environment. These efforts are unlikely to be sufficient to overcome the broad scale ecological processes external to, and acting on, the reserve. Small, isolated populations in remnant habitat are especially vulnerable to habitat disturbance or destruction and other external processes. The critical requirement of controlling these external influences on the remnant habitat at Yellingbo, with its small total area yet enormous edge, most likely cannot be met with the existing reserve condition and management level.

A considerable increase in reserve size and management activity will be required to overcome the deleterious influences of external ecological processes on the reserve, and 'insulate' the critical core riparian habitat. Achievement of this aim would require an increase in width (and therefore area) of at least 5-10 times. However, the single population at Yellingbo is still vulnerable to environmental disasters such as bushfires; to safeguard against this catastrophe requires the establishment of helmeted honeyeaters in other reserves of suitable habitat created elsewhere throughout the bird's former range. Given the considerable cost and effort involved in the present programme, such an ambitious, expanded programme would be a multi-million dollar proposition. This would be totally beyond the reasonable expectations of governments to implement, especially considering the limited financial resources generally available and the multiplicity of competing priorities from other endangered species' conservation programmes.

I suggest that successful management of remnant habitat at Yellingbo for the long term conservation of the helmeted honeyeater is unlikely to be successful given the present level of resource commitment. In conclusion, the helmeted honeyeater depends for survival entirely on remnant framented habitat within the Dandenong Ranges. Intensive management of this area may have stabilized population numbers in the short term. However, because of the small area and extensive ecological boundary of the reserve, considerable broad-scale external ecological processes are acting on the reserve and detrimentally affecting the remaining population of helmeted honeyeaters. Overcoming these problems is not likely to be achieved without the unrealistic commitment of further considerable resources, and even then success is far from guaranteed.

ACKNOWLEDGEMENTS

Richard Loyn (Fisheries and Wildlife Service, Department of Conservation, Forests and Lands) provided information and much useful discussion, and Ian Smales (Yellingbo State Nature Reserve, Department of Conservation, Forests and Lands) provided information and valuable discussion and comment on the manuscript. The manuscript was typed by Claire Taplin. I am grateful to them all.

REFERENCES

Ahern, L. D., 1982. Threatened wildlife in Victoria and issues related to its conservation. Fisheries and Wildlife, Victoria. Paper No. 27.

Barrett, C., 1933. Haunts of the Helmeted Honeyeater. *Vic. Nat.* **50**: 161-4.

Blakers, M., Davies, S. J. J. F. and Reilly, P. N., 1984. The Atlas of Australian Birds. Melbourne University Press, Melbourne.

Brown, A. M., 1983. Conservation genetics in Victoria. Fisheries and Wildlife, Victoria; Resources and Planning Branch Tech. Rep. Ser. No. 1, 50p.

Cooper, R. P., 1967a. Is the Helmeted Honeyeater doomed? *Aust. Bird Watcher* **3**: 1-13.

Cooper, R. P., 1967b. A centenary review of the Helmeted Honeyeater. *Vic. Nat.* **84**: 215-20.

Cooper, R. P., 1974. The Helmeted Honeyeater. *Aust. Wildlife Heritage* **31**: 972-5.

Crome, F. H. J., 1969. A preliminary study of the biology of the Helmeted Honeyeater (*Meliphaga cassidix* (Gould)) and its relationships with the Yellow-tufted Honeyeater (*M. melanops* (Latham)). B.Sc. (Hons.) Thesis, Monash University.

Crome, F. H. J., 1973. The relationship of the Helmeted and Yellow-tufted Honeyeater. *Emu* **73**: 12-8.

Dow, D. D., 1977. Indiscriminate interspecific aggression leading to almost sole occupancy of space by a single species of bird. *Emu* **77**: 115-21.

Hyett, J., 1976. Helmeted Honeyeater. V.O.R.G. Reports 6 and 7. *V.O.R.G. Notes* **12**: 14-32.

Lee, R. D. and Bryant, C. E., 1948. A count of Helmeted Honeyeaters. *Emu* **47**: 230-1.

Loyn, R. (in press). Bell miners: birds farming insect prey. *Nat. History* (USA).

Loyn, R., Runnals, R. G., Forward, G. Y. and Tyers, J., 1983. Territorial Bell Miners and other birds affecting populations of insect prey. *Science* **221**: 1411-3.

Mack, G., 1933. The Helmeted Honeyeater. *Vic. Nat.* **50**: 151-6.

McCulloch, E. M. and Noelker, F., 1974. Bell-miners in the Melbourne area. *Vic. Nat.* **91**: 288-304.

Moore, K. M., 1961. Observations on some Australian forest insects, 8. The biology and occurrence of *Glycaspis baileyi* Moore in New South Wales. *Proc. Linn. Soc. NSW* **86**: 185-200.

Petterson, B., 1985. Extinction of an isolated population of the Middle Spotted Woodpecker *Dendrocopos medius* (L.) in Sweden and its relation to general theories of extinction. *Biol. Conserv.* **32**: 335-53.

Pizzey, G., 1980. A field Guide to the Birds of Australia. Collins, Sydney.

Reader's Digest, 1976. Reader's Digest Complete Book of Australian Birds. Reader's Digest Services Pty Ltd, Surry Hills, NSW.

Schodde, R., 1975. Interim List of Australian songbirds; Passerines. R.A.O.U., Melbourne.

Simon, N. and Geroudet, P., 1970. Last survivors. World Publishing Co., New York.

Simpson, K. and Day, N., 1984. The Birds of Australia; A Book of Identification. O'Neil, Melbourne.

Slater, P., 1978. Rare and Vanishing Australian Birds. Rigby, Adelaide.

Smith, A. J. and Robertson, B. I., 1978. Social organization in bell miners. *Emu* **78**: 169-78.

Wakefield, N. A., 1958. The Yellow-tufted Honeyeater with a description of a new subspecies. *Emu* **58**: 162-87.

Ward, B. K., 1981. Eucalypt dieback in the Dandenong Ranges. Forests Commission, Victoria (unpublished report).

Wheeler, R., 1967. The helmeted honeyeater. *V.O.R.G. Notes* **5**: 2-4.

White, T. C. R., 1969. An index to measure weather-induced stress of trees associated with outbreaks of psyllids in Australia. *Ecology* **50**: 905-9.

Willis, E. O., 1974. Populations and local extinctions of birds on Barro Colorado Island, Panama. *Ecol. Monogr.* **44**: 153-69.

Wilson, F. E., 1933. The Helmeted Honeyeater at home. *Vic. Nat.* **50**: 157-60.

Wilson, F. E. and Chandler, L. G., 1910. The Helmeted Honeyeater (*Ptilotis cassidix*) *Emu* **10**: 37-40.

Woinarski, J. C. Z., 1981. The Helmeted Honeyeater at Cardinia Creek; a survey with implications for possible winter control activities. Report to Victorian Fisheries and Wildlife Service (unpublished).

Woinarski, J. C. Z. and Wykes, B. J., 1983. Decline and extinction of the helmeted honeyeater at Cardinia Creek. *Biol. Conserv.* **27**: 7-21.

Wykes, B. J. 1981. The ecology of the helmeted honeyeater and its relationships with potential competitors. Report to Victorian Fisheries and Wildlife Service (unpublished).

Wykes, B. J., 1982. Resource partitioning and the role of competition in structuring *Lichenostomus* honeyeater (and *Manorina melanophrys*) communities in southern Victoria. Ph.D. Thesis, Monash University.

CHAPTER 27

The Viability of Planning Control and Reservation as Options in the Conservation of Remnant Vegetation in New South Wales

W. B. Giblin[1] and S. King[1]

This chapter examines two options for the conservation of remnants of natural vegetation in New South Wales. The first alternative is reservation under the National Parks and Wildlife Act, 1974 (NPW Act). The second is land use planning control via the provisions of the Environmental Planning and Assessment Act, 1979 (EPA Act). The merits and constraints of each method are discussed then illustrated by case studies.

Reservation under the NPW Act achieves the objectives of nature conservation with the highest degree of security and greatest likelihood of appropriate management. Nevertheless, constraints such as funding for acquisition and management, the time involved in reaching government approval and the restriction of reservation to generally pristine environments mean that alternatives need to be pursued to achieve certain nature conservation objectives.

Potentially, the EPA Act has the powers to achieve these conservation objectives but as yet land use planning controls have not been satisfactorily applied to conserve areas of natural vegetation. This is particularly so at the local government level where there are few planning guidelines because of the lack of State and regional policies.

INTRODUCTION

BY the end of the first century of European settlement in New South Wales most of the State showed evidence of extensive vegetation clearing as the land was modified to realize its perceived maximum economic yield. The natural vegetation cover has to this day continued to shrink in area and the ecological value of remnant areas has increased as a consequence.

The traditional attitude of the rural dweller to the environment has been one of agricultural development taking precedence over conservation. Allied with this view has been the 'freehold rights' of private landowners to the land, interpreted to mean that the landowner can use the land as he/she desires. The concept of land use planning, and by inference the imposition of controls, has been considered an anathema to the philosopy of free enterprise in rural areas.

Community attitudes towards the environment started to become more oriented to conservation approximately 25 years ago. The formation of the National Parks and Wildlife Service (the Service) in 1967 occurred because of this increased public awareness of environmental matters in New South Wales. Community attention on quality of life issues became focused on the integration of local planning and environmental issues, evidenced in such activities as public discussion of sand mining in the Myall Lakes area.

It was not until the early 1970s that rural issues began to be addressed and the concept of environmental planning was canvassed. Existing rural planning was piecemeal and concentrated on local detail with little evidence of a regional perspective. Growing community concern eventually led to the enactment of a stronger legislative basis for the integration of planning and environmental protection

---

[1] Natural Resources Section, National Parks and Wildlife Service, Box No. 189, Grosvenor Street Post Office, Sydney 2000.
Pages 295-304 in NATURE CONSERVATION: THE ROLE OF REMNANTS OF NATIVE VEGETATION ed by Denis A. Saunders, Graham W. Arnold, Andrew A. Burbidge and Angas J. M. Hopkins. Surrey Beatty and Sons Pty Limited in association with CSIRO and CALM, 1987.

with the introduction of the Environmental Planning and Assessment Act, 1979 (EPA Act) on 1st September, 1980.

With the continuation of land clearing in inland New South Wales, this chapter examines two options for the conservation of remnants of native vegetation, namely:

1. reservation under the New South Wales National Parks and Wildlife Act, 1974 (NPW Act); and
2. land use control via the planning provisions of the EPA Act.

The merits and constraints of the two options are illustrated by way of case studies and evaluated.

THE LEGISLATION: AN OVERVIEW

National Parks and Wildlife Act, 1974

The NPW Act charges the Service with a statutory responsibility for the protection and care of native flora and fauna throughout New South Wales. In relation to the conservation of the State's flora, fauna and natural environments the principal aims and objectives of the Service are to:

1. preserve a full representation of the State's natural environments and more specifically to investigate and acquire land as national park and nature reserve so that a complete range of the State's natural environments are conserved;
2. conserve, manage and research flora and fauna of New South Wales;
3. manage the resources contained within the lands reserved under the NPW Act for conservation and appropriate use by the public; and
4. promote awareness, understanding and appreciation of wildlife, national parks and all aspects of conservation.

Clearly, these objectives require the Service to reserve land and to manage fauna and flora outside such reserves. Under the NPW Act the Service has a number of options for creating reserves. These are national parks, nature reserves, Aboriginal areas, historic sites, state recreation areas and state game reserves.

Only two of the above, national parks and nature reserves, are primarily aimed at the conservation of natural resources, although to a limited extent state recreation areas and state game reserves may afford a natural area or native species some degree of protection. In this chapter the term reservation applies only to lands reserved or dedicated under the former two categories of reserve.

The Environmental Planning and Assessment Act, 1979

Aside from the nature conservation provisions of the NPW Act, another option for the pursuit of nature conservation objectives is to introduce land use planning controls. With the advent of the EPA Act, New South Wales now has legislation which, in theory at least, provides the facility for controlling rural land use. Administered by the Department of Environment and Planning (the DEP), the EPA Act overrides most other State Acts and represents in theory a total environmental management process covering urban and rural planning, environmental protection and resource development functions.

The objects of the EPA Act are:

1. to encourage

    i. the proper management, development and conservation of natural and man-made resources, including agricultural land, natural areas, forests, minerals, water, cities, towns and villages for the purpose of promoting the social and economic welfare of the community and a better environment;

    ii. the promotion and co-ordination of the orderly and economic use and development of land;

    iii. the protection, provision and co-ordination of communication and utility services;

    iv. the provision of land for public purposes;

    v. the provision and co-ordination of community services and facilities; and

    vi. the protection of the environment;

2. to promote the sharing of the responsibility for environmental planning between the different levels of government in the State; and
3. to provide increased opportunity for public involvement and participation in environmental planning and assessment (Section 5, EPA Act).

The principal effects of this legislation were threefold. Firstly, it introduced new provisions relating to environmental planning to replace Part XIIA of the Local Government Act, 1919. Secondly, it introduced a statutory system of environmental impact assessment and thirdly, established the Land and Environment Court as the appellate and review tribunal in planning and environmental matters.

The EPA Act provides for the implementation of its objects principally through the making of plans. In general terms, major environmental issues and specific land uses are controlled within the parameters set by 'environmental planning instruments' prepared under Part III of the Act.

Environmental Planning Instruments. The EPA Act provides for three levels of environmental planning instruments; local, regional and State. The first is the responsibility of local government and the latter two, of State government. Usually a State or regional plan overrules a local plan.

A State environmental planning policy (SEPP) may be prepared for matters of State-wide significance or for issues where State-wide application of policy is necessary. Legoe (1985) noted that no SEPPs which relate to rural issues generally, or that of vegetation clearance, have yet been introduced although policies relating to the clearance of bushland in the Sydney region and littoral rainforest have been mooted.

The prime role of the regional environmental plan (REP) is to enable the resolution of State and regional issues separately from the local planning process. A region may be determined from a biophysical or environmental area, or general social and economic patterns. The procedures for the preparation of a REP are similar to that required for local environmental plans when a formal study is undertaken. A regional environmental study must be carried out, exhibited and submissions considered before the DEP can prepare a draft REP. The draft plan is then placed on public exhibition. A public inquiry may be held to resolve major issues of the plan. After consideration of submissions to the exhibited draft plan, and any inquiry, the final plan is prepared by DEP. The Minister then decides whether to accept the plan and, if approved, it is gazetted.

Only one REP covering rural lands, namely the Hunter Region, has been gazetted to date (Legoe 1985).

A local environment plan (LEP) is prepared by the local council. The procedure is similar to that for REPs, except that before a plan is exhibited or made, the Director of Environment and Planning has to be satisfied that the plan is consistent with State policies, regional plans, Section 117 Directions, and that the public participation procedures have been followed.

LEPs are by far the most common form of plans with over 1,000 being made each year. The large number of plans is due, in part, to councils making many minor LEPs rather than undertaking the preparation of a comprehensive plan for a whole local government area.

One of the current Section 117 Directions is the EPA Model Provisions 1980, containing definitions and provisions to be followed in the preparation of LEPs. The current Model Provisions contain no definitions or development control measures relating to vegetation clearance, with the exception of tree preservation orders.

Land Use Zones. Zonings are the basic land use control in LEPs. In 1976 the then New South Wales Planning and Environment Commission (PEC) introduced a range of zones for rural lands (PEC Circular No. 13, 1976). The zoning nomenclature is still used today. The circular established 10 environmental protection zones into which non-reserved land of environmental or cultural value could be placed for full or partial protection. The zones are wetlands (freshwater), estuarine wetlands, water catchment areas, scenic areas, escarpments, foreshore protection, archaeological sites, historic sites and wildlife refuges. Uses which could not pollute or damage the environment would be permitted.

The Service, by virtue of its statutory obligations to be involved in the planning process, provides certain information to local and State Government planning agencies. Matters addressed in Service submissions include maps of lands reserved under NPW Act, details of proposed national parks and reserves and identification of additional areas of conservation importance. In relation to these matters, the Service has in the past advanced zoning proposals for the protection of the environment. These are to identify areas of conservation importance, and to provide protection to these resources from land use changes that could detract from the value of the area.

With respect to conservation of natural environments, the Service has little or no interest in activities on lands that have been significantly altered from their pre-European condition except where the management practices in place have significant deleterious effects on native wildlife generally (e.g. the use of certain pesticides) or when practices affect reserved lands.

THE MERITS OF CONSERVATION OPTIONS

Reservation Under NPW Act

A number of options exist by which the Service can potentially achieve conservation objectives, other than under its own Act. Avenues include reservation or dedication for conservation purposes under the Forestry Act, 1916, the Fisheries and Oyster Farms Act, 1935 (aquatic reserve), and environmental protection zones under the EPA Act.

However, Hitchcock (1984a) has argued that land acquisition and reservation/dedication under the NPW Act is the most secure way to protect the nature conservation values of an area because national parks and equivalent reserves (nature reserves and flora reserves) have a high degree of security in that it requires an Act of Parliament before such a reserve can be revoked. Lands reserved or dedicated under the NPW Act are exempt from the Forestry Act. The Act also provides that operations under the various mining acts cannot be undertaken on Service managed lands without the concurrence of both Houses of Parliament and that leases under the Fisheries and Oyster Farms Act shall not be granted within national parks without the concurrence of the Minister. Although bound by the EPA Act there is provision under SEPP No. 4 for the Director of National Parks and Wildlife to be the determining authority for all development on Service managed land. All national parks and nature reserves are automatically placed on the Register of the National Estate and the Act permits the Service to formulate regulations which allow it to control use of lands reserved under the Act.

Merits of Planning Control

Conservation of native vegetation by planning control is clearly one of the objects of the EPA Act, however, land use planning decisions made since its inception, suggest the Act has not been fully applied. The benefits of planning control are that the conservation of wildlife may be only partially achieved by acquisition and reservation of lands. Many species of plant and animal require large contiguous areas of natural or semi-natural environments for their immediate and ultimate survival. For this reason, environmental protection zonings could be considered a suitable method, in conjunction with reservation, in achieving the goal of conserving natural habitats and wildlife while permitting other sympathetic land uses. Controls through planning instruments also provide the ability to constrain developments detrimental to nominated habitats either at State, regional or local levels. While rural councils might not always be sympathetic to representations for protective zonings, Giblin (1983) has suggested that the fact that they are informed of significant natural resources in an area should be seen as an important step in educating local decision makers of the merits of nature conservation. Planning control as a means of protecting the environment can avert the need for the State government to purchase land, thereby involving less disruption in the local community and a saving to the taxpayer.

THE CONSTRAINTS OF CONSERVATION OPTIONS

Reservation

Whilst the Service is the State Government department primarily concerned with nature conservation, there are a number of constraints which prevent it from achieving its identified objectives. Firstly, there is an inadequate identification of conservation objectives. One of the aims of nature conservation in New South Wales is to minimize the irreversible loss of representative examples of the natural environment and to adequately sample each species present, ideally over their entire geographic range. To date parks and reserves have been created generally in areas where land use conflicts have been minimal and where acquisition costs are low, or in areas where threats have been considerable and public pressure has created a political need to reserve or dedicate lands. To date unalienated Crown land has been the major source of land reserved under NPW Act (Fig. 1). However, it is now apparent that the remaining unalienated Crown land is of limited supply and there is increasing competition for such land, particularly as a result of the government's recent decision to legislate allowing for Aboriginal Land Councils to make land claims. Many wildlife communities and species of nature conservation value occur on alienated Crown land and freehold lands, and funding for land acquisition by the Service is limited. In real terms, acquisition funding has remained roughly at the 1973/74 levels (Hitchcock 1984b). Because of these factors it has become essential to establish conservation priorities and to consider alternative measures to achieve the Service's conservation objectives.

The Service initiated a major review of its conservation objectives in 1983. From the first stage of the review the Service has been able to identify areas where it needs to concentrate its efforts if it is to achieve its primary nature conservation objectives. For example, Benson (1985), in summarizing the conservation status of vegetation in New South Wales, has estimated that of the approximate 480 vegetation communities present within the State 51% are either extinct, endangered, vulnerable or inadequately conserved. Of these, 30 of the most threatened communities are present on the near western plains and western slopes.

One of the major constraints to nature conservation through reservation under the NPW Act is the need to secure agreement from a wide range of government bodies. As described by Hitchcock (1984a), once a nature conservation proposal has been formulated and reservation has been chosen as the most appropriate conservation option, the proposal then needs to be circulated to a range of other government departments for their approval before further action can be taken. Whilst rejection by another department does not automatically mean the proposal cannot proceed (although in some instances this may be so), it can delay it to the extent that the values of the proposal are placed in jeopardy due to competing, incompatible land uses, such as clearing.

A view expressed both within the Service and within other government departments is that the State has 'enough' national parks and nature reserves and that the aim in the future should be to consolidate rather than continue to expand. In relation to the Service's specific responsibility for the conservation of samples of vegetation types within New South Wales, this is clearly a misconception as 51% of the State's vegetation is at best inadequately conserved. For example, investigation has shown that remnant brigalow *Acacia harpophylla* and microphyll vine *Planchonella cotonifolia* var. *pubescens* thickets of the northwestern slopes and plains are very poorly conserved. Margules (1982), in assessing the conservation status of vegetation in the Murray-Darling Basin, clearly showed that the vegetation in the semi-arid and arid west were also

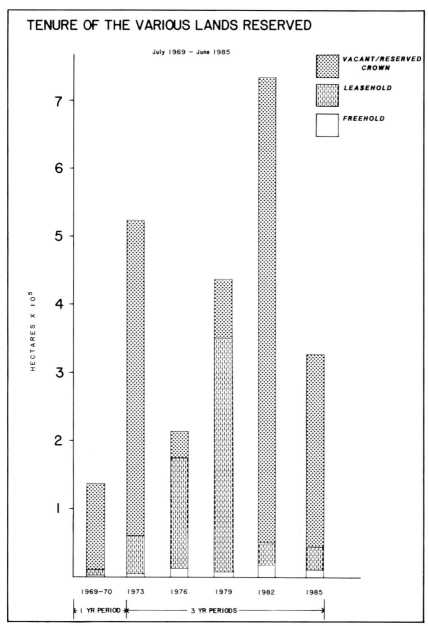

Fig. 1.

poorly represented in existing reserves. Another example is that of the 3 million ha of *Astrebla* spp. grassland which once occurred in New South Wales not one ha is represented in a conservation reserve (King 1983).

If other government bodies with an interest in Crown land become less receptive to the idea of the establishment of conservation reserves under the NPW Act, the Service will increasingly have to place more time and effort into achieving its conservation objectives in other ways and rely more heavily on other forms of environmental protection.

The Service utilizes Loan Funds for the acquisition of land and the annual allocation by Treasury for this purpose was $1.5 million in 1984-85. Acquisition obligations presently total some $10 million. Given this constraint, and the current price of land, the Service is severely limited in pursuing the option of reservation where it involves land purchase. At the present rate of funding many areas identified up to a decade ago as justifying reservation are likely to experience significant habitat modification before funds are ever made available.

In the area of staffing, levels have remained relatively constant over the past 5-10 years despite a doubling in the area of land now reserved under the NPW Act. This has limited the availability of staff for natural resources survey, and acquisition activities, not to mention land management.

Planning Control

The first step in the attempt to conserve remnant vegetation communities is identification of the relevant areas. There is a need for a computerised information system to provide comprehensive baseline information which is readily accessible and easily updated. Ideally such a system should link all relevant land use authorities. The present collection of data in New South Wales is incomplete and the data are difficult to locate and inefficient to use.

Consideration of environmental planning instruments has to occur within a limited time as dictated by the EPA Act. For example, submissions on a draft LEP must be made to the local council during the period of public exhibition. Whilst local studies and plans are usually exhibited for 30 to 40 days, the minimum statutory period for exhibition is only 14 days. Even at 30 days, the time limitation is a major constraint to an organization such as the Service, in terms of allocating resources to research and preparation of adequate responses.

Giblin (1983) has observed that the elected councillors in rural shires are usually agriculturalists and/or local businessmen who depend on rural endeavour for their livelihood. In addition, because rural councils tend to be involved only in local planning issues, they tend not to have State or regional perspectives. Councils are therefore often reluctant to include environmental protection zones in their LEPs, especially where there is no requirement by way of SEPP, Section 117 Direction or REP for them to include such zones. A council's decision to introduce such zones tends to be affected by the restrictive nature of such zones, particularly where it impinges upon the traditional use or perceived land use capability of private lands, thereby reducing its resale value. They may also believe that the acceptance of such a zoning proposal must be seen as a prelude to land acquisition by the Service.

If planning is to be properly implemented in rural areas, Legoe (1985) has suggested that it must overcome certain misconceptions about the planning process. These include the notions that planning only deals with problems, rather than encouraging better land use and protection of the environment, and that planning is always negative, concentrating on the use of controls and limitations to deal with problems, rather than incentives and guidelines that will achieve the planning objectives.

As agriculture is a use permissible in most zones without requiring the consent of council, existing use rights in the EPA Act permit the continuance of that use notwithstanding the gazettal of a new LEP which might otherwise require consent to be obtained. This means that in an area recently rezoned environmental protection instead of Rural 1(a), agricultural activities, including clearing, remain uncontrolled.

According to Legoe (1985), a number of problems with the administration of the EPA Act appear to be constraining the implementation of sound rural land use and management principles. These include the concentration of planning effort on large urban centres and the north coast of New South Wales as well as the lack of regional and State policies necessary to provide the planning guidelines. For example, until August, 1985, there was no Section 117 Direction that related to conservation or retention of vegetation. The Direction is, however, rather general, urging councils to adopt a responsible attitude with respect to the conservation of all facets of environmental heritage, including lands of ecological or aesthetic significance for the local area. Similarly, by simple amendment, the Model Provisions could specify vegetation clearance as a land use that requires consent of the local council. There also tends to be over-involvement in LEPs and the unwillingness and lack of authority to ensure that environment protection zones are appropriately applied in LEPs. Implementation of sound principles of land use and management is also affected by the apparent lack of priority given to rural planning, within the DEP. This issue has not been given the attention commensurate with its longer-term ecological significance. There is also the lack of prescription of activities under Part V of the EPA Act. The DEP could well investigate the activities that could appropriately be prescribed, and then ensure that government sets the example for land use and management in the State.

CASE STUDIES

Reservation

Three case studies are considered: brigalow, microphyll vine thickets and *Astrebla* spp. grasslands.

Brigalow. Service interest in the brigalow areas of the northwestern slopes and plains of New South Wales commenced with a report by Caughley (1979) on the black-striped wallaby *Macropus dorsalis*, a species associated with this vegetation type. As a direct result the conservation status of brigalow in New South Wales was examined (Pulsford 1984).

Past records indicated that there had been an extensive but fragmentary occurrence of brigalow in New South Wales. However, when the study by Pulsford was undertaken there were only a few remnant stands around Bourke, Narrabri and Moree. Given that the threat of clearing was highest around Moree and Narrabri, further systematic survey was conducted in this area to assess the extent and status of the brigalow remnants and to select areas that were suitable for reservation under the NPW Act.

In total, four areas were proposed for reservation. After discussions with various government land use departments the Service entered into negotiations with the respective owners in early 1983. To date negotiations over one area are nearing completion. Of the remaining three, two landowners are not willing to negotiate whilst one other owner has sold to another party.

Microphyll Vine Thickets. In 1979 the Service undertook a preliminary investigation of the vegetation remnants and associated avifauna on the basaltic soils of the northwestern slopes between Warialda and the Queensland border. This area, which formerly supported woodland of *Eucalyptus albens-E. melanophloia* with scattered occurrences of microphyll vine thickets, has been extensively cleared.

The microphyll vine thickets represent remnants of a once scattered but widespread vegetation type reportedly occurring as far south as the Liverpool Range and parts of the Upper Hunter. They are a scientifically interesting community being closely related to the monsoon rainforests of northern Queensland and the Northern Territory and the dry rainforests of southern Queensland and New South Wales.

The reservation effort focused initially on the areas north of Warialda as they appeared to have the highest diversity of plant species and were under constant threat from clearing and fire. A study in 1979 investigated a total of 10 areas, two of which contained contiguous vegetation types on sandy soil. The Service identified four of these areas for more detailed investigation in order to determine their suitability for reservation under the NPW Act (Pulsford 1983).

Only one remnant was recommended as a potential nature reserve. Given Service priorities, limited funding and the nature of the remnants, alternative conservation measures were recommended for the remaining three areas. The area proposed for reservation contains the largest known example of this vegetation type in New South Wales, a type which is not represented in an undisturbed form in any reserve in New South Wales or Queensland. The remnant community was brought to the attention of other land use authorites in 1983 and after a period of two years it was resolved that there were no objections to the proposal. Negotiations with the landholder are proceeding.

Given the limited information available on these communities a systematic comparative study is now needed before it will be possible to assess the conservation status of the remaining microphyll vine thickets or the flora or fauna associated with them.

Astrebla spp. grasslands. *Astrebla* grasslands were once widespread in Queensland, Western Australia, Northern Territory and New South Wales. In New South Wales they covered extensive areas in the midnorthwestern section of New South Wales, particularly on alkaline soils that are for the most part cracking or self-mulching clays. The dominant species in these communities is generally either *Astrebla lappacea* or *A. pectinata*. These grasslands are very productive both for cropping and as native pastures. They face continual threat from cropping and grazing.

Past distributions of these grasslands have been identified by Beadle (1948) and Pickard (1981). There has been no major attempt to map the present-day distribution of this vegetation type nor to assess its conservation status. However, it is known that the present areas used extensively for dryland cropping correspond closely with the past distribution of this grassland type. Indeed S. Jacobs (pers. comm.) suggests that there may be very little of the community left in its natural state.

The area has been identified as requiring urgent survey. Thus, in order to overcome this paucity of information, the Service has begun the initial stage of a survey programme for an area based on the known past distribution of *Astrebla* spp. grasslands. The information obtained from the survey should be sufficient to enable identification of priority areas for acquisition and determination of suitable land use strategies. However, it is likely to be three to seven years before any areas identified in such a survey might be reserved.

Planning Control

Moree Plains Shire Local Environmental Plan. In December, 1981, the Moree Plains Shire Council, following amalgamation of three former councils earlier the same year, resolved to prepare a LEP for the whole of the shire to standardize land use controls.

Moree Plains Shire is located in north-west New South Wales and covers an area of some 17,000 km^2 Geographically it is part of the western slopes and plains. The shire is largely cleared and developed for agriculture, particularly in the eastern section. It formerly carried box eucalypt woodlands which tended to grade into grasslands in the drier western parts. European settlement commenced in 1839 and by 1920 the wheat industry had been well established. Cotton, wheat, wool and mutton are the principal rural products. Eighty-five per cent of the shire is inundated during major flooding of the Gwydir and Macintyre Rivers, the major watercourses in the region.

In accordance with Section 82 of the EPA Act, the council consulted with the Service as an affected public authority and the Service provided relevant advice. The Service indicated that there were five nature reserves and part of one national park in the shire. There were also five wildlife refuges as well as areas of natural vegetation of known conservation value. In particular, two vegetation remnants were identified and described, namely brigalow and ooline *Cadellia pentastylis*. Natural areas associated with the Gwydir River floodplain were also noted as warranting further investigation with a view to determining their nature conservation values.

The Moree Plains Shire Local Environmental Study (1982) which preceeded the plan made the following comment regarding the impact of Europeans on the ecology of the area: 'The presence of European man since the early agricultural settlements of the 1830's has had profound and often catastrophic effects on the previously existing balance of nature. What was once a diverse and abundant fauna has been greatly reduced and simplified as a result of man's activities.' The document also noted that: 'due to extensive clearing for pastoral and agricultural uses, only small remnants of native vegetation remain within the shire, occurring mainly as corridors of riverine woodland along water courses and as occasional patches of uncleared woodlands in the east'.

The Service provided comment on the LES produced and further advice to be considered in the preparation of the draft LEP. The aims and objectives of the draft LEP adopted by council were, *inter alia*:

'3. To identify and conserve those areas of the shire which are especially significant as habitats for indigenous wildlife species.

4. To protect and enhance the visual amenity of the countryside and to encourage the protection and conservation of all items which form the environmental heritage of the shire.'

Consistent with these objectives, the Service suggested incorporation of environmental protection zones (Zone 7) over certain areas of remnant vegetation. These included wetlands, an area of scenic interest adjoining a national park and two areas of scientific interest (one ooline scrub and the other brigalow/belah *Casuarina cristata*/bimble box *Eucalyptus populnea*).

In October, 1983, the council advised the Service that it had accepted the Service's environmental protection zoning proposals. In addition the Service was made a concurring authority in respect of all new development proposed in Zone 7. This meant that no development application could be given the go-ahead without the approval of the Service.

The Moree Plains Shire Draft Local Environmental Plan which was placed on public exhibition in February, 1984, included; controls on the clear felling of timber, controls on the construction of earthworks (with respect to irrigation developments) and environmental protection zoning for some small, scattered areas of both freehold and Crown lands upon which proposed developments would require council consent. Table 1 shows the land use control table proposed for Zone 7.

Opposition to the draft plan was swift, strident and bitter. The local Member of State Parliament was quick to castigate the proposed plan, stating that the actions of the Service in regard to environmental protection amounted to resumption by proclamation and that it was totally unacceptable for freehold land to be the subject of such controls. In the face of the opposition the council reversed its earlier decision on the matter and sought to have the controls relating to earthworks, clear felling of timber and environmental protection zones removed from the plan.

Over 850 submissions to the draft LEP were received and the council approached the Minister for Planning and Environment for a Commissioner of Inquiry to be made available for the purpose of hearing objections and reporting to council. The Minister agreed and some 220 parties appeared before the public hearing held in April and May, 1984. During the first two days of the hearing the Service attempted to explain the basis for its interest in certain lands and also reasons for the zoning recommendations put to council. In his report to the Council of Moree Plains (July, 1984) the Commissioner made two recommendations in relation to land use controls pertaining to remnant vegetation. Firstly, that all environment protection zones be removed from the LEP and secondly, that all provisions relating to earthworks and clearfelling be similarly deleted with such lands remaining in a Rural 1(a) zone. [Rural 1(a) is the most general of all the rural zones. Agriculture does not require development consent and a wide range of broadacre uses are permissible, for example, irrigation development and forestry].

The recommendations of the Commissioner of Inquiry represent an endorsement for the continuation of the common law approach with respect to rural land use planning. It supports the traditional practice of a private landowner using the land as he/she wants. The decision also means that in a shire 80% devoid of its native vegetation cover, the opportunity has been forfeited by council to introduce constraints on further modification of those remaining vegetated lands which are considered to be of significant conservation value.

Unfortunately, from the conservation perspective, the Moree Plains Shire episode is not an isolated case; rather it is typical of the responses by local government. As a consequence of this response the Service has had to curtail its involvement in local planning, given increasing constraints on the expenditure of public funds. As a result of this decision, the Service now restricts its input to natural resource descriptions pertaining to lands reserved under the NPW Act and limits identification of specific areas that might be incorporated in environmental protection zones.

Table 1.

Column I	Column II	Column III	Column IV	Column V
Zone and colour or indication on the map.	Purposes for which development may be carried out *without* development consent.	Purposes for which development may be carried out *subject to* such conditions as may be imposed by the Council pursuant to Section 91 of the Act.	Purposes for which development may be carried out *only with* development consent.	Purposes for which development is *prohibited*.
7. ENVIRONMENT PROTECTION: (a) Environment Protection (Wetlands) Orange with heavy black edging and lettered 7(a).	—	Picnic grounds.	Agriculture*; earthworks; educational establishments; scientific establishment; soil conservation works.	Any purpose not included in Column III or IV.
(b) Environment Protection (Escarpment) Orange with heavy black edging and lettered 7(b).	—	—	Roadworks; drainage; public utilities; grazing.	Any purpose not included in Column IV.
(c) Environment Protection (Scientific) Orange with heavy black edging and lettered 7(c).	—	—	Scientific establishments.	Any purpose not included in Column IV.

*'Agriculture' is defined in the Local Government Act, Section 514(a) (applicable by reason of the model provisions).

Section 514(a)
'Agriculture and cultivation include horticulture and the use of land for any purpose of husbandry, including the keeping or breeding of livestock, poultry or bees, and the growing of fruit, vegetables, and the like, and agricultural and cultivate have a corresponding meaning.'

EVALUATION OF OPTIONS

Reservation

Reservation under the NPW Act achieves the objectives of nature conservation with the highest degree of security and the greatest likelihood of appropriate management. Nevertheless, reservation on its own cannot be expected to fully satisfy the Service's nature conservation objectives. This is particularly so where the economic cost is too high or where multiple use is considered necessary, for example, where the government decides that mining is in the State interest but adjacent vegetation has conservation values. Similarly, when considering the reservation of remnant communities, there is the possibility that the degradation of some communities may be so advanced (for example, the size of the remnant may be so small or the results of the surrounding land use so invasive) as to make their long-term survival doubtful even in a fully protected situation. Consequently, the Service must continue to pursue alternative options in order to fully satisfy its objectives.

Planning Control

The EPA Act has now been in operation for five years. With the competition for land ever increasing, it was thought the Act's wide powers would be used to determine major land-use and resource management issues. In particular, given the degree of land degradation and destruction of wildlife habitat, it was thought by some that controls would be implemented in the 'public interest' to restrict 'freehold rights' over agricultural activities.

Unfortunately, full implementation of the Act, particularly with respect to its potential to assist in the conservation of remnant natural areas, is yet to take place. For example, Section 26(e) of the EPA Act enables an environmental planning instrument to make provision for 'protecting or preserving trees or vegetation'. This section has yet to be fully applied. Because of the lack of State and regional plans to provide direction, the shortcomings in achieving the objective of protecting the environment are particularly noticeable at the local level.

As to the future, reference is made to comments made by the former New South Wales Minister for Planning and Environment, The Hon. T. F. Sheahan, M.P., on vegetation clearing and planning controls over agricultural activities. Mr. Sheahan has suggested that council approval should perhaps be required before land can be clearfelled. He has also acknowledged that sound conservation policies and efficient resource management are mutually supporting and has suggested that land use controls accepted in urban areas are now necessary in rural New South Wales. Such controls would override certain 'freehold rights' on vegetation clearance. There is no doubt the State has the legislative framework to implement such land management decisions should the government decide to embark on this course of action.

CONCLUSION

Providing native vegetation remnants in inland New South Wales meet the necessary criteria for the creation of viable reserves, they are best conserved by reservation under the NPW Act. However, because of certain constraints such as delays incurred in reaching agreement within government, and the provision of adequate funding for acquisition and management, not all areas identified for reservation can be secured. In addition, there are numerous areas which could not justify reservation but which still warrant conservation.

There is a need, therefore, for the pursuit of alternative means of conserving remnant areas. The EPA Act has the powers to achieve the Service's stated objectives but as yet land use planning controls have not been satisfactorily applied to conserve areas of natural vegetation. This is particularly so at the local council level where there are few planning guidelines because of the lack of State and regional policies.

REFERENCES

Beadle, N. C. W., 1948. The Vegetation and Pastures of Western New South Wales. Government Printer, Sydney.

Benson, J., 1985. The Conservation of Plant Communities in New South Wales. New South Wales National Parks and Wildlife Service Review of Nature Conservation Programmes (unpublished).

Caughley, J. W., 1979. Faunal Survey — North Western Plains. New South Wales National Parks and Wildlife Service (unpublished).

Giblin, W. B., 1983. The Achievement of Nature Conservation Objectives via the Environmental Planning and Assessment Act, 1979. New South Wales National Parks and Wildlife Service Review of Nature Conservation Programmes (unpublished).

Hitchcock, P. P., 1984a. Review of Nature Conservation Options for Implementation of the Service's Nature Conservation Objectives — Summary Paper. New South Wales National Parks and Wildlife Service. Review of Nature Conservation Programmes (unpublished).

Hitchcock, P. P., 1984b. Nature Conservation Review — A Situation Report. New South Wales National Parks and Wildlife Service (unpublished).

King, S., 1983. Grasslands. New South Wales National Parks and Wildlife Service. Review of Nature Conservation Programmes (unpublished).

Legoe, G., 1985. Planning for Vegetation Retention in Rural New South Wales. Masters Degree in Town and Country Planning Thesis, University of Sydney.

Margules, C. R., 1982. Conservation in the Murray-Darling Basin. Murray-Darling Basin project development study, stage 1. C.S.I.R.O. Division of Water and Land Resources, Canberra.

Pickard, J., 1981. Vegetation Map of North-West New South Wales. Royal Botanic Gardens, Sydney (unpublished).

Pulsford, I., 1983. Planchonella Nature Reserve New Area Proposal — Investigation Report. New South Wales National Parks and Wildlife Service (unpublished).

Pulsford, I., 1984. Conservation Status of Brigalow *(Acacia harpophylla)*. New South Wales National Parks and Wildlife vice Review of Nature Conservation Programmes (unpublished).

CHAPTER 28

Planning for Fire Management in Dryandra Forest

N. D. Burrows[1], W. L. McCaw[1] and K. G. Maisey[1]

The effective management of fire is fundamental to the management of nature reserves in the south-west of Western Australia. Fire management must take into account the influence of surrounding land uses and people who wish to visit the reserve but should allow processes of ecological succession within the reserve to continue with a minimum of disruption. This requires a sound understanding of the fuel and vegetation characteristics of the reserve that determine the behaviour and effects of fire, planned or unplanned.

This chapter discusses some of the factors that should be taken into account when preparing a fire management plan for a reserve or remnant using the Dryandra Forest as an example. Dryandra Forest is a particularly important area as it is one of the largest and most diverse reserves in the central wheatbelt of Western Australia, and contains a number of rare species of flora and fauna which may be affected by fire. The approach to data collection used here could be applied to many similar remnants.

INTRODUCTION

DRYANDRA Forest is one of the few remnants of natural vegetation found in the Narrogin district of the Western Australian wheatbelt. It lies to the east of the main jarrah *Eucalyptus marginata* forest and straddles the 500 mm rainfall isohyet (Fig. 1). Dryandra Forest is characterized by open woodlands of wandoo *E. wandoo*, powder bark wandoo *E. accedens* and marri *E. calophylla* with jarrah on specific sites. In addition to natural stands of brown mallet *E. astringens* there are extensive plantations of this species which were established between 1930 and 1950. The landscape is composed of remnant lateritic plateaux flanked by pediments and broad valley floors, with occasional granite outcrops. The landforms and soils of the district have been described in detail by McArthur *et al.* (1977).

Dryandra Forest is of major ecological significance. It provides an example of the original vegetation now almost vanished from this part of the wheatbelt, and is one of the few remnants large enough to maintain viable populations of certain rare animals. The total area of the forest is about 28,000 ha, consisting of a large central block and smaller outliers surrounded by cleared farmland. The Western Australian Department of Conservation and Land Management currently administers Dryandra as an area for the conservation of flora, fauna and landscape values.

Fire plays an important role in maintaining natural ecosystems in the south-west of Western Australia (Gardner 1957; Christensen and Kimber 1975; Baird 1977). The region experiences a Mediterranean type climate of cool wet winters, and hot dry summers that, together with accumulations of flammable vegetation and litter predispose the environment to fire (Underwood and Christensen 1981; Burrows 1984). Management of fire is therefore fundamental to the effective management of forests and nature conservation reserves. Fire management plans have been developed for individual forests and reserves in recognition of the importance of tailoring the use of fire to local ecosystems and human factors. Fire management must be based on research and information relevant to conservation objectives (Gill 1977a; Good 1981) and on the need to minimize unwanted wildfires. To use fire as a controllable management tool, the manager must not only understand the factors controlling fire behaviour but must also be aware of the ecological and social consequences of particular fire regimes. With few exceptions a detailed knowledge of ecological processes and of the role of fire does not exist for most nature conservation forests in this region. While certain general concepts can be applied in most situations significant local factors render it necessary to consider each reserve separately (Gill 1977a). In most cases, managers are not in a position to do nothing until researchers provide detailed

[1]Department of Conservation and Land Management, Research Branch, Brain Street, Manjimup, Western Australia 6258 Australia.
Pages 305-12 *in* NATURE CONSERVATION: THE ROLE OF REMNANTS OF NATIVE VEGETATION ed by Denis A. Saunders, Graham W. Arnold, Andrew A. Burbidge and Angas J. M. Hopkins. Surrey Beatty and Sons Pty Limited in association with CSIRO and CALM, 1987.

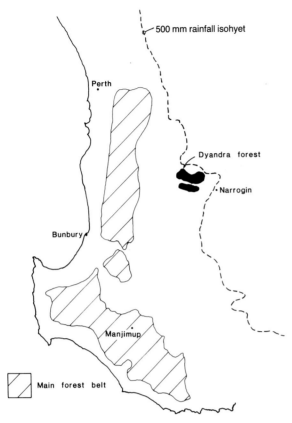

Fig. 1. Location of Dryandra Forest, an important remnant of wheatbelt vegetation in the south-west of Western Australia.

information as, if nothing else, they are often legally and morally obliged to take wildfire prevention measures.

We believe that an interim understanding of the role of fire in maintaining conservation values in forest remnants can be obtained at reasonable cost and in a relatively short time. Staff from the Manjimup Research Station of the Department of Conservation and Land Management conducted a survey of Dryandra Forest during mid 1984. The objectives of the survey were to:

1. determine the likely range of past fire regimes by examining characteristics of both individual species and communities;

2. examine the recent (last 50 years) fire history of the reserve in order to determine the threat posed by wildfires to both conservation and human values; and

3. recommend interim approaches for management of fire in Dryandra Forest that would not prejudice either human values or long term conservation values.

In this chapter we describe the techniques used in the survey, and discuss the implications of the key findings for future fire management of Dryandra.

METHODS

A variety of survey techniques were used to help establish the role of fire at Dryandra. For simplicity these are grouped into two categories:

1. methods to determine the 'fire environment' of the forest; and

2. biological indicators of the role of fire in maintaining ecological processes.

Methods Used to Determine the Fire Environment

The fire environment has been defined by Countryman (1972) as the conditions, influences and modifying forces that control fire occurrence and behaviour. Important factors determining the fire environment are weather patterns, fuel characteristics, topography and ignition source. We surveyed these factors at Dryandra to determine the fire environment of the forest and surrounding lands.

Climate and Weather. Climate, vegetation and fire have a long association in the south-west of Western Australia (Hallam 1975). Climate affects the fire proneness of an area by influencing the quantity of fuel available for burning, the surface burning conditions (such as fuel moisture, wind speed, temperature and humidity), the length of the fire season and the likelihood of ignition due to lightning (Gill 1984).

We examined monthly and yearly rainfall and mean daily maximum temperatures at Narrogin from 51 years (1933-1984) of records held by the Bureau of Meteorology. Wind and relative humidity data were not available.

Recent Fire History. We examined the records of prescribed burning and wildfires between 1938 and 1984 for the area within a 20 km radius of the fire lookout tower. This included the entire forest as well as a substantial area of adjoining agricultural land. The season, location and size of prescribed and wildfires was noted, and the cause of wildfires recorded.

Fuel Characteristics. Six blocks of natural forest were selected for detailed fuel sampling, each about 160 ha in area. Together these blocks contained the major landform and vegetation units found at Dryandra with most vegetation units represented at a range of ages since burning (from three to more than 46 years). Vegetation maps prepared in 1932 on the basis of overstorey and dominant understorey plant species were useful in selecting blocks for sampling, and in most areas these maps were surprisingly accurate over 50 years later. Sampling in natural forest was conducted systematically on a 200 × 100 m grid, providing a total of 380 points. Time since burning was initially determined using annual fire records, and these were later verified at each point on the basis of visual indicators that included the amount and condition of charcoal on logs and trees, diameter of coppice stems and the height of understorey regrowth.

Sampling in mallet plantations was stratified according to plantation age, as these areas had not been burnt since initial establishment. Sampling was conducted at 50 m intervals along transects with a total of 150 points in 10 ages of plantation.

In both natural forest and mallet plantation we identified three fuel arrays for measurement:

1. litter, consisting of leaves, twigs, bark and floral parts (less than 10 mm diameter) on the forest floor;
2. woody material (greater than 10 mm diameter) on the forest floor; and
3. live, and suspended dead vegetation up to 2 m above ground level.

Litter fuel was collected from 10 quadrats (each 0.04 m^2) located randomly within a 20 m radius of each point. Depth of litter in each quadrat was measured. Litter samples were subsequently oven dried, sorted into components and weighed. The quantity of dead material greater than 10 mm diameter on the forest floor was estimated using the line intercept technique of Van Wagner (1968), and included material from small branches up to large fallen logs.

In order to sample vegetation fuels we first classified the understorey using data on plant species composition and tree basal area collected at each sample point, except those in the mallet plantation. Vegetation was sorted into clusters on the basis of percentage similarity coefficient using a programme developed for the department's Perkin Elmer computer system (D. Ward pers. comm.). Detailed vegetation sampling was then undertaken at four or five points in each vegetation type. The procedure involved estimation of the height of each vegetation strata, estimation of percentage ground cover and collection of all vegetation up to 2 m tall on a 3 m^2 quadrat for determination of oven dry biomass.

Grass was common along the boundary of the farmland and forest. This fuel type, although important, was not examined in the initial survey but was dealt with in a subsequent survey of specific fire hazards along the reserve boundary.

Biological Indicators of the Role of Fire

Adaptive Traits. Plants have adaptive traits which enable them to survive and regenerate within specific regimes of fire frequency, season and intensity (Gill 1975, 1977b). During the survey we observed adaptive traits of particular plant species and attempted to interpret these as indicators of fire dependency or sensitivity to fire, thus constructing a picture of the likely range of fire regimes experienced by these species in the past.

All of the species recorded during vegetation assessments were classifed according to the following categories of life form:

1. woody plants with rootstock;
2. fire sensitive plants with soil stored seed;
3. fire sensitive plants with seeds stored in woody or papery capsules; and
4. fire sensitive herbs.

Fire sensitive means that mature plants subjected to full scorch of the foliage will die (Gill 1981). The contribution of each life form category was expressed as a percentage of the total number of species recorded.

In addition, four species were selected for more detailed examination; *Gastrolobium microcarpum, Dryandra nobilis, Eucalyptus astringens* and *Allocasuarina heugeliana*. These species have a restricted distribution outside Dryandra but are important within the reserve commonly growing in pure stands. The structure of a number of stands of each species was determined by measuring stem diameter overbark, and a sample of individuals were felled to facilitate counting of growth rings to estimate individual plant age. The origin (seed, lignotuber, coppice), size and abundance of regeneration within each stand was also recorded. This information was then compared to records of fire history to determine if the structure of the stand could be attributed to particular fire events.

The Fate of Seed in Poison Bush Thickets. Many species of Leguminosae are hard seeded (Gill 1981) and may regenerate prolifically following intense summer fires (Shea *et al.* 1979; Christensen 1980). During the survey we observed thickets of *Gastrolobium* spp. that had not regenerated to any degree following summer and autumn fires. We hypothesised that the paucity of regeneration may have been due to absence of seed beneath mature thickets caused by seed predation, as we frequently observed insect and bird (pigeons and quails) activity in these areas.

To estimate the availability of seed we counted the number of seeds collected by sorting through the litter and surface soil (to 20 mm depth) from 25 0.04^2 m quadrats taken from beneath individual plants at each of five sites. Initial sampling was done immediately after seedfall in February 1985 and samples were taken again in June and August.

RESULTS AND DISCUSSION

The Fire Environment

The mean annual rainfall at Narrogin (50 year average) was 508 mm with about 80% of the annual total falling in late autumn and winter. Drought years (rainfall < 400 mm) have occurred with a frequency of about one in five years although without any obvious pattern. Summer months (November-March inclusive) are dry, with only 2-3 wet days per month, and hot with mean daily maximum temperatures above 25°C. Conditions of high temperature, low humidity and strong winds often occur simultaneonsly during the summer (S. Gorton pers. comm.) but more detailed description of fire weather was prevented by the lack of humidity and wind data. In an average year forest fuels at Dryandra would be sufficiently dry to burn from October until April.

Grassland fuels and crops on land adjacent to the forest would also be flammable for a similar period, although harvesting and grazing would reduce the amount of fuel available in the later stages of the fire season.

In comparison with the main jarrah forest belt to the west, Dryandra experiences more severe fire weather conditions but in the last 50 years no major wildfires have burnt within the forest, and no wildfires have entered private property from the forest (Table 1, Fig. 2). However, records showed that in the past, fires commonly entered the forest from adjacent farmlands. Most of these fires were escapes from land clearing burns which were common in the decade 1950-60. In recent years fewer fires have escaped from farmland into the forest, and the proportion of the forest burnt by wildfires has substantially reduced (Fig. 2). The reduced frequency of wildfire in the forest can probably be attributed both to reduction in land clearing activity, and to improvements in the organization and equipment of fire suppression forces. Lightning was not a major cause of fire in the forest, causing only two fires between 1938 and 1984.

Fire prevention activities should continue to receive priority from forest managers. Important fire prevention activities include regular goodwill visits to adjacent landholders, good housekeeping along forest boundaries and co-operative fire control activities with local bushfire brigades to fully utilize the expertise and resources available.

For example, forest managers may assist with windrow or stubble burning on land adjacent to the forest.

Dryandra Forest is well suited to pre-fire suppression activities. Fuels are generally discontinuous and slow to accumulate compared with those of higher rainfall forests (Fig. 3). The potential for litter fuel accumulation was linked to the density of the forest overstorey. The light ground fuels, low open forest structure and absence of trees with fibrous or pendulous bark would tend to reduce the potential for long-distance spotting during wildfires. We consider that effective fire control in Dryandra Forest could be achieved by maintaining the extensive network of existing firebreaks (1 km of firebreaks per 40 ha of forest) and reducing fuels at strategic locations either by burning, herbicide spraying or ploughing. Fire breaks along the forest boundary, and areas of the reserve regularly used by people should receive priority. Natural firebreaks such as rocky outcrops, lateritic breakaways and areas of low fuel accumulation should be used to best advantage in both prescribed burning and wildfire control.

By examining the options for wildfire suppression in advance, managers should be able to develop suppression strategies which are compatible with conservation values. Fuel types at Dryandra are strongly linked to landforms so that accurate maps of fuel characteristics could be developed from air photographs to assist in suppression planning.

In general the highly flammable fuel types such as thickets of *Dryandra nobilis* are limited in area and are surrounded by less flammable vegetation with low fuel loads such as *E. accedens* forest.

Table 1. Summary of wildfires recorded within 20 kms radius of the Dryandra Forest fire tower between 1938 and 1984.

Location and total number of Wildfires	
Started in Dryandra Forest	9
Started in private property	69
Started in other lands	9
Burnt into Dryandra Forest from private property	20
Burnt into private property from Dryandra Forest	0
Fire size (ha)	
Average size of wildfire in Dryandra Forest	40
Average size of wildfire on private property	503
Largest wildfire which started in Dryandra Forest	260
Largest wildfire which burnt on private property	>2550
Cause of ignition (No. of fires)	
On Dryandra Forest — Escapes from prescribed burning	5
— Lightning	2
— Other	2
On all land within 20 kms radius	
— Private property	39
— Harvesting operations	18
— Locomotives	10
— Lightning	4
— Other cause (billy fires)	9
— Unknown cause	7

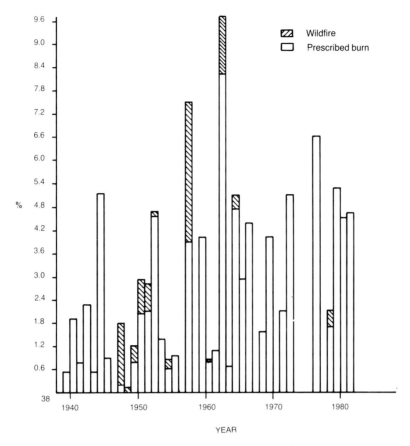

Fig. 2. Percentage area of Dryandra Forest burnt annually 1938-1984.

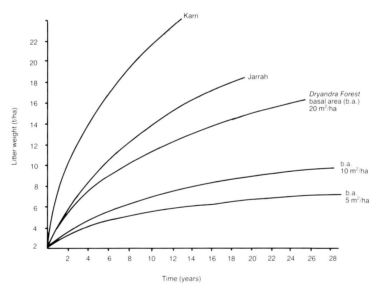

Fig. 3. Litter fuel accumulation for three levels of forest stand density at Dryandra (5, 10, 20 m²/ha of overstorey basal area) compared with jarrah and karri forest (data from Peet 1971).

Biological Indicators of the Role of Fire

Plants provide a useful guide to the fire regime most likely to have been experienced in the past and to that which is most likely to maintain natural processes in the future. We believe fire to have been an infrequent visitor to this forest in the past, based not only on the above-mentioned factors of fuel accumulation and fire history, but also on the occurrence of fire sensitive plant species (Fig. 4). Fire sensitive species not only comprise a high proportion of total species, but are important in that they form dense thickets and pure stands

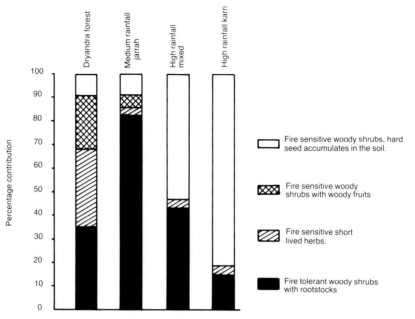

Fig. 4. Contribution of major plant life forms to species total of Dryandra (survey data) compared with three other forest types (data from Christensen and Kimber 1975).

over much of the landscape. In addition to the importance of thickets in maintaining the character of the forest vegetation, they may also be important habitat for some animals (Christensen 1980).

The ability of fire sensitive plants to regenerate in the absence of fire varied between species and between sites of similar species. In some cases *Gastrolobium* spp. appeared to have regenerated as even-aged thickets in response to past fires but we also found evidence of uneven-aged thickets where fire had been excluded for many years, indicating successful regeneration in the absence of fire.

Estimates of the age of individual *Gastrolobium* plants from growth ring counts suggested that some thickets were up to 50 years old. In contrast natural stands of *E. astringens* were for the most part even aged. *E. astringens* is often surrounded by heavy accumulations of bark fragments (up to 25 t/ha) so that even large trees do not normally survive after a fire. As *E. astringens* does not produce substantial quantities of seed for 12-15 years this species would not still be present in the forest if its particular niche had been subject to frequent burning over a long period.

Slow postfire response of rootstock species, poor ability of the trees for epicormic crown recovery, the high proportion of fire sensitive species and the slow rates of fuel accumulation are all indicative of the infrequency of fire. We believe that infrequent fires (20-60 year periodicity) may be important in maintaining the diversity of successional stages within the forest but that a regime of more frequent fires (less than 15-year rotation) would not be compatible with maintaining the structure and florisitics of the vegetation in their present form. Frequent burning of fire sensitive species that have a long juvenile period would inevitably lead to degeneration and even elimination of populations of these species.

The quantity of viable seed in litter and soil beneath *Gastrolobium* species thickets declined rapidly with time after seedfall (Fig. 5). Seed predation by bronzewing pigeons *Phaps chalcoptera*, quail *Coturnix pectoralis* and various insects is probably responsible for this. It appears that relatively high populations of seed eating birds and insects can be sustained in the region, even though much of the original vegetation has been removed for growing wheat. During the late spring and summer months there is an abundance of wheat grain for seed eaters but during the winter months the more mobile predators may concentrate on thickets within the forest. We observed many thickets in a declining state and suggest that this may be due to lack of viable seed for regeneration. To offset this we suggest that seed be collected from the bush to facilitate artificial regeneration where required.

Given that infrequent fires play an important role in maintaining natural processes, the question may then be asked 'what is the appropriate fire regime?' We suggest that prescribed burning for conservation values (biological burning) should be determined on the basis of clearly defined needs, rather than by committing blocks or areas to a long-term burning programme. Regular monitoring of carefully selected environmental components may help identify 'ecosystem health' in measureable terms such as the density and structure of thickets and fluctuations in animal numbers. We could perhaps construct and maintain an 'ecology index' which would

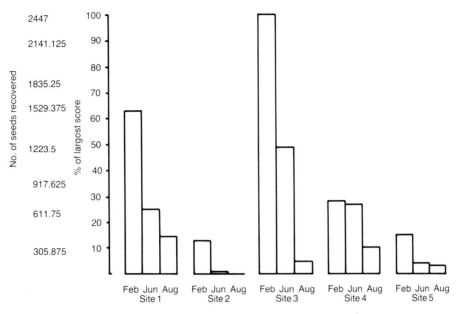

Fig. 5. Recovery of *Gastrolobium* seed from litter and surface soil beneath thickets in February, June and August 1985.

be a function of animal numbers and population age structure and of the density, cover and composition of important vegetation types.

Monitoring is an integral part of every management operation and the maintenance of an 'ecology index' could become part of routine operations. Further research and discussion with forest managers will be necessary to develop this concept. Other desirable traits of this approach to fire management include:

1. ensuring that the management dollar is spent at a time when it will be most effective;
2. providing a direct measurement of the performance of management in maintaining conservation values;
3. providing a historical perspective that enables managers and researchers to observe trends and cycles in the ecosystem; and
4. encouraging managers to become more aware of the biological aspects of reserve management.

It would be rather simplistic and, perhaps, arrogant to expect that the complex ecological interactions in Dryandra Forest should be actively managed or directed. Rather we should provide information and techniques that allow managers to intervene when these processes are disrupted by unnatural influences such as feral animals, weed invasion or frequent wildfires.

CONCLUSIONS

Reserve managers require information and ideas to properly plan and implement fire regimes which are in accordance with long-term conservation objectives and which meet the constraints of time, money, fire protection and legislation.

A survey such as the one briefly described here can provide a sound basis for developing an interim fire management plan for areas of remnant vegetation. Concurrent detailed research can modify fire management as results become available.

We consider that managers must take an active role in determining the fire regime necessary to maintain and protect important natural processes in Dryandra Forest. Infrequent firing (20-60 years) may be important in maintaining a diversity of successional stages but frequent firing could eliminate fire sensitive species and cause degeneration of certain habitats. The ability to protect the forest from wildfires is a vital prerequisite for more sophisticated management. Managers must continue to work in close harmony with adjacent land owners and forest users to maintain this valuable forest remnant.

ACKNOWLEDGEMENTS

We thank staff of the Department of Conservation and Land Management at the Manjimup Research Station who helped collect, collate and present data from Dryandra Forest. We also thank Steve Gorton and the departmental staff at the Narrogin District Office for their help with the survey.

REFERENCES

Baird, A. M., 1977. Regeneration after fire in Kings Park, Perth, Western Australia. *J. R. Soc. West. Aust.* 60: 1-22.

Burrows, N. D., 1984. Describing forest fires in Western Australia: A guide for managers. Forests Department of Western Australia. Tech. Pap. 9.

Christensen, P. E. and Kimber, P. C., 1975. Effect of prescribed burning on the flora and fauna of south-west Australian forests. *Proc. Ecol. Soc. Aust.* 9: 85-106.

Christensen, P. E. S., 1980. The biology of *Bettongia penicillata* Gray, and *Macropus eugenii* (Desmarest, 1817) in relation to fire. Forests Dept. of Western Australia Bull. 91.

Countryman, C. M., 1972. The fire environment concept. USDA Forest Service.

Gardner, C. A., 1957. The fire factor in relation to the vegetation of Western Australia. *West. Aust. Nat.* **5**: 166-73.

Gill, A. M., 1975. Fire and the Australian flora: A review. *Aust. For.* **38**: 4-25.

Gill, A. M., 1977a. Management of fire prone vegetation for plant species conservation in Australia. *Search* **8**: 20-6.

Gill, A. M., 1977b. Plant traits adaptive to fires in Mediterranean land ecosystems. *In* Proc. of the Symposium on the Environmental Consequences of Fire and Fuel Management in Mediterranean Ecosystems ed by H. A. Mooney and G. E. Conrad. USDA For. Serv. Gen. Tech. Rep. WO-3.

Gill, A. M., 1981. Adaptive responses of Australian vascular plant species to fires. Pp. 243-72 *in* Fire and the Australian Biota ed by A. M. Gill, R. H. Groves and I. R. Noble. Australian Academy of Science, Canberra.

Gill, A. M., 1984. Causes, frequency and extent of bushfires in Victoria. *In* Proc. of Symposium on Fuel Reduction Burning ed by H. M. Ealey.

Good, R. B., 1981. The role of fire in conservation reserves. *In* Fire and the Australian Biota ed by A. M. Gill, R. H. Groves and I. R. Noble. Australian Academy of Science, Canberra.

Hallam, S. J., 1975. Fire and hearth: A study of Aboriginal useage and European usurpation in south Western Australia. Australian Institute of Aboriginal Studies, Canberra.

McArthur, W. M., Churchward, H. M. and Hick, P. T., 1977. Landforms and soils of the Murray River Catchment area of Western Australia, C.S.I.R.O. Div. of Land Resource Management, C.S.I.R.O.

Peet, G. B., 1971. Litter accumulation in jarrah and karri forests. *Aust. For.* **35**: 258-62.

Shea, S. R., McCormick, J. and Portlock, C. C., 1979. The effect of fires on the regeneration of leguminous species in the northern jarrah (*Eucalyptus marginata* Sm.) forest of Western Australia. *Aust. J. Ecol.* 4:195-205.

Underwood, R. J. and Christensen, P. E. S., 1981. Forest fire management in Western Australia. Forests Dept.of Western Australia. A. Special Focus 1.

Van Wagner, C. E., 1968. The line transect method in forest fuel sampling. *For. Sci.* **14**: 20-6.

CHAPTER 29

Conservation Strategies for Human-Dominated Land Areas: the South Australian Example

S. G. Taylor[1]

Human-dominated land areas have a native vegetation cover which has been highly fragmented by urban and/or rural land use, creating a landscape pattern of small patches of remnant natural land scattered within a matrix of cultivated vegetation and human constructions. This landscape pattern presents unique problems for nature conservation. In South Australia, initial attempts to address these problems used assessments of conservation value to select remnant natural land for inclusion in nature reserves. More recently, assessments of conservation value have been used to select remnant natural land for native vegetation retention by private landholders as part of clearance control schemes. In both cases, these assessments of conservation value have been based on three different strategies for nature reserve design; namely an anthropocentric, a biocentric and a biogeographic strategy. Recent research on the ecological dynamics of remnant natural land suggests that these nature reserve design strategies do not provide appropriate nature conservation goals for human-dominated land areas. In the light of this research, two alternative strategies are recommended: one to conserve the productive/amenity value of remnant natural land and the other to conserve its biological value.

SOUTH AUSTRALIA'S HUMAN-DOMINATED LAND AREAS

IN South Australia, 150 years of European settlement have had relatively little impact on the State as a whole. Settlement is still exceedingly sparse throughout the north of the State in the arid province and the three pastoral provinces (Eastern Pastoral, Flinders Ranges, Western Pastoral) which contain approximately 85% of South Australia's total land area (Fig. 1, Table 1). Moreover, the settlement which has occurred in these environmental provinces is mainly associated with mining, tourism, Aboriginal social service centres and extensive livestock grazing using palatable native plant species rather than sown pastures. Thus, in the arid and pastoral provinces, the cultivated vegetation characteristic of settled land is confined to a few scattered points of permanent human occupation and the entire north of the State can be classed as a *natural land area* with a virtually continuous native vegetation cover (Laut *et al*. 1977a).

The remaining land area in the south of the State is divided into four agricultural provinces; South-East, Murray Mallee, Mt. Lofty Block, Eyre and Yorke Peninsulas (Fig. 1, Table 1). Here, in contrast with the arid and pastoral provinces, European settlement has involved widespread native vegetation clearance in order to establish cultivated vegetation for productive/amenity purposes. As a result of this clearance and other human impacts, only an estimated 18% of the land in the agricultural provinces is remnant natural land with a native vegetation cover (Dendy 1985).

From a conservation perspective, the amount of remnant natural land in the agricultural provinces is far less important than its geographic distribution. Because the settlement of South Australia has not been a random process in space or time, the extent of native vegetation clearance differs markedly from one agricultural province to another, and from environmental region to environmental region within each province. For example, approximately 15% of the Mt. Lofty Block Province is remnant natural land, but the amount of remnant natural land in the three environmental regions of this province (Fig. 1, Table 1) varies from 38% of the Kangaroo Island Region, to 16% of the Mid-North Wheatlands Region and 4% of the Peninsula Uplands Region (Dendy 1985).

---

[1]Department of Geography, University of Adelaide, Adelaide, South Australia 5000.
Pages 313-22 *in* NATURE CONSERVATION: THE ROLE OF REMNANTS OF NATIVE VEGETATION ed by Denis A. Saunders, Graham W. Arnold, Andrew A. Burbidge and Angas J. M. Hopkins. Surrey Beatty and Sons Pty Limited in association with CSIRO and CALM, 1987.

Table 1. Environmental Provinces and Regions of South Australia according to Laut *et al.* (1977b). Key to Figure 1.

Environmental Provinces	Environmental Regions
A. Agricultural Provinces 1. South-East	1.1 South Coast 1.2 Southern Coastal Plains 1.3 Mt. Gambier Volcanics 1.4 Southern Wetlands and Dune Ranges 1.5 Frances Plateau
2. Murray Mallee	2.1 Murray Lakes 2.2 Northern Calcarenite Ridges and Plains 2.3 South-East Mallee Heathlands 2.4 Upper Murray Lands
3. Mt. Lofty Block	3.1 Kangaroo Island 3.2 Peninsula Uplands 3.3 Mid-North Wheatlands
4. Eyre and Yorke Peninsulas	4.1 Southern Highland and Plains 4.2 West Coast 4.3 Central Mallee Plains and Dunes 4.4 Northern Myall Plains 4.5 Southern Yorke 4.6 Gulf Plains
B. Pastoral Provinces 5. Eastern Pastoral	5.1 Olary Plains 5.2 Olary Uplands 5.3 Lake Frome Plains
6. Flinders Ranges	6.1 Southern Basins and Ranges 6.2 Northern Complex
7. Western Pastoral	7.1 Gawler Uplands 7.2 Torrens Depression 7.3 Central Salt Lakes 7.4 Kingoonya Plains 7.5 Great Victoria Desert 7.6 Nullarbor Plains
C. Arid Province 8. Northern Arid	8.1 Northern Uplands 8.2 Western Sand Plains 8.3 Central Tablelands 8.4 Lake Eyre Basin

Notes. Environmental regions are further divided into environmental associations which are areas having a distinctive geologic/geomorphic history and thus a repetitive pattern of environmental units (i.e. distinctive landforms).

Environmental regions like Kangaroo Island and the Mid-North Wheatlands have a high proportion of remnant natural land because they contain environmental associations (Table 1) which are unsuitable for primary production and remote from South Australia's main urban centres. Either singularly or in combination these environmental associations support extensive (>10,000 ha) tracts of native vegetation; but they are the exception in the agricultural provinces. By far the majority of the 211 environmental associations in the agricultural provinces support a native vegetation cover which has been highly fragmented by urban and/or rural land use, creating a landscape pattern of small (mainly < 100 ha) patches of remnant natural land scattered within a matrix of cultivated vegetation and human constructions (Laut *et al.* 1977a).

This landscape pattern is by no means unique to South Australia's agricultural provinces. Instead, it is typically associated with the settlement of temperate wooded environments by sedentary agriculturalists and the consequent development of urban centres dependent on the intensive use of rural land for primary production. The result of this urban/rural development is a *human-dominated land area* where remnant natural land is largely an accidental artifact of the settlement process (Burgess and Sharpe 1981).

Most of South Australia's human-dominated land areas are of comparatively recent origin. Exceptions include the human-dominated land areas of the Mt. Gambier Volcanics, Gulf Plains, Mid-North Wheatlands and Upper Murray Lands Regions (Fig. 1, Table

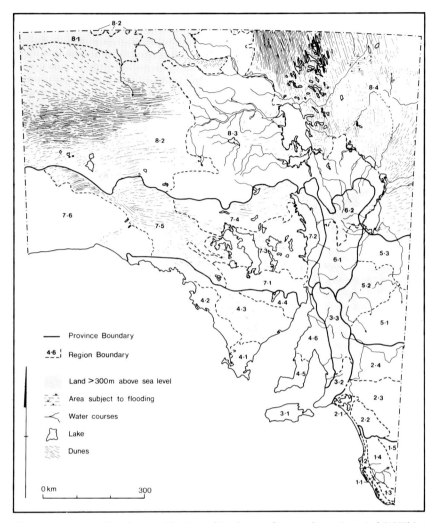

Fig. 1. Environmental Provinces and Regions of South Australia according to Laut *et al.* (1977b). A key to the number codes for these environmental provinces and regions is given in Table 1.

1), all of which have had a long history of intensive urban/rural development (Williams 1974). In other environmental regions, fragmentation of the native vegetation cover did not pass the critical point definitive of a human-dominated land area until the post-1945 period when South Australia's arable farming frontier was stabilized at its present northern limit.

Since 1945, northward expansion of the arable farming frontier has been replaced by the intensification of farming within this frontier, in close association with the introduction of leguminous sown pastures and superphosphate fertilization (Williams 1974). This change in rural land use, together wth the related suburban spread of South Australia's urban population, has greatly increased the fragmentation of the human-dominated land areas created before 1945 and has also caused the rapid fragmentation of the native vegetation cover in previously underdeveloped environmental regions such as the Peninsula Uplands, Kangaroo Island, Southern Highland and Plains, Southern Yorke Peninsula, South East Mallee Heathlands, and Southern Wetlands and Dune Ranges Regions (Fig. 1, Table 1) (Williams 1974).

With new human-dominated land areas being created at an accelerating rate, native vegetation clearance became a major public issue in South Australia during the 1940s and 1950s. Consequently, most of the State's nature conservation efforts were directed to the protection of remnant natural land in the agricultural provinces. At first these efforts involved the use of three different conservation strategies to select remnant natural land for protection in State nature reserves. More recently, however, assessments of conservation value have been used to select remnant natural land for native vegetation retention by private landholders as part of voluntary and enforced clearance control schemes. These assessments of conservation value are based on the assumption that South Australia's established nature reserve design strategies provide appropriate goals for the State's present clearance control strategy. The following review of nature reserve design and clearance control strategies in South Australia examines this assumption with particular reference to the formulation of appropriate goals for nature conservation in human-dominated land areas.

NATURE RESERVE DESIGN STRATEGIES

Anthropocentric Nature Reserves

During South Australia's expansionist phase of settlement prior to 1945, land with a native vegetation cover was not generally regarded as a resource worth conserving in its natural state, except for productive purposes (such as the protection of water catchments and the provision of native pastures) or as a 'savings bank' of natural resources for future productive use (e.g. timber reserves, stone reserves, etc.). Nevertheless, a small minority group, which Powell (1976) describes as an urban scientific élite, did persuade the State Government to reserve some remnant natural land within the arable farming frontier for biological conservation under the *National Pleasure Resorts Act* of 1914. Although the scientists who supported this Act were motivated by a concern for biological conservation, the Act itself reflected the same anthropocentric conservation strategy evident in the establishment of water catchment, timber, stone and other 'productive' reserves. Thus National Pleasure Resorts were areas of natural beauty and/or natural interest reserved primarily for their tourism and local recreation value.

After 1914, there were continuing appeals for the establishment of biocentric reserves to protect flora and fauna threatened by native vegetation clearance in the agricultural provinces. However, South Australia's official nature conservation strategy remained dominantly anthropocentric until the 1960s. As a result of this bias, the 28 State nature reserves established before 1960 include four reserves with a major tourism function (now classed as National Parks) and seven reserves designed to protect popular picnic sites in the Adelaide plains and hills area (now classed as Recreation Parks). The remaining 17 reserves are now classed as Conservation Parks; but 11 of these were established for tourism or local recreation purposes and have been managed primarily for biological conservation only in recent decades.

Biocentric Nature Reserves

The six explicitly biocentric reserves dating from South Australia's anthropocentric period of nature conservation were established at the end of the expansionist phase of settlement to prevent opportunistic clearing of remnant natural land with severe soil limitations for agricultural production on Eyre Peninsula and in the Murray Mallee Province (Williams 1974). From the mid-1930s, recognition of the soil limitations associated with most of the remnant natural land in the agricultural provinces gave concerned scientists reason to hope that their arguments for biocentric reserves would receive an increasingly positive government response, but this hope was frustrated by the agricultural advances of the 1940s and the need to settle servicemen returned from World War II. In combination, these factors provided the incentive for a number of government and private projects which 'opened-up' the last remaining large areas of undeveloped land in the agricultural provinces during the 1950s and early 1960s (Harris 1985).

By the end of the post-World War II development projects, approximately 75% of the land in the agricultural provinces was settled land and the continuing decline in the extent of remnant natural land had become the concern, not only of an urban scientific élite, but also of a populist nature conservation movement (Harris 1985). Since 1960, under pressure from the conservation lobby, the government has established 126 new State nature reserves on the mainland in the agricultural provinces and on Kangaroo Island (National Parks and Wildlife Service 1984). Of these reserves, three are National Parks, eight Recreation Parks and 115 Conservation Parks. It is indicative of the fragmented state of the native vegetation cover in the agricultural provinces by 1960 that only 7% of the new Conservation Parks are greater than 10,000 ha in area, while 46% are less than 1000 ha and 26% less than 100 ha.

Until the mid-1970s there was little evidence of a systematic strategy for selecting the remnant natural land included in biocentric reserves. The main selection criterion appears to have been the threat of native vegetation clearance, particularly where clearance would affect rare species or communities. As noted by Margules and Usher (1981, p.100), this criterion reflects a common characteristic of biocentric reserves; that they are advocated as a use for remnant natural land 'in response to a threat from some other land use' and are not usually suggested in their own right. In other words, the biocentric conservation strategy is typically a 'crisis-response strategy' in which the protection of rare species and communities has become the single most important goal of biological conservation (Margules and Usher 1981).

The political actvities of South Australia's conservation lobby still reflect a pre-occupation with the biocentric conservation strategy. The government, however, has become increasingly concerned about the long-term viability of its nature reserves and this concern has led to a drastic reduction in the rate of land acquisition for State nature reserves in the agricultural provinces since 1975 (National Parks and Wildlife Service 1984).

Biogeographic Nature Reserves

In Australia, the major stimulus for concern about the temporal dimension of biological conservation was the Specht Survey of the conservation status of the major plant alliances in each State (Specht *et al.* 1974). The results of this survey emphasized the need to replace the biocentric conservation strategy with a systematic attempt to protect the diversity and viability of Australia's native biological communities (Specht 1975). Subsequent to the Specht Survey, the Australian Academy of Science published guidelines for a national biogeographic conservation strategy designed to achieve this goal (Fenner 1975).

Australia's biogeographic conservation strategy was intended to promote the establishment of nature reserve systems which would include self-sustaining, representative examples of the range of biological communities found in the various environmental regions of each State (Specht 1975). The Australian Academy of Science guidelines identified two key criteria to be used in the selection of sites for these nature reserve systems; community diversity and area (Slatyer 1975).

The first criterion was derived from the concept of 'representativeness'. To be representative of a given environmental region, an ideal nature reserve system should include a portion of each of the environmental associations in the region, as well as the full range of environmental units characteristic of each association (Slatyer 1975). More realistically, it was assumed that the typical species and communities of an environmental region could be protected by establishing a few 'representative' reserves, each with a high level of community diversity. In addition, rare species and communities not included in the 'representative' reserves could be protected by establishing separate 'unique' reserves with much more flexible site requirements (Slatyer 1975).

The second criterion was only applicable to 'representative' reserves and was intended to ensure that they would be 'self-sustaining'. In practical terms, this meant that 'representative' reserves should be sufficiently large to maintain viable populations of their rarest species (where 'rarity' was measured by population density) and to meet the range requirements of their most mobile species (exclusive of long-distance migratory species) (Slatyer 1975). No minimum area was specified for 'representative' reserves in environments equivalent to those of South Australia's agricultural provinces, but 10,000 ha approximates the lower limit commonly used to identify 'self-sustaining' nature reserves in this State (Specht *et al.* 1974).

In his comments on the biogeographic conservation strategy, Frankel (1975) suggested that the limited State funds available for implementing the strategy should be directed to 'remote areas' where large nature reserves could be established at a reasonable cost, both in terms of land acquisition and subsequent reserve management. The establishment of additional State nature reserves in human-dominated land areas was not recommended, due to the necessarily small size of reserves in environmental regions and associations with a highly fragmented native vegetation cover, the high cost of purchasing 'scrub blocks' on freehold urban/rural properties and the need for intensive reserve management in the face of strong public pressure for recreational access to near-urban remnant natural land (Frankel 1975).

Although South Australia's present approach to land acquisition for nature reserves has been rationalized in the manner suggested by Frankel, the application of biogeographic principles to nature reserve design has not had an entirely negative impact on biological conservation in the State's human-dominated land areas. Instead, by highlighting the vulnerability of the existing small nature reserves in these areas to ecological isolation as a result of continuing native vegetation clearance, biogeographic principles have provided scientific support for attempts to promote native vegetation retention on private land.

The pressing need to involve private landholders in nature conservation became apparent in 1976 when the government received the report of an interdepartmental committee which it had commissioned to investigate the impact of native vegetation clearance on the agricultural provinces (Department for the Environment 1976). The recommendations of this committee were implemented in 1980 with the establishment of a Heritage Agreement Scheme providing incentives for private landholders who voluntarily agreed to manage their remnant natural land for nature conservation. The limited success of this voluntary scheme persuaded the government to introduce native vegetation clearance controls in 1983 (Dendy 1985).

THE CLEARANCE CONTROL STRATEGY
Assessment of Conservation Value

At present, South Australia's native vegetation clearance controls do not prohibit further clearing of specific environmental regions or associations within the agricultural provinces. They can, however, be used to prevent clearing throughout the agricultural provinces where the native vegetation concerned has a high conservation value. Here conservation value is measured in terms of both the human (productive/amenity) and biological benefits commonly attributed to native vegetation retention on private land. Unfortunately, the principles which have been established for processing clearance applications fail to make a clear distinction between the *potential benefits* of native vegetation retention and *practical criteria* for native vegetation retention.

This problem is particularly apparent with respect to the productive/amenity benefits of native vegetation retention. The existing clearance control legislation states that native vegetation should not be cleared if it contributes to the productivity or the amenities of an area by protecting water quality, reducing flood hazard, assisting soil erosion or salinity control, providing livestock shade and shelter, enhancing the scenic quality of the landscape, or preserving the natural character of heritage sites (Department of Environment and Planning 1983). Given the present extent of native vegetation clearance in the State's human-dominated land areas, it is arguable that all of the remaining native vegetation on private land in these areas has a very high conservation value with respect to most of the productive/amenity benefits identified by the existing legislation as grounds for refusing clearance applications.

Nevertheless, the government has not taken the step of identifying these human-dominated land areas within the framework of its State Development Plan in order to regulate native vegetation clearance in the same regional manner as other forms of public and private land development. In fact, it seems likely that a Native Vegetation Clearance Act will be drafted to entirely separate this form of land development from the general Planning Act (Dendy 1985). Consequently, applications to clear native vegetation are being (and probably will continue to be) processed on an 'individual-site' basis using assessments of conservation value which rely on biological benefits to add weight to sites which are otherwise equally suitable for retention in terms of productive/amenity benefits.

In this context, the present legislation again confuses potential benefits with practical criteria by stating that native vegetation should not be cleared if it provides important habitat for wildlife or wildlife 'corridors' and 'stepping stones' (Department of Environment and Planning 1983) — as, undoubtedly, does all native vegetation in human-dominated land areas. However, some specific criteria for assessing the biological conservation value of native vegetation are cited; namely plant species or community rarity, plant community diversity and plant community typicality (Department of Environment and Planning 1983).

These criteria reflect the combined goals of South Australia's biocentric and biogeographic strategies for nature reserve design; that is, to establish 'unique' (biocentric) reserves for the protection of biological rarity and to establish 'representative' (biogeographic) reserves for the protection of biological diversity/typicality. As has been demonstrated by the preceding discussion of the biogeographic conservation strategy, these goals are compatible when applied to an environmental region or association with a more or less continuous native vegetation cover, but the criteria used to assess the conservation value of potential sites for 'representative' reserves have little relevance to South Australia's human-dominated land areas.

In these areas, the few remaining very large (> 1000 ha) patches of native vegetation are, for the most part, on public land which is unlikely to be cleared, even where it is not being managed specifically for biological conservation. Patches of native vegetation in the 1000 to 100 ha range also tend to have a nucleus of public land. Thus the native vegetation most in danger of clearance occurs on private land, mainly as small (< 100 ha) isolated patches or as small sections of larger patches (Laut *et al*. 1977a). Under these circumstances, it might seem that rarity is the only valid biological criterion for retaining one small area of native vegetation on private land in preference to another.

Yet even this conclusion (which approximates present clearance control practice) can be questioned in the light of recent research on the ecological dynamics of remnant natural land in human-dominated land areas (Burgess and Sharpe 1981). Due to the effects of fragmentation and other forms of human-induced disturbance, it is unlikely that small areas of native vegetation on private land will have rarity value, in terms of either their species composition or community structure, and where they do have rarity value, it is unlikely that they will continue to maintain rare species or communities without intensive management intervention.

Human-Induced Disturbance and Conservation Value

When a patch of remnant natural land is created by fragmentation of the native vegetation cover and the external edge of the patch is maintained by cultivation or other forms of urban/rural land use, the remnant natural land in the zone along the external edge develops a physical environment which differs in microclimate and surface soil conditions from that of the patch interior (Ranney *et al*. 1981). Since the edge zone experiences higher levels of stress and more frequent disturbance (in both space and time) than the patch interior, it provides conditions favourable to plant species with ruderal tendencies (i.e. early successional native species and 'weedy' exotic species) (Grime 1979). Thus a combination of retrogressive succession within the edge zone and exotic invasion from the adjacent settled land eventually generates a distinctive type of edge vegetation. On the basis of its species composition and structure, this edge vegetation cannot be classed with either the mature or the successional native vegetation of the patch interior. Instead, in its South Australian form, edge vegetation is best described as human-induced vegetation with a degraded native canopy layer and an adventive ('weedy') native/exotic understorey (Laut *et al*. 1977b).

Where conditions in the patch interior approximate a natural *dynamic* disturbance regime (i.e. a low time frequency of spatially random vegetation perturbations), and in the absence of internal patch fragmentation (by fire breaks, roads and other minor clearings), the edge zone is merely a narrow ecotone between the patch interior and its maintained external edge (Ranney *et al.* 1981). Consequently, the proportion of edge vegetation to interior vegetation (the edge/interior ratio) is insignificant for large patches. As patch size decreases, however, the edge/interior ratio increases until, at some point, the entire patch becomes functionally edge.

In the Deciduous Forest Biome of the eastern United States, the edge zone is less than 20 m wide (under the ideal conditions cited above) and 4 ha has been identified as the theoretical critical minimum size (for a near circular patch) at which interior vegetation is still differentiated from edge vegetation (Levenson 1981). Although these figures are specific to a particular temperate wooded environment, they do indicate an order of magnitude for edge effects in similar environments, under ideal conditions.

Of course, real remnant natural land rarely satisfies these ideal conditions. In South Australia, for example, considerable attention has been focused on the conservation value of the narrow strips of native vegetation which commonly occur along water courses, fence lines and routeways in human-dominated land areas (Mollenmans 1982). These narrow strips may account for a high proportion of the remnant natural land within environmental associations which have experienced intensive urban/rural development and they are frequently described as providing 'wildlife' corridors between patches of remnant natural land in less fragmented environmental associations. Nevertheless, the conservation value of these narrow strips of native vegetation may be considerably less than has been assumed if, because of their linear shape, they are not native vegetation *per se* but merely two edges lacking an interior. Similarly, small (<100 ha) patches of remnant natural land often have such an irregular shape and such a high frequency of internal canopy gaps as to be, in reality, no more than a single, highly convoluted edge (Levenson 1981).

The significance of shape, in terms of edge effects, decreases as patch size increases, but even the native vegetation on very large patches can be degraded by edge effects if this land is internally fragmented. In South Australia, the internal fragmentation of large patches of remnant natural land is typically the product of their cadastral subdivision into a number of small public and private land tenure units. The impact of cadastral subdivision on native vegetation can be illustrated by the example of the Peninsula Uplands Region in the Mt. Lofty Block Province (Fig. 2).

There are 66 discrete patches of remnant natural land in this region with areas exceeding 100 ha, but only seven of these patches have areas in the 1000-5000 ha range. None of these seven very large patches are wholly public land and most are held under multiple public and private land tenure; the major public land holders being the National Parks and Wildlife Service, the Woods and Forests Department and the Engineering and Water Supply Department. Each of the public and private land tenure units within any given patch has its own system of access tracks, fire breaks, fence lines, power transmission corridors, residential or other facility sites, etc. These various minor clearings create internal edges which function in the same manner as the external edge of a patch. Thus all of the very large patches of remnant natural land in the Peninsula Uplands Region are actually composed of a sub-set of many much smaller patches separated by minor clearings and their edge zones.

Finally, remnant natural land in human-dominated land areas rarely has a dynamic disturbance regime because near-urban patches of native vegetation generally experience high levels of human-induced disturbance (other than fragmentation itself). Common forms of human-induced disturbance in South Australia include livestock grazing, deliberate or accidental burning, firewood collecting and a wide range of recreational usages. These disturbances tend to be frequent in both space and time. Consequently, as patch size decreases, the spatial frequency of human-induced disturbance approaches a 'whole-patch' disturbance situation. This, combined with frequent disturbance in time, creates a *static* (whole-patch/perpetual) disturbance regime, similar to the disturbance regime of an edge zone and with similar vegetation effects (Pickett and Thompson 1978).

In theory, large patches are more likely to have a dynamic disturbance regime than small ones; provided that the large patches do not contain small public or private land tenure units subject to high levels of human-induced disturbance (e.g. Recreation Parks and grazed farm wood lots). The presence of such static disturbance sites within a large patch may greatly reduce the amount of interior vegetation supported by the patch as a whole (Levenson 1981).

The concept of edge and interior vegetation implies that not all the native vegetation in human-dominated land areas is equally 'natural'. The edge vegetation generated by fragmentation and other forms of human-induced disturbance is commonly described as 'semi-natural' or 'sub-natural' in Europe and North America (Holzner *et al.* 1983). In Australia, where edge vegetation has a more significant exotic component than elsewhere (Hoehne 1981; Peterken 1981) due to invasion by exotic

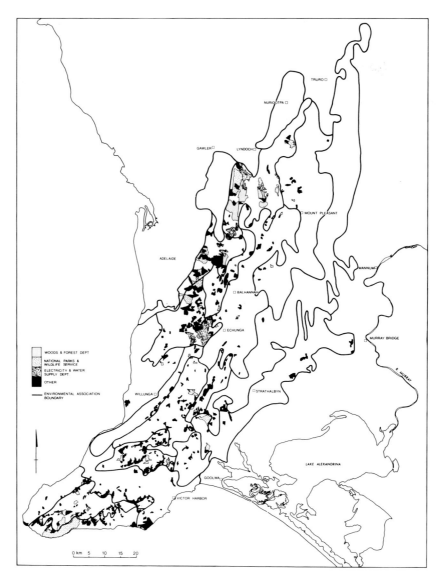

Fig. 2. Remnant Natural Land of the Peninsula Uplands Region showing land tenure.

rather than indigenous agricultural 'weeds', this vegetation is probably better described as 'human-induced adventive'. In addition to its exotic component, edge vegetation has a native component dominated by early successional shrubs and herbs capable of tolerating a static disturbance regime and therefore able to colonize a variety of 'waste' and 'derelict' sites on settled land. As noted by Peterken (1981), these early successional native species are the plant species least in need of protection within nature reserves.

Edge vegetation in human-dominated land areas does provide habitat for native animal species, but its animal assemblages are not ecologically equivalent to those of successional native vegetation in natural land areas where disturbance gaps (i.e. natural edges) in the mature native vegetation cover usually have a higher animal species richness than adjacent undisturbed habitats (Pickett and Thompson 1978). Because of their relatively small size and habitat uniformity, their ecological isolation from the surrounding settled land and their spatial isolation from each other, all patches of remnant natural land in human-dominated land areas tend to be impoverished with respect to a number of native mammal and bird guilds. These include species with habitat, range and life history attributes which make them prone to local extinction as a result of habitat fragmentation and other forms of human-induced disturbance (Matthiae and Stearns 1981; Whitcomb *et al.* 1981). Terborgh (1974) has identified these animal guilds as the ones most in need of protection within nature reserves and, in the context of this discussion, they can be identified as the ones least likely to be found in association with edge vegetation on remnant natural land. Instead, the animal assemblages of this edge vegetation, like the plant assemblages, are dominated by opportunistic native and exotic species capable of surviving in a broad range of habitats on settled land.

The ecological characteristics of edge vegetation and its associated animal species have been used to support the conclusion that the most appropriate goal for biological conservation in human-dominated land areas is to establish a few nature reserves to protect the largest available, relatively undisturbed patches of remnant natural land in a given area. It is argued that only these nature reserves will have the interior conditions essential for maintaining the ecological integrity of native vegetation and its ability to provide natural habitat for native animal species (Diamond 1975; Usher 1979; Levenson 1981; Peterken 1981).

In practice, this goal would be tautological because the diagnostic feature of a human-dominated land area is its highly fragmented native vegetation cover. Thus, where large, relatively undisturbed patches of remnant natural land do occur, it is generally because they have been, at least in part, acquired for some form of nature conservation on public land early in the settlement process. This suggests that it would be more realistic to define a biological conservation goal for human-dominated land areas in terms of the management of existing nature reserves, and any contiguous native vegetation, on large patches of remnant natural land. The management goal would be to co-ordinate the use of all land tenure units within large patches in order to minimize the effects of fragmentation and other forms of human-induced disturbance.

CONCLUSIONS

The preceding review of South Australia's nature conservation strategies has identified two areas of conservation concern with respect to continuing native vegetation clearance in human-dominated land areas; the loss of the productive/amenity benefits of remnant natural land and the loss of its biological benefits. The State's present clearance control strategy fails to distinguish between these concerns and attempts to assess the conservation value of individual sites proposed for clearance in terms of both kinds of benefits. As a result, the biological criteria cited by the clearance control legislation have proved to be the only practical basis for decisions to retain one small patch of native vegetation on private land in preference to another. These biological criteria have been derived from nature reserve design strategies which may be appropriate for South Australia's arid and pastoral provinces, and for extensive tracts of native vegetation within its agricultural provinces, but are not appropriate for its human-dominated land areas.

As an alternative to South Australia's present clearance control strategy, this chapter has suggested two different management strategies for human-dominated land areas. The first of these strategies involves the regional management of human-dominated land areas to protect the productive/amenity benefits of native vegetation retention on private land. The second strategy involves the local management of large patches of remnant natural land, with a nucleus of one or more public nature reserves, to protect the biological benefits of native vegetation retention on public land. Clearance controls would play an important role in both strategies, but not in their present form.

With respect to the first strategy, South Australia already has a mechanism for regulating regional urban/rural development in the form of its State Development Plan and the native vegetation clearance controls initially introduced as a supplement to this Plan. However, because clearance controls have been applied to individual land tenure units, rather than to specific environmental regions or associations within the agricultural provinces, an opportunity has been missed to use these controls for managing the biogeography of entire landscapes. As noted by Sullivan (1981) a regional management strategy for native vegetation clearance control can preselect a settlement pattern for extensively vegetated landscapes that will retain optimal productive/amenity values and can, in combination with rehabilitation measures, not only maintain but enhance the productive/amenity values of degraded landscapes.

With respect to the second strategy, the management priority would be to prohibit native vegetation clearance on private land tenure units within large patches of remnant natural land having a nucleus of one or more existing nature reserves. The next step would be to encourage private landholders to play an active role in the management of these land tenure units for nature conservation by providing government support for fencing and other rehabilitation measures through the Heritage Agreement Scheme or similar programmes. Finally, it is suggested that the only rational, long-term procedure for maintaining the biological conservation value of a large patch of remnant natural land under multiple public and private land tenure is co-operative management of the patch, by the various landholders, as a single multi-purpose nature reserve.

The present management of remnant natural land in South Australia's human-dominated land areas by a multiplicity of State or Local Government agencies and private landholders is an artifact of past nature reserve design strategies and other land use policies largely irrelevant to current biological conservation problems. Examples include Recreation Parks more than 1000 ha in area and Conservation Parks less than 4 ha in area; Woods And Forests Department holdings of mature and regenerating native vegetation which have no economic timber production potential; Engineering and Water Supply Department reserves which are the largest nature reserves in several environmental associations but are not managed specifically for biological conservation; and the multiple private tenure of native vegetation contiguous with many small Conservation Parks. These and other anomalies of South Australia's present approach to biological conservation in human-dominated land areas need to be re-evaluated in the light of current conservation research if these areas are to retain any ecologically viable elements of their pre-European native vegetation cover.

REFERENCES

Burgess, R. L. and Sharpe, D., 1981. Forest Island Dynamics in Man-Dominated Landscapes. Springer-Verlag, New York.

Dendy, T., 1985. South Australian Heritage Agreement Scheme and clearance controls. *S. Aust. Geographer* 1:16-26.

Department for the Environment, South Australia, 1976. Vegetation Clearance in South Australia. Govt. Printer, Adelaide.

Department of Environment and Planning, South Australia, 1983. Vegetation Clearance: Supplementary Development Plan. Govt. Printer, Adelaide.

Diamond, J. M., 1975. The island dilemma: lessons of modern biogeographic studies for the design of natural reserves. *Biol. Conserv.* 7: 129-46.

Fenner, F. (ed.), 1975. A National System of Ecological Reserves in Australia. Australian Academy of Science Report Number 19.

Frankel, O. H., 1975. Conservation in perpetuity: ecological and biosphere reserves. Australian Academy of Science Report Number 19: 7-10.

Grime, J. P., 1979. Plant Strategies and Vegetation Processes. John Wiley, Chichester.

Harris, C., 1985. Tree clearance in South Australia. *S. Aust. Geographer* 1: 2-9.

Hoehne, L. M., 1981. The groundlayer vegetation of forest islands in an urban-suburban matrix. Pp. 41-54 *in* Forest Island Dynamics in Man-Dominated Landscapes ed by R. L. Burgess and D. M. Sharpe. Springer-Verlag, New York.

Holzner, W., Wergner, M. J. A. and Ikusima, I., 1983. Man's Impact on Vegetation. Dr. W. Junk, The Hague.

Laut, P., Heyligers, P. C., Keig, G., Löffler, E., Margules, C., Scott, R. M. and Sullivan, M. E., 1977a. Environments of South Australia: Province Reports 1-4. CSIRO, Canberra.

Laut, P., Heyligers, P. C., Keig, G., Löffler, E., Margules, C., Scott, R. M. and Sullivan, M. E., 1977b. Environments of South Australia Handbook. CSIRO, Canberra.

Levenson, J. B., 1981. Woodlots as biogeographic islands in southeastern Wisconsin. Pp. 13-39 *in* Forest Island Dynamics in Man-Dominated Landscapes ed by R. L. Burgess and D. M. Sharpe. Springer-Verlag, New York.

Margules, C. and Usher, M. B., 1981. Criteria used in assessing conservation potential: a review. *Biol. Conserv.* 21: 79-109.

Matthiae, P. E. and Stearns, F., 1981. Mammals in forest islands in southeastern Wisconsin. Pp. 55-66 *in* Forest Island Dynamics in Man-Dominated Landscapes ed by R. L. Burgess and D. M. Sharpe. Springer-Verlag, New York.

Mollenmans, F., 1982. A Rapid Classification Scheme for Assessing the Conservation Significance of Roadside Vegetation. Govt. Printer, Adelaide

National Parks and Wildlife Service, South Australia, 1984. Area Statement as at June 1984. Govt. Printer, Adelaide.

Peterken, G. F., 1981. Woodland Conservation and Management. Chapman and Hall, London.

Pickett, S. T. A. and Thompson, J. N., 1978. Patch dynamics and the design of nature reserves. *Biol. Conserv.* 13: 27-37.

Powell, J. M., 1976. Environmental Management in Australia, 1788-1914. Oxford University Press, Melbourne.

Ranney, J. W., Brunner, M. C. and Levenson, J. B., 1981. The importance of edge in the structure and dynamics of forest islands. Pp. 67-95 *in* Forest Island Dynamics in Man-Dominated Landscapes ed by R. L. Burgess and D. M. Sharpe. Springer-Verlag, New York.

Slatyer, R. O., 1975. Ecological reserves: size, structure and management. Australian Academy of Science Report Number 19: 22-38.

Specht, R. L., 1975. The report and its recommendations. Australian Academy of Science Report Number 19: 11-21.

Specht, R. L., Roe, E. M. and Boughton, V. H., 1974. Conservation of major plant communities in Australia and Papua-New Guinea. *Aust. J. Bot. Supplementary Series* 7 pp.667.

Sullivan, A. L., 1981. Artificial succession — a feeding strategy for the megazoo. Pp. 257-66 *in* Forest Island Dynamics in Man-Dominated Landscapes ed by R. L. Burgess and D. M. Sharpe. Springer-Verlag, New York.

Terborgh, J., 1974. Preservation of natural diversity: the problem of extinction prone species. *Biosci.* 24: 715-22.

Usher, M. B., 1979. Changes in the species-area relations of higher plants on nature reserves. *J. Appl. Ecol.* 16: 213-5.

Whitcomb, R. F., Robbins, C. S., Lynch, J. F., Whitcomb, B. L., Klimkiewicz, M. K. and Bystrak, D., 1981. Effects of forest fragmentation on avifauna of the eastern deciduous forest. Pp. 125-205 *in* Forest Island Dynamics in Man-Dominated Landscapes ed by R. L. Burgess and D. M. Sharpe. Springer-Verlag, New York.

Williams, M., 1974. The Making of the South Australian Landscape. Academic Press, London.

CHAPTER 30

The Use of Fire as a Management Tool in Fauna Conservation Reserves

P. Christensen[1] and K. Maisey[1]

Fire is used to manage a fauna reserve for two macropod species, the woylie *Bettongia penicillata* and tammar wallaby *Macropus eugenii*. The reserve is divided into a number of 'burning blocks' which are burned off on a rotation system. The periodicity of burning is a compromise between that required for habitat maintenance and fire protection. The burning season is altered to provide the appropriate different fire regime required by the two species.

Although management in this reserve is based upon management for selected species, other species of fauna are monitored to ascertain the effects upon them of the current fire regimes. Monitoring is also important in assessing the effects and success of the recommended management practices.

This work has highlighted several principles important to management. These include the size of management units especially in relation to fire, grazing and weed invasion, the stability of populations of fauna within the reserve and, of particular importance, translating research findings into management practice. The latter requires some effort on the part of the research worker to present managers with clear management objectives, practical means of achieving the objectives and a simple method for monitoring the success of the management practices.

INTRODUCTION

It is widely recognized that Australian vegetation associations are well adapted to fire and respond in different ways to different fire regimes (Gill *et al.* 1981). It was thought that by combining information on the habitat requirements of particular species of fauna with knowledge of the responses of vegetation to fire regimes it would be possible to promote the vegetation complexes and seral stages most suitable for the selected species of fauna by prescribing the appropriate fire regime. A programme designed to put these ideas into practice was initiated in the Perup Forest of Western Australia in 1972 (Christensen 1974).

The Perup Forest is designated a fauna management priority area (MPA) and is a fauna conservation reserve within state forest. An MPA is a land management unit which is part of the multiple use system in operation in the state forests of Western Australia (Anon. 1977). The major management consideration in the Perup MPA is the conservation of fauna and flora; however, the effect of any management practice on other land uses in the area must also be considered. The guidelines which were developed for the management of this MPA were based on the concept of selected species management, here tailoring management to suit the woylie *Bettongia penicillata* and tammar wallaby *Macropus eugenii*.

Two management objectives were set.
1. Conservation of the rare and endangered fauna; maximizing populations whilst at the same time attempting to cater for other species of fauna and flora.
2. Research; use of the area for biological research with particular emphasis on species biology and habitat requirements in relation to fire.

This chapter presents a description of the attempts which are being made to meet these objectives. It is not possible to present the development and evolution of the practical management plan together with all the data upon which it is based in a chapter of this nature and so these data are presented elsewhere and relevant publications have been referred to in the text.

---
[1]Department of Conservation and Land Management, P.O. Box 104, Como, Western Australia 6152.

DEVELOPMENT OF THE MANAGEMENT PLAN

As part of a state forest the Perup MPA was burned every 5-7 years to reduce the fuel loadings to a level that would not support an uncontrollable fire (Underwood and Christensen 1981). This provided a well organized fire control system which was available to be developed as a habitat management tool. The schedule of fuel reduction burning was maintained initially with the addition of an area left as an unburnt control (Fig. 1). Research into the biology and habitat requirements of the woylie and tammar gave some indication of woylie and tammars' use of their habitat and adaptations to fire in their environment.

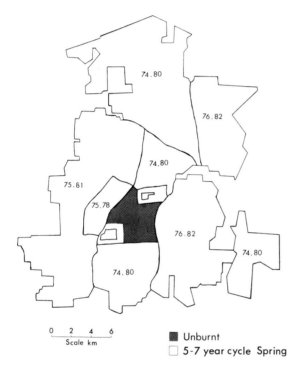

Fig. 1. The standard fuel reduction burning schedule was maintained (with the addition of an area to remain unburned) while fauna research was carried out.

The Woylie

Ground cover seems to be an important factor in the selection of habitat by the woylie. Low, clumped vegetation with between 50 and 80 percent cover density and with 20 to 40 percent bare ground in the vicinity of nest sites is suitable habitat (Christensen 1980a). This type of vegetation structure is attained within three to four years of burning in the Perup reserve.

Woylies have a strong home range site fidelity and when an area is burnt the local residents usually disappear. Although they survive the actual flames, they will not leave their home range area, and a large proportion of the population fall victim to predators. Starvation is not a problem as their diet consists very largely of hypogeous fungi, which are still plentiful in the burnt area. Individuals whose home range overlaps adjacent unburned areas tend to continue to use and survive in the burnt area. Low numbers persist within the burnt area and a few may survive to breed (Christensen 1980a).

Following a fire there is a significant increase in the number of non-adults and the proportion of males caught on the burnt area. Repopulation of the burnt areas was almost exclusively by sub-adults. A similar pattern of repopulation has been observed to occur in an area where the residents had been removed. Young animals come from the adjacent unburnt populated areas (the progeny of the established population), rather than from further afield, suggesting recolonization depends more upon fecundity than immigration. Although woylies breed rapidly at a rate of three young per year and live approximately four to five years (Christensen 1980a), there is a high mortality among young colonizers, and this slows the rate of repopulation.

Based on this data the burning plan for the Perup was altered to one designed to promote habitat suitable for the woylie. The size of each area to be burned was reduced, so that a smaller proportion of the population would be affected; the rotation time was extended to 8-12 years to allow two to three generations between burns; and the burn blocks were arranged so that the blocks adjacent to the one due to be burned were at least four years of age (Fig. 2).

A few years later a west/north-west and a central buffer were added for safety reasons. These buffers are burned every five to seven years, for protection,

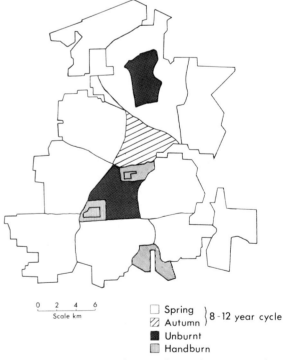

Fig. 2. The first burning plan was designed to accommodate woylie habitat requirements. The burning rotation was lengthened and the size of each burn block reduced.

Fig. 3a. Tammar wallaby *Macropus eugenii* in a *Melaleuca viminea* thicket. *3b.* A heartleaf poison *Gastrolobium bilobum* thicket. This thicket was generated by a hand-lit experimental burn and at seven years of age the structure is suitable for tammar habitation.

as it is usually north-west winds which contribute to the worst wildfire situations. The boundaries adjacent to farmland are also burnt more frequently to form an effective fire break. These changes were initiated because of three wildfires: one deliberately lit, another due to an escape from farmland and the third caused by a lightning strike.

The Tammar

The tammar presented a different set of management problems. These animals inhabit dense vegetation provided in the Perup area by thickets of heartleaf poison *Gastrolobium bilobum* and *Melaleuca viminea*. The former species grows in sandy loam valleys and the latter in seasonally inundated creek systems. They are, therefore, patchily distributed and comprise a very small total area within the reserve. The presence of grassy areas in association with the thickets is also important as a food source for the tammar (Christensen 1980a). In this case it was important to gather detailed information on the fire responses of the thickets themselves as well as on the ecology of the tammars (Fig. 3).

Tammar wallabies live in groups, each group having a home range area rather than having individual home ranges. Although some tammars, usually males, will travel considerable distances (several kilometres) between thickets, the group will usually stay in the vicinity of their home range for some time after it has been burned. They remain in unburnt patches or in marginal habitat and some individuals may transfer to the edge of neighbouring thickets. Mortality due to predation and lack of food is usually high. The tammars cannot begin to use the new thickets for shelter until the crowns provide approximately 20-40 percent overhead cover and the canopy is high enough to allow the wallabies to move about below it (Christensen 1980a). This structure is not achieved until five to six years after the burn. Therefore, burning all of a thicket will affect the entire population of that thicket for up to six years after the burn. Tammars do not breed as rapidly as woylies, having only one young per year. Thus the population would take longer to become re-established if severely depleted during the period when the thicket is unsuitable for habitation (Christensen 1980a).

Thickets form as a result of mass germination following wildfires of high intensity under dry conditions (total fuel consumption, some crown scorch) and the thicket structure is maintained for 25-30 years before the plants become senescent and break down. Laboratory trials confirm that heartleaf seed required intense heat treatment for good germination. This type of fire does not often occur during fuel reduction burning which is predominantly carried out in spring. A high intensity fire is more likely to occur under autumn or summer fire weather conditions.

It was thought that the natural variation in flammability between the valley and ridge vegetation could be utilized to achieve the appropriate fire regime. A combination of spring and autumn burning was devised. The first rotation burn in a block would be carried out in spring when the fuels in the valley are still comparatively moist and will not burn. The fire would burn only the understorey on the ridges. The second or third rotation burn would be carried out in autumn under conditions of low fuel moisture and dry weather conditions when the fire will burn the scrub on the ridges together with the valley thickets, thus regenerating the thickets every 18-27 years (Christensen 1982). The change of some spring burns to autumn did not affect the woylie (Fig. 4).

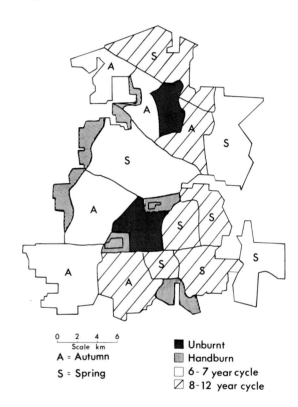

Fig. 4. The current burning plan incorporates spring and autumn burns, shorter rotation burns on the western side of the reserve for protection and a block to remain unburned. This plan seeks to satisfy the requirement needs of tammars, woylies and safety.

It was found that although the valley vegetation was being burned during the autumn fuel reduction burns, the fires were not consistently of a high enough intensity to produce the thicket structure. Hand-lit experimental burns were carried out to identify the parameters of a 'thicket regeneration burn' and showed that the type of fire required would have to be carried out in very dry fuel and soil conditions. This would achieve total litter fuel consumption and heating of the soil needed to trigger germination of the heartleaf seeds or release of *Melaleuca* seed from the dehiscent capsules. These conditions occur during the season of highest fire danger and therefore are not practical on a large scale. Smaller controlled burns of tammar thickets may have to be considered from the safety point of view as well as that of tammar management.

The suitable tammar thickets in the Perup reserve are some distance apart and often separated by areas of cleared land. Burning an entire thicket at one time may be detrimental to the tammar population and it may be necessary to have smaller hand-lit burns within the thickets, burning only one section each rotation.

The experimental burns carried out so far have demonstrated that this is feasible; however, these small burns pose other major problems. On small areas there is considerable grazing pressure on small heartleaf seedlings. Western grey kangaroos *Marcropus fuliginosus* and the tammars themselves invade the burn from the surrounding area to feed on the fresh young growth which follows the fire. In extreme cases all the heartleaf seedlings may be eliminated.

In addition there may be problems with achieving the desired balance between heartleaf thicket and grassed areas. If the fire is not intense enough, too large an area may be occupied by grasses. Grasses only remained where the fire intensity was insufficient to remove the compacted litter or 'duff' and expose the soil.

Future research will investigate the possibility of producing a localized high-intensity burn, required for thicket regeneration, within one of the autumn fuel reduction burns by manipulating fuels and carrying out the burns in dry but stable weather conditions, for example, by burning at night-time. Research into fire behaviour is continually evolving new techniques, increasing the manager's ability to prescribe certain types of fire and to predict fire behaviour which broadens the safety limits under which burning may be carried out (Sneeuwjagt and Peat 1985).

OTHER SPECIES

Although the management plan for the Perup is specifically tailored to suit the woylie and the tammar, enough is known of the fauna and flora of the reserve to be confident that the current burning regime is unlikely to cause any major or irreversible changes to the ecosystems of the area in the short term. Nevertheless, because of the uncertainty of long-term effects of prescribed fire regimes, work is also being done on other species within the area.

Some work has been carried out on the numbat *Myrmecobius fasciatus* (Christensen *et al.* 1984), the short-nosed bandicoot *Isoodon obeselus,* the brushtail possum *Trichosurus vulpecula,* the ringtail possum *Pseudocheirus peregrinus* (Inions 1985), the chuditch *Dasyurus geoffroii* and the brushtailed phascogale *Phascogale tapoatafa.*

Results to date suggest that the current burning regime may suit these species reasonably well. It has become clear, however, that there may be conflict at times when trying to manage an area for a specific purpose such as maintaining tammar populations at stable levels. It was found that high intensity burns, during autumn, intended to regenerate thickets of heartleaf, were particularly devastating on the local possum population (Inions 1985). Individuals were killed during the fire or fell prey to predators. This situation is similar to that which occurs following a wildfire and the possum population soon re-establishes by migration of young animals from adjacent unburnt forest. A high intensity fire also tends to create as many or more nesting hollows than it destroys (Inions 1985).

A monitoring programme which includes trapping and spotlighting is providing data on population levels of the woylie, Western grey kangaroo, Western brush wallaby *Macropus irma,* the brushtail and the ringtail possum, and to a lesser extent the numbat, chuditch and short-nosed bandicoot. Information to date suggests that populations of the small to medium sized species may undergo fluctuation in numbers.

RE-ESTABLISHMENT

Trial reintroductions of woylies to forest areas from which the species disappeared earlier this century have proven successful in some instances. Much has been learned about such ventures which makes this technique a realistic option in areas where the species formerly occurred. A major problem in this work has been predation by foxes *Vulpes vulpes* and it is now thought that fox control may be necessary in some areas if animals like the woylie are to be successfully re-established (Christensen 1980b; King *et al.* 1981; Kinnear *et al.* 1984).

Re-establishment may be an option in tammar wallaby management. Thickets of heartleaf are being planted on abandoned farmland at the Perup Forest Ecology Field Centre with the intention of establishing a colony of tammar wallabies in these thickets as soon as they provide enough cover. Fire is a factor in both instances and essential in the maintenance of habitat structure in the long term.

IMPLICATION FOR RESERVE MANAGEMENT

The work at the Perup MPA has highlighted some fundamental principles which we believe are basic to the management of other reserves, particularly small isolated remnants of native vegetation.

1. It is necessary to have special management plans for certain species in reserves which are regularly burned. The tammar is an example of such a

species. It is unlikely to survive indefinitely either under a frequent burning regime, regardless of whether it is carried out in spring or autumn, or under full protection from fire.

2. The biology of the species in relation to reserve sizes. For example, species such as the tammar may survive in comparatively small reserves, whilst more 'territorial' species such as the woylie or numbat need larger reserves. This needs to be recognized or we may waste valuable time and effort on 'lost causes' when such effort might be better spent elsewhere.

3. The interaction between burn size and grazing is of critical importance. On small burns, palatable plant species, particularly legumes, may be wiped out by grazing at the seedling stage. The level of grazing appears to be related to rainfall so that larger burns are needed in the lower rainfall areas, where the rate of recovery of vegetation after fire is slower. In the Perup, areas need to be larger than 500 ha. Burning of small reserves is even more of a problem where grazing animals are present. In such cases culling may have to be considered or certain plants may be eliminated by selective grazing.

4. Weed invasion and establishment following fire; this is especially important in small reserves which are surrounded by farmland. Little factual data exists but we know that many weeds invade recently burned areas which are adjacent or in close proximity to farmland. The extent of this problem may depend on rainfall, soil type and other factors. In our opinion the burning of small reserves is a matter which should be treated with extreme caution. Small reserves surrounded by farmland or adjacent to other weed sources should not be burned unless absolutely necessary.

5. Monitoring of the population levels of key species is critical. Woylie, numbats and other species undergo periodic and possibly cyclic fluctuations in numbers. Burning needs to be timed so that critical areas are not burned when population numbers are down. At these critical times when burning and population lows coincide, fox predation becomes a significant factor and special measures such as poisoning may need to be implemented.

6. It has been demonstrated that reintroduction of woylies is possible and this may be a viable option for other species. There are many reserves with suitable habitat for species which no longer occur in the area but which might flourish if they were reintroduced. Fire may be essential in the maintenance of these new populations.

7. There is an urgent need for education as the vast majority of people do not understand the role of fire in natural ecosystems.

INTEGRATION OF RESEARCH AND MANAGEMENT

The practical and flexible use of prescribed fire in reserve management is something which both scientists and the general public need to face up to. There is far too much hesitancy by scientists to get involved. Fear, perhaps of being criticized, condemned or simply of making mistakes too often results in a lack of action. It is easier and more comfortable for us to sit back, produce more papers and criticize, justifying the lack of action with assertions about the lack of 'essential basic data'.

There needs to be more involvement in the practical issues. Further research is needed at a more practical grass roots level in close co-ordination with actual reserve management programmes. Often the practical problems are relatively simple and easy to solve. Too often in the past such research has been avoided, downgraded or labelled as 'unscientific'. It has failed to attract researchers or funds perhaps because it does not fit the 'hi-tech' image of our time where computers and ecosystem modelling are the 'in thing'.

The work at the Perup has demonstrated to us that one cannot simply take research results and apply them in the field. It is just not possible even with comparatively simple information let alone more complex data. It is therefore necessary for the scientist to provide an extension link with the reserve manager.

In the Perup example the close links which have been developed between the management, research and education have led to the development of a reserve management strategy based on four main criteria:

1. management using the best available data;
2. simplicity;
3. monitoring; and
4. on-the-spot integration of management, research and education.

Reserves must be managed, they do not remain static simply because there is no planned influence. Things are happening to reserves every day; visitors are camping, hiking and picnicking, roads are being planned and built, people are removing fence posts and flowers, dumping rubbish and starting fires. The manager must deal with all of these problems daily. The reserve will be managed whether or not scientists supply the necessary information. However, we believe that most managers prefer to manage their reserves using scientific data made available in a useable form. But they need it now, not tomorrow or the day after. The successful transfer of scientific data into management practice depends on presenting the best available data in simple form for the manager's use. There is more management information available than is generally realized. However, it needs to be recognized that few managers have the capacity or the time to interpret research findings for themselves let alone adapt them for management purposes. This is a role in which increasingly the scientist should become involved. We need to encourage the use of the best data available and using it in the simplest way possible.

The way to start is not to present the manager with a complex ecosystem model; that may come later. We should make every effort to simplify the complex results whenever possible. Success should not be measured by the degree of complexity of the management system, but be gauged by definable tangible results in the field. For this, monitoring is necessary but it is only practical if it is relatively cheap and simple, given the constraints we face with finance, materials and manpower.

Generally, the simpler the management objectives and the management system the simpler will be the monitoring of the operations and the more likely the chances of successful application. Feedback from the monitoring operation may result in changes to management practices and a need for further research. The interaction between the manager and the scientist is an ongoing commitment.

Finally, the role of education cannot be stressed too strongly. There is a high level of interest in ecology amongst the general public. Unfortunately, this is not matched by an equal level of understanding. Knowledge if often superficial having been gained through watching television documentaries or reading the press. This combination of a high level of interest and a low level of understanding can result in problems for the reserve manager. Scientists and reserve managers together have the task of educating the public so there is understanding and acceptance of reserve management practices.

In conclusion, the message is that scientists should become actively involved in the dissemination and application of research findings. No one will volunteer to do it for us, the task falls squarely on our own shoulders.

ACKNOWLEDGEMENTS

The authors would like to thank the past and present staff of the Manjimup Forest Research Station for their work and assistance throughout the duration of this project.

REFERENCES

Anon, 1977. Forests Department General Working Plan No. 86. Forests Dept. of Western Australia.

Christensen, P., 1974. The Concept of Fauna Priority Areas. 3rd Fire Ecol. Symp. Victorian Forests Commission, 1974.

Christensen, P. E. S., 1980a. The Biology of *Bettongia penicellata* (Gray, 1837) and *Macropus eugenii* (Desmarest 1917) in Relation to Fire. Forests Dept. Western Australia Bulletin No. 91.

Christensen, P. E. S., 1980b. A Sad Day for Native Fauna. Forests Dept. Western Australia *Forest Focus* No. 23.

Christensen, P., 1982. Using prescribed fire to manage forest fauna. Forests Dept. Western Australia *Forest Focus* No. 25.

Christensen, P., Maisey, K. and Perry, D. H., 1984. Radiotracking the Numbat, *Myrmecobius fasciatus*, in the Perup Forest of Western Australia. *Aust. Wildl. Res.* 11: 275-88.

Gill, A. M., Groves, R. H. and Noble, I. R. (eds.), 1981. Fire and the Australian Biota. Australian Academy of Science, Canberra.

Inions, G., 1985. Interactions between Possums, 'habitat trees' and Fire. Honours Thesis Australian National University, 1985.

King, D. R., Oliver, A. J. and Mead, R. J., 1981. Bettongia and Fluoroacetate, A Role for 1080 in Fauna Management. *Aust. Wildl. Res.* 8: 529-36.

Kinnear, J. E., Onus, M. and Bromilow, R., 1984. Foxes, Feral Cats and Rock Wallabies S.W.A.N.S. 14: 3-8.

Sneeuwjagt, R. G. and Peat, G. B., 1985. Forest Fire Behaviour Tables for Western Australia. Dept. of Conservation and Land Management.

Underwood, R. J. and Christensen, P. E. S., 1984. Forest Fire Management in Western Australia. Forests Dept. Western Australia. Special Focus No. 1.

CHAPTER 31

The Impact of Tree Decline on Remnant Woodlots on Farms

F. R. Wylie[1] and J. Landsberg[2]

TREE decline in rural areas, and the associated hazards of soil erosion and salinity, are among the foremost conservation issues in Australia. About 30 percent of Australia's forests and woodlands have been cleared or severely modified since European settlement (Wells *et al.* 1984). Since the late 1960s and early 1970s there has been a marked increase in the rate of decline of remnant native vegetation on farmland in many parts of Australia. Tree loss in rural areas involves three elements. The first is the deliberate removal of trees as part of farm management. The second is the death of trees due to old age, coupled with a paucity of recruitment of tree seedlings because of active suppression, grazing and competition with improved pastures. The third element, rural dieback, is the premature and relatively rapid decline and death of native trees on farms, apparently as a consequence of interacting environmental stresses. Rural dieback affects many different species of trees, of all ages, in most states (Old *et al.* 1981).

A number of features characterise 'healthy' remnants of native woodland. These remnants are usually substantial (several hectares or more) and have been minimally disturbed by man or grazing livestock. There is a high degree of diversity; in vegetation structure, in plant and animal species present and in the age classes of trees and shrubs represented. In such remnants insect grazing pressure on trees is generally light to moderate. Most trees have normal, full-leaved crowns with very few dead or leafless branches and little obvious epicormic growth. Only the occasional tree shows symptoms of disorder and there are relatively few dead trees.

In contrast, remnant woodlots with dieback are usually small and have been severely disturbed. They lack diversity of vegetation structure, floristics, age classes and fauna. In such woodlots the pressure on trees from grazing by insects is usually severe and sustained. Many trees have sparse crowns with some death of both minor and major branches and a high proportion of epicormic growth. Tree disorder is widespread and tree death common.

Research in Queensland (Wylie and Johnston 1984) has shown a direct relationship between the extent of modification of original tree cover and the severity of rural dieback. Deliberate tree clearing, and the intensification of land use associated with it, are pivotal factors in the development of rural dieback.

Remnant woodlots on farms represent ecosystems which are precariously balanced. Once a dieback sequence is initiated, positive feedback can cause an originally stable woodlot to rapidly regress through a series of unstable states to treeless grassland (Fig. 1). Small or highly modified woodlots have least buffering capacity and are therefore most at risk. In the long term both smaller woodlots and woodlots which have been structurally or floristically modified have little chance of survival unless

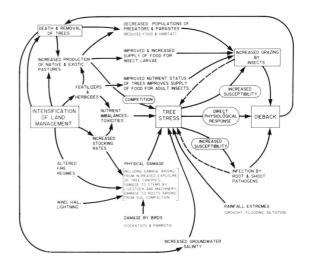

Fig. 1. A conceptual model of the development of rural dieback.

---

[1]Queensland Department of Forestry Biology Laboratory, Indooroopilly, Queensland, 4068 Australia.
[2]Research School of Biological Sciences, Australian National University, P.O. Box 475, Canberra City, Australian Capital Territory, 2601 Australia.
Pages 331-2 *in* NATURE CONSERVATION: THE ROLE OF REMNANTS OF NATIVE VEGETATION ed by Denis A. Saunders, Graham W. Arnold, Andrew A. Burbidge and Angas J. M. Hopkins. Surrey Beatty and Sons Pty Limited in association with CSIRO and CALM, 1987.

supplemented by replanting or by natural regeneration. As land use intensifies dieback will accelerate, and will continue to erode the value of remnants of native woodlands on farms as a conservation resource, unless there is positive intervention to restore ecosystem complexity. Research to establish the parameters which define a stable woodland remnant is urgently needed.

Rural tree decline has major consequences for conservation of biota, soil and water, at a scale which affects both rural and urban communities. Successful maintenance and rehabilitation of existing farm woodlots, and the establishment of new ones, require an understanding of the dynamics of rural dieback, if its spread is to be arrested and its resurgence prevented.

REFERENCES

Old, K. M., Kile, G. A. and Ohmart, C. P., 1981. Eucalypt Dieback in Forests and Woodlands. CSIRO, Melbourne.

Wells, K. F., Wood, N. H. and Laut, P., 1984. Loss of forests and woodlands in Australia; A summary by state, based on rural local government areas. *CSIRO Div. Water and Land Res. Tech. Mem.* **84/4**.

Wylie, F. R. and Johnson, P. J. M., 1984. Rural tree dieback. *Qld. Ag. J.* **110**: *3-6.*

CHAPTER 32

The Conservation and Study of Invertebrates in Remnants of Native Vegetation

J. D. Majer[1]

INTRODUCTION

THERE are just over 1900 species of terrestrial vertebrates in Australia but well in excess of 108,000 species of insects (Taylor 1976). One might expect that the research and management priorities of State and Commonwealth conservation agencies would reflect this statistic but a recent telephone survey (September 1985) revealed that there are, in fact, only three entomologists or invertebrate zoologists on the permanent payroll of any Australian conservation agency. These three scientists are the legacy of amalgamations between conservation agencies with what were formerly forestry organizations. This situation contrasts with numbers of scientists working on plants and other animals, and with the situation in other countries such as Britain or the Netherlands where the Institute for Terrestrial Ecology and the Rijkinstituut voor Natuurbeheer, respectively, employ a wide range of invertebrate zoologists on projects considered to be of significant national importance.

A lack of awareness of the importance of terrestrial invertebrates has also been reflected in various Acts governing Australian wildlife conservation. For instance, until 1976 terrestrial invertebrates were not included in the definition of fauna used in the Western Australian Fauna Conservation Act, 1950-70. The definition of fauna has now been expanded in the Western Australian Wildlife Conservation Act, 1950-1980 and the Conservation and Land Management Act, 1984, to include all native animals. However, at present only the Buprestidae, a beetle family, and *Nothomyrmecia macrops,* a primitive ant, are included on the list of protected terrestrial invertebrates.

The attention which has been paid to terrestrial invertebrates by conservation agencies in Australia has generally covered studies on groups, such as trapdoor spiders, which are of interest to local researchers (Anon. 1977), pollination studies (e.g. Stoutamire 1983), insectivorous vertebrate feeding studies (e.g. Chapman and Dell 1985), rare or supposedly threatened insects such as the wingless stonefly, *Riekoperla darlingtoni* in Victoria (Neumann and Morey 1984) or has taken the form of biological surveys of nature reserves (e.g. Bailey and Richards 1975). Such surveys are usually reported in an extremely superficial manner in comparison with the vertebrate and plant sections of the reports. Some work is also performed in universities and colleges under the sponsorship of conservation agencies although such support is relatively limited. It is encouraging to note that the Australian National Parks and Wildlife Service has recently sent out a questionnaire (J. D. Ovington pers. comm.) to over 600 Australian entomologists soliciting information on threatened species of terrestrial and aquatic invertebrates, and the International Union for Conservation of Nature and Natural Resources has now published an Invertebrate Red Data Book (Wells *et al.* 1983). Reviews of the conservation status of Australian insects have been written by Key (1978) and New (1984).

REASONS FOR LACK OF CONSIDERATION BY CONSERVATION AGENCIES

Why is it that there is so little interest in terrestrial invertebrates from Australian conservation agencies when over 20 years ago scientists were emphasizing their place in ecosystems (Serventy 1961)? It appears that the reasons fall into three categories.

1. Of the large number of insect species in this country, about 55% have not yet been described. The taxonomic knowledge of certain other invertebrate groups is probably even less. There is such a lack of taxonomic, ecological and

---

[1]School of Biology, Western Australian Institute of Technology, Bentley, Western Australia 6102
Pages 333-5 *in* NATURE CONSERVATION: THE ROLE OF REMNANTS OF NATIVE VEGETATION ed by Denis A. Saunders, Graham W. Arnold, Andrew A. Burbidge and Angas J. M. Hopkins. Surrey Beatty and Sons Pty Limited in association with CSIRO and CALM, 1987.

distribution baseline data that the inclusion of terrestrial invertebrates in ecological studies is considered by some to be virtually meaningless. This viewpoint would be invalidated if invertebrate studies became an accepted component of biological investigations in conservation areas.

2. Invertebrates are diverse, numerous, inconspicuous and often strongly seasonal in overt behaviour or occurrence. As a result, conservation scientists tend to be overwhelmed by the task of sampling and studying invertebrate communities.

3. Most invertebrates, with some important exceptions due to nuisance value (e.g. bushflies), arouse little public interest in comparison with the large vertebrates. Their ecological and economical importance is not realized by many people, and consequently there is little pressure on conservation agencies to consider them.

Obviously we should question whether these are valid arguments. Invertebrates play such a large role in soil aeration and drainage, litter decomposition and nutrient cycling, pollination, seed distribution and survival, plant predation and also in the provision of food for insectivorous invertebrates, that their importance should surely be appreciated by conservation agencies. This role is possibly of more critical dimensions in small areas than in large ones where biological functions are spread or performed through a diversity of species. Also, because vertebrates frequently require a larger habitat area than many invertebrates, the biological load carried by invertebrates in restricted habitats is increased.

In order to demonstrate how baseline data can be collected on terrestrial invertebrates, at the same time as collecting baseline data on flora and other fauna, I will describe a simple survey method which has been sponsored by Worsley Alumina Pty Ltd (Worsley Alumina Pty Ltd 1985). This technique integrates flora, vertebrate and invertebrate surveys, and is designed so that baseline data in individual habitat units may be collected in a relatively cost-efficient manner. The data may be used to compare areas throughout Australia.

AN INVERTEBRATE SURVEY SCHEME

First, representative habitats which are to be assessed are selected. Within each habitat an intensive study location (ISL) is staked out (Fig. 1). This includes a 200×200 m bird observation quadrat which forms the basic sampling unit for plants and vertebrates as described in Worsley Alumina Pty Ltd (1985). In linear habitats, such as riparian vegetation, the quadrat may be laid out as a series of four 200×50 m transects.

One side of the ISL is then chosen for invertebrate sampling. Each invertebrate sampling transect (IST) is located so that, as far as possible:

1. it is representative of the physiographic and ecological characteristics of the habitat; and

2. sampling will cause minimal interference to botanical work being carried out in the same habitat.

Three sampling methods are then applied in order to provide an estimate of the relative abundance of invertebrates on: (a) the ground (pitfall-traps); (b) herbs and shrubs (sweep-samples); and (c) trees (beat-samples). It is not possible using these methods to obtain absolute population estimates, although each technique is useful for ranking ISL's in terms of their species richness or the relative abundance of the various taxa.

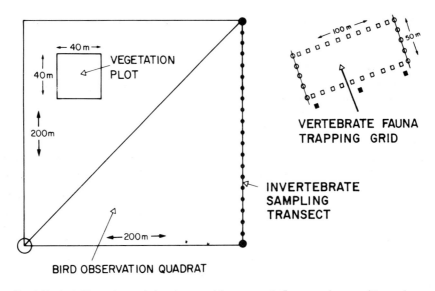

Fig. 1. Design of intensive study locations used for systematic flora, vertebrate and invertebrate surveys by Worsley Alumina Pty Ltd (1985).

Along each IST, 20 pitfall-traps are located at 10 m intervals. At each trap location, one 21 mm (internal diameter) PVC pitfall-trap holding-cylinder is sunk into the ground. At the commencement of each sampling period an opened pitfall-trap (18 mm internal diameter Pyrex test tubes containing 10 ml 70/30 v/v alcohol/glycerol preservative) is placed in each holding cylinder. This trap preservative is non-attractive to ants, although some other invertebrates may be attracted. The trap is left open for one week.

A wire stake marked with flagging tape is placed at each pitfall-trap site. These markers also constitute the starting point for each sweep sample, which runs at right angles to the IST and towards the ISL interior. Herb and shrub-sweeps are performed using a quadrilateral-shaped sweep-net measuring 32 cm across the opening. Each sweep consists of a 40 m walk into the interior of the ISL from the pitfall-trap site, and a 40 m return walk offset 2 m from the original traverse. Eighty sweeps are performed over the entire 80 m distance. In each ISL, 20 trees are also flagged for beat-sampling. Generally, the tree nearest to each pitfall-trap is selected, although in some instances a representative mix of tree species is marked for sampling: for instance, if the trees in the ISL comprise an equal mix of jarrah *Eucalyptus marginata,* marri *E. calophylla,* bull banksia *Banksia grandis* and *Persoonia,* five trees of each species are selected for beating. Beating is performed by jarring the trees with a stout stick and collecting dislodged animals on a 1 m² calico tray beneath each tree. The tray is moved to four positions around the tree canopy in order to sample each side of it. All animals are hand-picked from the tray and placed in vials of 70 percent alcohol.

In order to account for the strong seasonality of most invertebrate groups the survey should be repeated four times during the year, in the middle of each season. Samples may be sorted to the order or family level, although at least one taxa should be identified to species level in order to provide information on species richness.

CONCLUSIONS

The significance of invertebrates in small remnants of native vegetation is deemed to be as, or even more, important than in large areas. The scheme described here is presented to show a simple survey method that could be adopted as a standard for future studies. By using these methods, data on a wide range of species may be collected. The invertebrates which are sampled come from all trophic levels and from a range of strata. There is, therefore, the potential to obtain information relevant to the whole gambit of ecosystem processes. Certain taxa may be suitable for use as bio-indicators. The value and choice of such taxa is discussed elsewhere in this volume in the workshop reports by T. J. Ridsdill-Smith and J. D. Majer.

REFERENCES

Anon., 1977. Trapdoor spider study. *SWANS* 7: 3-4.

Bailey, W. J. and Richards, K. T., 1975. A report on the insect fauna of the Prince Regent River Reserve, North-west Kimberley, Western Australia. *Wildl. Res. Bull. West. Aust.* 3: 101-12.

Chapman, A. and Dell, J., 1985. Biology and zoogeography of the amphibians and reptiles of the Western Australian wheatbelt. *Rec. West. Aust. Mus.* 12: 1-46.

Key, K. H. L., 1978. The conservation status of Australia's insect fauna. *Australian National Parks and Wildlife Service, Occasional Paper* No. 1.

Neumann, F. G. and Morey, J. L., 1984. A study of the rare wingless stonefly, *Riekoperla darlingtoni* (Illies), near Mt Donna Buang, Victoria. *For. Comm. Vic., Res. Branch Rep.* No. 253.

New, T. R., 1984. Insect conservation — an Australian perspective. Dr. W. Junk, The Hague.

Serventy, D. L., 1961. Fauna conservation in Australia and in Australian-controlled New Guinea *in* Tenth Pacific Science Congress — 1961 (abstract).

Stoutamire, W. P., 1983. Orchids and wasps. *SWANS* 13: 8-9.

Taylor, R. W., 1976. A submission to the enquiry into the impact on the Australian environment of the current woodchip industry programme. *Hansard* 12 August 1976: 3724-31.

Wells, S. M., Pyle, R. M. and Collins, N. M., 1983. The IUCN invertebrate red data book. International Union for Conservation of Nature and Natural Resources, Gland.

Worsley Alumina Pty Ltd, 1985. Worsley Alumina Project. Flora and Fauna Studies, Phase Two. Worsley Alumina Pty Ltd, Perth.

CHAPTER 33

A Monitoring System for Use in the Management of Natural Areas in Western Australia

A. J. M. Hopkins[1], J. M. Brown[1] and J. T. Goodsell[1]

INTRODUCTION

In Western Australia, the Department of Conservation and Land Management exercises direct control over some 16.4 million ha of land including National Parks, Nature Reserves and State Forest. In addition, it is called on to give advice on management of other areas of natural land. Proper management of natural areas imposes a requirement to predict likely effects of any management decision. To make effective management decisions, it is desirable to know:

1. the distribution and abundance of the various species of plants and animals over the landscape; and
2. the dynamics of the communities and ecosystems. This state of knowledge is seldom achieved for any particular area of land.

One approach to improving the information base involves the establishment of a system of monitoring sites throughout the State. Such a system is currently being developed with the following objectives:

1. to provide data on long-term changes in natural communities; such data should be based on standardized observations so that some interpretation is possible;
2. to complement and supplement the biological survey function;
3. to provide a simple and effective method for assessing effects of present management decisions in order to improve subsequent decisions; and
4. to increase the interest of departmental staff and the public in particular areas by involving them in the monitoring programme.

It is anticipated that the monitoring programme may provide some early indications of management problems such as over-grazing, epidemics, local extinctions, etc.

Observations spanning many years (at least five) are considered essential to enable one to distinguish long-term trends in the biota from year-to-year variability and changes resulting from short-term cycles. These long-term studies will also provide a sound basis for interpreting short-term inferential studies of community dynamics (i.e. those based on one-off observations of sites with different disturbance histories).

THE MONITORING METHOD

The method selected and tested through a pilot study incorporates the following features:

1. repeatability — plots are permanently marked, and methods for recording are standardized;
2. flexibility/adaptability — the design of the monitoring quadrat enables it to be used for a range of organisms that differ in size and abundance;
3. user-friendliness — data recording sheets are being designed for simplicity and ease of use. A system of reporting back to observers is also planned; and
4. regularity of recording — it is proposed to structure the system so sites are visited at least annually.

A permanently marked, square quadrat is used (Fig. 1). A photopoint is located 10 m north of the south-west corner of the quadrat; the photograph is to be taken with that corner peg in the centre of the field of view. The quadrat can be expanded away from the south-west corner, as indicated in Figure 1, to suit the the organism of interest. In the case of vascular plant species in the South-West Botanical Province of Western Australia, a 20 m × 20 m quadrat is used. To facilitate recording of presence/absence data, this quadrat can be marked off in a series of nested quadrats. Canopy cover is measured as

[1]Department of Conservation and Land Management, Western Australian Wildlife Research Centre, P.O. Box 51, Wanneroo, Western Australia 6065.
Pages 337-9 in NATURE CONSERVATION: THE ROLE OF REMNANTS OF NATIVE VEGETATION ed by Denis A. Saunders, Graham W. Arnold, Andrew A. Burbidge and Angas J. M. Hopkins. Surrey Beatty and Sons Pty Limited in association with CSIRO and CALM, 1987.

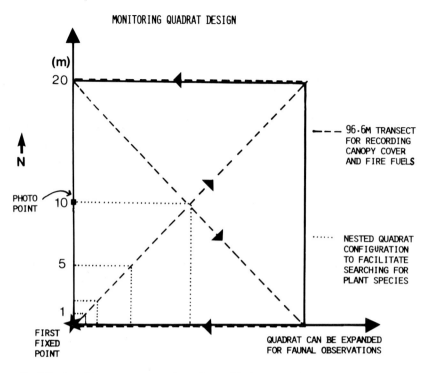

Fig. 1. Design of the permanent quadrat to be used in the monitoring programme.

intercepts along two sides and two diagonals of the quadrat; this provides a repeatable indicator of abundance and biomass. Observations on life history attributes of species can be recorded simultaneously with either of the two preceding steps. Measures of other attributes such as tree DBH and faunal observations over larger areas can be referenced back to the initial quadrat. For example, systematic observations can be made of birds using a 1 ha square that incorporates the 400 m^2 square used for vascular plants and shares the common south-west corner. Note that the quadrat must be located *within* a community; the techniques outlined are not appropriate for measuring changes at a boundary or ecotone. Any number of quadrats can be used but, at the moment, the programme is based on a single quadrat, representative of the particular community of interest, at each site.

Each monitoring site is assigned an identifying code. At the time of establishment, basic information to be recorded includes identity of the observer and date, purpose/status of site, locality data, and geomorphological and vegetation descriptions. Monitoring procedures to be undertaken on a regular basis are graded from a photograph (plus observer and date), through notes to supplement the photograph, a plant species list, vegetation structure and habitat data, cover values along the transect lines to fire fuels, faunal observations, etc. Observations can be recorded directly on to computer data entry sheets that are suitable for immediate key punching. Data processing will be fully automated with provision for feedback to each observer indicating how his/her observations have contributed to the knowledge at that site.

It is proposed to involve departmental field staff and members of the general public in the monitoring programme to the greatest extent possible. This has the potential to increase the amount of information being collected; it is also efficient and educational. However, a large programme would still require careful supervision and management to maintain interest and continuity of observations.

STUDIES USING THE MONITORING METHOD

During the evaluation phase, some 24 monitoring sites have been established in the south-west of Western Australia using the methods described above (Fig. 2). Objectives for these studies are:

1. to assess rates of change in natural communities in relation to major environmental gradients (climate, microclimate, soil type, fertility) e.g. York gum *Eucalyptus loxophleba,* jam *Acacia acuminata* woodlands on loamy sands across a rainfall gradient;

2. to assess rates and patterns of change in different communities e.g. pilot study at Clackline Nature Reserve; and

3. to assess the effects of management actions e.g. the effects of prescribed fires on wandoo *E. wandoo* and York gum woodland communities at Dryandra State Forest and Mockerdungulling Nature Reserve (south-west of Tutanning).

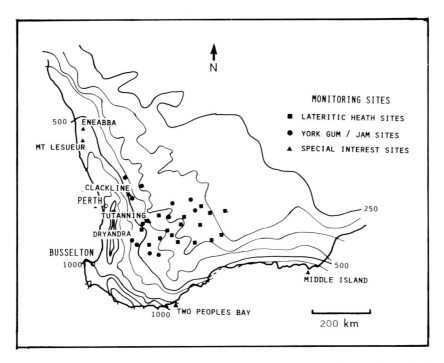

Fig. 2. Locations of the monitoring sites established in southwestern Australia using the standard methods together with other permanent study sites yet to be incorporated in the monitoring programme. Annual rainfall isohyets are also shown.

In addition, about 1000 permanently marked sites already used for ecological survey are now being re-evaluated for incorporation into the monitoring programme.

BIBLIOGRAPHY

Austin, M. P., 1981. Permanent quadrats: an interface for theory and practice. *Vegetatio* **46**: 1-10.

Beeftink, W. J., 1979. Vegetation dynamics in retrospect and prospect. Introduction to the proceedings of the Second Symposium of the Working Group on Succession on Permanent Plots. *Vegetatio* **40**: 101-5.

Collins, S. L. and Adams, D. E., 1983. Succession in grasslands: thirty-two years of change in a central Oklahoma tall grassland prairie. *Vegetatio* **51**: 181-90.

Grimsdell, J. J. R., 1978. Ecological Monitoring. Handbook No 4. African Wildlife Leadership Foundation, Nairobi.

Noble, I. R., 1977. Long-term dynamics in an arid chenopod shrub community at Koonamore, South Australia. *Aust. J. Bot.* **25**: 639-53.

Rogers, G. F., Malde, H. E. and Turner, R. M., 1984. Bibliography of repeat photography for evaluating landscape change. University of Utah Press, Salt Lake City.

Williams, E. D., 1978. Botanical composition of the Park Grass plots at Rothamsted 1856-1976. Rothamsted Experimental Station, Harpenden.

CHAPTER 34

Bird Dynamics of Foster Road Reserve, near Ongerup Western Australia

B. J. Newbey[1] and K. R. Newbey[1]

ROAD reserves are considered important for birds by providing habitat, refuges and corridors for various species. However, there is no published work to support this belief. Our study of a short section of a road reserve was an attempt to quantify the use that various species of bird made of such a reserve.

Foster Road reserve is situated in wheat and sheep farmland near Ongerup (33°54'30"S, 118°26'00"E) Western Australia. The climate is Mediterranean with 375 mm average annual rainfall, though during the study period of April 1978 to March 1979 only 292 mm was recorded. The vegetation was senescent low woodland (>75 years old) with a few open areas invaded by perennial grasses. Shrubs were few, herbs rare and the litter was 2-5 cm thick and almost continuous under trees. The upper stratum consisted of *Eucalyptus annulata* or *E. gardneri* 7-9 m high with 15-50% canopy cover. One section (350 m) consisted of *E. platypus* 2.5-4 m high with 50-80% canopy cover. *E. platypus* was also present in some places under *E. annulata*.

The area under study was 2.0 km long and 20 m wide, with a central graded road, and was part of a similar road reserve system around the district. Connected to this network was grazed but uncleared vegetation along salt creeks. One end of the study area was connected to a 4 ha patch of uncleared but grazed low woodland. Similar areas of 1 to 10 ha were scattered in surrounding farmland. The nearest large area of natural vegetation connected to Foster Road via the system of road reserves, was 200 ha of mallee/heath, 6 km from the study area.

Once every week, commencing two hours before sundown, two observers with binoculars walked slowly along the road, taking 40-50 minutes. The location and behaviour of all birds sighted was recorded.

Forty-four species were recorded (Table 1) during the study. The number of individuals in a single recording period varied from 35 (March) to 146 (December). Lowest monthly mean of individuals was 44.0 (March) and the highest 101.8 (June). Monthly number of species varied from 19 (March and September) to 27 (August). Ten species were present every month and eight species bred in the road reserve.

Bird usage of the study area was broadly categorized (based also on our detailed local knowledge of distribution and behaviour) as:

A. the major part of their range (4 spp.);

B. an important part of their range for 4-12 months (18 spp.);

C. corridor (9 spp.); reliably present in nearest different habitat of *E. occidentalis* woodland, *Acacia acuminata* dwarf woodland or mallee/heath (2-6 km); or uncommon in district;

D. minor but not critical part of their range (10 spp.); and

E. uncertain (3 spp.). These data are shown in Table 1.

Some variations in species records and population numbers were positively correlated with eucalypt flowering (Table 2). Yellow-throated miner *Manorina flavigula* and red wattle bird *Anthochaera carunculata* numbers rose sharply and peaked during the flowering of *Eucalyptus gardneri*. Purple-crowned lorikeet *Glossopsitta porphyrocephala* did likewise with *E. annulata* and *E. platypus* flowering.

The management implications of the study are that senescent low woodlands are an important habitat for birds in well-cleared areas. Not all vegetation should be replanted once it has become senescent. The loss of low woodlands can be partially offset by encouraging farmers to plant shelterbelts of mixed local species.

ACKNOWLEDGEMENTS

For commenting on the manuscript, our thanks to Mr Mike Brooker and Dr Graeme Smith of CSIRO, Division of Wildlife and Rangelands Research.

REFERENCES

Blakers, M., Davies, S. J. J. F. and Reilly, P. N., 1984. The Atlas of Australian Birds. Melbourne University Press: Melbourne.

Table 1. Birds of Foster Road (Monthly mean, number of recordings and usage). Nomenclature follows Blakers *et al.* (1984). N=Sum of weekly recordings. U=Usage — see text for explanation of code. *=Breeding.

Bird Species	April	M	J	J	A	S	O	N	D	J	F	March	N	U
Little eagle														
Hieraaetus morphnoides			0.3								0.3		2	D
Australian hobby														
Falco longipennis								0.2					1	D
Brown falcon														
Falco berigora	0.3	0.3			0.3		0.3			0.3		0.2	6	D
Australian kestrel														
Falco cenchroides								0.3					1	D
Common bronzewing														
Phaps chalcoptera	0.5				0.3			0.3		0.3			5	D
Crested pigeon														
Ocyphaps lophotes	0.5	0.5			0.8	1.2	1.3	1.8	1.6*	1.7*	3.8	2.4	59	B
Purple-crowned lorikeet														
Glossopsitta porphyrocephala					4.0	0.6	0.5	4.0	19.8				136	B
Regent parrot														
Polytelis anthopeplus									3.8				19	C
Western rosella														
Platycercus icterotis									1.2				6	C
Port Lincoln ringneck														
Barnardius zonarius	2.0	1.0	2.8	0.8	1.0	1.2	6.0	7.5	10.6	2.0	6.8	4.0	195	B
Elegant parrot														
Neophema elegans	3.0	6.5	5.0	0.6	0.5	0.4	1.8	3.8	6.8	5.3	3.0		149	B
Pallid cuckoo														
Cuculus pallidus				0.3	0.5								3	D
Welcome swallow														
Hirundo neoxena	0.8	0.3	0.3	0.8		0.6	0.3	0.5	0.4		0.5		19	B
Tree martin														
Cecropis nigricans	7.8	0.8						4.8	1.8	16.3	1.3		116	B
Richard's pipit														
Anthus novaeseelandiae	0.5					0.2	0.5	0.5	0.4				9	D
Black-faced cuckoo-shrike														
Coracina novaehollandiae		0.3		0.2	0.3	1.2	0.3	0.5					12	D
Red-capped robin														
Petroica goodenovii	0.3	0.3									0.3	0.2	4	C
Rufous whistler														
Pachycephala rufiventris		0.3		0.4						0.3			4	E
Grey shrike-thrush														
Colluricincla harmonica	0.3	0.3	1.0	0.4	1.0	0.8	1.3*	0.5	0.2	0.3	0.3	0.6	29	A
Restless flycatcher														
Myiagra inquieta	0.3	1.5	1.0	0.6	0.3		0.3				0.3	0.6	20	B
Grey fantail														
Rhipidura fuliginosa			0.3		0.3								2	C
Willie wagtail														
Rhipidura leucophrys	3.8	3.5	5.0	3.4	5.0	6.0	7.5	5.8	5.0	7.0	5.5	4.8	259	A
White-browed babbler														
Pomatostomus superciliosus	9.0	8.5	9.0	14.0	11.0	12.0	3.3	14.5	9.8	12.3*	17.5	3.8	530	A
Brown songlark														
Cinclorhamphus cruralis					0.2								1	D
Western gerygone														
Gerygone fusca	0.3	2.3	0.5	0.8	1.3								18	E
Weebill														
Smicrornis brevirostris			0.3								0.5		3	C
Yellow-rumped thornbill														
Acanthiza chrysorrhoa	1.8	15.3	10.0	4.4	1.8				0.8		2.0	5.8	168	B
Varied sittella														
Daphoenositta chrysoptera	0.8				1.8								10	C
Red wattlebird														
Anthochaera carunculata		5.8	13.3	15.2	15.5	9.4	4.5*	5.3	4.0	0.7			322	B
Yellow-throated miner														
Manorina flavigula	2.5	1.0	21.5*	18.2	13.0	1.6	1.8	2.8	5.2*	0.3	1.0	2.4	312	B
Singing honeyeater														
Lichenostomus virescens	2.5	5.3	4.3	2.0*	0.8			1.0	0.3			0.2	68	B
Purple-gaped honeyeater														
Lichenostomus cratitius	3.0	4.3	4.3	2.6	4.8	2.6	2.8	1.5	1.4	0.3	1.0	2.2	131	B
Yellow-plumed honeyeater														
Lichenostomus ornata	14.3	11.0	11.5	22.0	18.0	13.4	13.5*	16.5*	15.0*	18.6*	6.5	10.2	725	B
Brown-headed honeyeater														
Melithreptus brevirostris			3.3						0.2				14	E
Brown honeyeater														
Lichmera indistincta		2.8	1.3	0.6	0.3				0.8	0.3	0.3		15	B
New Holland honeyeater														
Phylidonyris novaehollandiae				0.2									1	C
Striated pardalote														
Pardalotus striatus	0.5	0.3			0.8								6	C
Silvereye														
Zosterops lateralis		0.5											2	C
Australian magpie-lark														
Grallina cyanoleuca	1.3	1.0	2.0	1.0	1.5	0.8	2.5	0.8	1.0	1.7	2.3	0.8	78	B

Table 1. — Continued.

Bird Species	Month												N	U
	April	M	J	J	A	S	O	N	D	J	F	March		
Dusky woodswallow														
Artamus cyanopterus	6.0	2.0	2.0	1.4	5.8	4.8	5.3	3.3	5.8*	3.3	4.0*	2.4	195	A
Grey butcherbird														
Cracticus torquatus			0.3				0.3			0.3		0.4	5	D
Australian magpie														
Gymnorhina tibicen	3.0	0.5	2.0	1.0	1.0	0.8	2.8	1.5	3.0	5.0	4.8	1.0	106	B
Grey currawong														
Strepera versicolor	1.3	0.5		0.6					0.2	1.3	1.0	1.4	25	B
Australian raven														
Corvus coronoides	0.5	0.8	0.5	1.8	0.5	2.6	1.3	0.3		0.3	1.3	0.6	46	B
Total	66.9	77.5	101.8	93.2	92.2	60.4	58.2	76.8	100.0	78.2	64.3	44.0	Total spp.	
No. of spp.	26	28	24	24	27	19	21	21	25	22	22	19	seen: 44	

Table 2. Flowering and recordings of flower feeders (monthly mean) along Foster Road Reserve. F=few flowers. M=many flowers.

Bird Species	Month											
	April	M	J	J	A	S	O	N	D	J	F	March
Purple-crowned lorikeet					4.0	0.6	0.5	4.0	19.8			
Red wattlebird		5.8	13.3	15.2	15.5	9.4	4.5	5.3	4.0	0.7		
Yellow-throated miner	2.5	1.0	21.5	18.2	13.0	1.6	1.8	2.8	5.2	0.3	1.0	2.4
Singing honeyeater	2.5	5.3	4.3	2.0	0.8				1.0	0.3		0.2
Purple-gaped honeyeater	3.0	4.3	4.3	2.6	4.8	2.6	2.8	1.5	1.4	0.3	1.0	2.2
Yellow-plumed honeyeater	14.3	11.0	11.5	22.0	18.0	13.4	13.5	16.5	15.0	18.6	6.5	10.2
Brown-headed honeyeater			3.3						0.2			
Brown honeyeater		2.8	1.3	0.6	0.3				0.8	0.3	0.3	
New Holland honeyeater				0.2								
Plant Species												
Eucalyptus annulata		F	F	F	F	F	M	M	M			
Eucalyptus gardneri		M	M	M	M	F						
Eucalyptus platypus									M	M	F	
Total of above nine species of bird	22.3	30.2	59.5	60.8	56.4	37.6	23.1	30.1	47.4	20.5	8.8	15.0

CHAPTER 35

Relevance, Accountability and Efficiency of Research for Management

Workshop Leader: R. McKellar[1]

IN the groups discussing the 'Relevance, accountability and efficiency of research for management' there was general agreement that there were failings with regard to these aspects within virtually every conservation programme. There was also general agreement regarding ways by which the failings could be overcome, even though specific failings identified by those involved with management differed from those identified by research workers.

In this review of the discussion groups, I first recount the perceived failings, and then outline ways by which they might be overcome. It should be noted that in most discussions, 'relevance', 'accountability' and 'efficiency' were not discussed as separate issues, but rather as parts of a larger question pertaining to how the different aspects of many existing nature conservation programmes could be integrated.

Some major failings identified by managers were:

1. much research is only peripherally relevant to the needs of management;

2. some research that could be relevant to management is not being taken far enough to be useable;

3. often research findings are not made available in a form that is easily accessible to managers;

4. research workers are seldom interested in filling information gaps found during the development or, implementation of management programmes, and, in particular, are usually not interested in monitoring the effects of management activities; and

5. research appears often to be oriented to personal areas of interest, rather than areas where present information needs for management have been demonstrated.

Research workers identified major failings as:
1. research findings are often ignored when management plans are prepared and implemented, and are unknown to politicians and other decision-makers; and
2. managers often do not appreciate that apparently simple management questions may involve scientific uncertainties which are difficult to resolve.

Several research workers stated that all well conducted research is relevant to management, but others saw this as a facile view when limited funds require that research be undertaken on some priority basis to achieve maximum benefits.

These failings were classified into three issues which were separately definable but inter-related: the perception of nature conservation; communication; and the setting of objectives. These three issues provide a framework for defining both the problems and some of the solutions.

The perception of nature conservation affects legislation, research, management as well as public opinion and support. The two most common perceptions are that nature conservation is a problem for which a solution must be found or that it is an opportunity that should be exploited. Rather than being a problem or an opportunity, nature conservation should be viewed as a process which has long-term objectives which are difficult to achieve, and strategic and programme objectives which may be more easily achieved. Many workers in the field of nature conservation lose sight of the overall objectives, which are often only enunciated in legislation or general policies, and concentrate on the clearer and more achievable objectives of the programmes in which they are involved. The real success of nature conservation, however, lies in the success of the overall process. By maintaining an awareness of their place in the overall process, each worker will be aware of

---

[1]Darradup Environmental Services, Nannup, Western Australia 6275.

the need to work towards long-term objectives, and to serve and communicate with workers in other parts of the process who are also working towards the same objectives.

Each worker must also make his work interesting and useful to other workers. The manager must define uncertainties and problems and implement his management plans in a way that facilitates research, and he must communicate with research workers, if he wishes their involvement. Research workers must recognize that applied research is more likely to be used in management than pure research, because of its greater relevance to immediate management decisions.

Communication, or lack of it, contributes to many problems, including the duplication of research, research information not being used by managers, and information needed by managers not being produced by research workers.

Better communication between those involved in or interested in nature conservation is needed on all levels, from the interpersonal to the geographic. This is at present often frustrated by institutional arrangements, which can physically separate researchers, managers and policy-makers by placing them in different locations. This may lead to a lack of opportunity for individuals to meet others in the same or a related field. There is also the lack of national directories within which the summaries of objectives, programmes and findings in the field of nature conservation could be widely published.

Communication could be improved to some extent by integrating nature conservation activities within a single group within a regional office. The potential drawbacks of spreading researchers and research facilities would, however, have to be addressed. Other ways to improve communication include providing more opportunity for workers to meet, both formally and informally, and the production of national directories.

Objective setting is essential for success in any endeavour, including nature conservation. Several different types and levels of objectives are required.

First, since nature conservation is a process, there should be explicitly stated objectives pertaining to the development, maintenance and change to the process. These are at present usually unstated, and discernible only by investigating the organization and objectives of the agencies and departments involved in nature conservation in a region.

Second, the overall objectives for nature conservation in a region (such as a state) must be explicitly stated. At present, such objectives are usually stated in the relevant legislation, but not within a policy document which has been developed through public and professional discussion.

Third, strategic objectives which elaborate on the overall objectives and set priorities for nature conservation within a region must be defined. They also should be developed with public and professional involvement, rather than by an isolated policy group.

Finally, each management programme and action must have clearly stated objectives. There was much discussion on this point, with support for an 'experimental management' approach in which objectives and evaluative criteria would be set for each management programme and each action prior to implementation. It was noted that if management had a well-defined experimental structure then research workers would be more willing to monitor the effects of management actions. It was also noted that this would require greater integration of management and research expertise.

SUMMARY

There was clear agreement in both discussion groups that improvement could be made, and should be made, with regard to the relevance, accountability and efficiency of research for management in nature conservation. The workshops defined the ways by which this could be accomplished as being:

1. enlarging the viewpoint of those involved in nature conservation from their own goal-oriented tasks to include the larger objectives of nature conservation and therefore the needs and interests of other workers in nature conservation;

2. facilitating communication between those involved in or interested in nature conservation; and

3. developing and aiming to achieve a set of long-term objectives, as well as shorter-term objectives, goals and strategies.

CHAPTER 36

Achieving a Balance Between Long and Short Term Research

Workshop Leader: K. L. Tinley[1]

TWO separate groups discussed the topic 'Achieving a balance between long and short term research' specifically in relation to the theme ECOLOGICAL STUDIES AS THE BASIS FOR MANAGEMENT of remnants of native vegetation. The main points raised during the discussions are listed below.

1. It was generally agreed that short term research is of less than 2-3 years duration and long term research is more than this.

2. As well as being of immediate value, short term research should contribute to long term objectives. Short term research includes: reconnaissance, survey, inventory, identification of key components, processes and relationships (see (8) below and Table 1).

3. The time factor is determined by: the subject; the object or purpose of the study; adequacy of financial, technical and material backup; or whether the study is done by a single researcher or a team.

4. The first step in ecological research should be a biophysical reconnaissance or survey of the subject area. This synoptic analysis should be integrated to form the basic framework for both extensive and intensive detailed research of longer time spans, and updated at the stage of synthesis (Fig. 1).

5. As long term research is the most difficult to conduct for many reasons, a staged approach should be used. This should be made up of semi-discrete units which form part of a progressive series each augmenting and enhancing the next stage in the programme. Each stage should be sufficiently discrete to be a goal in its own right and provide material proof of its value both to the workers and the funding and support agencies. Any discontinuity due to loss of funding or other reasons does not mean the loss of the entire programme.

6. Monitoring should be the responsibility of all levels of research. Monitoring combines the roles of both short and long term research as studies or analyses of short duration are repeated and reassessed at intervals over the long term. It is thus liable to the same constraints as long term research where financial constraints may determine the frequency of analysis rather than natural parameters such as seasonal or catastrophic events. Each analysis should be of value in its own right yet form part of a time-space continuum.

7. The most effective means of monitoring change in systems is by fixed point photography both from the ground and the air. High 'tech'/high

Table 1. Targets in Field Research.

1. Dominants (in size and/or number).
2. Unique components (e.g. endemics, rare and endangered species, biogeographic outliers or refugia).
3. Prime mover components (e.g. geomorphic and edaphic factors causing changes in soil moisture and/or relief; fire; pollinators and seed dispersers).
4. Identify indicators of various conditions, trends or change.
5. Key processes and patterns (identify (a) those operating under present circumstances, (b) trends or tendencies towards future change, and (c) processes which are inexorable and those which can be damped or ameliorated by management).
6. Structure and physiognomy (in plan and in profile) (including changes wrought by regrowth to maturation).
7. Successional relationships (geomorphic and biotic in time and space (Drury and Nisbet 1971).
8. Ecotones (changes in position, dimensions and cover; analysis of status/condition and trend by monitoring. The ecotone of today can become the biome of tomorrow/yesterday's ecotone is today's biome.
9. Phenology (ideally long term monitoring of numbers of species plus numbers of each species).
10. Relationship of an area or site to (a) its arena of interaction (hydrologic unit or process compartment), and (b) its position in the landscape.
11. Identify buffer systems and habitats (e.g. floodplains) or abundant species which protect other less durable systems or susceptible species from overuse (overgrazing, overpredation). Physical buffers include water bodies, gibber surfaces, rock outcrops, bare dunes.
12. Record historical information from oldtimers.
13. Develop strategies whereby landowners and local communities surrounding protected areas take part in protection as part of their resource and life support system.
14. Map data and rank on a most-least importance scale derived from single or combined values.

[1]Department of Conservation and Environment, 1 Mount Street, Perth, Western Australia 6000.
Pages 347-50 in NATURE CONSERVATION: THE ROLE OF REMNANTS OF NATIVE VEGETATION ed by Denis A. Saunders, Graham W. Arnold, Andrew A. Burbidge and Angas J. M. Hopkins. Surrey Beatty and Sons Pty Limited in association with CSIRO and CALM, 1987.

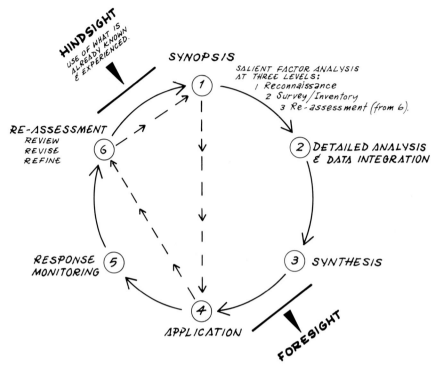

Fig. 1. The continuum of ecological research. Many situations can be adequately handled by going directly from 1 to 4 to 6 and 1 again.

Table 2. Hierarchy of Salience (or of key and master factors).

1st Level:	REGIONAL ECOSYSTEM e.g. ocean, continent, island, desert, mountain, river basin, biome. Natural processes of landscape evolution, climate, hydrography, geomorphic and edaphic controls, plant formations and succession. The regional ecosystem as a whole remains primary no matter how important one or more of its components may be; it is the contextual setting or process arena in which everything interrelates.
2nd Level:	MAJOR ELEMENTS Elements or components with the greatest impact, most importance or largest space requirements; one or more of these are derived from the other Levels e.g. malaria mosquito. Examples: 1. Man (hunter-gather, fisherman, pastoralist, cultivator, beekeeper, industrial man). 2. Large mammals (migration, overgrazing, etc.). 3. Representation of the full spectrum of ecosystems. 4. Unique elements (e.g. scenery, aquifers, endemics, rare or endangered species). 5. Dominants and prime mover components.
3rd Level:	INDIVIDUAL ECOSYSTEMS (and communities)
4th Level:	MACRO-COMPONENTS e.g. ungulates, flora and fauna
5th Level:	MICRO-COMPONENTS e.g. insects, fungi micro-organisms, chemicals.

(INCREASING LEVEL OF IMPORTANCE / INCREASING LEVEL OF COMPLEXITY)

cost equipment is most liable to disruption from changes in direction of policy and funding and this should be borne in mind before becoming overly reliant on satellite imagery.

8. Synoptic ecological assessments using salient (key) factor analysis should be undertaken first as this most rapidly provides the arena for deciding what the directions, goals and objectives should be of detailed research. It also allows quicker access to identifying priorities and ways of resolving environmental problems (see Tables 1 and 2).

9. Once the first stages of ecological research are completed, short duration research (including monitoring) should work in tandem with long term

research each interacting in the progressive development of information to the benefit of both. Neither short or long term research should be done to the exclusion of the other. They should be complementary, reciprocal and incremental.

10. Continuity and fresh initiatives must be maintained by updating and reviewing the findings of each stage and the cascading effect this may have on priorities, goals and objectives in research approaches and directions (Fig. 1).

11. Like short term research the incremental stages of long term research need to be integrated and presented in such a way that they facilitate and stimulate multidisciplinary and cross sectoral correlations and new combinations or aspects to explore. Methods and techniques must be repeatable and streamlined so that valid comparisons can be made.

12. All ecological studies should relate to the unit of interaction and interrelationship (process arena) e.g. a drainage basin or other process unit. Too many studies continue to be unrelated to an ecosystem or whole catchment context. The dynamic or kinetic (Drury and Nisbet 1971) status and trend of both the protected area *and* its surroundings should be assessed in a hydrologic ecosystem context (Tinley 1986). The physical, especially geomorphic and edaphic, process variables and trends require far more attention and monitoring than is usually accorded them, as they play a primary role in regulating the biotic components (e.g. Tinley 1982). The regional ecological matrix is primary (Table 2).

13. For any given region or political area first *the ideal* distribution, size and shape of representative examples of ecosystems (or specific components) on a geographic and altitudinal gradient should be mapped. Only after the ideal has been identified should the second step be undertaken — how close to the ideal can be practically attained (and maintained) as determined by various constraints (e.g. availability of land, government support, etc.).

14. Once a series of ecological reserves have been established it must be clearly identified at the outset for what purpose *each area* has been protected so that ensuing research and management is unequivocal in its objectives and activities.

15. At all times first principles of conservation should be applied at the beginning *to prevent* damage or degradation to ecosystems and resources, while waiting for long term research to refine minimum requirements for the maintenance of system viability and process relations.

Appended here is an example of how the application of first principles would maintain connectivity and diversity in the landscape, as well as providing a natural framework for separating those parts of the landscape most easily protected and those better suited for sustainable farming and other activities.

16. There should be a co-ordinated approach by all land managing authorities to ensure that first principles of conservation (soil, water, forestry, ecosystems) are applied by all land owners as an ethical obligation of individual responsibility for the health of the land. 'The tendency is to relegate to government all conservation activities that land owners fail to perform. The lack of a conservation ethic amongst land users relegates to government many functions eventually too large, too complex, or too widely dispersed for it to efficiently perform' (Leopold 1949). Closer adherence of all land users to the Soil Conservation Act would have gone a long way to mitigate the present environmental predicament.

17. National Parks and Nature Reserves should be core areas for each ecological region where field stations can provide basic facilities for the maintenance of long term research.

18. Theoretical research should have practical objectives, by providing answers for management dilemmas (Cherfas 1985), e.g. minimum size of reserves in different situations and for different purposes, and the minimum widths of connecting bands of native vegetation.

19. As bureaucracies, politics and universities are constantly changing policies and directions, there is no guarantee of long term funding. A main requirement is thus to explore alternative ways of undertaking both short and long term research (e.g. through Trusts, Endowments or programmes such as the Bicentennial National Time Capsules, and the possibility of funding from Conservancy Groups of land owners who obtain direct benefits from management oriented research).

20. Any re-appraisal of research programmes, directions and priorities needs to be sure that research resources are aimed at the right targets for management and are not squandered on irritating environmental problems of high political priority which can better be resolved by application of first principles or salient fsctor analysis (e.g. unnecessary compilation of massive scientific tomes as 'proof' for 'resolving' environmental issues).

REFERENCES

Cherfas, L., 1985. The biology of conservation. *New Scientist* 107: 43-5.

Drury, W. H. and Nisbet, I. C. T., 1971. Inter-relations between developmental models in geomorphology, plant ecology and animal ecology. *General systems* 16: 57-68.

Leopold, A., 1949. A Sand County Almanac. Oxford University Press, New York.

Tinley, K. L., 1982. The influence of soil moisture balance on ecosystem patterns in southern Africa. Pp. 175-92 *in* Ecology of Tropical Savannas ed by B. J. Huntley and B. W. Walker. Springer-Verlag, Berlin.

Tinley, K. L., 1986. Ecological Regions of Western Austrlia: The Basis for Co-ordinated Planning and Management of Conservation and Development. Pp. 219-33 *in* Towards a State Conservation Strategy for Western Australia: Invited Review Papers ed by R. G. C. Chittleborough. Department of Conservation and Environment, Western Australia, Bulletin 251.

APPENDIX

An Example of the Application of First Principles in Conservation

The maintenance of native vegetation connections along natural linkages in the landscape — the West Australian pattern.

Natural Sites (parts of the landscape along which native vegetation should be left or restored):

1. along both sides of perennial and seasonal drainage lines (riverbanks, streams, creeks and washes);
2. around margins of both permanent and temporary wetlands (e.g. marshes, swamps, lakes, salt and claypans, estuaries);
3. on escarpments and slopes steeper than 25°;
4. on and around rocky isolates of all dimensions (from local outcrops to inselbergs — mesas, buttes, breakaways *et al.*);
5. on coast foredunes (first line of vegetated dunes landwards to the first trough, or in the absence of definite relief, 100 m landward from the seaward edge of the shrub cover);
6. on inland dunes (whole of crest and slopes); and
7. on low rainfall sandplains.

Human Artifacts

1. alongside roads and railways.

Key Requirements

1. Width of the native vegetation bands will vary with landform dimensions, but minimum viable widths need to be researched and practically applied.
2. As coastal flux is not only to and fro but lateral, a setback line must be identified that is related to local coastal processes and long term management objectives.
3. Native vegetation bands along roads and railways can only be effective as habitat corridors so long as they are protected in perpetuity along only one side of the track and widening or other route changes are confined to the opposite side, otherwise the effort is pointless. In addition bands confined to one side of the road will help minimize road kills of wildlife crossings.

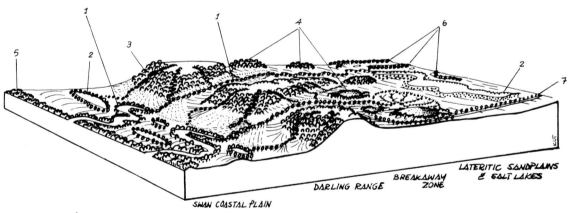

Fig. 2.

CHAPTER 37

Nutrient Cycles: Their Value in Devising Management Strategies

Workshop Leader: A. R. Main[1]

MANAGEMENT strategies in the sense used here are taken to relate to the art of preparing plans with a view to success in retaining remnant vegetation and its included biota. Developing a strategy involves a study of the nature of the biotic resources within the remnant, the relationships of the biota to each other and to primary resources such as nutrients. Finally, it is necessary to consider the natural forces which may affect successful retention of the remnant being conserved.

In looking at an ecosystem or reserve it is perceived at an instant in time. Past events have shaped the present ecosystem but it is possible to model or visualize the likely happenings of the future. Being aware of both the past and the future we wish to maximize the possibility of persistence of what is now present — not fossilized as it is, but kept within the ambit of what would have happened if nature alone were operating, yet modified to take account of what are visualized as being unavoidable perturbations.

The basis of persistence lies in the continuation of the nutrient cycling process. In particular it is necessary to consider what projections can be made about the continued effects of past events (extinctions, isolation of reserve) or of future happenings on nutrient cycling.

Thus in devising management strategies it is necessary to consider the following.

1. The need to cycle as evidenced by: nutrient-deficient soils, long-lived shrubs and trees, nutrients held in the standing crop, nutrients bound to minerals in the soil.

2. The function of diversity: in time and place; for example, the possibility of cycling under all conceivable (likely) conditions. But we do not know seral stages or their duration, or whether a facilitation model of succession applies, or roles of biotic elements in:

    (a) present situation, (b) naturally changing situation, (c) situations altered by man.

    And finally, we cannot predict incidence of climatic alteration.

3. The workings:

    Abstraction — what makes nutrients accessible or available (root systems, mycorrhiza, proteoid roots, micro-organisms).

    Cycling communities, e.g., arthropods, fungi.

    Scavenging or conserving mechanisms, for example plants, especially tuberous ones, or those with rapidly developing root systems or ephemerals, which grow after fire or break of season, and the hyphae of fungi.

    Soil turnover — large animals — echidna, bandicoots; birds, lizards, and invertebrates [e.g., mygalomorphs, lycosids, zodarids (spiders); Scarabaeoidea (beetles)].

    Nitrogen fixation — *Rhizobium,* actinomycetes; blue-green algae (Cyanobacteria).

4. Balances:

    In: from soil cycled, scavenged, N-fixed, animals.   Standing crop   *Out*: animal dispersal flying termites, ants, birds, kangaroos, seed and fruit dispersal.

5. The changed and possibly simplified system in remnants or small reserves where for example imports can come from macropods grazing adjacent farmland introducing phosphorus. Moreover in small remnants gains and losses may not balance as they may do in much larger areas.

Thus in remnants it is especially important to know whether the area is large enough for seral changes and natural processes to continue. For example within remnants it is easy to detect the loss of larger conspicuous animals, but this may not be as important in recycling nutrients as the bacteria,

[1]Zoology Department, University of Western Australia, Nedlands, Western Australia 6009.
Pages 351-2 *in* NATURE CONSERVATION: THE ROLE OF REMNANTS OF NATIVE VEGETATION ed by Denis A. Saunders, Graham W. Arnold, Andrew A. Burbidge and Angas J. M. Hopkins. Surrey Beatty and Sons Pty Limited in association with CSIRO and CALM, 1987.

fungi and small invertebrates with fast growth rates and rapid turnover. Thus in devising management strategies it is necessary to know whether the remnant situation is affecting the diversity, abundance or effectiveness of this largely unseen part of the ecosystem on which cycling depends.

The above approach relates the whole of the biota to the currency of persistence, viz., nutrients.

In the context of nutrient cycles, it:

(i) gives a functional role for diversity, viz., ensuring the capacity or flexibility to respond to natural changes in the future; and

(ii) relates changes in biota present, e.g., during seral stages, drought, flooding, seasonality of rainfall, and other perturbing events, to maintenance of nutrient levels and thus continued functioning in the short and long term.

However, in order that a strategy can be developed, it is necessary that groups be classified according to their role or function in nutrient cycling, especially whether it is related to: abstraction, short-term conservation preventing loss, holding in standing crop, returning (cycling strictly), soil turnover, accelerates downward leaching (opening soil) for water percolation, redistribution, importation (macropods), export (dispersing birds, breeding flights of termites and ants, hepialid and cossid moths), facilitation in seral stages.

Thus with respect to nutrients it is possible to consider *plants* as:

(a) abstractors, especially those with special root systems; (however while mycorrhizae are important in making nutrients available to plants we are ignorant of their precise contributions to nutrient cycling);

(b) nitrogen fixers, both symbionts and free living;

(c) scavengers or conservers leading to short-term retention;

(d) withholding nutrients in the standing crop, particularly in very long-lived elements;

and *animals* as:

breaking down plant material and so initiating recycling, redistributing resources or opening soil for water and nutrient percolation.

Having considered the above, the workshop discussion group concluded the following.

1. The maintenance of nutrient cycles is fundamental and should always be considered when devising management strategies for reserves.
2. To do this it is necessary to define:
   (a) the nutrient characteristics of a reserve in terms of what and how much is there; and
   (b) the factors controlling and maintaining the cycling.
3. Management incorporating nutrient cycling needs an approach which picks indicator species and monitors changes.

CHAPTER 38

Invertebrates as Indicators for Management

Workshop Leader: J. D. Majer[1]

INTRODUCTION

THIS workshop discussion commenced with a review of reasons why terrestrial invertebrates may act as good indicators of habitat composition and what may generally be referred to as 'habitat quality'. The criteria include:

1. invertebrates are abundant and therefore readily found;
2. they are relatively easy to study and sample;
3. habitats generally contain many invertebrate species. Thus invertebrate samples usually contain material which yields much information;
4. many invertebrate species occupy specialized niches or higher trophic levels and may therefore indicate the status of particular habitat factors or food resources; and
5. many invertebrate species are responsive to changing environmental conditions.

Although there are many cases of invertebrates being used as bio-indicators of water quality, terrestrial examples are not so common. The few terrestrial examples include the use of springtails to predict future land degradation by recreational pressures (Mahoney 1976), epigaeic fauna to elucidate changes in the environment in areas of high industrial pressure (Puszkar 1979) and ants to follow the success of open-cut minesite reclamation (Majer 1983).

Invertebrates may serve as valuable indicators for the management of native vegetation remnants when the following questions are being posed:

1. Is the area worth conserving in terms of its biological diversity?
2. What is the condition or 'quality' of the area?
3. Are there any pristine or degraded areas in the remnant?
4. Is there a need to rehabilitate degraded areas?
5. How effective is the rehabilitation procedure?
6. Is the remnant deteriorating as a result of edge effects?
7. What is the impact of invasive plant or animal species?
8. Is grazing affecting the condition of the remnant?
9. Is the controlled burning programme, if in existence, appropriate for the region?
10. Are there any rare invertebrate species in the remnant?

INVERTEBRATES v PLANT INDICATORS

What are the advantages and disadvantages of using plants or invertebrates as indicators of habitat quality and change? Plants may be particularly appropriate as they reflect the outcome of an integration of environmental variables to a greater extent than the mobile animals, which may be less restricted by edaphic and man-influenced factors. Plants may reflect different environmental factors to those indicated by animals and this might lead to different conclusions, depending on which group was studied. One disadvantage of using plants is that many species are not present or are difficult to identify throughout the year. On the other hand, there are groups of invertebrates which may be sampled continuously throughout the year. There are obviously advantages in using groups of plants or animals which have a proven value for such work, but there is no information available about the number of taxa needed to be considered to provide meaningful information on environmental quality.

Invertebrates provide additional information about habitats which complement, and reinforce, that provided by the study of plants. However, extensive and unqualified species lists of plants and invertebrates are of little value to reserve managers. It is clearly desirable to analyse such data to provide information on groups of species which share common resources and thus serve as indicators of

---

[1]School of Biology, Western Australian Institute of Technology, Bentley, Western Australia 6102.

habitat type and quality. Appropriate species groups should be defined, and understood, by managers involved in monitoring and surveys. If this task is achieved it will have significant importance in the interpretation of data from ecological surveys, impact assessments and monitoring programmes. The actual choice of animals and/or plants used in these monitoring programmes would be dependent on the questions being posed and by their occurrence in the areas being evaluated. Where invertebrates are selected it is necessary to define the reasons for choosing particular species or groups.

CHOICE OF INVERTEBRATES

Which invertebrate groups are most appropriate for measuring environmental quality or change? There is a tendency for the taxa to be selected on the basis of the particular interests of the investigators involved. This may be reasonable in Australia where the taxonomic and ecological knowledge of many orders is severely limited but caution should be taken to avoid taxa which are inappropriate for the questions being raised.

A number of candidate taxa suggested have potential value as indicators. These include:

1. certain soil-dwelling taxa such as springtails and mites which are 'K'-strategists and are present in the soil throughout the year. Surveys may therefore be performed at various times of the year and are more likely to produce comparable data than those taken from seasonally active taxa;

2. social insects such as ants and termites which are able to buffer variations in climate and food availability and are therefore available for sampling throughout the year;

3. Diptera (flies) which may act as good indicators because they are abundant, diverse and include representatives from all trophic levels and from a wide range of available niches;

4. large invertebrates with localized distribution which have been used effectively as indicators of habitat status in New Zealand; and

5. rare species which could be useful for indicating the status of specific habitats on which they depend in some way. However, in Australia the knowledge of terrestrial invertebrate distribution is so poor that it is generally not possible to pinpoint which species are actually rare.

PROBLEMS OF USING INVERTEBRATES AS INDICATORS

Three potential problems concerning the mechanics of using invertebates as indicators were highlighted by the discussion groups. First was the extreme seasonality of occurrence of many groups. This might pose problems in repeating a survey or in interpreting the response of indicator species to environmental change. This problem may be lessened by performing surveys four times throughout the year, once in each of the seasons. Following from this, the precision provided by invertebrate data was questioned. This is one area which needs to be investigated. Finally, the validity of applying particular invertebrate data to other groups of species was also questioned. The minesite data of Majer (1983) suggests that the variation in ant fauna between sites reflects that of several other invertebrate groups. Therefore, for this group at least, ants may be used as indicators of the diversity and species composition of other invertebrate taxa.

REFERENCES

Mahoney, C. T., 1976. Soil insects as indicators of use patterns in recreation areas. *J. For.* 74: 35-7.

Majer, J. D., 1983. Ants: bio-indicators of minesite rehabilitation, land-use and land conservation. *Environ. Manage.* 7: 375-83.

Puszkar, T., 1979. Epigeal fauna as a bio-indicator of changes in environment in areas of high industrial pressure. *Bull. Acad. Pol. Sci. Ci. II Ser. Sci. Biol.* 28: 925-31.

CHAPTER 39

Modelling — its Role in Understanding the Position of the Remnants in their Ecosystems and the Development of Management Strategies

Workshop Leader: G. R. Beeston[1]

THE topic was introduced under the headings of ecosystem and management models. The concepts and aspects proposed as covered by each type are shown in Table 1.

The major points to emerge from the discussion are listed below.
1. What need is there for models?
2. What precision is needed in the model?
3. While traditionally models have been thought of as mathematical they can also be thought of as a conceptual framework in which to place processes.
4. Modelling is a process of simplification which enables the structuring of our knowledge requirements and a drawing together of our existing knowledge.
5. All models have flaws and for any to work properly they need an adequate data base and validation by results being applied. Models may enable inadequacies in a data base to be exposed.
6. Expert systems were raised as a 'model' type, which may provide some solutions to managers in the future. These systems will also allow data from traditional owners and managers to be integrated into future models.
7. In developing models for management, monitoring will be important as for every action taken there is a non-action, which could be taken. Also, the recording by managers of the results of particular management practices, possibly by simple check lists is the only way that models can be developed past the testing phase.

Table 1.

ECOSYSTEM MODELS

1. SPATIAL
   Position
   Relationship to other remnants and land uses
2. TERRAIN AND GEOLOGICAL
   Position in landscape
   Underlying and surrounding geology
3. HYDRAULIC: WHAT FORM DOES IT TAKE
       Run on Run off
       Underground movements
       Flooding
       QUALITY How do Pollutants
                   and Nutrients affect
4. LIGHT   Normal situation
             Derived situation
5. AIR   TYPE AND MAGNITUDE OF MOVEMENTS
        Normal situation
        Present How crucial to survival
        QUALITY
        Pollutants — Industrial
                  — Railways
                  — Farming

MANAGEMENT MODELS

1. SINGLE SPECIES LIFE CYCLE
   Every species present
   Development — Growth — Stagnation — Decline
   Redevelopment — Migration
2. PREDATOR — PREY INCLUDING PESTS
   Relationship between species
   Resource partitioning (Sharing — Symbiosis).
   Sharing — Symbols
   Pests, Invasion of Aliens

[1] Department of Agriculture, Baron-Hay Court, South Perth, Western Australia 6151.

Table 1. Continued.

3. VEGETATION — SOIL
   ANIMAL — SOIL
   IN BOTH NEED TO CONSIDER
   Physical
   Compaction Problems
   Nutrients
   Changes/Removal of species and its effect
   Invasion and its effect
   Runoff from Urban Areas
   Pollutants
   Role in Food Chains
4. OTHER EXTERNAL INFLUENCES
   MAN — Fire
       — Grazing by different species
       — Weeds
       — Water Harvesting
5. HOLISTIC
   Do we aim for this?

8. To avoid the needless repeating of experiments and trials there needs to be a good communication path between researchers and managers: and the use of model structures could formalize this process.

9. The use of models in the management of remnants would need:-

    (a) the clear definition of management objectives;

    (b) the structuring of the decision-making process;

    (c) the development of selective detailed models; and

    (d) a clearly defined monitoring and reporting system with the ultimate goal a good communication model.

10. Economic models were common-place and should that not be also the case in conservation planning and management?

CHAPTER 40

The Value of Corridors (and Design Features of Same) and Small Patches of Habitat

Workshop Leader: T. Dendy[1]

INTRODUCTION

IN any consideration of the role of native vegetation remnants, reference is often made to their value as wildlife corridors or stepping stones between other remnants or larger habitat areas. However, there is little published research for Australian environments supporting this concept, let alone providing assistance in determining design features for such areas [see Henderson *et al.* (1985) for a Northern Hemisphere example]. Not surprisingly, little attempt has been made within Australian parks and reserves systems to provide for or facilitate the retention of wildlife corridors between reserves.

This summary of the views of two discussion groups on the topic suggests basic advantages and disadvantages of wildlife corridors, provides some suggestions regarding design features (having considered a case study), and proposes that support be given to scientific research.

DEFINITION

A corridor for wildlife may be described as a narrow strip, stepping stone or series of stepping stones of hospitable territory traversing inhospitable territory providing access from one area to another. A corridor will almost certainly not be suitable for all wildlife species, particularly discontinuous corridors (stepping stones) which are likely to be unsuitable for the smaller and/or more cryptic species.

ADVANTAGES

Corridors are considered to facilitate the movement of the biota (both plants and animals) between otherwise isolated areas. As such, corridors:

1. increase the effective size of plant and animal populations;

2. maximize heterozygosity — genetic variability which is considered to enhance long-term viability of organisms through greater resilience or adaptiveness to changing conditions; and

3. facilitate recolonization of an area affected by some disturbance (e.g. bushfire, clearance, etc.).

Corridors may also have a side value as:

1. linear or island reserves, supporting viable plant and animal communities in their own right;

2. refuges or alternate habitat areas for animals that are able to utilize cleared landscapes; this is particularly valuable to a landowner when the species are insectivorous and utilize agricultural insect pests (e.g. Australian magpie *Gymnorhina tibicen*);

3. assist in the maintenance of species diversity in largely cleared landscapes;

4. domestic stock shade and shelter belts although their value as wildlife habitat will be considerably lessened if they are not appropriately protected from degradation by stock;

5. windbreaks (ameliorating climatic extremes on crops and pastures);

6. significant landscape features (particularly where they are retained in harmony with the environment by following such natural features as creeklines and rivers); and

7. a source of renewable timber for such things as fenceposts and firewood.

DISADVANTAGES

It is recognized that there may be detracting features of corridors. Specifically they may, among other things:

1. facilitate the spread of wildfire, disease and pests (plants and animals);

2. provide a harbour for plant and animal pests; and

3. occupy potentially arable land.

---

[1]National Parks and Wildlife Service, Department of Environment and Planning, G.P.O. Box 667, Adelaide, South Australia 5001.

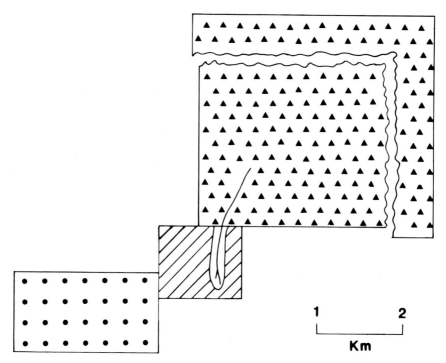

Fig. 1. Illustration of clearance proposal. The question is: should an area be retained as a corridor between Knee-bone and Thigh-bone Parks?

The data available for considering the decision:
● Knee-bone Park. 1000 ha with vegetation associations A and B.
▲ Thigh-bone Park. 100,000 ha with vegetation associations A, B, C (creekline) and D.
/ Private land proposed for clearance. 500 ha with vegetation associations A, B and C.

Rare fauna: Species W is restricted to vegetation association C (creekline); Species X observed in all vegetation associations and cleared land; Species Y observed in all vegetation associations (known to cross cleared areas up to 50 m); and Species Z observed in all vegetation associations (known to be cryptic and unlikely to cross even short distances of cleared land greater than 20 m).

DESIGN FEATURES

These workshop discussions on corridors were set within a realistic frame by using an example from South Australia where State legislation provides that clearance proposals must be assessed with regard to, *inter alia,* the need to retain native vegetation for/or as wildlife corridors. Workshop participants were invited to take the role of the decision making body with regard to a specific clearance proposal (see Fig. 1). The conclusion of both workshop groups was as follows (having reluctantly accepted that a decision was required on the information provided).

1. *A corridor should be retained.*
   Despite limited information, it appears that a corridor between Knee-bone and Thigh-bone parks would facilitate the movement of (at least) species y and z. Should future evidence prove the contrary, an option is still available to clear more land. Should the alternative decision be taken, and subsequent evidence indicate that a corridor was necessary, in reality an option to reinstate the corridor probably does not exist.

2. *The corridor should be as wide as possible.*
   The corridor in this example should include the whole of the area proposed for clearance — the logic follows on from 1 above. Additionally, the corridor should be as wide as possible to minimize the effect of fertilizer drift, weed invasion, etc. and fenced to prevent degradation by stock. This may be facilitated by retaining vegetation along property boundaries incorporating roadside vegetation or vegetation on a neighbouring property.

3. *The corridor should incorporate natural features.*
   In this example, the corridor should include the creekline for part of its length as well as different habitat types. Preference should be afforded to denser habitat to facilitate the movement of the smaller and/or more cryptic species, whereas the less cryptic species may migrate through fairly open vegetation or even cleared land.

CONCLUSION

The workshops that discussed the value of corridors were certain that they were positive features despite the lack of supporting research. Nevertheless it was recognized that landowners and politicians will expect a more scientific justification to retain wildlife corridors at the expense of cleared agricultural land. To this end it is essential that research be directed towards obtaining appropiate data. Indeed, it is expected that such research will demonstrate that past clearance practices have been excessive for wildlife conservation and that the re-creation of corridors by re-planting may be an appropiate remedial direction for the future.

REFERENCES

Henderson, M. T., Merriam, G. and Wegner, J., 1985. Patchy environments and species survival: chipmunks in an agricultural mosaic. *Biol. Conserv.* 31: 95-105.

CHAPTER 41

Single Large or Several Small Reserves?

Workshop Leader: C. R. Margules[1]

GIVEN the existing patchiness of much habitat throughout the world, research should be concentrated on the management problems of small patches, e.g. minimum viable population sizes, genetic consequences of isolation, species' habitat requirements, rather than with an unhelpful academic debate on the size and shape arrangements of reserve networks. Although that debate is entrenched in ecological literature, so rarely is there an opportunity to plan reserves prior to habitat fragmentation, it is not really a major conservation issue.

There is no doubt that both single large reserves and groups of small reserves have a role to play. In some parts of the world the opportunity still exists to conserve entire functioning ecosystems, like rainforests, where fragmentation would destroy ecological relationships and reduce biological diversity. Groups of small reserves are not likely to conserve many of the large vertebrates adequately, nor will they always provide adequate buffering against adverse external effects.

On the other hand, groups of small reserves might divide extinction risk among several populations, and in some environments, networks of small reserves mimic natural patchiness. Many species and communities are only found today in small remnants of previously widespread habitat.

Options for reserve selection tend to be severely constrained by the existing sizes, shapes and geographic arrangement of habitat patches so that the distributions of species among them becomes the major consideration.

Strategies for selecting reserves should be based on species' distributions and abundances, or better still, where they are known, population dynamics and habitat requirements. In practice, due to a lack of appropriate detailed information, this might mean selecting reserves to maximize the range of environments they represent.

Studies on species populations and habitat requirements, and understanding mechanisms of species distribution and abundances are more likely to serve the cause of nature conservation than arguments over single large or several smaller reserves.

---

[1]CSIRO, Division of Water and Land Resources, P.O. Box 1666, Canberra City, Australian Capital Territory 2601.

CHAPTER 42

Ecotones, Patchiness and Reserve Size

Workshop Leader: Eleanor Russell[1]

An Ecotone is an artificial construct imposed on environmental gradients, and for management purposes it is more profitable to think in concrete terms, such as edges: natural edges as between two plant communities (an enviromental gradient) or two different successional stages, or human induced edges. 'Edge-effects' are well-documented in the literature, and these effects may be beneficial or deleterious. In fact, small remnant areas of native vegetation may be all edge.

Environmental gradients, whether abrupt or across a continent, are important as places for observing the dynamics of interactions between species and communities and for observing evolutionary events. For example, the flora of the southwest of Western Australia is richest in the transitional rainfall zone between the relatively species poor high rainfall forests and arid zone communities, with many recently evolved species (Hopper 1979).

When examining ecological gradients, the problem of scale needs careful consideration. Different species perceive a change in the physical environment differently, depending on taxon, size, mobility and overall life history strategy. The sort of moisture gradient which would be significant for a snail could be irrelevant to an insectivorous bird. Are the ecotones or gradients which ecologists see between habitat types of relevance to other organisms? For example, Friend and Taylor (1985) showed that, in the Northern Territory, the habitat types perceived by various species of small mammals usually overlapped several of the habitat types delineated on the basis of vegetation structure or floristics. It may be that the small mammal's scale of perception is different from the human's scale as expressed on a vegetation map.

Time scale is as important as spatial scale. The use of different habitat patches may vary with time of day or season. The loss of one particular habitat may affect not only species which are specific to that habitat, but also species which depend on that habitat at one particular time.

'Environmental patchiness' must be distinguished from 'successional patchiness.' Environmental patchiness is an important consideration in the maintenance of genetic diversity in species. If a species occurs with isolated populations in a patchy environment, then a reserve large enough to include several patches will allow migration between patches. Reserves sited *along* an environmental gradient would do more to promote maximum genetic diversity than reserves sited in the more homogeneous environment *across* a gradient. The scale of environmental patchiness is an important factor to be considered when determining reserve size.

The steepness of environmental gradients determines the steepness of ecoclines, as in Western Australia, where, in the south, such things as soil type and annual rainfall change rapidly with distance from the coast. A small area near the west coast would include a greater variety of habitat types than would a similar area near the eastern border.

Although it may be desirable to acquire large blocks of land for reserves with the greatest possible range of habitat types and transitions between them, that is, the greatest environmental patchiness, we are rarely able to do so. Instead, we already have many very small reserves which we have to learn how to manage. It is desirable to think in terms of a system of small reserves which provides a diversity of habitats found in no single reserve. This system may sample across an environmental gradient which no single small reserve could span. From this point of view, the many small reserves in the Western Australian wheatbelt have a significance as a whole which is greater than that of any individual reserve. Therefore management of individual reserves should be integrated into a policy for the whole system.

---

[1] CSIRO Division of Wildlife and Rangelands Research, L.M.B. No.4, P.O. Midland, Western Australia 6056.

The purpose of a particular reserve is usually neglected in many acquisition and management decisions. Aims which relate to conservation of habitat diversity, of communities or of single species may have quite different management implications, and must be translated into what is realistically possible for each single small reserve or system of reserves.

REFERENCES

Friend, G. R. and Taylor, J. A., 1985. Habitat preferences of small mammals in tropical open-forest of the Northern Territory. *Aust. J. Ecol.* 10: 73-185.

Hopper, S. D., 1979. Biogeographical aspects of speciation in the southwest Australian flora. *Ann. Rev. Ecol. Syst.* 10: 399-422.

CHAPTER 43

Viability of Small Populations of Plants and Animals and the Value of Introductions and Translocations

Workshop Leader: J. Kinnear[1]

THE question relating to the concept of a minimum viable population (MVP), which in the literature ranges from 50 to 500 individuals, was discussed. This range appears to be a reasonable estimate having been mathematically derived from a number of approaches, but this range should be used only as a guide. The discussion groups concluded that research and management should strive to maintain populations within the above range.

The value of translocations was debated in some detail in both sessions. It was concluded that translocations had merit given that one had identified and rectified the factor(s) responsible for the species' decline in range and numbers (e.g. fox predation).

It was generally agreed that it would be desirable to increase the variability of inbred populations that normally outcross. The point was made that one should avoid mixing populations from extreme ecotypes. No agreement was reached on the number of individuals which might be introduced nor were guidelines made regarding the size of founding populations.

Mutation rates were discussed by one group particularly with respect to ameliorating the negative consequences of inbreeding. No conclusions were reached because estimates of mutation rates in most eukaryote species have not been reliably quantified.

Introductions and translocations were considered to represent good conservation strategy in that the risks of extinction of endangered species would be reduced. There was general agreement that introductions or translocations should not be undertaken without adequate research, and that an ongoing commitment from management was essential.

---

[1] Department of Conservation and Land Management, Western Australian Wildlife Research Centre, P.O. Box 51, Wanneroo, Western Australia 6065.
Page 365 *in* NATURE CONSERVATION: THE ROLE OF REMNANTS OF NATIVE VEGETATION ed by Denis A. Saunders, Graham W. Arnold, Andrew A. Burbidge and Angas J. M. Hopkins. Surrey Beatty and Sons Pty Limited in association with CSIRO and CALM, 1987.

CHAPTER 44

Use of Surveys and Data Bases for Conservation

Workshop Leader: M. P. Austin[1]

INTRODUCTION

PRELIMINARY discussion led to the following definitions:

Survey: a field study where observations are made at separate locations

Data Base: any organized data set maintained in a permanent form, usually on a computer

for use during the workshop discussions.

The introductory remarks drew attention to the problems of inconsistent data recording on surveys and of the dangers of GIGO — garbage in and garbage out — in the use of data bases. The need to set clear objectives for the use of both surveys and data bases was emphasized. Depending on the objectives, a sequence of operations linking surveys to data bases and their subsequent use for conservation management could be identified. A brief review of some recent developments in the use of surveys and data bases relevant to conservation was made. These included the survey design alternatives to stratified random sampling, the collation of data from different surveys and the use of generalized linear modelling (GLM) for analysis of survey data.

The issues then raised were:

1. whether the purpose of surveys was well-defined;
2. the adequacy of sampling methods;
3. the appropriateness of analysis methods;
4. the purpose of data bases;
5. the reliability of the contents of data bases; and
6. whether data bases were used effectively.

In order to structure the discussion and to summarize the participants' contributions, a questionnaire on the issues was distributed. The participants were grouped as either researchers or managers. Their responses, however, showed no major differences. Most answers were either yes or no, but frequently answers were qualified as depending on the context of the particular study. Answers were therefore scored as yes, qualified yes, equivocal, qualified no, no and don't know.

SURVEY PURPOSE

There was marked disagreement as to whether conservation surveys had well-defined purposes; of 23 responses, 10 answered yes and 10 no! While there was general agreement that general-purpose surveys were worthwhile, there was no consensus on whether special-purpose surveys were more cost-effective than general-purpose surveys. The majority of participants felt that a vegetation map of communities and a composite list of species was not sufficient for most conservation purposes, though a few thought that it could be useful in certain circumstances. There was agreement that surveys should be on a regional basis, rather than restricted to reserve areas. One point made was that the major value of surveys may lie in their overall contribution to knowledge of distribution and status of the flora and fauna in the longer term. This is seldom recognized in assessing the usefulness of a survey at the time it is carried out.

SAMPLING

There was general agreement that stratified random sampling was not necessary or cost-effective. However, the need for consistent procedures and more carefully designed surveys was accepted. Many participants also thought that examples of effective surveys could be identified.

ANALYSIS

Multivariate and other analytical techniques were seen as important for obtaining maximum information from surveys. No consensus existed, however, regarding the relative amount of time which should be spent on field work or data analysis.

---

[1] CSIRO, Division of Water and Land Resources, G.P.O. Box 1666, Canberra, Australian Capital Territory 2601.
Pages 367-8 in NATURE CONSERVATION: THE ROLE OF REMNANTS OF NATIVE VEGETATION ed by Denis A. Saunders, Graham W. Arnold, Andrew A. Burbidge and Angas J. M. Hopkins. Surrey Beatty and Sons Pty Limited in association with CSIRO and CALM, 1987.

DATA BASE

A clear division was apparent between those who thought that existing data bases had well-defined purposes and those who did not. Many participants felt that they had insufficient experience of data bases to give definitive answers and this was reflected in the number of 'don't know' answers to the questionnaire. For example, many participants were unsure of the cost-effectiveness of data bases (nine yes, 10 don't know, four no), or of examples of successful data bases (12 yes, eight don't know, three no). Reliability of items in the data bases were not seen to be a problem as it was at least equal to that from other sources, nor was the maintenance of data bases seen as a problem. The predictive capability of data bases was thought to be acceptable though many people felt that insufficient validation had been done. There was complete agreement that data bases have a major use in conservation, and that future use will increase dramatically.

CONCLUSIONS

Several conclusions were reached by the participants though these were not unanimous in the workshop discussions:
1. surveys are essential for conservation;
2. more consistent methods of survey are needed;
3. attention should be given to ensuring that the purpose of surveys and data bases is clearly defined;
4. data bases have great potential in conservation work but experience of their use is limited;
5. reliability of data in, and maintenance of data bases was not seen as a current problem; and
6. doubts existed about the value of statistical design and analysis methods for use in conservation surveys.

These conclusions should be used as a basis for more intensive workshops on the use of surveys and data bases. The recent workshop proceedings on Survey Methods for Nature Conservation by K. Myers, C. R. Margules and I. Musto and these current proceedings offer a firm basis for planning such activity.

CHAPTER 45

Monitoring of Management Practices

Workshop Leader: G. R. Friend[1]

INTRODUCTION

THE standard dictionary definition of monitoring — 'to check, observe or record the operation of [something] without interfering with its operation' (Macquarie Dictionary 1981) tends to assume one is dealing with 'steady state' or 'closed' systems, and allows some appreciation of why the monitoring of ecosystem changes in complex, naturally variable ecosystems is not a simple matter. Indeed, in Australia at least, the monitoring of ecosystem changes associated with the management of nature reserves and national parks is rather poorly developed. Management usually takes place in the absence of relevant research information, and its effects are frequently not documented within the framework of a formal monitoring system.

To be useful, monitoring must provide long-term records of the status of various ecosystem components (e.g. soil, vegetation, fauna) and facilitate measurement of the *direction* and *rate* of change. In particular, it must enable the manager to distinguish changes attributable to use and management from those attributable to climatic variation or natural succession (see Macdonald and Grimsdell 1983; Wilson *et al.* 1984).

These two discussion groups concentrated on aspects of the monitoring of management practices. Discussion was based on a conceptual monitoring framework (adapted from Macdonald and Grimsdell 1983, and Wilson *et al.* 1984), which was further modified in the light of the workshop discussion (Fig. 1). Two principal questions, central to the successful operation of a monitoring system, were addressed. These were:

1. how can we measure changes in ecosystem components due to a particular management practice, as distinct from those associated with natural processes such as climatic phenomena or succession?; and
2. how can we facilitate the integration of management, planning and research personnel to form an optimal feedback system?

MEASUREMENT OF CHANGE IN ECOSYSTEM COMPONENTS

In an optimal monitoring system the difficulty of distinguishing management-induced changes from natural changes is reduced by a substantial research input prior to the monitoring system being designed (Fig. 1). This research should aim at identifying the major components and pathways in the ecosystem, and adopt an experimental approach (small scale?) to provide some understanding of the nature and dynamics of change in the system. The development of simple models and predictions should also be part of this work. Unfortunately, in most situations such research has either not been carried out, or is too short a term to provide useful insight.

Both discussion groups considered that long-term work with adequate controls or multiple reference areas was essential. There is also the need for good experimental design involving consultation with statisticians at an early stage in the design. The interpretation of the effects of any management action could be greatly enhanced by maintaining accurate, site-specific and detailed management records (i.e. of who did what, where, when and why). This would allow flexibility in personnel and ensure long-term continuity.

The problem of how to find and set up 'real controls' arose. For example, to exclude fire from a control area in a system designed to monitor various burning regimes is unnatural in the longer term: the control cannot remain as a constant reference point, and may change more than the treatment areas. In such cases it was considered that the 'control' should be that area subjected either to the present management regime or a natural disturbance regime (see Hobbs, this volume), and 'treatment' areas are those that are manipulated for a long term and assessed against the land management objectives previously specified (see Fig. 1).

---

[1]Western Australian Wildlife Research Centre, P.O. Box 51, Wanneroo, Western Australia 6065.
Pages 369-71 *in* NATURE CONSERVATION: THE ROLE OF REMNANTS OF NATIVE VEGETATION ed by Denis A. Saunders, Graham W. Arnold, Andrew A. Burbidge and Angas J. M. Hopkins.
Surrey Beatty and Sons Pty Limited in association with CSIRO and CALM, 1987.

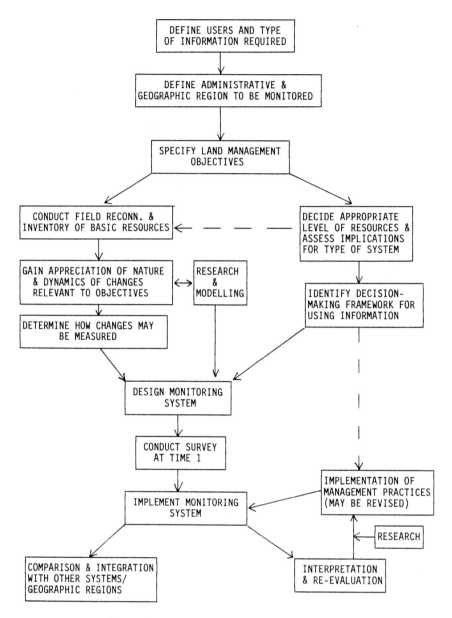

Fig. 1. A monitoring framework.

The components to be measured, and the scale, frequency and precision of the monitoring are important considerations if the effects of various management practices are to be clearly elucidated. These factors are determined primarily by the land management objectives (Fig. 1) which must be clearly stated at the outset. When coupled with research (as in lower right of Fig. 1), the interpretation of results and re-evaluation of the system allows managers to narrow down the number and type of components that need to be measured.

INTEGRATION OF MANAGEMENT, PLANNING AND RESEARCH PERSONNEL

Discussion on this problem initially centred on the respective roles of research and management in the monitoring process, and thus overlapped some deliberations from question one. It was considered that the issue of strategic and tactical research was a red herring; there need be no sharp distinction between the two, nor between the researcher and manager. Research and monitoring are interdependent and both contribute to the establishment of sound management principles. Ideally, research input is required at two stages:

1. prior to the design and implementation of monitoring; and
2. in the interpretation of results and re-evaluation of management practices (Fig. 1). In the latter phase of an optimal system, management and planning personnel are involved in ongoing research, by necessity.

The question of who is responsible for monitoring arose as an important issue. It was generally agreed that the initial responsibility was with the research scientist to design and set up the system, ideally based on some research findings already gathered (part 1, as above). Researchers have an obligation to point out the management implications of their findings (preferably by publication), and it becomes the responsibility of the manager to apply the findings to the relevant situation and carry out the monitoring process. Researchers should have further input in the interpretation of the results from monitoring (part 2 as above). Thus monitoring systems are designed and initiated by researchers (through an experimental approach) but the actual long-term monitoring is carried out by managers.

If proper integration is to be achieved, managers, planners and researchers should be physically together (e.g. same building, and working on the same sites), and a forum for critical scrutiny developed. A 'challenge' system involving the public as well as professional colleagues, and the establishment of working groups to ensure consistent methodologies across different organizations and regions were also considered desirable for facilitating integration. Government bodies responsible for research and management of conservation areas must remain accountable to both the public and private sectors of the community. By making results and plans available for public scrutiny, the public is given the opportunity to become involved in conservation matters, thereby reducing tensions and misunderstandings and ensuring true integration.

The use of an 'EXPERT' system was considered by at least some members of the discussion groups to be a major positive contribution towards achieving integration. Such a system assembles all classes of information into a structured framework for decision-making, and is able to rapidly identify gaps in our knowledge. Researchers, managers and planners all have the opportunity to participate in the decision making process. Furthermore, no knowledge is lost with a change of personnel, and the process can be applied even if not done on a formal basis with a computer.

Finally, all agreed that a lack of *communication* was the most basic problem in management of conservation areas. Workshop sessions such as the present may make a major contribution towards reducing this problem, provided that the results of discussions are widely disseminated.

REFERENCES

Macdonald, I. and Grimsdell, J., 1983. What causes change — getting at the facts. The role of ecological monitoring. Pp. 77-95 *in* Guidelines for the Management of Large Mammals in African Conservation Areas ed by A. A. Ferrar. *S. Afr. Nat. Sci. Prog., Rep.* No. 69.

Wilson, A. D., Tongway, D. J., Graetz, R. D. and Young, M. D., 1984. Range inventory and monitoring. Pp. 113-27 *in* Management of Australia's Rangelands ed by G. N. Harrington, A. D. Wilson and M. D. Young. C.S.I.R.O., Melbourne.

CHAPTER 46

Measuring and Monitoring Dynamics of Remnants. Types of Organisms that should be Monitored: Why and How

Workshop Leader: T. J. Ridsdill-Smith[1]

INTRODUCTION

THE topic of this workshop presented two questions:
1. what types of organisms should be monitored; and
2. what methods should be used to study them?

Monitoring of communities was considered, but most of the discussion was concerned with the selection of species for ecological studies in remnants. A set of criteria was proposed to select species or groups of species the study of which would provide most information about the dynamics and health of remnants.

DISCUSSION

Autecology versus Community Studies

Autecological studies were perceived to give more information about remnant functioning than community ecology studies. Since an important aim of conservation is to conserve the diversity of species it is surprising that researchers showed a preference for concentrating on the study of single species. The general expectation was that insights into the functioning of the ecosystem were better from a study of individual species, provided the correct species is chosen. Studies must be long term.

Organisms to Study

Umbrella species. Frequently one species in a remnant attracts public attention, but may be of little use in developing an understanding of the remnant. Their value is that they provide the opportunity to study more meaningful species. These species are called 'umbrella species' by P. R. Ehrlich (this volume). They are usually spectacular, unusual or endangered species, but may also be a common pest species. The remnant may have been set aside for the species. Examples would be an orchid, the noisy scrub-bird *Atrichornis clamosus*, the numbat *Myrmecobius fasciatus* or the grey kangaroo *Macropus fuliginosus*. The aim of the study would usually be to determine how to manage the remnant so that populations of the umbrella species be maintained and preserved.

Representative species. One representative species can be selected from one or all of the trophic levels in the system. The species would probably be dominant in its group and possibly characteristic of the habitat. Species selected might include a tree, a small herbaceous plant, an aquatic plant, a large vertebrate, a bird, an insect, etc.

A set of characteristics were suggested for selecting a plant which would be reasonably vulnerable to disturbance and thus a good indicator of the health of the remnant. Plant characteristics considered to be useful in remnant studies include: (a) a woody perennial; (b) an obligate seed regenerator; (c) a serotinous species; (d) long-term seed viability; and (e) limited seed fertility.

Criteria for selecting animal species considered to be useful in studies of remnants include: (a) widespread so it can be studied at several sites with contrasting environments; (b) abundant so it can be effectively sampled and changes in population size detected; (c) sampling methods should be available which are suitable for the aims of the study; (d) taxonomy should be known (especially important for invertebrates) so that the species can be easily separated from closely related species; (e) scale of activity (its normal feeding range) should be known so that the area of sampling can be determined. For example trapdoor spiders can be sampled within one remnant, but black cockatoos *Calyptorhynchus* spp. and grey kangaroos need to be studied both in the remnant and adjoining cultivated land where they feed; and (f) responsive to changing environmental conditions.

Although larger animals are usually studied, it was suggested that studies of insects in remnants had many advantages over those of the larger animals. The biomass of the small invertebrates in a remnant may exceed that of the larger animals and their contribution to processes may be as great if not greater. The study of insects for conservation management is discussed by J. D. Majer in this volume.

Processes and nutrient cycling. Since management of remnants is about the maintenance of processes, species which indicate the continuing successful functioning of a process should be studied. However, participants agreed that the selection of such species was difficult, and no real guidelines were available. Ants and spiders as dominant predator groups were considered as useful indicators of predatory activity and are relatively easy to study. Keystone species would be valuable to study if they can be identified. They are the species whose absence would cause a major upset to the process. Species important in cycling nutrients would also be useful to study. A. R. Main discusses the importance of studying nutrient cycling elsewhere in this volume.

Study of Dynamics

To study the dynamics of systems in a remnant it will usually be necessary to undertake long-term studies on the abundance of several species. Long-term studies are needed to be able to distinguish permanent changes in the system from short-term temporary changes caused by management, climate or disturbance. These may be monitoring studies or ecological studies (see below).

However, there is a quicker short-term approach to study dynamics by correlating differences in species abundance and diversity at different sites with habitat differences between sites, or when habitats at different sites are similar with management or environmental differences between sites. In this monitoring approach little understanding of mechanisms is involved but a general answer can often be obtained in a short time. Examples of site differences which could be studied in this way are: (a) habitats and climates; (b) perturbations such as fire, mining or road building; and (c) manipulation by humans such as control of weeds, removal of feral animals such as foxes *Vulpes vulpes* and cats *Felis catus* by poisoning, grazing by domestic animals. There are problems in this approach, in that spatial heterogeneity may obscure temporal patterns, so that care must be taken in interpreting data.

Study Methods

Monitoring studies use the least specific techniques. The abundance of a number of species and species diversity are usually estimated. Distribution and biogeography of species can be studied using data from a large number of monitoring sites. The methods are usually easy to use, and sets of methods are put forward in this volume by A. J. M. Hopkins, J. M. Brown and J. T. Goodsell and by J. D. Majer. This approach is the easiest and quickest for managers to use, and is cheap to support. Analysis is usually by correlations between variables in relation to environmental factors or time. This approach provides a general picture of the health of a remnant. The disadvantage is that it provides little understanding of the functioning of the system.

Long-term ecological studies are needed to provide more detailed information on the factors influencing the abundance of species, the interrelations between them, and with the environment. Techniques will be more complex and frequently methods to estimate population size, such as mark-recapture, will be needed for each study.

Community ecology involves the study of many species in a remnant and their interactions, and therefore this type of study obviously represents the approach to studying the ecosystem which most closely mirrors the system. However, while this approach provides data on factors influencing the structure of the community, and the relative abundance of species, it is hard to obtain much understanding about interactions between species, because so many interactions are involved.

Autecology involves the study of a single species. Its ecology can be studied in detail, including its interactions with other species, and the mechanisms influencing its abundance and occurrence. Methods can be developed relatively easily for one species, and will be highly specific. The species will usually be at one or a limited number of sites, and the study will be long-term. This is the most expensive method to obtain answers about managing remnants, because research resources will be committed for a long time on a single species. The approach is used by specialist researchers. An autecology study can lead to a reasonable understanding of the functioning of a whole remnant with time. Conserving rare and endangered species concerns the conservation of a population gene pool, and this can be investigated with single species studies as described in this volume for butterflies (P. R. Ehrlich) and trees (G. F. Moran and S. D. Hopper).

An option that lies between these two approaches is to study guilds of species. These are groups of species which partition the same niche axis, e.g. food, shelter. Similar methods can be used. Guilds suggested for study include flies (Diptera), ants (Hymenoptera) (see J. D. Majer in this volume) and spiders (Arachnida).

When the various key responses to the system are understood, a predictive model can be developed. This may be a simple word model or a complex mathematical model. Data from limited studies can then be used to plan management of remnants on a more general scale.

CONCLUSIONS

Three types of study were discussed.

1. Monitoring of communities. This is the easiest and quickest to carry out. It is therefore likely to be used most, particularly by managers who have many other commitments. It will provide managers with some information on which they can act while waiting for the results of longer-term studies.

2. Community ecology. This involves long-term studies of a number of species. Sampling methods are difficult to devise, and interactions between species hard to interpret. This approach is probably most valuable when using field experiments to study management practices.

3. Autecology. This involves long-term studies of one species. A detailed analysis of mechanisms can be carried out. Many participants considered this approach to give the best information about the functioning and health of the remnant. However, because it is the most specialized approach and involves long-term commitment of resources to a single species, it will probably be used least. Guidelines to the selection of the most suitable species to study were developed.

ACKNOWLEDGEMENTS

Many people have contributed to this summary. I would especially like to thank Chris Margules, Angas Hopkins and Jon Majer for their comments.

CHAPTER 47

The Integration of Survey and Monitoring

Workshop Leader: R. W. Braithwaite[1]

MUCH resource inventory type of information continues to be collected throughout the world. This information can be used as baseline data for monitoring programmes but the problem is how to measure change through time.

The biotic and environmental attributes to be measured will depend on the objectives of the monitoring programme and these should be clearly defined. Ideally, monitoring should be part of an experimental process of management. The research worker does the initial survey, poses the initial hypotheses and designs methods to test them using management. The manager then applies management prescriptions and the results of these tests of the hypotheses are identified using the monitoring programme. Thus the transition from survey to monitoring will generally involve a reduction in effort (fewer sites, less measurement) as objectives for management are defined and the monitoring tailored to fit. However, as most management problems are persistent, monitoring is potentially forever. A monitoring programme may be refined and modified as part of an iterative experimental management process but it does need to be simple and inexpensive to ensure its continuity.

The design of the management and monitoring programme needs to be an interactive process involving both managers and research workers. Who does the monitoring? Many research workers felt that park rangers were not to be relied upon to undertake monitoring. However, in places where rangers were given sufficient time in their routines they had proved successful in this role. The research worker should at least be involved in the interpretation of the monitoring results but for reasons of logistic efficiency the managers on site must take the major responsibility for the collection of data. However, rangers doing monitoring usually have not been rewarded with feedback on the results of their labours.

The discussion groups felt that surveys and monitoring not only need carefully standardized methodology, but need to be performed on precisely defined sites. These sites need to be permanently and prominently marked so that they can be readily relocated well into the future.

If the data collected are to be used, computer data bases must be installed. If the rigour of survey and monitoring is to be increased, greater provision for and encouragement of publication must be provided. Finally, if management is to work, both researchers and managers must be committed to an interactive partnership for the duration of the monitoring programme.

---
[1]CSIRO, Division of Wildlife and Rangelands Research, P.M.B. 44, Winnellie, Northern Territory 5789, Australia.
Page 377 in NATURE CONSERVATION: THE ROLE OF REMNANTS OF NATIVE VEGETATION ed by Denis A. Saunders, Graham W. Arnold, Andrew A. Burbidge and Angas J. M. Hopkins. Surrey Beatty and Sons Pty Limited in association with CSIRO and CALM, 1987.

CHAPTER 48

The Role of Government and the Community

Workshop Leader: G. J. Syme[1]

INTRODUCTION

REMNANTS of vegetation will only be retained if private landowners and private individuals and groups are involved with the retention and management of small areas of bush. Farmers need to be guided, or encouraged, perhaps, if necessary, by legislative means, if desirable bush is to be retained on private property. Voluntary input to management is also required from both groups and individuals. The relevant State Government authorities in Australia are unlikely to ever have the resources to adequately manage either public or private remnants of vegetation. The major issue confronting planners, therefore, is to what degree we should rely on legislation or compulsion as opposed to persuasion.

LEGISLATIVE ALTERNATIVES

Those seeing the advantages of legislation did so because they considered that too little time was left for effective persuasion in areas where there is already little bush and clearing still continues by private landholders. The South Australian legislation in which clearing of bush requires a permit was regarded by those supporting legislation to be a successful model. Compensation can be provided for the farmer in terms of rate relief and assistance with fencing costs where clearing is forbidden. The farmer has provision for appeal if he/she feels that a permit was unjustifiably withheld.

A second argument for legislative control was that for ecological reasons specific areas (both in size and location) may be required. In this case provision for buying land, consolidating areas for conservation interests and reselling appropriate land for agriculture would be a great advantage.

Arguments against legislation centred around the fear that if farmers felt that legislation was imminent they would clear land in anticipation of clearing restriction. This phenomenon has been observed in Western Australia before clearing ban legislation was introduced to catchments. In addition, there was discussion of bulldozer operators who were thought to be crossing from South Australia to western Victoria offering their services in view of rumoured clearing legislation in that state.

ALTERNATIVES FOR ENCOURAGEMENT

A wide variety of suggestions were received for this category and these are summarized below. It should be emphasized, though, that most participants in the workshop group saw a variety of techniques or strategems being used simultaneously.

Local Pride and Regional Significance

Programmes should be undertaken to foster pride in the local flora and fauna and an appreciation of its regional significance. Competitions could be run by voluntary organizations such as Greening Australia and farm-based groups such as Western Australia's Land Management Society within each locality with appropriate prizes. Local media could be used to extensively promote these events so that conservation minded citizens are seen as desirable models.

Television

Television should be used extensively in all areas to foster understanding of the importance of remnants of vegetation. The development of high credibility 'personalities' was thought by some to be a particularly effective means of promoting awareness. Emphasis should be placed on the economic as well as ecological benefits of conservation.

Economic Incentives

Provision of rate relief or tax concessions for farmers retaining areas of natural vegetation was considered. While there was general support for this approach some difficulties were foreseen. About

---

[1]CSIRO, Division of Groundwater Research, Private Bag, P.O. Wembley, Western Australia 6014.

one quarter of Australian farmers paid no tax and, therefore, equity issues could arise. Any taxation or rate concession was seen as unlikely to be large enough to motivate the farmer to forego production. In the case of incentives there may be problems in the farmer attempting to claim for relief for land which would not have been cleared in any event.

Rural-Urban Relations

Assistance should be given to urban-based conservation groups to improve their appreciation of rural views on conservation and to enhance their assistance to rural communities in managing remnants of vegetation.

The Role of Women

Rural women were seen as a largely untapped resource in terms of their potential to encourage conservation or assist in the management of existing remnants. Government should work through groups such as the Country Women's Association to encourage womens' involvement. Extension on individual farms should be directed to women as well as men.

Direction of Extension Officers

Government extension officers should be directed to talk about conservation opportunities to all their clients rather than to only those who were already sympathetic. This procedure had reportedly been successfully used in the United Kingdom and had enhanced the rate of adoption of conservation ethics.

Children: Investing in our Future

Increasing the awareness and knowledge of children in local reserves was regarded as an important priority. Ongoing future commitment to maintenance and preservation of our reserves can only be maintained by increasing resources to environmental education now.

CHAPTER 49

Management Options, Practical Constraints and the Establishment of Priorities

Workshop Leader: Andrew A. Burbidge[1]

OPTIONS

OPTIONS for the management of remnants range from no interference through varying levels of interference to a high level of manipulation of the several components which make up or affect the biota. No intervention is an ideal less attainable as the size of the protected area diminishes and the area becomes a remnant. No intervention is not an option today unless the management agency's lack of resources make it the only option by default — a situation unfortunately only too common. Even where an agency cannot afford to intervene it is essential that there is sufficient surveillance to identify major problems and prevent avoidable degradation.

Where some form of manipulation is possible it is vital to identify the objectives of management. Management can be for the benefit of a particular species or a particular ecosystem, it can be for maximum diversity or for the benefit of people visiting the area, or it can be for combinations of these. When considering objectives and strategies for management in remnants it is desirable that groups of small areas be considered as a system or network, thus allowing greater flexibility and the recognition of complementary values.

CONSTRAINTS

Constraints often vary with the region being considered. Time, possibly the greatest constraint, affects all others. The previous history of a remnant may affect management options, so may adjoining land-uses. A significant constraint is size, which together with shape greatly affects the rate of environmental degradation and limits a manager's ability to carry out procedures on the boundary, which may increase the magnitude of edge-effects.

The lack of scientific data on a particular remnant is often a constraint since the management agency cannot make rational decisions without knowing their consequences. Of equal importance is the lack of technical procedures to carry out the necessary manipulation. Scientific studies which do not make the transition from theory to practice are often of limited value to the manager. Community expectations are also important. Socio-political expectations concerning remnants are changing rapidly and different sectors see different values ranging from zero to high. Since management actions taken now will usually affect a remnant for some time, an idea of future expectations is desirable, but often is difficult to obtain. Finally a constraint which can only be overcome to a degree is the level of resources available to the management agency.

There are ways of limiting or removing constraints. These include research and planning, education and communication, and lobbying. Public involvement in management is important and the level of public group involvement in the actual management of remnants in Australia seems to be much lower than that prevailing in some other countries, e.g. the United Kingdom.

In reality managers are having to make decisions without sufficient information. In order to improve the standard of decision-making in the future it is vital that managers record how they reached such decisions and what they did.

DEVELOPMENT OF PRIORITIES

How do management agencies develop priorities when they are faced with many areas to manage and few data and resources to use? Firstly, they should consult the community and debate the issues. For consultation to be effective a minimum data-base on the remnant or network of remnants is needed. Gaps in the data-base need to be identified and targeted for future work. Community groups need access to

---

[1] Department of Conservation and Land Management, Western Australian Wildlife Research Centre, P.O. Box 51, Wanneroo, Western Australia 6065.
Pages 381-2 in NATURE CONSERVATION: THE ROLE OF REMNANTS OF NATIVE VEGETATION ed by Denis A. Saunders, Graham W. Arnold, Andrew A. Burbidge and Angas J. M. Hopkins. Surrey Beatty and Sons Pty Limited in association with CSIRO and CALM, 1987.

all the information about an area; presentation of selected data will not result in the co-operation and useful debate which is needed.

In developing priorities, options and constraints should be reviewed together with global and local conservation objectives, from the world and national conservation strategies to local conservation problems. The experience of local people and managers should be tapped, possibly through 'expert' computer-based systems, and, most importantly, there should be regular review of objectives, methods and results.

One method of developing priorities and procedures in the face of varying options and numerous constraints is the public participation management planning process outlined by Wallace and Moore (this volume). Already this has shown itself to be an effective method of bringing the community and the management agency together to try to manage remnants of both local and national value.

CHAPTER 50

Creation of Ecotones and Management to Control Patch Size

Workshop Leader: E. M. Mattiske[1]

An 'Ecotone' is a transitional zone between two dissimilar ecosystems and it may be natural or man-made. It is a region of interaction of the two ecosystems and may include species which are specific to the ecotone.

A 'Patch' is defined as an ecosystem with uniform physiognomy or floristic composition.

The creation or modification of ecotones may have conservation value by increasing species diversity (in terms of increased diversity and densities of species in ecotones), increasing maintenance of heterogeniety and genetic integrity, increasing habitat diversity and by providing advantages to some species which may be affected by seral stages in nearby ecosystems.

It is idealistic to try and accommodate all species in a small remnant of native vegetation. Any expansion of available niches in an area through the creation or modification of ecotones would be predicted to increase the habitat and species diversity, but may raise the interaction between the ecotones and nearby ecosystems. Ecotones may provide buffers between ecosystems; particularly when management options are considered at the level of patch size. The concept of creating or modifying ecotones either within or near smaller remnants provides greater flexibility of management. The management of ecotones and modification of patches provides possibilities for research (see Christensen and Maisey, this volume) whose findings could be applied by extrapolation, to larger remnants, particularly where funds for management and research are lacking. This approach could be used in the case of the extensive 'linear' remnants along transport corridor systems in Western Australia.

The potential disadvantages of creating and modifying ecotones include the risk of misguided management through lack of understanding of native communities and possible detrimental effects which may follow. This risk could be reduced by increased assessment and monitoring of modifications to ecotones and patches. There is also the possibility for the misguided use of introduced or alien species in any alteration of ecotones; emphasis should be placed on the need to maintain and expand the diversity of the native landscape.

The methods of creating or modifying ecotones and the options available to control patch size include the introduction of native plants and animals along the edges of remnants and natural corridors (e.g. creek beds, gullies, etc.), or introducing species which have a site specificity to ecotones. Smaller remnants could be expanded by encouraging adjacent landowners to recreate ecotones. Other methods for recreating ecotones and modifying the size of patches within remnants include the use of fire or replanting in disturbed areas, but the options for management of patch size and creation of ecotones may be less in the smaller remnants.

There is undoubtedly a need for positive management of remnants but any management actions must have flexibility to incorporate future options should re-evaluation by research workers or managers indicate the need. There is a great deal to be learnt from research into the biological communities of smaller remnants and their management. In some instances the fragmentation of native vegetation has removed background factors particularly in the smaller remnants and made it realistic to manage areas of remnant plant and animal communities by the manipulation of ecotones and patch size.

There is a real need in future for increased funding and research on the management of ecotones and patch sizes in the smaller remnants, as well as a need to increase the communication between adjacent landowners and managing authorities (local and state authorities).

[1] E. M. Mattiske and Associates, P.O. Box 437, Kalamunda, Western Australia 6076.

# CHAPTER 51

Management Theory and Optimum Feedback

Workshop Leader: A. J. M. Hopkins[1]

WHAT IS BEING MANAGED?

THEORIES of management have been developed to cover a wide variety of subject areas from people to property, traffic to tourists. Several of these subject areas are relevant to the overall workshop theme (the role of remnants in nature conservation). In order to decide which theories are relevant, it is first necessary to identify what is being managed.

1. *The Conservation Reserve System.* The designated parks and reserves, together with remnants in other forms of ownership, are the basic resource to be managed. However, because these form only a small proportion of the landscape, other parts of the biophysical system often require attention too.

2. *The Personnel of the Conservation Agency.* The employees can be deployed on the basis of themes (e.g. administration, wetland management, etc.) or on a regional basis (e.g. responsible for management of particular conservation lands within a region). Because of the nature of the conservation task, it is imperative that dedicated, enthusiastic and committed staff be encouraged through effective personnel management programmes.

   Staff are often despatched to field offices before they have had the opportunity to acquire the necessary background in the legislative and administrative policy of the agency or they may lack experience in expounding on the ramifications of these as they apply in dealing with the public. The development of a Manager's Kit for field staff could resolve this problem. Such a kit should contain essential information as well as providing a guide to sources of information of relevance to management. A list of items which could be included in the Manager's Kit is given in Appendix 1.

3. *The Public* (i.e. the non-agency personnel). Important groups are the reserve neighbours and near-neighbours, owners of remnants, other visitors/users, members of the public who wish to retain a user-option, local government (councillors and staff) and staff of other State and Commonwealth instrumentalities. The goodwill and support of the public for conservation objectives is essential for the achievement of these objectives. A variety of strategies is required so that agency personnel can interact with the diverse groupings within the public to foster this goodwill and support. All planning and management activities should include a high degree of liaison and consultation but this is particularly important at the local community level (see Wallace and Moore, this volume).

4. *Finances.* Most nature conservation agencies are substantially under-resourced relative to the job to be done. Financial management should provide for efficient use of existing resources (including such things as incentives for private conservation initiatives).

5. *The Legal Framework.* Conservation activities generally occur within a complex legal and statutory framework. Some Acts outline the justification for these activities and provide powers to perform them; other Acts impose constraints on the activities. This framework is central to the functioning of any conservation agency and requires constant attention.

THE LAND-MANAGEMENT PROCESS

A suggested method of developing and implementing management programmes for conservation lands is illustrated in Figure 1. It has the following features.

1. Once management objectives are known, it is desirable to construct a model, albeit a simple one, of the reserve or group of reserves. This pre-planning modelling provides a means of synthesizing existing information and may highlight gaps in the knowledge that require attention before further planning proceeds.

[1] Department of Conservation and Land Management, Western Australian Wildlife Research Centre, P.O. Box 51, Wanneroo, Western Australia 6065.

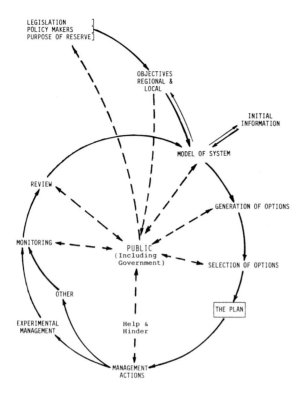

Fig. 1. Diagrammatic representation for developing and implementing management programmes for nature conservation lands. The process is an iterative one with feedback of information gained during management into the subsequent planning stages. A high level of public involvement is indicated.

2. Some of the management activities should be designed as experiments (hypothesis testing). This approach provides many benefits which include cost-efficient research.

3. The effects of management should be monitored and the management programme should be subjected to periodic review on the basis of the results of the monitoring.

4. The results of the experimental management programmes, the monitoring and the review will contribute to a gradual (iterative) improvement in the model developed during the pre-planning stage and thus to subsequent management plans.

Optimum Feedback. The states of the groupings identified in the section 'What is Being Managed?' change in the following ways: the appearance of the reserves changes following biological sequences or climatic or other changes; personnel change with staff movements; finances expand or are reduced; laws are amended; and public perception of the values of reserves and their acceptance of them alters over time. It follows that management must be dynamic and take into account any of the possible changes that eventuate.

Public participation is an important avenue through which changes in public attitudes can be established. It should be a regular occurrence so that public education is sustained. It can (and ideally should) involve all parts of the planning and management process. In the short term, organizing public participation may appear unproductive. However, when viewed as part of an educational process, its value is more obvious. The central nature of public participation in the development and implementation of management programmes for nature conservation lands is indicated in Figure 1.

ACKNOWLEDGEMENTS

All the participants in the workshop contributed to a lively and stimulating discussion which I have tried to report here. However, I particularly thank Professor Main for his vigorous input and subsequent assistance in preparing this summary and Claire Taplin for typing the manuscript.

APPENDIX 1

The Manager's Kit

1. The legislation.
2. Organizational policies.
3. Principles of management.
4. Other sources of information.
5. List of priorities/workplan (including description of the resources to be managed).
6. List of the resources available for implementation of management decisions.
7. Knowledge of the social process.

The Manager's Kit should be improved, expanded and revised on a regular basis.

CHAPTER 52

The Role of Remnants of Native Vegetation in Nature Conservation: Future Directions

D. A. Saunders,[1] G. W. Arnold,[1] A. A. Burbidge,[2] and A. J. M. Hopkins[2]

INTRODUCTION

THE Oxford English Dictionary defines the word 'remnant' as being 'that which remains after the removal of a portion; the remainder, rest, residue'. The dictionary adds that among drapers and clothiers a remnant is 'an end of a piece of goods left over after the *main* portion has been used or sold'. The emphasis is ours.

It is the fact that the main portion has been removed that has created the remnant or remnants and it is logical that any owner of such scraps would want to know if they are useful. With most of the fabric removed, is there enough substance in a single piece or group of remnants to make anything, particularly if they are from different cloth? The issue confronting managers dealing with remnants of native vegetation is basically the same: in the long term are they going to be of any practical nature conservation value? If so, how must they be managed to retain this value?

Remnants of native vegetation have been created over much of the world, for a variety of reasons and over a wide range of time scales. In the majority of cases, primary production (agriculture, grazing or intensive forestry) has been the cause of the fragmentation of native vegetation. In some cases the fragmentation happened gradually many hundreds of years ago. Furthermore, often the residuals have been altered to suit particular needs and are no longer representative of pre-existing vegetation associations. In Europe, for example, humans have been actively modifying natural systems for thousands of years. Many of the present remnants consist of secondary vegetation (for example, heathlands, alpine meadows, coppice woods and hedgerows) which do not represent pre-existing vegetation associations. They are regarded, nevertheless, as valuable patches for conservation today.

In many cases, landowners are encouraged to continue traditional land management practices in order to maintain what are essentially unnatural systems. Another extreme is typified by the agricultural area of the south-west of Western Australia where the fragmentation was rapid, extensive and very recent. In this region 54% (over 160,000 km^2) of all land developed for agriculture was cleared in only 37 years (between 1945 and 1982). Now clearing has almost ceased. In yet other parts of the world such as in the forests of the Amazon Basin in South America or in the rainforests of South-east Asia fragmentation of native vegetation is still proceeding rapidly.

The result is that, over much of the world, nature conservation agencies are left with small isolated fragments of communities and ecosystems in which to conserve the biota of the region. To achieve the goal of conservation, the agencies must devise appropriate management strategies that are based on a proper understanding of the biota and the processes that sustain it.

CHARACTERISTICS OF REMNANTS

As discussed earlier, the word 'remnant' is used to define any patch of native vegetation around which most or all of the original vegetation has been removed. This definition covers patches of many shapes and sizes, soil types, vegetation associations, degrees of isolation and types of ownership.

Typically, remnants are small and not necessarily representative of the pre-existing range of plant or animal communities. As stated by Usher (Chapter 9), the process of fragmentation is usually not random. Remnants often occupy the less arable soils or areas totally unsuitable for agriculture, perhaps consisting of rock outcrops or saline flats. Since plant communities are generally closely linked with soil types, certain communities or species have been selected for or against (depending on your point of view). For example, salmon gum *Eucalyptus salmonophloia* and York gum *E. loxophleba* woodlands were once widespread through the transitional rainfall zone of the south-west of Western Australia. Woodlands of these species were regarded by the early settlers as indicators of soils with good agricultural potential. Consequently, salmon gum and York gum are poorly represented on dedicated conservation remnants throughout the agricultural area (as are other woodland species, see Smith, Chapter 24) although many small areas of degraded woodland remain on farms.

Although remnants are generally isolated from one another, some are connected by corridors of native vegetation. The value of these corridors in the Australian landscape is, as yet, poorly documented but Bennett (Chapter 4), Recher *et al.* (Chapter 14), Bridegewater (Chapter 15) and Saunders and Ingram (Chapter 22) all discuss the role of corridors in the context of the management of remnants for conservation.

Ownership is an important facet of remnants which needs to be considered. In some countries the majority of remnants are privately owned; this can present a problem for conservation managers particularly where species exist solely or predominantly on such remnants. In the U.K. most of the declared National Nature Reserves are owned not by the Nature Conservancy Council but by private individuals and so a large proportion of the Council's budget goes to those landowners in compensation for the forgone development rights.

In an area of 1760 km^2 near Kellerberrin (Western Australia) there are 531 remnants totalling 5.7% of the area and ranging in size from less than 1 ha to 1100 ha (see LANDSAT photograph on frontispiece of this volume). Only 16 of these are vested in the Crown and only six are reserves set aside for the conservation of flora and fauna (20% of the area of uncleared vegetation). By far the greatest proportion of naturally vegetated land (particularly woodlands) is in private ownership so that effective conservation will hinge on the co-operation of the farmers.

These 531 patches also demonstrate the different treatment accorded to remnants in the one district: only 60 do not show obvious signs of deterioration as a result of grazing by domestic stock or other disturbances whereas the remainder show varying degrees of degradation. The 600 km of roads in the district have verges that range from possessing no native vegetation to well vegetated corridors 20 m wide linking remnants.

Remnants have a high edge to area ratio and, in some cases the entire remnant may be edge [for example, Yellingbo State Nature Reserve (Victoria) has an area of only 500 ha but 50 km of boundary (see Backhouse, Chapter 26)]. Corridors, too, are mainly edge. These high edge to area ratios pose many special problems for management.

Remnants have human neighbours. The ecosystems (or parts thereof) functioning on remnants may affect hydrology, nutrient cycling and the rate of erosion on surrounding land. Remnants are also repositories of organisms that can help control outbreaks of pests on the surrounding lands. Conversely, neighbouring land uses often have a direct impact on remnants by altering fire regimes, increasing nutrient input, especially from fertilizers (in nutrient-poor systems), changing water run-of, affecting ground water levels, increasing salinity, introducing domestic stock which graze, trample (breaking down soil structure), defecate and introduce exotic seeds via faeces or body surfaces, introducing domestic pets, particularly cats *Felis catus*, dumping rubbish and selectively removing species (for example, shooting native animals or digging up native shrubs).

Remnants have aesthetic, educational, recreational and scientific values. They contribute an additional visual dimension to man-modified landscapes because they retain their original appearance. They are valuable indicators of what the landscape used to be like. And, because remnants are generally dispersed across the landscape, they are relatively accessible to local communities. The simplicity of the biological systems on many remnants makes them ideal for research and interpretation. Taken together, these characteristics of remnants create a potential for education which must be exploited.

THE ROLE OF REMNANTS IN NATURE CONSERVATION

The perception of the role of remnants for nature conservation has suffered until recently because of the emphasis that has been given (correctly) to the acquisition of large national parks and nature reserves. This role now needs to be re-emphasized. Where it has not been possible to establish systems of large, interconnected reserves, conservation objectives must be achieved on remnants. In areas where there are large national parks and nature reserves, the small, dispersed remnants have an important complementary role.

Remnants have to serve the purpose that was previously performed by the whole landscape (preservation of landscape and ecosystem functions and conservation of biota). Unlike the very large or contiguous areas in which natural processes may continue to function as before, remnants are subject to a variety of exogenous disturbances (see Hobbs, Chapter 20) which may disrupt normal processes and often result in a decline in the number of species present. Thus, management will always be necessary to maintain the conservation values of the remnants.

In order to achieve nature conservation objectives it is often necessary to consider reserves as a group rather than individually. The selection and management of single reserves for the conservation of rare and/or endangered species, while worthwhile, may not be sufficient to conserve all species. Vagile species, for example, may require a network of patches of native vegetation. A network might be regional, as has been shown to be necessary for the conservation of Carnaby's cockatoo *Calyptorhynchus funereus latirostris* (Saunders and Ingram, Chapter 22), or it may need to be international to preserve the resources essential to all stages of the life cycle of migratory species (Lynch, Chapter 10). If wetland remnants on the Swan Coastal Plain in Western Australia had been viewed in the context of their importance for transequatorial migratory waders (or shorebirds), the large-scale destruction of those wetlands might not have occurred.

The specific role of remnants in nature conservation is most clearly perceived when it becomes necessary to identify objectives and priorities for management. This is normally done in the planning process and involves establishing a hierarchy from global down through continental, regional and local to those for individual remnants. However, some major problems still exist in developing these objectives and priorities especially at the regional to individual remnant level. Information is often lacking: how can we set priorities when we scarcely know what is present, let alone how the system works and, therefore, what parts of it need most attention? Furthermore, the methodology for rational priority setting is not well developed. Managers often focus on lower level objectives when developing work priorities because these are seen to be achievable (for example, conserving a particular species on a particular reserve). However, it is essential that managers, planners and research workers maintain a total perspective on their work by recognizing the higher level objectives.

A remnant may have several roles. Perhaps the best way to identify the role(s) is to ask 'why should that particular remnant be retained?' The reasons may be:

1. process oriented;
2. example oriented;
3. species oriented;
4. social values oriented; or
5. a combination of the above.

Process oriented reasons include the value of the remnants in the maintenance of evolutionary opportunities and successional states, as functional links or corridors between other remnants (or within and between landscapes as stepping stones for more vagile species i.e., values of connectivity), or for their roles in nutrient cycling, groundwater movements, effects on salinity and erosion and so on.

Example oriented reasons include the value of the remnants as representatives of ecosystems/communities, ranges of habitats or vegetation associations. Remnants can be important for the conservation of the regional biota solely because they provide representative or selected samples of communities that were once more widespread. This importance is enhanced when the sample is the best remaining example of the particular community. Other special features of remnants that justify their retention include the presence of benchmark sites or important research sites.

Species oriented reasons include those which involve conserving individual species and their gene pools, and the right of all species to exist. As a result of fragmentation of native vegetation, remnants now contain many species of plant and animal that occur nowhere else and are thus a repository of irreplaceable genetic material.

Social values oriented reasons relate to recreation, aesthetics, and, most importantly, public education. Remnants provide a focus for the development of an improved relationship between man and nature that is central to conservation.

Once the values of each remnant are identified then it is possible to fit it into a regional framework and thence to develop objectives and priorities. It should be remembered, however, that the values may change with time either as the biophysical attributes of the region change or as new information comes to hand. This is one of the many sound reasons for regularly reviewing management plans and programmes.

FUTURE DIRECTIONS

We have defined remnants, pointed out their characteristics (and some of the attendant problems), stated what the role of remnants should be and pointed to the need for management. We have also highlighted a fundamental problem: the lack of information necessary for the setting of management objectives and priorities and thus for prescribing management actions. Where should we, as concerned research workers, planners and managers, now direct our meagre resources?

Research should include autecological studies, studies of population genetics and research on the dynamics of communities and ecosystems. Autecological studies are clearly important in the case of rare and/or endangered species but should also be conducted on 'umbrella species' (see Ehrlich and Murphy, Chapter 16): species of popular or political appeal. A better understanding of umbrella species often engenders public empathy and political action that conserves not only the large and spectacular plants and animals, but many other species that coexist with them as well as the ecosystems that sustain them.

Most statements on nature conservation highlight the need to conserve gene pools and genetic diversity. It appears, however, that in relation to conserving remnant populations, too little is known about the effects of inbreeding, minimum viable population sizes, genetic drift, 'bottlenecks', mutations, etc., (see Usher, Chapter 9; Boecklen and Bell, Chapter 11; Moran and Hopper, Chapter 12; Ehrlich and Murphy, Chapter 16). There is a clear need to improve the genetic theory relating to small populations and to document actual genetic events in such populations of species across a broad phylogenetic spectrum.

There is also a need to define and refine knowledge on the current status of species and communities in remnants. Recent developments in computerized biogeographic data bases have provided a valuable tool to facilitate this process.

Studies of plant and animal communities should be integrated as far as possible with a view to identifying keystone species; those whose demise would cause at least partial collapse of the system, as well as indicator and vulnerable species. Once these ecologically signifcant species are identified and the reason for their significance is known then autecological studies and management and monitoring programmes can be set up.

Two types of process studies are relevant to management of remnants. The first, which includes such things as nutrient cycles and hydrological regimes, is difficult and time consuming but should not be overlooked. The second, involving community dynamics, should focus on the changes to disturbance regimes brought about by fragmentation and the biological consequences of such changes. Information on edge effects such as the conditions under which exotics invade a remnant, succession in relation to such things as fire and requirements for the regeneration or rehabilitation of degraded remnants will have immediate and direct application to management.

Relevance and accountability are key issues in the minds of conservation planners, managers and administrators. Since resources available for research are usually limited, these issues should be of paramount importance to researchers too. Study design and hypothesis testing are two areas that have received insufficient attention in field-based ecological research to date. Improved rigour will allay some of the non-reseachers' concerns but much will depend on better liaison between research and management personnel, a process that we emphasize later in this chapter. At the same time as pursuing applied research goals, however, scientists should maintain a strong theoretical basis to their work because it is here that the longer-term solutions to management problems lie.

Planning and Management. As mentioned earlier, a hierarchy of clearly defined objectives for the management of remnants needs to be established. Management programmes should be developed for groups of related reserves using a systems approach; this will provide a context into which objectives for each individual remnant can be set. At present there is little information available to produce management programmes so research is needed on resource effective methods for gathering relevant data and using it in assigning priorities to individual remnants in the network.

Modelling is becoming increasingly important as a means of structuring research as it assists in identifying key questions. Models can also aid communication. They can provide a means to make the most of available knowledge and experience. But models also have application in the planning process and in day-to-day management. To be appealing, however, the modelling systems must be cheap to set up and easy to run. The same caveats apply to computerized resource management information systems. While further development of modelling and information systems is required, it is essential that planners and managers show a willingness to use such systems in order to enhance their performance.

It is a relatively straightforward task to produce a management plan for an individual remnant particularly when the management objective is to conserve a rare species. But it is a major task to draw up management plans for each and every individual patch in a network of remnants. It is feasible, however, to produce a single plan for the network as a whole that takes into account the values of each remnant. The objective for management of these networks should be to ensure that the constituent plant and animal populations fluctuate over time within the expected changes of density and succession. Unfortunately, little is known about such fluctuations so research is needed. Some information on the fluctuations should come from studies centred on the extensive areas of native vegetation in which ecological processes are operating without the influence of fragmentation and its attendant disturbances. These studies should be complemented by work in the remnants themselves.

Management must also take into account special case areas where the purpose is to ensure that natural changes do not take place, as, for example, a heath reserve which may be deemed a best example reserve and managed to ensure that it does not change to woodland with time.

Invariably, conflicts will arise between different user groups and their needs and nature conservation objectives. The resolution of such conflicts is an accepted part of the management planning process. Where compromise is necessary we suggest the bias should always be towards nature conservation because remnants are both precious and vulnerable.

Liaison. Today it is difficult for even the specialists to keep up with the flood of literature in relevant scientific journals. It is completely unrealistic to expect managers to wade through the literature, interpret results and weave the findings into their management activities. It is the responsibility of *all* conservation scientists to interpret their results and, where they have a significant application to management, to provide those interpreted results to land managers in an accessible form. This means that research workers should be producing for an audience that is wider than one comprising merely their scientific peers. Research workers must communicate their findings, not only in the scientific literature, but also in popular publications and the daily press. It is only by educating the public and politicians that sufficient resources may be allocated to conservation. The need to communicate to a wider audience must be recognized by the institutions in which research is carried out, and included in the work statements of relevant workers.

It is easy to generate enthusiasm and resources for special conservation efforts to which people can relate. 'Save the numbat' or 'Save the helmeted honeyeater' are war-cries which have rallied resources around them in Australia, particularly as one is Western Australia's mammal emblem and the other is Victoria's bird emblem. 'Save our remnants' does not raise the same emotions. Thus, the 500 ha nature reserve set aside for the helmeted honeyeater has one ranger/naturalist and two support staff (i.e. one person per 167 ha) (Backhouse, Chapter 26) but the wheatbelt of Western Australia has one person per 64,000 ha of nature reserve or one person per 39 reserves (Wallace and Moore, Chapter 23) without taking into account the need for management of thousands of privately owned remnants as well as road/railway verges.

The magnitude of the task of managing remnants of native vegetation for nature conservation is clearly beyond the resources of most nature conservation agencies. To succeed in the long term a management plan must involve the general public *actively*. In particular, it should draw members of each local community into the management of *their* local network. Every remnant has one neighbour; many are privately owned. In some cases, remnants are purchased by specialist conservation groups (e.g., the Royal Society for the Protection of Birds in the U.K.). These neighbours and owners must be encouraged to participate in all aspects of remnant conservation from research through planning and management to monitoring. To achieve this level of involvement, however, will require a concerted public education campaign that must begin with a clear demonstration of the important role of remnants. If a catch-cry is necessary, perhaps 'Adopt a remnant today' might appeal.

Monitoring. At present, most management proceeds in the absence of adequate knowledge. Not enough is known about the way ecosystems function and particularly about the way ecosystems respond to disturbance. A great deal of insight can be gained by instituting carefully designed monitoring programmes. With neighbours and owners, there are potentially many monitors available to undertake relatively simple monitoring procedures (e.g. taking repeat, standardized photographs of weed invasion from surrounding land). The work of such volunteers can complement the more detailed monitoring programmes of agency staff.

Monitoring has for too long been regarded as third-rate science and therefore the province of 'also rans'. Monitoring *is* good science but it may take a long time to get results as many important ecological processes can only be detected over a scale of decades or even centuries (see Hobbs, Chapter 20). Appropriate design and excecution combined with hypothesis testing can make long-term monitoring attractive to a wider scientific community. Funding bodies also must recognize the need for studies to be long-term. Monitoring must not be too expensive or complicated; it must be relevant and some of the studies must be designed so that they can be carried out by local managers including neighbours. As stated above, it is important that local people are involved in the management process and monitoring is one simple, inexpensive way to give them hands-on involvement. Most importantly, there must be a mechanism for incorporating the results of monitoring programmes into the planning and management process. This reinforces the view that managers, research workers and the public must work closely together to achieve the objectives of any management plan.

PROTECTION AND TITHING

Protection of remnants. Throughout the world natural vegetation is being destroyed at an alarming rate. In many places, only small scattered fragments remain, most without legal security to protect their conservation values. The workshop held at Busselton concluded that more effort must go into protecting remnants and, wherever possible, into expanding existing protected reserves by acquisition of adjacent areas. Education of the public is essential to enable politicians to achieve this goal.

There is a perception in the community that remnants are merely unproductive areas of native bush. It is desirable that this perception be changed to one which recognizes the importance of these areas as part of a world-wide conservation system. It would be a major advance if a new approach to land management could be developed — one that is based on balanced and sustainable production and that takes into account the values of remnants to production (landscape protection, source of biological control organisms, etc.). Under such a system, farmers particularly would accept a greater responsibility for ensuring that their local remnants are conserved.

Ecological tithing. The workshop held at Busselton challenged scientists and naturalists who understand the values of native vegetation to spend a significant part of their time (say 10%) helping the public and politicians understand the role of remnants of native vegetation in nature conservation so that these valuable repositories of the world's heritage will not be lost forever.

ACKNOWLEDGEMENTS

We thank Dr Richard Hobbs who constructively criticized an early draft and Mrs Claire Taplin who typed the chapter.

Appendix 1

LIST OF PARTICIPANTS

Robyn Adams, Victoria College.
Jeni Alford, Western Australian Department of Conservation and Land Management.
Graham Arnold, CSIRO.
Ken Atkins, Western Australian Department of Conservation and Land Management.
Mike Austin, CSIRO.
Gary Backhouse, Victorian Fisheries and Wildlife Service.
Joe Baker, James Cook University.
Greg Beeston, Western Australian Department of Conservation and Environment.
David Bell, University of Western Australia.
Andrew Bennett, Arthur Rylah Institute, Victoria.
David Bennett, Muresk, Institute of Agriculture, Northam, Western Australia.
John Blyth, Western Australian Department of Conservation and Land Management.
William Boecklen, Northern Arizona University, USA.
Louise Boscacci, Western Australian Department of Conservation and Land Management.
Dick Braithwaite, CSIRO.
Peter Bridgewater, Bureau of Flora and Fauna.
Michael Brooker, CSIRO.
Judith Brown, Western Australian Department of Conservation and Land Management.
Rod Buckney, New South Wales Institute of Technology.
Allan Burbidge, Western Australian Department of Conservation and Land Management.
Andrew Burbidge, Western Australian Department of Conservation and Land Management.
Neil Burrows, Western Australian Department of Conservation and Land Management.
Harry Butler, Dinara Pty Ltd, Western Australia.
Stuart Chape, Western Australian Department of Conservation and Environment.
Per Christensen, Western Australian Department of Conservation and Land Management.
Sally Collins, Social and Ecological Assessment Pty Ltd.
Gabby Corbett, Western Australian Main Roads Department.
Tim Dendy, South Australian National Parks and Wildlife Service.
Peter Durkin, Conservation Council of Victoria.
Anne Ehrlich, Stanford University, USA.
Paul Ehrlich, Stanford University, USA.
Anthony Ferrar, CSIR, South Africa.
Pam Ferrar, Australian National University.
Gordon Friend, Western Australian Department of Conservation and Land Management.
Tony Friend, Western Australian Department of Conservation and Land Management.
Peter Goodman, Natal Parks Board, South Africa.
John Greenslade, CSIRO.
David Hamilton, Ballarat College of Advanced Education.
Joe Havel, Western Australian Department of Conservation and Land Management.
Ted Henzell, CSIRO.
Ian Herford, Western Australian Department of Conservation and Land Management.
Richard Hobbs, CSIRO.
Angas Hopkins, Western Australian Department of Conservation and Land Management.
Stephen Hopper, Western Australian Department of Conservation and Land Management.
Jocelyn Howell, New South Wales Nature Conservation Council.
Greg Keighery, Western Australian Department of Conservation and Land Management.
Jack Kinnear, Western Australian Department of Conservation and Land Management.
Jill Landsberg, Australian National University.
Brett Loney, Western Australian Main Roads Department.
Richard Loyn, Victorian Department of Conservation, Forests and Lands.
Jim Lynch, Smithsonian Environmental Research Centre, USA.
Bert Main, University of Western Australia.
Barbara York Main, University of Western Australia.
Karen Maisey, Western Australian Department of Conservation and Land Management.
Jonathan Majer, Western Australian Institute of Technology.
Chris Margules, CSIRO.
Bernie Masters, Westralian Sands Ltd.
Libby Mattiske, E. M. Mattiske and Associates.
Alex McDonald, South Australian National Parks and Wildlife Service.
Michael McGrath, Conservation Council of Western Australia.
Richard McKellar, Private Consultant.
Norman McKenzie, Western Australian Department of Conservation and Land Management.
John McLennan, DSIR, New Zealand.
Keiran McNamara, Western Australian Department of Conservation and Land Management.
Susan Moore, Western Australian Department of Conservation and Land Management.
Susan Mopper, Northern Arizona University, USA.
Gavin Moran, CSIRO.
Barry Muir, Western Australian Department of Conservation and Land Management.
Ken Newbey, Private Consultant.
Owen Nichols, ALCOA of Australia.
Colin Ogle, Department of Internal Affairs, New Zealand.
Graeme Olsen, ALCOA of Australia.

Geoff Park, DSIR, New Zealand.
Harry Recher, Australian Museum.
Viv Read, University of Western Australia.
James Ridsdill-Smith, CSIRO.
George Rothschild, CSIRO.
Ian Rowley, CSIRO.
Eleanor Russell, CSIRO.
Denis Saunders, CSIRO.
Nick Sheppard, New South Wales National Parks and Wildlife Service.
Jim Shields, Forestry Commission of New South Wales.
Dianne Simmons, Monash University.
Graeme Smith, CSIRO.
Geoff Syme, CSIRO.
Sandy Taylor, University of Adelaide.
Richard Thackway, Australian National Parks and Wildlife Service.
Susan Timmins, Department of Lands and Survey, New Zealand.
Ken Tinley, Western Australian Department of Conservation and Environment.
Michael Usher, University of York, UK.
Brian Walker, CSIRO.
Ken Wallace, Western Australian Department of Conservation and Land Management.
Craig Whisson, South Australian Department of Environment and Planning.
Gary Whisson, Western Australian Department of Conservation and Environment.
Robert Whitcomb, US Department of Agriculture.
Peter Williams, DSIR, New Zealand.
Barry Wilson, Western Australian Department of Conservation and Land Management.

INDEX

Abies lasiocarpa, Rocky Mountain Fir 115
Aborigines
    environmental modification 234
    fire 5
    fire based hunting technology 5
    interaction with landscape 197
abundance
    of birds 65-77
    regulation of by animals 5
    relative changes in 11
Acacia 10, 152, 279
    as habitat for trapdoor spiders 33, 34, 38
    A. acuminata, Jam 34, 37, 338, 339
    A. harpophylla, Brigalow, conservation in NSW 298, 300
    A. papyrocarpa, regeneration 237
    parasitism by mistletoe, *Amyema preissi* 84
Acanthiza
    A. lineata, Striated Thornbill 188, 189
    A. pusilla, Brown Thornbill 188
Acanthorhynchus tenuirostris, Eastern Spinebill 188, 189
accountability of research for management 345
Acer spp., Maple 163, 164
    A. saccharum 235
Acrobates pygmaeus, Feathertail Glider 185, 186
Acrotriche patula 92
Actaea spicata, Herb Christopher 109, 110
Adirondak Mountains, New York, USA 165
Aegiceras corniculatum, River Mangrove 228
Aepyceros melampus, Impala 202
Aepyprymnus rufescens, Rufous Bettong 42
aerial spraying of fertilizer 237
aestivation, trapdoor spiders 36, 37
Aganippe 31-34, 36, 38
age structure 168
agriculture(al) 65-77, 387
    development, WA 270, 271
Agropyron cristatum, Crested Wheatgrass 168
Aimophila aestivalis, Bachman's Sparrow 165
Aland Islands, Finland 105, 114
Albany, WA 250, 251
alleles 142, 143, 145-148
    classes of 155
    loss of 159
allelic
    evenness 154, 157-159
    frequencies 154
        shifts in 160
    richness 154-157, 159
    variants, classes of 155
Allocasuarina (see also *Casuarina*) 5
    as habitat for spiders 33, 34, 38
    A. acutivalvis 37
    A. heugeliana 307
    A. humilis 216
    A. verticillata 91, 92
alpine meadows 387
altitude(inal)
    and bird communities 66-68, 70, 71, 73, 76, 77
    range, forest patches 80, 82, 83
Amazon Basin 201, 387
American Avocet, *Recurvirostra americana* 173

American Chestnut, *Castanea dentata* 163
American Heath Hen, *Tympanuchus cupido* 293
American Museum of Natural History 279
American Redstart, *Setophaga ruticilla* 166
Ammodramus
    A. bairdii, Baird's Sparrow 173
    A. henslowii, Henslow's Sparrow 173
    A. savannarum, Grasshopper Sparrow 165
Amyema
    A. miquelii 84
    A. preissii 84
Amytornis textilis, Thick-billed Grass-wren 271
Anagallis arvensis 92
Aname 32, 34
    A. armigera 32, 34
Andropogon
    A. gerardi 169
    A. virginicus, Broom Sedge 165
Anidiops villosus 29-38
animal populations
    and landscape fragmentation 236
    and vegetation structure and composition 236
Anolis
    A. gingivinus 209
    A. wattsi 209
Antechinus
    A. stuartii, Brown Antechinus 46, 186
    A. swainsonii, Dusky Antechinus 186
Anthochaera carunculata, Red Wattlebird 185
Anthus novaeseelandiae, Richard's Pipet 275
ants, Formicidae
    and nutrient balance 351, 352
    as prey of trapdoor spiders 38
    and redistribution of resources 5
    indicators of nutrient cycling 374
    indicators of processes 374
    suggested for study of guilds of species 375
Aphelocephala leucopsis 271
Appalachian Mountains, USA 105, 163, 165
aquifer recharge, Barrow Island, WA 281
Aquila audax, Wedge-tailed Eagle 202
Arachnida, spiders, suggested for study of guilds of species 375
Arbanitis 32, 34
arboreal marsupials 181, 182, 185, 186, 190
Arcadian Flycatcher, *Empidonax virescens* 129, 166
Archaeidae, relics of Tertiary environment 30
Argentine Ant, *Iridomyrmex humilis* 236
aridity gradients 171
Arum maculatum, Cuckoo-pint 111
Asplenium viride, Green Spleenwort 111
assessment of conservation value
    diversity 317, 318
    natural beauty and/or interest 316
    productive/amenity value 318
    rarity 316-318
    representativeness typicality 317, 318
    threat 316
Astrebla spp. grassland, conservation of remnants in NSW 299, 301
Athysanella spp. 171, 172
atlas, breeding birds, Maryland 166
Atrichornis clamosus, Noisy Scrub-bird 269-276, 373

Austral Bracken Fern, *Pteridium esculentum* 66, 77
Australasian Bittern, *Botaurus stellaris* 79, 83-85
Australasian Gannet, *Sula bassana* 80
Australia 234-236
Australian Magpie, *Gymnorhina tibicen* 70
Australian National Parks and Wildlife Service 333
Australian Raven, *Corvus coronoides* 185
autecology 373-375
    contribution to mammal conservation 50, 51
    integration with biogeographic studies 80
    Long-nosed Potoroo, *Potorous tridactylus* 41, 43, 46-50
authorities, statutory, and management of remnants 7
avian community 185
Avicennia marina, Grey Mangrove 228

Bachman's Sparrow, *Aimophila aestivalis* 165
Baird's Sparrow, *Ammodramus bairdii* 173
bandicoots and soil turnover 351
Banksia spp. 72, 77
    B. ericifolia 235
    food for Carnaby's cockatoo 251
    in mixed woodlands 22
Barro Colorado Island, Panama 118
Barrow Island, WA 17
    management 279-85
Bartramia longicauda, Upland Sandpiper 165
base-line data, invertebrates 334
Baudin's Cockatoo, *Calyptorhynchus baudinii* 24, 251
Bay Checkerspot, *Euphydryas editha* 202-204, 207-209
Beautiful Firetail, *Emblema bella* 72
Bell Miner, *Manorina melanophrys* 72, 74
    competitor with Helmeted Honeyeater 289, 292
    "farming" psyllids 292
Bell's Vireo, *Vireo bellii* 165
Beltsville Agricultual Research Centre, Maryland 166, 172
Bettongia
    B. gaimardi, Tasmanian Bettong 42
    B. lesueur, Boodie 6
    B. penicillata, Woylie 42, 217, 323, 327, 328
    recolonization 324
Betula spp. birch 163, 164
    regeneration 236
Bewick's Wren, *Thryomanes bewickii* 165
Big Bend National Park, Texas 167
Big Bluestem, *Andiopogon gerardi* 169
Bilby, *Macrotis lagotis* 6
bio-indicators 353
    invertebrates 335, 353
bioclimatic zones, south-west WA 271
biogeographic theory, application in New Zealand 80-85
bioindicators, invertebrates 30
biological diversity 353
biota 387
    conservation of 388, 389
Birch woods in Scotland 237
Birch, *Betula* spp. 163, 164
Bird Observers Club 290

bird(s) 227, 230-232
   abundance 65-77
   and nutrient balance 351, 352
   and soil turnover 351
   community 185
   dynamics 341-343
      corridors 341-343
      Foster Road Reserve, WA 341-343
      in senescent low woodland 341-343
      Ongerup, WA 341-343
      road reserves 341-343
   forest 183-185
   generalized species 183, 184
   open country 183-185
   population 65-77
   survey and census procedures 180, 181
Bitter Lake National Wildlife Refuge, New Mexico 167, 173
Black Cockatoos, *Calyptorhynchus* spp. (see under Cockatoos)
Black Stilt, *Himantopus novaezelandiae* 80
Black-and-white Warbler, *Mniotilta varia* 166
Black-faced Cuckoo-shrike, *Coracina novaehollandiae* 188
bladderwort, *Utricularia* spp. 85
Blue Grama, *Bouteloua gracilis* 168-170
Blue Grosbeak, *Guiraca caerulea* 165
Blue-winged Warbler, *Vermivora pinus* 165
Bombala, NSW 167, 180, 190
Bondi State Forest, Bombala NSW 180, 190
Boodie, *Bettongia lesueur*, and disturbed soil 6
Borya nitida, pincushion plants 32
Botaurus stellaris, Australasian Bittern 79, 83-85
Bouteloua gracilis, Blue Grama 168-170
Bowdleria punctata veleae, Fernbird 80
brackish swamp 229, 230
breeding
   bird atlas, Maryland 166
   season of Carnaby's Cockatoo 252-256
   sites, required by Carnaby's Cockatoo 255, 257
   success
      Carnaby's Cockatoo 249, 252, 254-257
      Helmeted Honeyeater 292
Brigalow, *Acacia harpophylla* 298, 300
Bristlebirds, *Dasyornis*
   Rufous, *D. broadbenti* 271, 276
   Western, *D. longirostris* 271-276
Bronzewing Pigeon, *Phaps chalcoptera* 307, 310
brood parasitism 215-217
Broom Sedge, *Andropogon virginicus* 165
Brown Antechinus, *Antechinus stuartii* 186
Brown Goshawk, *Accipiter fasciatus* 70
Brown Mallet, *Eucalyptus astringens* 305, 307, 310
Brown Quail, *Coturnix australis* 71
Brown Thornbill, *Acanthiza pusilla* 188, 189
Brown Treecreeper, *Climacteris picumnus* 72
Brown-headed Cowbird, *Molothrus ater* 135, 172
Brush Cuckoo, *Cuculus variolosus* 189
Brush Wallaby, *Macropus irma* 327
Brush-tailed Phascogale, *Phascogale tapoatafa* 327
Brushtail Possum, *Trichosurus vulpecula* 84, 185, 186
Buchloe dactyloides, Buffalograss 168
Buffalograss, *Buchloe dactyloides* 168
Bunbury, WA 250, 251
Buprestidae, Jewell Beetles 333

Burgan, *Leptospermum phylicoides* 66, 77
burn, burning (see also fire)
   a procedure for hazard reduction 7
   consequences of frequent 7
   fuel reduction 326
   persistence of fauna 29
   rotation 324
   time since 30
burrowing 234
Bush Rat, *Rattus fuscipes* 186
But-But, *Eucalyptus bridgesiana* 77
butterflies, Lepidoptera 113-114, 202-209

Cabin John Island, Washington, D.C. 167
Cacatua
   C. galerita, Sulphur-crested Cockatoo 24
   C. leadbeateri, Major Mitchell's Cockatoo 24
   C. pastinator gymnopis, Little Corella 24
   C. pastinator pastinator, Long-billed Corella 272
   C. roseicapilla, Galah 236, 256, 269, 272
   diet 23, 24
cacti and cattle grazing in south west USA 237
Cadellia pentastylis, Ooline 301
Calamanthus, *Sericornis fuliginosus* 275
calcium carbonate
   amount in Eyre Peninsula soils 90
   classification of layers 91, 96
calcrete 90, 91
California 235
Callaeas cinerea, Kokako 79-82, 85
Callitris 32
Callocephalon fimbriatum, Gang-gang Cockatoo 185
Calluna vulgaris, Heather 243, 244, 246
Calothamnus quadrifidus 216
Calpatanna Waterholes Conservation Park 91, 98, 100
Calyptorhynchus spp., Black Cockatoos (see under Cockatoos)
Canadian Forest 163
canopy continuity 163
Cape Kidnappers, NZ, gannet colony 80
Cardinia Creek, Vic. 288, 289, 292
Carnaby's Cockatoo, *Calyptorhynchus funereus latirostris* 19, 21-24, 138, 236, 249-255, 257, 389
carrying capacity 54, 59, 61, 62
Carya spp., Hickory 163, 164
Castanea dentata, American Chestnut 163
Casuarina (see also *Allocasuarina*)
   C. cunninghamiana, distribution 157
   C. huegeliana 84
   litter as habitat for spiders 29
Casuarinaceae (see also *Allocasuarina*, *Casuarina*) and symbionts 5
Cat, Feral, *Felis catus*, removal from remnants 374
catastrophic loss, multiple reserves 178
central place theory 179, 185, 186, 189-191
Cercartetus nanus, Pygmy Possum 185, 186
cereal cropping areas 4
Chafer Beetle, *Prodontia lewisi* 80
changes
   following perturbation 11
   interpretation of 11
characteristic species 119
chemicals, Barrow Island, WA 281
Chenistonia tepperi 32, 34
Chestnut-rumped Hylacola, *Sericornis pyrrhopygius* 72
Chestnut-sided Warbler, *Dendroica pensylvanica* 164
China, People's Republic 205

Chihuahuan Desert, USA 164, 170, 171, 173
Chlorotettix spp. 171
Chrysococcyx basalis, Horsfield's Bronze-cuckoo 216
Chuditch, *Dasyurus geoffroii* 327
Cinclosoma punctatum, Spotted Quail-thrush 72
Cirsium helenioides, Melancholy Thistle 109, 110
Clackline Nature Reserve, WA 339
   monitoring sites 338
clearing
   for agriculture 249, 251, 254-256
   of forests 65-77
   of natural vegetation 235
Clematis vitalba, Old Man's Beard 242-246
Climacteris leucophaea, White-throated Treecreeper 188, 189
climate(ic) 2, 273, 274
   change 230-232
      as determinant of ecosystem properties 164
   gradient 173, 174
clubmoss, *Lycopodium serpentinum* 85
clutch size 167
Coast Range Road, Vic. 180
coastal
   plain, USA 163, 166
   sandplain, Vic. 66
Cockatoo, Vic. 288, 289
cockatoos
   changes induced by clearing 5
   corridors of native vegetation 138
   diet 21-24
   feeding along roads 180
   hollows suitable for 191
cockatoos, *Cacatua*
   Galah, *C. roseicapilla* 236, 256, 269, 272
      diet 23, 24
   Little Corella, *C. pastinator gymnopis* 24
   Long-billed Corella, *C. pastinator pastinator* 272
   Major Mitchell's, *C. leadbeateri* 24, 272
   Sulphur-crested, *C. galerita* 24
cockatoos, *Calyptorhynchus*
   Baudin's, *C. baudinii* 24, 251
   Carnaby's, *C. funereus latirostris* 19, 21-24, 138, 236, 249-255, 257, 272, 389
   Inland Red-tailed Black, *C. magnificus samueli* 24, 272
   Forest Red-tailed Black, *C. magnificus naso* 24
cold fronts 2, 3
colonial nesters 189
colonization 113-114
Colorado, USA 115
Colluricincla harmonica, Grey Shrike-thrush 185, 188, 189
Common Brushtail, *Trichosurus vulpecula* 327
   population recovery 54
Common Heath, *Epacris impressa* 77
Common Ringtail, *Pseudocheirus peregrinus* 327
Common Starling, *Sturnus vulgaris* 71
Common Yellowthroat, *Geothlypis trichas* 130, 164
common, commoness as opposed to rare, rareness 34
communication
   between researchers and managers 345, 346
   improvement of 391
   in management of conservation areas 371
community(ies)
   attitudes to conservation 237

community(ies) — cont'd
  ecology 373-375
  expectations 381
  spiders in 29
compensation in lieu of development rights 388
Congo basin 201
conifers of north America 157
Connecticut, USA 166
connectivity 389
  definitions 196
conservation 72, 74, 387-392
  agencies 333, 334
  birds 273
  contribution of autecology 50, 51
  contribution of insular biogeography 50
  ethic 349
  forest birds 171-172
  genetics 142, 143, 146
  grasslands and associated insects 172-174
  in situ 152, 157, 159, 160
  in situ strategies 152, 155, 160
  insects 172, 173
  invertebrates 333, 334
  of biological communities 151
  of *Eucalyptus* 151-161
  of gene pools 160
  of genetic resources 151-161
  of remnant vegetation 151-161
  purpose 367
  reserve system
    as a resource to be managed 385
    vegetation in NSW 298
  stratagy, introductions and translocations 366
  value
    definition 89, 96, 100
    evaluation 91, 96, 100, 102
    native vegetation patches 79-85
constraints, research 347-349
continuous sampling 353
continuum of research 348
control, in management context 38
controls
  as natural disturbance regime 369
  for monitoring of management 369
Cook Strait, NZ, barrier to species' dispersal 80
Coolangubra State Forest, NSW 180, 185
Coomallo Creek, WA 249-257
coordination of action 349
coppice woods 387
Coprosma 85
Coracina
  C. novaehollandiae, Black-faced Cuckoo-shrike 188
  C. papuensis, White-bellied Cuckoo-shrike 72
Corcovax melanorhamphos, White-winged Chough 72
corridors 165, 178, 188, 190
  advantages 357
  between habitat fragments 114
  bird dynamics 341-343
  definition 357
  design features 358
  disadvantages 357
  habitat value 357
  in landscape
    function and form 195-198
    management 198
  of native vegetation 249, 257, 388
  pest plants and animals 357
  research requirements 359
  roadside vegetation 358
  role of 388, 389

corridors — cont'd
  shade and shelter for stock 357
  stepping stones 318, 319
  width 358
  wildfire spread 357
  windbreaks 357
Corvus
  C. coronoides, Australian Raven 185
  C. orru, Torresian Crow 271
cossid moths, and nutrient cycling 352
cost of travel, foraging animals 179, 190
Costa Rica 205
Coturnix pectoralis, Stubble Quail 310
coupe size, logging unit 177, 190
creek reserves 178, 179, 181, 185
Cresent Lakes National Wildlife Refuge, Nebraska 167, 173
Crested Shrike-tit, *Falcunculus frontatus* 185
Crested Wheatgrass, *Agropyron cristatum* 168
Crimson Rosella, *Platycercus elegans* 185
criteria for judging conservation value
  diversity 89
  representativeness 89, 90, 98, 100
criterion, 1%, 104
Cromwell, NZ, chafer beetle reserve at 80
cryptic annual species, corridors for wildlife 358
Cuculus
  C. pyrrhophanus, Fan-tailed Cuckoo 188, 189
  C. variolosus, Brush Cuckoo 189
Cumbria, UK 108
Custer State Park, South Dakota 173
cutting cycle, Eden NSW 177
cyanobacteria
  binding soil surface 5
  nitrogen fixation 5, 351
Cyclodina spp. 79, 84, 85
  predation by rats 84, 85
cyclones, tropical 2, 3

Dacelo novaeguineae (= D. gigas), Laughing Kookaburra 71, 185
Dalgyte, *Macrotis lagotis* 6
Dandenong Ranges, Vic. 287, 288, 292, 293
Daphoenositta chrysoptera, Orange-winged Sitella 189
Dasyornis
  D. broadbenti, Rufous Bristlebird 271, 276
  D. longirostris, Western Bristlebird, 271-276
Dasyurus geoffroii, Chuditch 327
data
  analysis 367
  base 367, 368
  mapping need 349
decline
  faunal, local 53, 54, 57-59, 62, 63
  Helmeted Honeyeater 287-292
defoliation, by insects 71, 74
deforestation 165, 167
degradation 65-77
  of native vegetation 256, 257
demographic stochasticity 148
Dendrocopus
  D. villosus, Hairy Woodpecker 166
  D. medius, Middle-spotted Woodpecker 293
Dendroica
  D. discolour, Prairie Warbler 165
  D. kirtlandii, Kirtland's Warbler 172, 202
  D. pensylvanica, Chestnut-sided Warbler 164
  D. petechia, Yellow Warbler 165

density, of wrens 214, 215
Department of Conservation and Land Management, WA 305
desert plains 168
desertification 170, 171
design of nature reserves 141-143, 145, 148
diapause 170
Dicaeum hirundinaceum, Mistletoe-bird 189
Dickcissel, *Spiza americana* 165
dieback 190
  of Eucalypts, due to insect attack 65, 68, 71, 74
Diomedea epomophora, Royal Albatross 80
Diptera, flies, suggested for study of guilds of species 375
dispersal 167, 217
  barriers to 80, 84, 85
  of juvenile trapdoor spiders 30, 32, 34-37
  of young numbats 58-62
  mechanisms, weeds in NZ reserves 243-245
  natural 53, 54, 58-62
  rates in land snails 84
distance of isolation 67, 70, 76, 77
distribution
  birds 271, 272, 276
  Helmeted Honeyeater 288
  of alleles 155, 157
disturbance 233, 234, 237, 319
  and bird communities 67, 68, 76, 77
  bird habitat 273, 274, 276
  human induced 233, 234
  natural 163, 165
  of soil 233, 234
  regime 233, 238
disturbed areas
  edges of tracks 38
  gravel pits 38
diversity 317, 318
  and roles in ecosystem 12
  biological 353, 354
  functional role 352
  invertebrates 353
  of spiders 29, 30
dominance hierarchy 171
Dorset, UK 105-106, 119
Douglas Fir, *Pseudotsuga menziesii* 191
Drepanotermes
  D. niger, prey of trapdoor spiders 38
  D. tamminensis, prey of trapdoor spiders 38
Dromaius novaehollandiae, Emu 17
  dispersal of *Santalum* spp. 6
drought 35, 38, 170, 174, 179, 180, 191, 234
  effects on mammal populations 54
  frequency 3, 6, 7
Dryandra Forest, WA 54, 58, 59, 63, 84, 305, 306, 339
  monitoring sites 338
Dryandra spp. 133, 216
  food for Carnaby's Cockatoo 251
  D. nobilis 307, 308
Dryocapus pileatus, Pileated Woodpecker 166
Dumetella carolinensis, Gray Catbird 164
Dunedin, NZ, Royal Albatross colony at 80
Dusky Antechinus, *Antechinus swainsonii* 186
dust, Barrow Island 281
dynamic equilibrium
  model 141-143, 147
  of landscapes 235

East Boyd State Forest, Eden NSW 180
eastern forest, USA 163-169

Eastern Rosella, *Platycercus eximius* 71
Eastern Spinebill, *Acanthorhynchus tenuirostris* 188, 189
Ecdeicolea monostachya 33, 34, 37
Echidna, *Tachyglossus aculeatus* 7
   and soil turnover 351
   interaction with trapdoor spiders 36, 38
ecocline 363
   definition 198
ecological
   isolation 317, 320
   need for management 15, 17, 18
   survey 104, 119, 354
   tithing by scientists 392
ecology index 310, 311
economic incentives 379-380
ecosystem(s) 351, 387-391
   function and resilience 10
   changes
      management induced 369
      measure of direction and rate 369
      monitoring of 369
      natural 369
      re-establishment of 257
      research on nature and dynamics of 369
      spiders as part of 29
   models 355
ecotone 363
   creation 383
   definition 383
   modification 383
edaphic and geomorphic influences 348
Eden, NSW 177, 180
   national parks and reserves at 178, 191
edge
   effects 134, 190, 318, 319, 363, 390
   effects on birds 134
   of pasture and forest 67, 70, 71, 76, 77
edge to area ratio 206, 232
Edgewood Park, California, USA 204
education 379-380, 381
   Barrow Island workforce and visitors 280, 283
   role of planning in 266-268
educational process 386
effective population size 57, 142
   fluctuations in 159
efficiency of research for management 345
elephants in African savannas 236
elevation 163, 170, 173, 174
Emex australis, as food for cockatoos 23
Empidonax virescens, Arcadian Flycatcher 166
Emu, *Dromaius novaehollandiae* 17, 72, 73
   dispersal of *Santalum* spp. 6
Enchylaena tomentosa 92
endangered species 53
endemic species 104
endemism 171, 174
   New Zealand 79-81, 84, 85
Eneabba, WA 19-21
Enhydra lutris, Sea Otter, translocation 62
environmental
   changes, birds 271
   factors 2, 3
   gradients 363
   planning in NSW 295, 300-303
   protection zones 297, 298, 300, 302
Environmental Planning and Assessment Act, NSW 296, 300-304
Eopsaltria australis, Yellow Robin 188, 189
equilibrium theory 107
   of island biogeography 73
Equus przewalskii, Przewalski's Wild Horse 115

erosion 65, 74
establishment 113-114
Eucalyptus 151-161
   as habitat for trapdoor spiders 34
   E. accedens, Powderbark Wandoo, in mixed woodlands 22
   E. annulata low woodland, bird dynamics 341-343
   E. annulata, feeding by honeyeaters 341, 343
   E. astringens, Brown Mallet 305, 307, 310
   E. caesia 151-153, 155, 157-161
   E. calophylla, Marri 211, 271, 305
      as food for cockatoos 24
      food for Carnaby's Cockatoo 255
   E. cloeziana 157
   E. cypellocarpa, Mountain Grey Gum 77
   E. delagatensis 153, 157, 160
   E. diversicolour, Karri 271
   E. diversifolia 91, 92
   E. dumosa 91, 92
   E. doliqua 42
   E. gardneri
      low woodland, bird dynamics 341-343
      feeding by honeyeaters 341, 343
   E. gracilis 91, 92
   E. grandis 157
   E. johnsoniana 153-154, 157-160
   E. lateritia 152-154, 157-160
   E. loxophleba, York Gum 34, 37, 291, 338, 339
   E. marginata, Jarrah 153, 160, 206, 271, 305
   E. oleosa 91, 92
   E. ovata, Swamp Gum 42, 287, 290, 292
   E. pendens 154-157, 159-161
   E. platypus
      low woodland, bird dynamics 341-343
      feeding by honeyeaters 341, 343
   E. radiata, Narrow-leaf Peppermint 66, 77
   E. saligna 157
   E. salmonophloia, Salmon Gum 271, 272, 388
   E. socialis 92
   E. todtiana, Pricklybark, in mixed woodlands 22
   E. viminalis, Manna Gum 42, 287, 290
   E. wandoo, Wandoo 21, 22, 84, 211, 271, 305
   forest 235
   localized 152-160
   monitoring sites established 338
   nesting hollows for Carnaby's Cockatoo 255
   of Western Australia 151, 152-154, 160
   preferential clearing in WA 238
   regional distribution 152-155, 157, 159, 161
   subpopulation structure of 159-160
   widespread 151-154, 157, 161
Euphydryas editha, Bay Checkerspot 202-204, 207-209
European
   colonization, disruption of disturbance regimes 234
   settlement 71
Eutaxia microphylla 92
evolution 115-116
exotic invasion 318-320
experimental management 25, 26, 346, 386
expert systems 355, 382
Explorers Club, New York 280
exponential model, of bird species/area relationships 69, 70, 73

extinction(s) 141-145, 207, 237, 269-276
   coefficient 143-145
   local 53, 54, 63
      of Carnaby's Cockatoo 251, 253, 257
   of large marsupials 234
   rate 116-118
extirpation 167, 170, 172, 174
exudates 71

Fabaceae 5, 10
   and symbionts 5
faecal accumulation rate
   variation with vegetation 221, 222
   Western Grey Kangaroo 221-223
   pellet number, relationship with pellet weight 224
   pellet weight, variation with season 223
      Western Grey Kangaroo 222, 223
Fairfields farm, Tammin, WA 32, 34, 37
Falcunculus frontatus, Crested Shrike-Tit 185, 188, 189
Fan-tailed Cuckoo, *Cuculus pyrrhophanus* 188, 189
Fantail, *Rhipidura fuliginosa* 81
farm dams 66, 67, 77
farmland 65-77
   regenerating 77
   birds 65-77
Fauna Management Priority Areas (MPAs), WA 232, 324
fauna(1)
   assemblage, Barrow Island, WA 280-283
   collapse models 141-145, 148
   conservation legislation, invertebrates 333
   decline, local 53, 54, 57-59, 62, 63
Feathertail Glider, *Acrobates pygmaeus* 185, 186
feeding habitat of Carnaby's Cockatoo 249
Felis catus, Feral Cat 374, 388
fencing 67, 74
fens of eastern England 237
feral vertebrates, Yellingbo State Nature Reserve 291
Fernbird, *Bowdleria punctata vealeae* 80
Feral Cat, *Felis catus* 374
fertilizers, as perturbation 6
fescue grassland, Montana 169
Festuca ovina, Sheep's Fescue 114
Ficus 279
Field Sparrow, *Spizella pusilla* 165
finances 385
fire (also see burn, burning) 2, 6, 7, 16, 20, 23, 38, 164, 165, 233-236
   accidental 235
   adaptive traits 307
   and bird communities 67, 70, 73, 76, 77
   and kangaroo grazing 236
   and vegetation response 234
   as a management tool 323
   Barrow Island 282
   climate 306
   controlled burning 235
   effects on mammal populations 54
   environment 306
   frequency 3, 214, 215, 218, 235, 273-285
   hazard 7
      reserves perceived as 7
   history 306, 309
   loss of Helmeted Honeyeater 289, 291, 293
   management 305, 306, 311
      guidelines 11
      in remnants 237
   Maori-induced, in New Zealand 79
   natural catastrophe 38

fire — cont'd
  persistence of fauna 29
  prevention 308
  protection 67
  regime 190, 305-307, 309-311, 326, 327
  suppression 235
  temperature 216
firebreaks 308
firewood 74
first principles
  preventative conservation by application 349
  use and application 349, 350
Fisheries and Wildlife Department, Vic. 289-291
Flame Robin, *Petroica phoenicea* 72, 185
fledgling weight, Carnaby's Cockatoo 252
Flexamia spp. 171, 172
flies, Diptera, suggested for study of guilds of species 375
flood 6, 233-235
folded left wing, length in Carnaby's Cockatoo 252, 257
food
  Helmeted Honeyeater 287-292
  required by Carnaby's Cockatoo 255, 257
foraging
  animals, cost of travel 179
  strategies, trapdoor spiders 32, 34-37
forbs 170
forest(s) 271-273
  birds 183-185
    abundance and numbers of species 65-77
    insectivorous 68
  dependent fauna 178
  dry sclerophyll 66
  eucalypt 65-77
  fragmentation
    at Naringal, Vic. 42
    composition of mammal assemblages 45, 46, 49
    mammal species richness 41, 44, 45, 48, 49
    responses of birds 123-140
  interior bird species 166
  islands, Eden NSW 179
  loss in New Zealand 79, 80, 84, 85
  mature 164, 166
  mixed-species 66
  mosaic 178
  patches
    Britain 133, 134
    New Zealand 134
  riparian 185
  successional 164, 165
  time required to regenerate 191
    species in different sized patches 81, 82
  wet 71, 72
  wildlife, management at Eden NSW 190
Forest Red Gum, *Eucalyptus tereticornis* 72, 77
Forest Red-tailed Black Cockatoo, *Calyptorhynchus magnificus naso* 24
Forest Wire-grass, *Tetrarrhena juncea* 77
forestry 382
Formicidae, ants
  and redistribution of resources 5
Foster Road Reserve, WA, bird dynamics 341-343
founder effect 55, 56
fox(es) 37, 54, 327, 328, 374
fragmentation 65-77, 166, 318, 319
  at Naringal, Victoria 42
  cause of food shortage for cockatoos 249, 250, 257
  composition of mammal assemblages 45, 46, 49

fragmentation — cont'd
  dividing extinction risk 361
  effect on biological diversity 361
  forests 65-77
  mammal species richness 41, 44, 45, 48, 49
  natural patchiness 361
  of habitats 103-121
  of native vegetation 233, 235, 238, 387-390
  patterns of 17
  research priorities 361
  responses of birds 123-140
fresh initiatives 344
freshwater birds, species in different sized patches 83, 84
frogs and reptiles 187, 190
frugivorous species and plant regeneration 236
fuel
  accumulation 307-310
  reduction burning 326
  sampling 306, 307
funding problems 349

Gahnia lanigera 92
Galah, *Cacatua roseicapilla* 236, 256, 257, 269, 272
  diet 24
  dietary preference 23
game reserves in Africa 118
Gang-gang Cockatoo, *Callocephalon fimbriatum* 185
gap-phase species 235
Gastrolobium spp. 307, 310, 311
  G. bilobum, Heartleaf 326, 327
  G. microcarpum 307
Gazella dorcas, Dorcas Gazelle 115
gene
  flow 58, 148
    of eucalypts 157, 159
  pools, manipulation of 53, 62
generalized
  linear models 92
  species, birds 183, 184
genetic(s)
  bottlenecks 56, 57, 62
  distance 160
  diversity
    allelic
      evenness 154, 157-159
      richness 154-157, 159
    distribution within populations 157-159
    effects of
      gene flow 159
      genetic drift 159
      selection on 159
    geographic distribution of 151
    in plant species 152
    in small populations 159
    measurement of 154
    migration effects on 159
    patterns of 152, 159
    relationship with population size 159
  drift 56, 57, 62, 63, 142, 143, 145, 207
  of populations
    bottleneck (founder) effects 115-116
    effective population size 115, 116
    genetic drift 115, 116
    heterozygosity 115
    inbreeding 115
    mutation rate 115, 116
  management 142
  of isolated populations 57, 63
  resources
    assessment of 154, 160
    conservation of 151-161

genetic(s) — cont'd
  maintenance of 151, 159, 160
  retention 192
  within and between populations 151, 161
  structure
    and gene flow 157, 159
    and migration 159
    of populations 151, 152, 154
    of subpopulations 159-160
    spatial array of individuals 159, 160
  theory of small populations 390
  variability 142, 143
  variation
    assays of 152
    loss of 55-58, 63
    patterns of 152
genotypes
  identical 160
geographic(al)
  population structure 152
  range
    of eucalypts 152-154, 161
    outliers 160
  races 157
geological history 170, 171
geomorphic and edaphic influences 348
Geothlypis trichas, Common Yellowthroat 164
Geraldton, WA 250
Germanium robertianum, Herb Robert 111
Gerygone igata, Grey Warbler 81
Gilbert's Whistler, *Pachycephala inornata* 271
Gippsland Yellow-tufted Honeyeater, *Lichenostomus melanops gippslandica* 288
Gippsland, Vic. 288
Gladiolus aureus, Golden Gladiolus 202
GLIM 92
Glossopsitta porphyrocephala, Purple-crowned Lorikeet 341, 343
Golden Gladiolus, *Gladiolus aureus* 202
Gombe Stream Reserve, Tanzania 205
Gophers, *Thomomys* spp. 207
gradients, ecological 118
grasses
  introduced 66, 77
  native 77
Grasshopper Sparrow, *Ammodramus savannarum* 165
grasshoppers and elimination of seedlings 7
grassland(s)
  biome 234
  north America 168-174
gravel pit, establishment of ants 38
Gray Catbird, *Dumatella carolinensis* 130, 164
grazing 43, 65-77, 168, 174, 233, 234, 236, 387, 388
  and cactus establishment 237
Great Basin, USA 170
Great Smoky National Park, Tennessee 167, 172
Greater Glider, *Petauroides volans* 185, 186
Greater Prairiechicken, *Tympanuchus cupida* 173
greenhouse effect and nature reserves 2
Grevillea
  G. bipinnatafida 216
  G. paradoxa as trapdoor spider habitat 38
  G. spp., food for Carnaby's Cockatoo 251
Grey Butcherbird, *Cracticus torquatus* 71
Grey Fantail, *Rhipidura fuliginosa* 185, 188, 189

Grey Kangaroo, *Macropus fuliginosus* 373
Grey Shrike-thrush, *Colluricincla harmonica* 185, 188, 189
Grey Warbler, *Gerygone igata* 81
Grizzly Bear, *Ursus horribilis* 148
Ground Parrot, *Pezoporus wallicus* 271
ground water table 4
growth rate, weeds in NZ reserves 243
guilds 170
 study of 374
Guiraca caerulea, Blue Grosbeak 165
gullies 66, 72, 77
 heathy 72
gymnosperms 155, 157

habitat (see also microhabitat) 32, 34, 37
 birds 65-77, 271-273, 275, 276
 change(s)
  birds 275
  climatic 230-232
  man-made 230
  south-west WA 270-272
 corridors for wildlife 357
 diversity 70
 destruction 256
  bird 269, 271, 272, 276
 discrimination 165
 dividing extinction risk 361
 dynamics, birds 274-276
 effect on biological diversity 361
 fragmentation 53, 54
  cause of food shortage for cockatoos 249, 250, 257
 Helmeted Honeyeater 287-292
 heterogeniety, birds 275
 indices 66-74
 islands 73, 235, 249
 isolates 251
  management, birds 272-276
  mimic natural patchiness 361
 patches 89
  conservation evaluation 90, 96-98, 100, 102
  mallee 91, 102
 recolonization, birds 274
 research priorities 361
 size, persistence of trapdoor spiders 37, 38
 specificity 166
 trees 191
Hairy Woodpecker, *Dendrocapus villosus* 166
Hakea spp. 216
 as habitat component for trapdoor spider 38
 food for Carnaby's cockatoo 251
Hart's-tongue Fern, *Phyllitis scolopendrium* 109, 111
Heartleaf, *Gastrolobium bilobum* 326, 327
heath (see also heathland) 249, 254-257, 271-276
 Scottish 236, 237
Heather, *Calluna vulgaris* 243, 244, 246
heathland (see also heath) 211, 216, 217, 218, 387, 391
Hebe 85
hedgerows 195-198, 387
Heitman's Scrub, Tammin, WA 30
Heliconius ethilla, longwing butterfly 205
Helmeted Honeyeater, *Lichenostomus melanops cassidix* 287-293
 breeding success 292
 competition from Bell Miner 289, 292
 decline 287-289, 292
 distribution 288
 food and feeding 287-292

Helmeted Honeyeater — cont'd
 habitat 287, 289-290, 292
 loss to bushfires 289, 291, 293
 movements and territoriality 287, 292
 publicity 289
 research 291
 surveys 288
 Victoria's State avifaunal emblem 289
Helmitheros vermivorus, Worm-eating Warbler 166
Hemlock, *Tsuga* spp. 163
Henslow's Sparrow, *Ammodromus henslowii* 173
hepialid moths, and nutrient balances 357
herbicides, as perturbation 6
heterozygosity 142, 143, 145-147
 expected panmictic 154, 157, 159, 160
 observed 154, 159, 160
Hickory, *Carya* spp. 163, 164
hierarchy of salience 348
Himantopus novaezeliandiae, Black Stilt 80
hole-nesting birds 71, 272
hollows 185, 191
 in trees 251-253, 257
 suitable for arboreal marsupials 191
 suitable for cockatoos, owls 191
holocene faunal changes and extinctions 5
home range 166
 Potorous tridactylus 47, 49
honeyeaters 287-293
 changes related to eucalyptus flowering 341, 343
 feeding on *Eucalyptus annulata* 341, 343
 feeding on *Eucalyptus gardneri* 341, 343
 feeding on *Eucalyptus platypus* 341, 343
Hooded Warbler, *Wilsonia citrina* 166
Horsfield's Bronze-cuckoo, *Chrysococcyx basalis* 216
House Wren, *Troglodytes aedon* 130
human use of reserves 273
human-dominated land areas
 characteristics of remnants 318
 definition 314
Hunter River, NSW 227, 230
hurricane 233, 234
hyaenas 202
hydorologic(al)
 ecosystem unit 348
 regime 3
Hymenoptera, ants 114
hypogean fungi 324

Icteria
 I. virens, Yellow-breasted Chat 165
 I. galbula, Northern Oriole 164
ideal versus attainable reserves 349
Idiosoma nigrum 32, 34, 37
Illinois, USA 105
immigration 113-114, 141, 142
 model 53, 59-62
Impala, *Aepyceros melampus* 202
inbred populations, mixing from extreme ecotypes 366
inbreeding 207
 coefficient 57, 59
 depression 56, 57, 142, 145, 148
 plants 160
incentives
 for conservation 74
 for private conservation initiatives 385
incubation, in Carnaby's Cockatoo 252
indentification of status and trend 349
Indiangrass, *Sorghostrum mutans* 169
indicator species 152, 161, 354
Indigo Bunting, *Passerina cyanea* 130, 165

industry 65, 74
information transfer 174
Ingleborough, North Yorkshire, UK 108-113
Inland Red-tailed Black Cockatoo, *Calyptorhynchus magnificus samueli* 24
insect(s) 65-67
 conservation 172, 173
 defoliating 74
 infestation 164
 in intertidal wetlands 227, 230, 231
 natural control 74
 parasitic and predatory 74
insular biogeography
 contribution to mammal conservation 50
 of mammals at Naringal, Victoria 41-46, 48-50
integrated logging 177, 190
integration
 management and research 346
 of research, management and planning 370, 371
 survey and monitoring 378
 using "expert" systems 371
interconnecting corridors 141
 conservation strategy 366
introduced birds 71
introductions, Barrow Island, WA
 Mus musculus 281, 284
 Rattus rattus 281, 284
invasibility, of natural communities 237
invasion
 by non-native species 236, 238
 invertebrates 353
 weeds into NZ native communities 242
invasive species 353
invertebrates 71
 as bioindicators 30
 conservation legislation 333
 ecological role 334
 of intertidal wetlands 232
 persistence in small areas 29, 30
 significance in ecosystems 30
 terrestrial 353
Iridomyrmex
 I. humilis, Argentine Ant 236
 I. purpureus, prey of trapdoor spiders 38
Irwin Botanical District, WA 18-20
island(s)
 biogeography 141, 142
 biogeography theory 11, 73, 249
 distribution of trapdoor spiders 30
 forest islands, Eden, NSW 179
 of native vegetation 256
 populations, of birds 138
 species 257
isolated populations 53, 55, 58, 59, 62, 63
isolating mechanisms 170
isolation 141-144, 166, 167, 170
 native vegetation patches in New Zealand 80, 85
Isometroides vescus, scorpion, predator of trapdoor spiders 36
Isoodon obesulus, Brown or Short-nosed Bandicoot 46, 327
Isoptera, termites, and redistribution of resources 5
isozymes 151, 154, 161
 techniques 154, 161
 variation 151, 154
iteroparity/iteroparous, trapdoor spider 35, 37
IUCN 333
Izu Islands, Japan 107

INDEX

Jacksonia sericea, shelter for spiders 29
Jam, *Acacia acuminata* 338, 339
Jarrah, *Eucalyptus marginata* 206, 271, 305
Jasper Ridge Biological Reserve 202-204, 208
Juncus kraussii 228, 229
Junegrass, *Koeleria cristata* 169

Kaka, *Nestor meridionalis* 81
Kakapo, *Strigops habroptilis* 80
kangaroos 17, 219-225, 327, 373
　and grazing after fire 236
　and nutrient balances 351, 352
Karri, *Eucalyptus diversicolour* 271
Kellerberrin, WA 388
Kentucky Warbler, *Oporornis formosus* 125, 127, 130, 131, 166
Kenya 205
keystone
　mutualists 202
　species 390
King Country, NZ, kokako in 12, 32, 345
Kings Park, Perth, WA, spiders of 29
Kirtland's Warbler, *Dendroica kirtlandii* 172, 202
Koeleria cristata, Junegrass 169
Kokako, *Callaeas cinerea* 79-82, 85
Konza Prairie, Kansas 167, 169, 173
Kooragang Island, NSW 227-232
Krakatau, Indonesia 113-114
kwongan 3, 5, 18, 22
　effect of clearing 5
Kwonkan sp. 32
　K. eboracum 32

Laestrygones 29
Lampropholis guichenoti 187
Lancashire, UK 108
land
　salt affected 4, 9
　management, adjacent to remnant 6
　tenure 319, 321
land snail
　dispersal 84
　New Zealand 84
　Paryphanta busbyi 79, 84, 85
　predation by wild pig 84
landscape
　ecology 138
　　definition 195
　heritage 174
　mosaics 237, 238
　stability 237
　unit 3, 6
laterite 32, 33
latitude 163, 170, 174
Latrobe Valley, Vic. 65-77
Laughing Kookaburra, *Dacelo novaeguineae (= D. gigas)* 71, 185
Laura Bay Conservation Park 91, 100
Leaden Flycatcher, *Myiagra rubecula* 72, 188, 189
legal framework of conservation 385
legislation 333, 379
Leiopoa ocellata, Mallee Fowl, and disturbed soil 6
Leptospermum scrub in NZ 85
Lesser Prairiechicken, *Tympanachus pallidicinctus* 173
liaison between reserve workers and managers 390, 391
Lichenostomus
　L. melanops cassidix, Helmeted Honeyeater
　　breeding success 292
　　competition from Bell Miner 289, 292

Lichenostomus — cont'd
　decline 287-292
　distribution 288
　food and feeding 287-292
　habitat 287, 289, 290, 292
　loss to bushfires 289, 291, 293
　movements and territoriality 287, 292
　publicity 289
　research 291
　surveys 288
　Victoria's State faunal emblem 289
　L. melanops gippslandica, Gippsland Yellow-tufted Honeyeater 288
　L. melanops, Yellow-tufted Honeyeater 72, 288
lichens
　binding soil surface 5
　fixing nitrogen 5
life cycle, as a strategy, trapdoor spiders 30, 32, 35
life form and height, weeds in NZ reserves 242-243
life history
　of trapdoor spiders in relation strategy 166, 167, 171, 173
　traits, development of 6
life span, weeds in NZ reserves 243
lifetime distribution 143
limestone pavement 109, 111-113
Limotettix spp. 171
linear environments, wildlife conservation in 178, 179, 189
litter, as habitat of trapdoor spiders 36, 38
Little Bluestem, *Schizachyrium scoparium* 170
Little Corella, *Cacatua pastinator gymnopis* 24
livestock 236
　and tree regeneration 237
lizards
　and soil turnover 351
　in WA wheatbelt 73
local endemism, New Zealand's flora and fauna 79, 80, 84, 85
local environment plan, Moree Plains Shire, NSW 301-303
locus(i)
　number of alleles per locus 154
　polymorphic 154
log-normal distribution 143
logging
　and bird communities 67, 68, 70, 76, 77
　at Eden, NSW 177, 178
　unit-coupe size 177, 188, 189
　visual impact 177
long and short term reseach, definition 347
Long-billed Corella, *Cacatua pastinator pastinator* 23, 24, 272
　diet 24
　dietary preference 23
Long-billed Curlew, *Numenius americana* 173
Long-billed White-tailed Black Cockatoo (see Baudin's Cockatoo)
Long-nosed Potoroo, *Potorous tridactylus,* population ecology 41, 43, 46-50
longevity, of trapdoor spiders 36, 37
longitude 163, 170, 171
Loranthaceae, mistletoes 84
Louisiana Waterthrush, *Seiurus motacilla* 166
low woodland, senescent, bird dynamics 341-343
Loy Lang, Vic. 66
Lycopodium serpentinum, clubmoss 85
lycosid spiders, and soil turnover 351
Lyperobius huttoni, speargrass weevil 80

Macclesfield Creek, Vic. 288
Macropus
　M. eugenii, Tammar 323, 326, 327
　M. fuliginosus, Western Grey Kangaroo 219-225, 327, 373
　　faecal accumulation rate 221-223
　M. irma, Brush Wallaby 327
　M. rufogriseus, Red-necked Wallaby 46
　M. rufus, Red Kangaroo 17
Macrotis lagotis, Dalgyte or Bilby, and disturbed soil 6
Macrozamia reidlei 216
Maireana trichoptera 92
Major Mitchell's Cockatoo, *Cacatua leadbeateri* 24
Mallee Fowl, *Leiopoa ocellata,* and disturbed soil 6
mallees of WA 151-152, 160
Mallet, Brown, *Eucalyptus astringens* 305, 307, 310
Malurus
　M. cyaneus, Suberb Fairy-wren 185, 188, 189
　M. splendens, Splended Fairy-wren
　　egg laying period 212
　　response to fire 211, 215-217
mammals
　arboreal 73
　assemblages
　　composition in remnant forest patches 45, 46, 49
　medium sized 53, 54
　predation by Red Fox 54
　species numbers 65, 69, 71-73
　survey and census procedures 180, 181
management 6, 8, 10, 74, 355
　additional costs 3
　and communication 371
　as an experimental process 378
　birds in senescent low woodland 341
　constraints 238, 381
　corridors in landscape 198
　crisis 265
　decisions in absence of knowledge 23
　definition 16
　fire 217
　forest wildlife at Eden, NSW 190
　goals 7, 8
　integration with research 370, 371
　interpreting effects of 369, 370
　intertidal wetlands 232
　knowledge required 17
　methods 262-265
　minimum goal of 3
　models 355
　monitoring of practices 369
　objectives 262, 385, 386
　options 381
　people 263, 264
　philosophy 262-265
　planning for 389-391
　plans (see also planning) 265, 386
　　Barrow Island 280, 282, 283
　　for remnants 390, 391
　pre-emptive 265
　priorities 381
　process 262, 263
　programmes 390
　records of 369
　recreation 264
　remnant communities 383
　requirements for successful 263-265
　setting objectives and priorities 389
　strategies 321, 322, 351
　theory 385
　to promote regeneration 38, 174

management — cont'd
  use of research and technical data 263
  value of monitoring in management 337
  Yellingbo State Native Reserve 289-291, 293
Manmanning, WA 249-257
mangroves 227-232, 279
Manna Gum, *Eucalyptus viminalis* 66, 77, 287, 290
Manorina melanophrys, Bell Miner
  competitor with Helmeted Honeyeater 289, 292
  "farming psyllids" 292
Maori firing of vegetation in New Zealand 79
Maple, *Acer* spp. 163, 164
marine species 250, 256
mark-recapture
  butterflies 209
  endangered insects 209
Marri, *Eucalyptus calophylla* 271, 305
  as food for cockatoos 24
  food for Carnaby's Cockatoo 255
marsupials 236
  extinction of 236
Maryland, USA, bird communities 124-131, 136, 138
Mascarene Archipelago 142
mating systems of plants 152
matriarch, female trapdoor spiders 35, 36
maturation, of trapdoor spiders 36
mature forest, role in nature conservation, NSW 177, 190
McCown's Longspur, *Rhynchophanes mccowni* 173
measurement, of ecosystem change 369
Megaleia rufa (= *Macropus rufus*), Red Kangaroo 17
Melaleuca
  M. acuminata 92
  M. adnata 92
  M. lanceolata 91, 92
  M. uncinata 37
  M. viminea 326
Melanerpes erythrocephalus, Red-headed Woodpecker 165
Meliphaga
  M. chrysops, Yellow-faced Honeyeater 185, 188, 189
  M. leucotis, White-eared Honeyeater 185, 188, 189
Mercurialis perennis, Dog's Mercury 111
Mesomelaena tetragona 216
Messmate, *Eucalyptus obliqua* 77
methodology, for monitoring 370, 371
methods and techniques 349
microclimate 191
microhabitat 207
  as an environmental factor 30
  of trapdoor spiders 34, 36-38
Micropholcomma (Micropholcommatidae) 29
Microphyll Vine, *Planchonella cotonifolia* var *pubescens* 298, 301
Middle-spotted Woodpecker, *Dendrocopos medius* 293
migrant birds 72
migration 145-148, 159
  strategy 166
Mimetus, Mimetidae 29
Mimosaceae 5
minimum
  area
    for landscape stability 238
    for population 116-118
    foraging distance 190, 191
    viable population 57, 59, 62, 116-118

Mirounga angustirostris, Northern Elephant Seal 115, 116
Missulena hoggi 32, 34, 35
mistletoe, Loranthaceae 77, 79, 84, 85
  possum browsing in New Zealand 79, 84
Mistletoe-bird, *Dicaeum hirundinaceum* 189
mixed prairie, USA 171, 173
Mniotilta varia, Black-and-white Warbler 166
mobile links 202
Mockerdungulling Nature Reserve, WA, monitoring sites 338
modelling 355
  role in planning 390
models
  animal-soil 355, 356
  as part of management process 385
  effect of fragmentation 172
  geological 355, 356
  holistic 355, 356
  hydraulic 355, 356
  of communities, ecosystems 18, 20, 26
  of ecosystems 386
  of minimum area 116-117
  predator-prey 355, 356
  single species 355, 356
  spatial 355, 356
  terrain 355, 356
  vegetation-soil 355, 356
Mojave Desert, USA 170
Molothrus ater, Brown-headed Cowbird 172
monitoring 7, 9, 10, 24-26, 32, 104, 249, 250, 252, 253
  Barrow Island, WA 283
  effects of management actions 346
  integration with survey 378
  involvement of public 391
  involvement of staff and public 338
  methodology 370, 371
  methods/procedures 337, 338
  need for scientific involvement 391
  of ecosystem changes 369
  of management practices 369
  population densities of kangaroos 219
  remnants as islands 206
  responsibility for 370, 371
  role of rangers 378
  role of research and management 370, 371
  role of research workers 378
  scale, frequency and precision 369
  sites 24, 378
    data to be recorded 338
    need to be established 337
  studies 374, 375
  system/programme 337
Monkey Gum, *Eucalyptus cypellocarpa* 191
Monongahela National Forest, West Virginia 167
Monte Carlo simulation 145
Monterey Pine, *Pinus radiata*
  plantations 177, 190
Moornaming, WA 249-252, 255-257
Moose, in boreal forest 236
Moree Plains Shire, local environmental plan 301-303
Morgan Hill, California, USA 204
mortality, of trapdoor spiders 35
mosiac, forest and prairie, USA 164, 169
Mountain Brushtail Possum, *Trichosurus caninus* 185, 186
Mountain Grey Gum, *Eucalyptus cypellocarpa* 77

movement
  and territoriality, Helmeted Honeyeater 287, 292
  corridors 190
Mt Gardner, WA 272, 273
Mt Lesueur, WA 19-23
Mt Lofty Ranges, SA 313-315, 319, 320
mudfish, *Neochanna diversus* 85
multiple
  non-interchangeable resources 180, 186, 190
  reserves, catastrophic loss 178
mycorrhizae 5, 351, 352
Mygalomorphae, trapdoor spiders
  and soil turnover 351
  persistence in small areas 29-39
Myiagra rubecula, Leaden Flycatcher 188, 189
Myrmecobius fasciatus, Numbat 7, 11, 327, 328, 373
  decline and recovery 54, 59, 63
  dispersal of young 58-62
  home range 58, 59
  minimum viable population 57, 59
Mrytaceae 5

NSW Environmental Planning and Assessment Act 296, 300-304
NZ protected natural areas
  problem weeds 241-247
Nadgee State Forest, Eden NSW 180
Naringal, Victoria
  ecology of *Potorous tridactylus* 41-43, 46-50
  forest fragmentation 42
  insular biogeography of mammals 41-46, 48-50
Narrogin, WA 305-307
Narrow-leaf Peppermint, *Eucalyptus radiata* 66, 67
national grasslands, USA 168, 174
national nature reserves, UK 388
national parks 151
  at Eden, NSW 178, 191
  Badgingarra, WA 155
  New Zealand 80
National Parks and Wildlife Service, NSW 295, 297-302
National Parks and Wildlife Act, NSW 296-304
native vegetation 254-257, 388-390
  clearance/clearing 259
    extent, SA 313
    for agriculture 250, 251, 254, 255
    history, SA 314, 315
    legislative controls, SA 317
  loss of in New Zealand 79, 80, 85
  retention
    benefits 317
    criteria for 318
    Heritage Agreement Scheme, SA 317
natural
  catastrophes 148
  disturbance regime, as control 369
  processes in small reserves 190
naturalness 318, 319
Nature Conservancy Council UK 143, 388
nature reserve(s)
  area 317
  biogeographic principles 317
  community diversity 317
  design 119, 316, 317
  history, SA 316, 317
  illegal use of 263
  in Yorkshire 104
  land acquisition for 260, 268

nature reserves(s) — cont'd
  national nature reserves, UK 388
  rare species and communities 316, 317
  recreation on 264
  representativeness 317
  Royal Society for the Protection of Birds 106
  Specht report 316, 317
  system of 260
  threat 316
  tourism/recreation 316
Nebraska Sand Hills region, USA 168, 171, 173
Needlegrass, *Stipa* spp. 169
Neochanna diversus, Black Mudfish 85
neotropical migrant bird species 165-168, 172
Nereeno Hill, WA 249-252, 255-257
nest
  height 166-167
  type 166, 167
nesting
  habitat of Carnaby's Cockatoo 249
  sites, for birds 70, 73
nestling growth rate in Carnaby's cockatoo 249, 252, 257
Nestor meridionalis, Kaka 81
network of reserves 381
New England Tablelands, NSW 74
new world warblers, Parulinae 125
New Zealand
  biogeographic theory 79-81, 84, 85
  endemism in flora and fauna 79-81, 84, 85
  fauna 79-81, 84, 85
  flora 79-81, 84, 85
  forest loss in 79-81, 84, 85
  national parks 79-81, 84, 85
  native vegetation 79-81, 84, 85
  Northland 79-81, 84, 85
  tussock grasslands 79-81, 84, 85
New Zealand Wildlife Service 80, 81
niche shifts 165
Niobrara Prairie, Nebraska 167, 173
nitrogen fixation 5, 351, 352
  lichens and cyanobacteria 5
nocturnal birds 77
Noisy Friarbird, *Philemon corniculatus* 72
Noisy Miner, *Manorina melanocephala* 65-77
  abundance 66-77
Noisy Scrub-bird, *Atrichornis clamosus* 269-276, 373
North Bungulla Nature Reserve, WA 29-39
North Yorkshire, UK 108
Northern Elephant Seal, *Mirounga angustrostris* 115, 116
Northern Oriole, *Icterus galbula* 164
Northern Parula, *Parula americana* 166
Nothomyrmecia macrops 333
Numbat, *Myrmecobius fasciatus* 7, 11, 327, 328, 373
  decline and recovery 54, 59, 63
  dispersal of young 58-62
  home range 58, 59
  minimum viable population 57, 59
Numenius americana, Long-billed Curlew 173
nutrient(s) 4
  as resources 351
  availability 5, 6
  balances 351
  cycles 351, 352
    roles of biota 352
  cycling processes and persistence 351
  principal roles 10

nutrient(s) — cont'd
  recycling
    contribution of plants and animals 6
    invertebrates and micro-organisms 5
    transfer from agriculture to remnants 237
Nuytsia floribunda, WA Christmas Tree 216

oak regeneration 236
Oak, *Quercus* spp. 163-165
objectives
  achievable 345, 346
  for management 16, 25
  long term management 345, 346
  management program/action 346
  need for explicit objectives 346
  overall objectives 345, 346
  strategic programme 345, 346
Okapia johnstoni, Okapi 115
Old Man's Beard, *Clematis vitalba* 242-246
Olearia muelleri 92
oligophagy 170
onus of land use 349
Ooline, *Cadellia pentastylis* 301
open country birds 183-185
Oporornis formosus, Kentucky Warbler 125, 127, 130, 131, 166
optimal foraging theory 179, 190
optimum feedback 385, 386
Orange-winged Sitella, *Daphoenositta chrysoptera* 189
orchids 373
Orchis simia, Monkey Orchid 114
organic matter, transfer from agriculture to remnants 237
Orsolobidae, relics of Tertiary environment 30
Oryctolagus cuniculus, Rabbit 46, 216
  effect on trapdoor spiders 36
Ottawa, Canada 114
outliers of widespread species, eucalypts 160
Ovenbird, *Seiurus aurocapillus* 129, 166
over-reliance on high 'tech' 348
overgrazing 168, 173
owls, hollows suitable for 191

Pachycephala
  P. inornata, Gilbert's Whistler 271
  P. rufiventris, Rufous Whistler 185, 188, 189
Panama Canal, Panama 118
Panthera tigris, White Tiger 115
Pardalotus punctatus, Spotted Pardalote 188, 189
park rangers, role in monitoring 378
Parula americana, Northern Parula 166
Parulinae, new world warblers 125, 127, 133
Paryphanta busbyi, land snail 79, 84, 85
Passerina cyanea, Inigo Bunting 165
pasture(s) 228, 230
  as bird habitat 66, 67, 70, 71, 73, 74
patch(es) 387, 388
  area 65-77
  definition 383
  in landscape 196, 198
  of remnant vegetation 249, 255, 257
  size
    control 383
    management 383
    native vegetation in New Zealand 79-85
    small and large coupes 181, 188
  structure 170, 173
  wildlife conservation in 178, 179, 189

patchiness
  created by fire 234
  environmental 363
  successional 363
pattern 114-115, 118
Patuxent Wildlife Research Centre, Maryland 166, 172
Pawnee National Grasslands, Colorado 167, 174
peninsulas of forest 71
Pennsylvania, USA 127
Perameles
  P. gunnii, Eastern Barred Bandicoot, habitat selection 198
  P. nasuta, Long-nosed Bandicoot 46
persistence
  and nutrient cycling processes 351
  biological 12
  capacity to persist of taxa 30, 36
  life history of trapdoor spiders 30
  of invertebrates in small areas 29, 30
  of trapdoor spiders 29-39
personnel, of agency, as a resource to be managed 385
Perth, WA 250
perturbations 351
  and roles 10, 11
  response modes 4, 10, 11
Perup Forest, WA 54, 55, 63, 323, 324, 327
pest
  control, forest 192
  plants and animals 357
  status 171
pesticides, transfer from agriculture to remnants 237
Petauroides volans, Greater Glider 185, 186
Petaurus
  P. australis, Yellow-bellied Glider 185, 186
    hollow suitable for 191
  P. breviceps, Sugar Glider 185, 186
Petrogale, rock-wallabies 11
Petroica
  P. macrocephala toitoi, Pied Tit 82, 83
  P. multicolour, Scarlet Robin 188, 189
  P. phoenicia, Flame Robin 185
Pezoporus wallicus, Ground parrot 271
Phaps chalcoptera, Bronzewing Pigeon 307, 310
Phascogale tapoatafa, Brush-tailed Phascogale 327
Pheucticus ludovicianus, Rose-breasted Grosbeak 164
philopatry 168
phosphorus 4
  toxicity 5
Phyllitis scolopendrium, Hart's-tongue Fern 109, 111
Phyllocladus trichomanoides 85
Picea spp., spruces 163, 164
  P. engelmannii, Engelmann's Spruce 115
Pied Currawong, *Stepera graculina* 72
Pied Tit, *Petroica macrocephala toitoi* 82, 83
Piedmont, USA 166, 172
pig, predation of land snails 84
Pileated Woodpecker, *Dryocopus pileatus* 166
Pilotbird, *Pynoptilus floccosus* 188, 189
pine, *Pinus* spp. 163-165
  plantations 65-77
  pinewoods in Scotland 237
Pink Cockatoo (see Major Mitchell's Cockatoo)

Pinus spp., pines 163-165
   P. ponderosa 157
   P. radiata, Monterey Pine 157
      plantations 178, 190
   P. resinosa 157
   P. virginiana, Scrub Pine 165
Pipilo erythrophthalmus, Rufous-sided Towhee 164
Pittosporum
   P. obcordatum, reserve for 80
   P. turneri, reserve for 80
Planchonella cotonifolia var *pubescens*, Microphyll Vine 298, 301
planning
   as education 266-268
   for management 389-391
   process 265
   public participation in 267
   with inadequate resources 265, 266
plant
   associations
      dynamics of and perturbations 6
      resilience after perturbation 6
      stability after perturbation 6
   disease 164
   dispersal mechanisms 235
Platalea flavipes, Yellow-billed Spoonbill 70
Platycercus elegans, Crimson Rosella 185
Pleistocene 164, 171
   faunal changes and extinctions 5
pollination, birds 273
population
   changes 273-276
      birds 273-276
   colonization 113-114
   densities
      Bitterns 84
      Kokako 82
      of birds 65-77
   decline
      in Carnaby's Cockatoo 249
   disjunct 155, 160
   ecology
      of *Potorous tridactylus* 41, 43, 46-50
   effective size 82
   equation for 113
   establishment 113-114
   estimation
      lincoln index method 219
   evolution 115-116
   extinction 116-118
   fluctuation 53, 54, 57
   genetic structure of 151, 152, 154
   genetics 115-116
   immigration 113-114
   initiation 113-114
   isolated 53, 55, 58, 59, 62, 63
   Macropus fuliginosus, Western Grey Kangaroo
      capture-recapture method 220, 221
      faecal transect method 221
      walking transect method 220, 221
   minimal viable size 57, 62, 151, 159, 161
   minimum 116-118
   minimum area 116-118
   neighbourhoods of 159
   of birds 65-77
   resilience or stability of 6
   size 146, 159
      effective 57
      Potorous tridactylus 43
   stability 114-115
   stationarity 114-115
   structure 207
   viability 114-116
   viable size 116-118

Potorous
   P. longipes, Long-footed Potoroo 42
   P. tridactylus, Long-nosed Potoroo, population ecology 41, 43, 46-50
poultry, effect on trapdoor spiders 37
Powderbark Wandoo, *Eucalyptus accedens* 305, 308
Powerful Owl, *Ninox stenua* 72
prairie 201
   north America 168-174
Prairie Warbler, *Dendroica discolor* 130, 165
precipitation 163, 170, 173
predation 54, 57
   by Red Fox 54, 327, 374
   methods, techniques of trapdoor spiders 36, 37
predators
   introduced 236
   of trapdoor spiders 36
prescribed burning 190, 306, 308
presentation technique 349
preservation (see conservation)
Primula vulgaris, Primrose 111
priorities 381
   indentification 348, 349
private property 66
problem weeds (see also weeds)
   characteristics 241-247
   in NZ reserves 241-247
process
   arena 349
   of nature conservation 345, 346
   unit 349
Prodontia lewisi, chafer beetle 80
productivity, of wrens 212, 214, 216
progeny "leakage" 168
Proteaceae 5, 236
protected natural areas, NZ
   problem weeds 241-247
   representativeness 246
proteoid roots 5
Prunus serotina 235
Przewalski's Wild Horse, *Equus przewalskii* 115
Pseudocheirus peregrinus, Common Ringtail 46, 185, 186, 327
Pseudomys shortridgei, habitat selection 196
Pseudotsuga menziesii, Douglas Fir 191
Psophodes nigrogularis, Western Whipbird 271-275
psyllids, Psyllidae, Insecta 74
   "farmed" by Bell Miner 292
Pteridium
   P. aquilinum, Bracken 114
   P. esculentum, Austral Bracken Fern 66, 77
Ptilonorhynchus violaceus, Satin Bowerbird 72
public
   land 65, 66
   participation 382, 386
      in planning 267
   scrutiny
      of management plans 371
      of research results 371
pulpwood, Eden NSW 177
Purple-crowned Lorikeet, *Glossopsitta porphyrocephala* 341, 343
purpose of reserves 349
Pycnoptilus floccosus, Pilotbird 188, 189
Pygmy Possum, *Cercartetus nanus* 185, 186

Quail, interaction with trapdoor spiders 36
Queensland 201, 205

Quercus spp., oaks 163-165
   Q. petraea, regeneration 236

Rabbit
   and disturbed soil 6, 236
   effect on trapdoor spiders 36, 38
   grazing of Santalum seedlings 6
radio-tracking 58
rainfall, heavy 3
rainforest(s) 201
   of South-east Asia 387
rainshadow 71
raptors 70
rare species 10, 354
   at Mt Lesueur, Eneabba, WA 19
   birds 272, 273, 276
   local endemics 79, 80, 84, 85
   relictual 79-81, 84, 85
   value in setting management objectives 16
rare, rareness, as opposed to common, commonness 34
rarity 171, 174, 316-318
   species and site conditions 85
rat predation of *Cyclodina* skinks 84, 85
Rattus fuscipes, Bush Rat 46, 186
recolonization
   after decline 54
   of disturbed areas by trapdoor spiders 38
recovery
   after fire 215-217
   mammal populations 53-57, 59, 62, 63
recreation 174
   management of 264
recruitment, whether annual or episodic 7
Recurvirostra americana, American Avocet 173
recycling 351
Red Fox, *Vulpes vulpes* 54, 327, 328, 374
   predation on medium-sized mammals 54
Red Kangaroo, *Megaleia rufa* (= *Macropus rufus*) 17
Red Wattlebird, *Anthochaera carunculata* 185
Red-eyed Vireo, *Vireo olivaceous* 129
Red-headed Woodpecker, *Melanerpes erythrocephalus* 165
Red-necked Wallaby, *Macropus rufogriseus* 46
Red-tailed Black Cockatoo (see Forest and Inland Red-tailed Black Cockatoos)
regeneration 114-115
   Betula spp. 236
   initiation of 2
   lack of in native vegetation 256, 257
   oak 236
   of native species 237
   of remnants 390
   of vegetation 38
regime
   disturbance 23
   fire 16
regional
   distribution of habitat 72
   system, primary position of 348, 349
rehabilitation 353
   invertebrates 353
   of remnants 390
   studies at Eneabba 20
relaxation 142-144
relevance of research for management 345
relictual distributions of New Zealand's flora and fauna 79-81, 84, 85
remnant(s) 218, 387-392
   affect of surrounding land 388
   and management strategies 351
   as a repository of genetic material 389

remnant(s) — cont'd
  as wildlife habitat 318, 320
  characteristics of 387, 388
  communities, management 383
  conservation value 160
  corridors 318, 319
  critical minimum size 319
  definitive 387
  degradation of 388
  disturbance regime 319
  ecological isolation 317, 320
  edge effects 318, 319
  edge/interior ratio 319
  effect on surrounding land 388
  exotic invasion 318-320
  genetics of species in 152
  heirarchy of 389
  land tenure 319, 321
  management 17, 390
    plans 390
    strategies 321, 322, 351
  native vegetation 249, 250, 254-257
  naturalness 318, 319
  neighbours of 388, 391
  network of 389, 390
  of native woodland, characteristic features 331
  of outliers 160
  ownership 388, 391
  patches of forest 65-77
  protection 391
  regeneration 390
  rehabilitation 390
  research on 389, 390
  retention by planning control 298-304
  role of in nature conservation 152, 159, 388, 389, 392
  shape 319, 381
  size 2, 314, 318, 381
  subject of disturbance 388
  values 388, 389, 391
  viable populations 180
  woodlots with dieback, characteristic features 331
remnant populations, genetic information of 390
representativeness/typicality 317, 318
reproduction, interoparous, of trapdoor spiders 35, 37
reproductive
  behaviour, wandering of male trapdoor spiders 35-37
  effort 167
  strategies, trapdoor spiders 32, 35-37
reptiles and frogs, survey and census procedures 180, 181
research
  and management approach 347-349
  constraints 347-349
  corridors for wildlife 359
  grants, Barrow Island, WA 283
  Helmeted Honeyeater 291
  integrating management and planning 370, 371
  interrelations 349
  on dynamics of ecosystem change 369
  publishing results of 371
  relationship with management 263
  review 349
  role-centred questions 11
  strategic versus tactical 370
  workers, role in monitoring 378
reserved coupes, stepping stones between reserves 191
reserve(s)
  at Eden, NSW 178, 191
  optimal size 135-137

reserve(s) — cont'd
  purpose of 2
  selection
    importance of habitat requirements 361
    options 361
    population dynamics 361
    species abundances 361
    species distribution 361
  size 17, 138, 173, 174, 232, 363
    effect on immigration 61, 62
    large versus small reserves 80, 82, 84, 85
    representativeness 79, 80, 85
    wheatbelt, WA 271
resilience
  and persistence 10
  interpretation of 6, 10-12
resource(s)
  dedication of 267
  extraction, Barrow Island, WA 280, 282
  management 266
  management with inadequate 265, 266
  staff and financial 263, 264
resprouting after fire 216
restoration, Barrow Island, WA 280-283
retained patches, unlogged forest 177
revegetation
  to reduce erosion, salinity, etc. 257
  Yellingbo State Nature Reserve, Vic. 290, 291
Rhagodia
  R. crassifolia 92
  R. preissii 92
Rhipidura
  R. fuliginosa, Grey Fantail 81, 185, 188, 189
  R. leucophrys, Willy Wagtail 216
  R. rufifrons, Rufous Fantail 188, 189
Rhizobium 351
Rhynchophanes mccownii, McCown's Longspur 173
Richard's Pipit, *Anthus novaeseelandiae* 275
Riekoperla darlingtoni, stonefly 333
Ringtail Possum, *Pseudocheirus peregrinus* 185, 186
riparian forest 185
River Red Gum, *Eucalyptus camaldulensis* 72, 77
riverine plain, creation of 256
road(s)
  affect on drainage 282
  as barriers 282
  Barrow Island, WA 281, 282
  construction 281, 282
  dividing forest patches 66, 77
  effect of dust 281
  reserves, as bird habitat 341-343
  verges 254, 255
roadside vegetation 358
rock-wallabies, *Petrogale* spp. 11
role of reserves 349
roosting sites for birds 70, 73
Rose Robin, *Petroica rosea* 72
Rose-breasted Grosbeak, *Pheucticus ludovicianus* 164
Rothamsted grassland plots 209
Royal Albatross, *Diomedea epomophora* 80
Royal Society for the Protection of Birds, UK 391
Rue-leaved Saxifrage, *Saxifraga tridactylites* 109
Rufous Bristlebird, *Dasyornis broadbenti* 271, 276
Rufous Fantail, *Rhipidura rufifrons* 188, 189

Rufous Whistler, *Pachycephala rufiventris* 185, 188, 189
Rufous-sided Towhee, *Pipilo erythrophthalmus* 164
rural dieback
  conceptual model 331
  conservancy groups 347, 349
  definition 331
  ecosystem 65, 74
  link with modification of tree cover 331
  tree decline, causes and consequences 331-332
Ryukyu Islands, Japan 107

salience, hierarchy of 348
salinity 4, 65, 74
Salmon Gum, *Eucalyptus salmonophloia* 271, 272, 388
salt
  affected land 4, 9
  marsh 227-232
Samburu-Isiolo Game Reserve 144, 145
sampling
  continuous 353
  invertebrates 334, 354
  methods 367
Samuel Ordway Prairie, South Dakota 167, 173, 174
San Francisco Bay region 204
sandhills, Nebraska 168
Santalum spp.
  dispersal of by emu 6
  grazing of seedlings by rabbits 6
Sarcocornia quinqueflora 228
Satin Bowerbird, *Ptilonorhynchus violaceus* 72
savanna, USA 164, 169, 171
Saxifraga tridactylites, Rue-leaved Saxifrage 109
Scaraboeoidea and soil turnover 351
Scarlet Robin, *Petroica multicolor* 188, 189
Scarlet Tanager, *Piranga olivacea* 129
Schizachyrium scoparium, Little Bluestem 170
scincid lizard 190
Sclerolaena diacantha 92
scrub, *Leptospermum*-dominated, in New Zealand 85
Scrub Pine, *Pinus virginiana* 165
Sea Otter, translocation 62
seed longevity, weeds in NZ reserves 244-245
Seiurus
  S. aurocapillus, Ovenbird 166
  S. motacilla, Louisiana Waterthrush 166
selection effects on genetic diversity 159
selective logging, Eden NSW 177, 188
Senecio turneri, reserve for 80
Sericornis
  S. frontalis, White-browed Scrubwren 185
  S. fuliginosus, Calamanthus 275
  S. pyrrhopygius, Chestnut-rumped Hylacola 72
serpentine
  barrens, USA 164, 174
  rock, endemic plants on 85
Sesleria albicans, Blue Moor-grass 110
Setophaga ruticilla, American Redstart 166
shade
  and shelter belts for stock 357
  and trapdoor spiders 38
shale barrens, USA 164, 174
shape 319
sheep, habitat disturbance of trapdoor spiders 37

Short-billed White-tailed Black Cockatoo (see Carnaby's Cockatoo)
short-grass prairie 168, 171
Short-nosed Bandicoot, *Isoodon obesulus* 327
shrubland 249, 255
Sierra Nevada Mountains, California 235
Silver Wattle, *Acacia dealbata* 72
Silvereye, *Zosterops lateralis* 81
Simla Research Station, Trinidad 205
sites, monitoring 378
Sitta carolinensis, White-breasted Nuthatch 166
size 314, 318, 319
    of reserve 232
skink, *Cyclodina* spp. 79, 84, 85
SLOSS ("single large or several small") 73, 141-144, 147, 148
small reserves, natural processes occurring in 190
social
    behaviour 148
    or gregarious animals 180
Society for Growing Australian Plants 290
soil(s) 3
    compaction 74
    fertility 85
    type 186
    dwelling species 354
Sonoran Desert, USA and Mexico 170
Sorghostrum nutans, Indiangrass 169
South Australia 237
    Calca 90
    Ceduna 90
    extent of native vegetation clearance 313
    Eyre Peninsula 89-91
    Heritage Agreement Scheme 317
    history of native vegetation clearance 314, 315
    history of nature reserves 316, 317
    legislative controls of clearance 317
    Mt Lofty Ranges 313-315, 319, 320
    Poochera 90
    Port Kenny 90
south west botanical province, WA 269, 271
    size of monitoring quadrats 337
south west of Western Australia 250
Southern Emu-wren, *Stipiturus malachurus* 71
spatial processes 114-5, 118
Speargrass Weevil, *Lyperobius huttoni* 80
species
    widespread tree 157
    with localized distributions 152-155, 157
    with regional distributions 152-154, 157
    distribution
        Casuarina cunninghamiana 157
        localized species 152-160
        of eucalypts 152-154
        outliers of widespread species 160
        regional eucalypts 152-155
        widespread eucalypts 152, 155
    diversity
        birds 74-77, 272, 273
        plants 273
    numbers
        birds 65-77
        mammals 65, 69, 71-73
    richness 168
        correlates of mammal species richness 41, 44, 45, 49
        mammals in forest remnants 41, 44, 45, 48, 49
    incidence in relation to patch size 80-83
species-area relationships 44, 49, 73-74, 104-107, 109, 111-112, 141-143, 148

Sphenomorphus tympanum 190
spiders, Arachnida
    in intertidal wetlands 227, 230, 231
    indicators of nutrient cycling 374
    indicators of processes 374
    mygalomorphs and soil turnover 351
    suggested for study of guilds of species 375
Spinifex longifolius 279
Spiza americana, Dickcissel 165
Spizella pusilla, Field Sparrow 165
Sporobolus virginicus 228
Sporodanthus 85
Spotted Pardalote, *Pardalotus punctatus* 188, 189
Spotted Quail-thrush, *Cinclosoma punctatum* 72
Spruce, *Picea* spp. 163, 164
stability, interpretation 6, 11
staged approach 347
Starling, *Sturnus vulgaris* 202
statistical models
    logistic regression analysis 92-94, 101
    probability estimation 95-98, 101
    regression 94, 95
statutory authorities and management of remnants 7
stepping stones 165, 389
    corridors 257
    reserved coupes 191
Stipa spp., Needlegrass 169
    S. eremophila 92
    S. setacea 92
Stipiturus malachurus, Southern Emu-wren 71
stock
    effects on trees 74
    shelter for 74
Striated Pardalote, *Pardalotus striatus* 71
Striated Thornbill, *Acanthiza lineata* 188, 189
Strigops habroptilis, Kakapo 80
Strzelecki Ranges, Vic. 288
Stubble Quail, *Coturnix pectoralis* 310
Sturnus vulgaris, Starling 202
Suaeda australis 228
subdivided populations 143, 145-148
subpopulation genetic structure 159-161
succession 118
    vegetation 276
    stages within remnants 238
Sugar Glider, *Petaurus breviceps* 74, 185, 186
Sula bassana, Australasian Gannet 80
Sulphur-crested Cockatoo, *Cacutua galerita* 24
Superb Fairy-wren, *Malurus cyaneus* 185, 188, 189
Superb Lyre Bird, *Menura novaehollandiae* 72
supersaturation, absence of 2
"supertramps", insects 173
survey(s)
    and census procedures
        birds 180
        mammals 180, 181
        reptiles and frogs 180, 181
    data 367
    design 367
    Helmeted Honeyeater 288
    invertebrates 334, 354
    methods 368
        environmental gradients 91
        stratification 91
        stratified random 91
    periods for monitoring Carnaby's Cockatoo 252, 253
    purpose 367, 368

Survey Cassidix 288
survival, wrens 212, 214-217
swamp 67, 77
    brackish 229, 230
Swamp Gum, *Eucalyptus ovata* 77, 287, 290, 292
Swamp Paperbark, *Melaleuca ericifolia* 71
Swan Coastal Plain, WA 389
synoptic analysis 347-349

Tachyglossus aculeatus, Echidna 7, 36, 38
    interaction with trapdoor spiders 36
Takahe, *Notornis mantelli* 80
tall-grass prairie 165, 168, 173, 174
Tamias striatus, Chipmunk 114
Tammar, *Macropus eugenii* 323, 326, 327
Tammin Shire, WA 30
Tanzania 205
targets in field research 347
Tarwonga, WA 249-252, 255-257
Tasmanoonops, relics of Tertiary environment 30
taxonomy of weeds in NZ reserves 241
technical data, use in management 263
termites, Isoptera
    and nutrient balances 351, 352
    as prey for trapdoor spiders 38
    destroying fallen logs 7
    harvesting litter and plant debris 7
    redistribution of resources by 5
terrestrial birds 183-185
territories
    interspecific defence by Noisy Miners 65-77
    interspecific defence of Bell Miners 74
    of Powerful Owls 72
Tertiary environment, relics of 30
Teyl 34-38
    T. luculentus 34
the manager's "kit" 385, 386
Themeda australis 216
Theodore Roosevelt National Park, North Dakota 173
theoretical research 349
theory
    of island biogeography 106, 107
    for management 15, 18, 23, 26
Thick-billed Grass-wren, *Amytornis textilis* 271
thicket 271-273, 275
Thomomys spp., Gophers 207
    T. bottae, Western Pocket Gopher, interpopulation genetic differences 58
Three Springs, WA 272
Threlkeldia diffusa 92
Thryomanes bewickii, Bewick's Wren 165
thunderstorms 2, 3, 30
tidal flats 228, 229
Timbillica State Forest, Eden NSW 180
time required for clear-felled forest to regenerate 191
tithe to conservation 209, 392
Todea barbara, reserve for 80
tornadoes 2, 3
Tornidirrup National Park, WA, spiders of 30
Toxopidae 29, 30
Toxops, a relic of tertiary environment 30
Trachydosaurus (= *Tiliqua*) *rugosus* 216
Tradescantia fluminensis, Wandering Jew 243, 245, 246
traditional attitude of rural community 295, 300, 302, 303
trampling by animals 234
translocation 53, 57, 62

trapdoor spiders, Mygalomorphae
and shade 38
    case studies, persistence 29-39
    distribution on islands 30
    disturbance by sheep 37
    effect of rabbits 36
    Grevillea paradoxa, habitat 38
    interaction with echidna 36, 38
    life history and persistence 30
    persistence in small areas 29-39
    predation techniques, strategies 36, 37
    prey 38
    scorpion predator 36
    Varanus gouldii as predator 36
    York Gum as habitat 37
treatments for monitoring of management
    practices 369
Tree Martin, *Cecropis nigricans* 71
treefalls 234
trees
    decline, Dandenong Ranges, Vic. 292
    genetic resources of 151-161
    health 65, 74
    population size of 159
Trelease Woods, Illinios 172
Trichosurus
    T. caninus, Mountain Brushtail Possum
        185, 186
    T. vulpecula, Common Brushtail 185,
        186, 327
        mistletoe distribution 84
        population recovery 54
Triglochin striata 228
Trinidad 205
Triodia
    T. augusta 279
    T. irritans 92
    T. pungens 279
    T. wiseana 279
tropical rainforest 205
Tsuga canadensis, Hemlock 163
Turnix, quail, interaction with trapdoor
    spiders 36
tussock grasslands, New Zealand 80
Tutanning Nature Reserve, WA 58
Two Peoples Bay Nature Reserve, WA 272, 273
Tympanachus
    T. cupida, Greater Prairiechicken 173
    T. cupido, American Heath Hen 293
    T. pallidicinctus, Lesser Prairiechicken
        173
Tyrannus tyrannus, Eastern Kingbird 164

Ulex europaeus, habitat role 197, 198
ultramafic rock, endemic plants on 85
Ultricularia
    U. delicatula, bladderwort 85
    U. lateriflora 85
"umbrella" species 202, 205, 373, 389, 390
unburnt reserves 184, 190
understorey 65-77
univoltinism 171
unlogged forest, retained patches 177, 188
Upland Sandpiper, *Bartramia longicauda*
    165
Ursus horribilis, Grizzly Bear 148

Valentine National Wildlife Refuge,
    Nebraska 167
Varanus gouldii, as predator of trapdoor
    spiders 36
variability of habitat 232
vegetation
    classification, numerical 91, 92, 100, 101
    dynamics 272, 274-276
    remnants, causes of 1

Vermivora pinus, Blue-winged Warbler 165
viability of small populations
    minimum viable populations 366
    mutation rates 366
viable populations, remnants 180
Victoria, Australia 65-77
    State avifaunal emblem (see Helmeted
    Honeyeater)
Vireo
    V. bellii, Bell's Vireo 165
    V. flavifrons, Yellow-throated Vireo 166
    V. gilvus, Warbling Vireo 165
    V. griseus, White-eyed Vireo 165
Virginia, USA 127
visitors, control of in reserves 38
visual
    impact, logging 177
    isolation of remnants of native
        vegetation 249, 255, 257
volcanic eruptions 233, 234
voluntary conservation 379-380
vulnerable species, value as a short cut to
    management 23, 26
Vulpes vulpes, Red Fox 37, 46, 54, 327, 328,
    374
    predation on medium-sized mammals 54

waders
    Australasian 231
    migratory 227, 231
    palearctic 231
Wandoo, *Eucalyptus wandoo* 271, 305
    monitoring sites 338
WAPET (West Australian Petroleum) 279,
    283, 284
Warbling Vireo, *Vireo gilvus* 165
water regime, of remnants 237
waterbirds 65-77
Wedge-tailed Eagle, *Aquila audax* 202
weeds
    dispersal mechanisms in NZ reserves
        243-245
    effect on habitats of trapdoor spiders 37
    growth rate in NZ reserves 243
    in protected natural areas, NZ 241-247
    life span in NZ reserves 243
    lifeform and height in NZ reserves 242,
        243
    persistence in reserves 243
    rapid spread in NZ reserves 242
    seed longevity in NZ reserves 244-245
    taxonomy, NZ 241
    usurp native vegetation 243
    vegetative reproduction, NZ 243, 244
    Yellingbo State Nature Reserve, Vic. 291
Wellington, New Zealand
    occurrence of *Cyclodina whitakeri* 84, 85
    speargrass weevil reserve at 80
West Australian Petroleum Pty Ltd 279, 283,
    284
West Virginia, USA 127, 165
Western Australia 235, 236, 238
    eucalypts 152
    flora 363
    landscape 161
    size of wheatbelt reserves 363
Western Australian Museum 279
Western Grey Kangaroo, *Macropus
    fuliginosus* 219-225, 327
    faecal accumulation rate 221-223
Western Pocket Gopher, interpopulation
    genetic differences 58
Western Whipbird, *Psophodes nigrogularis*
    271-275
    habitat 273-275
Westernport Bay, Vic. 288

Westringia rigida 92
wetlands
    birds 227
    insects 227
    insects, spiders in 227, 230, 231
    intertidal 227, 230, 231
    loss of, in New Zealand 79, 85
    mangrove 227-232
    New Zealand reserves 80
    saltmarsh 227-232
Whangamarino wetland, NZ 84
wheatbelt, WA 16, 23, 26, 53, 54, 58, 63, 73,
    119, 143, 148, 250, 251, 257, 271, 272, 387,
    388
    birds 133, 136, 137
    desciption of 261
    landscape 38
    mistletoes in 84
    reserves WA 30
    resources 264
White-bellied Cuckoo-shrike, *Coracina
    papuensis* 72
White-breasted Nuthatch, *Sitta carolinensis*
    166
White-browed Scrubwren, *Sericornis
    frontalis* 185
White-eared Honeyeater, *Meliphaga
    leucotis* 185, 188, 189
White-eyed Vireo, *Vireo griseus* 165
White-throated Treecreeper, *Climacteris
    leucophaea* 188, 189
White-winged Chough, *Corcovax
    melanorhamphus* 72
wildfire (see also fire, burning) 305, 306,
    308, 309, 311
    and corridors for wildlife 357
wildlife conservation
    in linear environments 178, 179, 189
    in small patches 178, 179, 189
wildlife corridors (see corridors for
    wildlife)
Wilsonia citrina, Hooded Warbler 166
wind-storm, as perturbation 6
windbreaks 357
windthrow 190
Wittelbee Conservation Park 91, 100
Wandering Jew, *Tradescantia fluminensis*
    243, 245, 246
wodjil 33, 34
Wongan Hills, spiders of 29
Wood Thrush, *Hylocichla mustelina* 129
woodland(s) 18, 249, 254-257, 272, 388, 390
    Box 72
    degradation of 256, 257
    in England 116, 117, 119
    in WA 272
    regeneration of 7
World Conservation Strategy 382
World Wildlife Fund 205
Worm-eating Warbler, *Helmitheros
    vermivorus* 166
Woylie, *Bettongia penicillata* 323, 327, 328
    recolonization 324

Xanthorrhoea preissii 216
Yambarra Conservation Park 98
Yarra River, Vic. 287, 288
Yarram, Vic. 288
Yellingbo State Nature Reserve, Vic. 287,
    288, 290, 293, 388
    advisory committee 290
    feral vertebrates 291
    management 289-291, 293
    revegetation 290, 291
    weeds 291
Yellingbo, Vic. 287-290, 292, 293

Yellow Robin, *Eopsaltria australis* 188, 189
Yellow Warbler, *Dendroica petechia* 165
Yellow-bellied Glider, *Petaurus australis* 185, 186
Yellow-billed Spoonbill, *Platalea flavipes* 70
Yellow-breasted Chat, *Icteria galbula* 130
Yellow-breasted Chat, *Icteria virens* 165
Yellow-faced Honeyeater, *Meliphaga chrysops* 185, 188, 189
Yellow-tufted Honeyeater, *Lichenostomus melanops* 72, 288
Yertchuk, *Eucalyptus consideniana* 77
York Gum, *Eucalyptus loxophleba* 271, 338, 339, 388
  as habitat for trapdoor spiders 37

zodarid spiders, and soil turnover 351
Zosterops lateralis, Silvereye 81
Zygophyllum
  Z. aurantiacum 92
  Z. ovatum 92